T0350924

Pumps and Compressors

Wiley-ASME Press Series

Corrosion and Materials in Hydrocarbon Production: A Compendium of Operational and Engineering Aspects
Dr. Bijan Kermani, Don Harrop

Design and Analysis of Centrifugal Compressors
Rene Van den Braembussche

Case Studies in Fluid Mechanics with Sensitivities to Governing Variables
M. Kemal Atesmen

The Monte Carlo Ray-Trace Method in Radiation Heat Transfer and Applied Optics
J. Robert Mahan

Dynamics of Particles and Rigid Bodies: A Self-Learning Approach
Mohammed F. Daqaq

Primer on Engineering Standards, Expanded Textbook Edition
Maan H. Jawad, Owen R. Greulich

Engineering Optimization: Applications, Methods and Analysis
R. Russell Rhinehart

Compact Heat Exchangers: Analysis, Design and Optimization using FEM and CFD Approach
C. Ranganayakulu, Kankanhalli N. Seetharamu

Robust Adaptive Control for Fractional-Order Systems with Disturbance and Saturation
Mou Chen, Shuyi Shao, Peng Shi

Robot Manipulator Redundancy Resolution
Yunong Zhang, Long Jin

Stress in ASME Pressure Vessels, Boilers, and Nuclear Components
Maan H. Jawad

Combined Cooling, Heating, and Power Systems: Modeling, Optimization, and Operation
Yang Shi, Mingxi Liu, Fang Fang

Applications of Mathematical Heat Transfer and Fluid Flow Models in Engineering and Medicine
Abram S. Dorfman

Bioprocessing Piping and Equipment Design: A Companion Guide for the ASME BPE Standard
William M. (Bill) Huitt

Nonlinear Regression Modeling for Engineering Applications: Modeling, Model Validation, and Enabling Design of Experiments
R. Russell Rhinehart

Geothermal Heat Pump and Heat Engine Systems: Theory And Practice
Andrew D. Chiasson

Fundamentals of Mechanical Vibrations
Liang-Wu Cai

Introduction to Dynamics and Control in Mechanical Engineering Systems
Cho W. S. To

Pumps and Compressors

Marc Borremans
Erasmus University College Brussels
Anderlecht, Belgium

This Work is a co-publication between ASME Press and John Wiley & Sons Ltd.

This edition first published 2019

This Work is a co-publication between John Wiley & Sons Ltd and ASME Press

Registered Offices
John Wiley & Sons, Inc., 111 River Street, Hoboken, NJ 07030, USA
John Wiley & Sons Ltd, The Atrium, Southern Gate, Chichester, West Sussex, PO19 8SQ, UK

Editorial Office
The Atrium, Southern Gate, Chichester, West Sussex, PO19 8SQ, UK

For details of our global editorial offices, customer services, and more information about Wiley products visit us at www.wiley.com.

Wiley also publishes its books in a variety of electronic formats and by print-on-demand. Some content that appears in standard print versions of this book may not be available in other formats.

Library of Congress Cataloging-in-Publication Data

Names: Borremans, Marc, 1951- author.
Title: Pumps and Compressors / Marc Borremans, Erasmus University College Brussels, Anderlecht, Belgium.
Description: First edition. | Chichester, West Sussex : John Wiley & Sons Ltd, [2019] | Series: Wiley-ASME press series | Includes index. | Identifiers: LCCN 2019009784 (print) | LCCN 2019012441 (ebook) | ISBN 9781119534204 (Adobe PDF) | ISBN 9781119534082 (ePub) | ISBN 9781119534143 (hardcover)
Subjects: LCSH: Pumping machinery. | Compressors.
Classification: LCC TJ900 (ebook) | LCC TJ900 .B67 2019 (print) | DDC 621.6/9–dc23
LC record available at https://lccn.loc.gov/2019009784

Cover design: Wiley
Cover image: ©Franco Nadalin/EyeEm/Getty Images, ©esemelwe/Getty Images

Set in 10/12pt WarnockPro by SPi Global, Chennai, India

Printed and bound by CPI Group (UK) Ltd, Croydon, CR0 4YY

10 9 8 7 6 5 4 3 2 1

Contents

Preface

When I studied electro-mechanical engineering at the University of Brussels, my professor of applied mechanics explained how turbopumps and turbines work. He proved the equation of Euler: he drew two curved lines on the blackboard (yeah, that was 1970) and proved Euler's law, a proof of one page. Then he said, "This law applies to all pumps and turbines." That was it. I wondered, "What do I know about a pump or turbine?" I had no answer. It was only when I became a professor and had to teach courses like "Pumps and compressors" and "Steam and gas turbines" and started to read magazines, brochures, and books on the subject that I saw what those devices looked like without a casing and how they worked. I also think that a beautiful or detailed picture can explain much more than text alone or dry formulas. That opinion informs this book. There are more than 700 drawings and pictures in this book. I hope you like them!

I worked a lot as a professor. I started in 1973 giving courses in electricity, electrotechnics, electronics, and high-frequency techniques for four years. Then I became a professor in mechanics, giving courses in thermodynamics, applied thermodynamics (pumps and compressors, combustion engines, steam and gas turbines, refrigeration techniques, heat techniques), materials science, fluid mechanics, strength of materials, pneumatics and hydraulics, CAD2D, CAD3D, CNC, CAM, and so. For these subjects, I designed detailed courses, first on the typewriter (do you know what that is?) and then, from 1983, on the computer. My first computer had an 8-bit processor with 16 kB of RAM and the printer was a matrix printer. In total I offered 45 subjects. When I ended my career, my courses comprised 2857 pages per year.

When I retired in 2007, I started collecting information for this book, beginning with my own course, and spent a year constantly writing on and researching the subject. I wrote in my mother tongue: Dutch.

In 2007, a search for "pumps" on Google elicited no fewer than 94 million references. The word "compressor" had a hit rate of 18 million. This is because both devices account for a significant part of the infrastructure of buildings, houses, and factories. It is reckoned that in a petrochemical plant there is one pump installed for every employee.

With this knowledge and 35 years' teaching experience, I started to draw up my own sort of encyclopedia. I collected as much information as possible from the literature, company brochures, and the Internet in order to catalogue as many of the pumps and compressors on the market as possible, writing a description of them, including their properties. It is left to the reader to find out which pump or compressor is the best for a certain job. That choice will not be distilled immediately from the book. The choice of a pump or compressor is not an exact science; it is an assessment of the pros and cons before a definitive choice can be made. Reasons for choosing a pump for a specific job are based on price, maintenance, lifecycle, regulation, type of fluid, etc. A reference list at www.wiley.com/go/borremans/pumps provides a lot of information including videos and animations that can be found on the internet for most types of pumps and compressors.

Much later, in 2018, I translated it into English, not without doing more research over several months. What you hold in your hands is the result. Maybe you could do the same work. But by buying this book you spare yourself a lot of time and money. If you now search Google using the word "pump", you will get a lot of hits, but a lot of these will concern "shoe pumps." For the word "compressor" you'll get 21 million hits.

Beyond this, most pictures in the book are not available anymore on the Internet. Nowadays, nearly all companies just show the casings of their pumps and compressors. Somebody told me it is because they don't want their ideas to be stolen. But to me that is pointless: rival companies can just buy a pump, dismantle it, and reverse engineer it. Just like the Japanese did after World War II, but they added the concept of constant quality and so made many improvements.

This book also uses concepts of fluid mechanics and thermodynamics, two subjects I taught. In fact, pumps and compressors apply the concepts of these two basic branches of engineering science. Don't worry if this isn't your area of expertise: the information you will need to understand these branches is given in this book. When I started my career the subject I taught was called "Applied mechanics and thermodynamics" and later it was separated into "Pumps and compressors," "Combustion engines and turbines," and "Refrigeration techniques."

This book is intended for technical high school students, college students, plant engineers, process engineers, and pump and compressor sales reps. Of course, in high schools, one has to make abstract on the mathematical framework. The book is also a kind of encyclopedia of the greater part of pumps and compressors on the market. It is impossible for a teacher or professor to go through the whole book in one course.

I use simple language to explain everything and in the hope that it will be easy for the reader to follow the reasoning. I oppose writing that forces the reader first has to make a grammatical analysis of every sentence.

Marc Borremans
borremans.m@telenet.be
https://www.borremansengineering.com

Acknowledgment

This colorful book wouldn't have been possible without the contribution of 67 enthusiastic companies, all over the world. They allowed me to use their pictures. They are, in alphabetical order:

Aerzen
Agilent
Alfa Laval
Alup
Allweiler
Andritz
Ateliers François
Atlas Copco
Begeman
Bitzer
Boge
Burhardt
Börger
Bornemann
Bosch-Rexroth GmbH
Cameron Compressor Systems
Clasal
Cornell
Dab
Daikin,
Daurex
Dewekom Engineering
Dura
Ensival Moret
Ebbm-Papst
Egger
Eriks
Flowserve
Friotherm
Johnson
Gardner Denver
Gea
Glynwed/Reinhütte pumpen
Grasso
Hibon
Hoerlinger Valve
Hermetic pumps GmbH
Hus (Verder)
Ingersoll Rand
iLLmVARC
ITT Goldpumps
SPXFlow
Kaeser
Klima
Knf
Koerting
KSB
LewVac
Leybold Vacuum
PDC Machines

Pfeiffer
Psg Dover
Ritz
Rotojet
Seepex
Sepco
Sihi-Sterling (Flowserve)
Speck Pumpe
Scam-Torino
Shamai
SKF
Systemair
Sulzer
Stork
Verderair
Viking Pump
Warman
Watts
Weir
Wemco

I would like to thank some people from Wiley, as well: Eric Willner for believing in the project, Steve Fassioms, Tim Bettsworth (the man with the sharp eye) for the editing, and in advance the typesetters, for their excellent work.

Used Symbols

Symbol	Meaning	Unit
a	Acceleration	m/s^2
$a^{'}$	Acceleration in suction line	m/s^2
A	Section piston	m^2
A	Cross section channel	m^2
A	surface	m^2
$A^{'}$	Section suction line	M^2
$A^{'}$	Perpendicular surface	m^2
$c_{\perp}$	Velocity perpendicular on surface	m/s
$c_{\parallel}$	Velocity along surface	m/s
c	Absolute velocity at impeller	m/s
c	Absolute at inlet rotor	m/s
c_l	Velocity lower surface hydrofoil	m/s
c_p	Specific heat at constant pressure	J/kg.K
c_v	Specific heat at constant volume	J/kg.K
c_u	Velocity upper surface hyrdofoil	m/s
c_{1r}	Radial component	m/s
C_L	Lift factor	-
C_D	Drag factor	-
D	Diameter	m
D	Drag force	N
D	Diameter impeller	m
D_H	Hydraulic diameter	m
F	Force	N
F_C	Centrifugal force	N
g	Gravity acceleration	m/s^2
H	Height	m
h	Depth impeller	m

Symbol	Meaning	Unit
h	Specific enthalpy	J/kg
H_{geo}	Geodetic height	m
H_{man}	Manometric height (head)	m
H_p	Geodetic press height	m
H_s	Geodetic suction height	m
$H_{s,max}$	Maximum suction head	m
H_f	Friction loss head	m
k	Absolute roughness	m
l_1	Distance covered during suction stroke	m
l_2	Distance covered during press stroke	m
L	Length piston rod	m
L	Lift force	N
L	Length pipe line	m
L'	Fictive length suction line	m
m	Mass	kg
m^*	Displaced mass per cylinder	kg
M	Molar mass	kg/kmol
n	Polytropic exponent	-
N	Speed	rmp
N_s	Specific speed	$m^{3/4} \cdot s^{-3/2}$
N_{ss}	Suction specific speed	$m^{3/4} \cdot s^{-3/2}$
N_q	Dimionless specific speed	-
N_ω	Dimensionless specific speed	-
$NPSH_a$	Available net positive suction head	m
$NPSH_{ss}$	Suction net positive suction head	-
$NPSH_r$	Required net positive suction head	m
O_{cd}	Surface under curve cd	J
O_{ab}	Surface under curve ab	J
p_v	Vapor pressure	Pa
p	Static pressure	Pa
p_a	Atmospheric pressure	Pa
p_{abs}	Absolute pressure	Pa
p_{eff}	Effective pressure	Pa
p_{dyn}	Dynamic pressure	Pa
p_{man}	Manometric (feed) pressure	Pa
p_{geo}	Geodetic (feed) pressure	Pa
$p_{v,p}$	Vapor pressure press vessel	Pa
$p_{v,s}$	Vapor pressure suction vessel	Pa

Symbol	Meaning	Unit
$p_{r,p}$	Pressure press chamber	Pa
$p_{r,s}$	Pressure suction chamber	Pa
$p_{f,s}$	Friction pressure suction pipe	Pa
$p_{f,p}$	Friction pressure press pipe	
$p_{l,p}$	Friction loss in pump	
p_s	Static feed pressure	Pa
$P_{tot,s}$	Total pressure suction side	Pa
$P_{tot,p}$	Total pressure press side	Pa
P_t	Technical power on drive shaft	W
q	Specific heat	J/kg
Q_M	Mass flow	Kg/s
Q_V	Volumetric flow	m^3/s
$Q_{v,a}$	Average volumetric flow	m^3/s
Q_{Veff}	Effective volumetric flow	m^3/s
$Q_{n,nom}$	Nominal effective flow	m^3/s
u	Peripheral speed	m/s
U	Voltage	V
R	Universal gas constant	J/kmol.K
R, r	Crankstroke, radius	m
s	Stroke length	m
T	Absolute temperature	K
u	Circumpheral velocity impeller	m/s
v	Specific volume	m3/kg
V	Total volume	m3
w	Relative velocity	m/s
w	Specific work	J/kg
w_c	Specific compression work	J/kg
w_t	Specific technical work	J/kg
W_c	Total compression work	J
W_t	Total technical work	J
x	Position	m
z	Number of vanes (channels)	-

Greek Symbols

α	Absolute angle impeller	° or rad
β	Relative angle impeller	° or rad
γ	Isentropic exponent	-

n	Polytropic exponent	-
δ	Correction factor laminar flow	-
ε	Factor dead volume	-
ρ	Specific mass	kg/m^3
θ	Angle of incidence	° or rad
Δ	Difference	-
Δpot	Specific change of potential energy	J/kg
Δkin	Specific change of kinetic energy	J/kg
ΔH	Geodetical height	m
λ	Hydraulic resistance factor	-
λ	Volumetric efficiency	-
λ	Mean free length	m
τ	Formation time	s
$\varphi(..)$	Function of	-
ω	Angular velocity	rad/s
ζ	Correction factor laminar flow	-
ν	Kinematic viscosity	m/s^2
v	Axial velocity	m/s
η	Dynamic viscosity	Pa·s
ξ	Hydraulic resistance factor	-
Γ	Reaction degree	-
Σ	Sum	-
Π	Product	-
Π_1	1st number of Rateau	-
Π_2	2nd number of Rateau	$m^{-1} \cdot s^2$

Indexes

1	State 1, inlet impeller
2	State 2, outlet impeller
3	State 3
a	State a
b	State b
c	State c
d	State d
d	Dead
s	Stroke
a	allowable
c	compression

t	technical
ab	State change ab
ac	State change ac
cd	State change cd
12	State change 12
13	State change 13
23	State change 23

Upper indexes

‘	1st stage
“	2nd stage

About the Companion Website

This book is accompanied by a companion website:

www.wiley.com/go/borremans/pumps 

The website includes:

- References
- Videos

Scan this QR code to visit the companion website.

Part I

Pumps

1 General Concepts

CHAPTER MENU

1.1 Hydrostatics

Consider an incompressible liquid at rest. The law of Pascal applies (Figure 1.1):

$$p = p_a + \rho \cdot g \cdot H$$

where

- p: the static pressure at the considered point [Pa = N/m^2]
- p_a: atmospheric pressure (ca. 1 [bar] = 10^5 [Pa])
- ρ: specific mass of the liquid [kg/m^3]
- g: gravitational acceleration (9.81 [m^2/s])
- H: height beneath the liquid surface [m]).

The standard pressure at sea level amounts to 1.013 [bar]. This pressure is the *absolute pressure.*

In practice the *relative*, or *effective*, pressure p_{eff} is of importance. This is the difference between the absolute pressure p_{abs} and the atmospheric pressure p_a:

$$p_{eff} = p_{abs} - p_a$$

Pumps and Compressors, First Edition. Marc Borremans.

Companion website: www.wiley.com/go/borremans/pumps

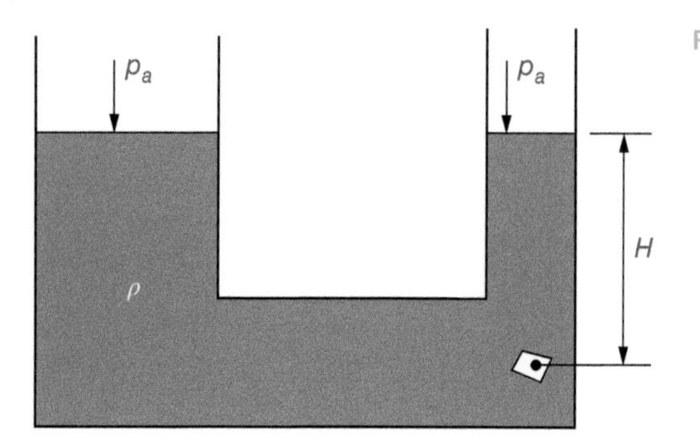

Figure 1.1 Law of Pascal.

One can distinguish the following pressure measurement apparatus:

- A *manometer* usually measures the effective pressure; this is an overpressure (mostly with respect to the atmospheric pressure).
- A *vacuum meter* measures an underpressure (with respect to the atmospheric pressure).
- A *barometer* measures the absolute pressure of the atmosphere.

Concerning the unities nowadays a consistent notation is held. One notes [bar(g)] of [bar_g] for effective pressure ("g" stands for gauge) and [bar(a)] or [bar_a] for absolute pressure.

1.2 Flow

Consider a pipe with variable section (Figure 1.2). At section 1 a fluid (gas or liquid) possesses a velocity c and a specific mass ρ. The cross-section there is A_1. Use an analogous notation for section 2.

The *mass flow*, i.e. the amount of mass that per unit of time flows through the section, also flows through section 2 (conservation of mass).

The mass flow Q_M is given by:

$$Q_M = \rho_1 \cdot c_1 \cdot A_1 = \rho_2 \cdot c_2 \cdot A_2 \text{ [kg/s]}$$

In the case of an incompressible fluid (liquids) ρ is constant. One can use the *volumetric flow* Q_V instead, i.e. the amount of volume that flows per unit of time through a section:

$$Q_V = c_1 \cdot A_1 = c_2 \cdot A_2 \left[\frac{m^3}{s}\right]$$

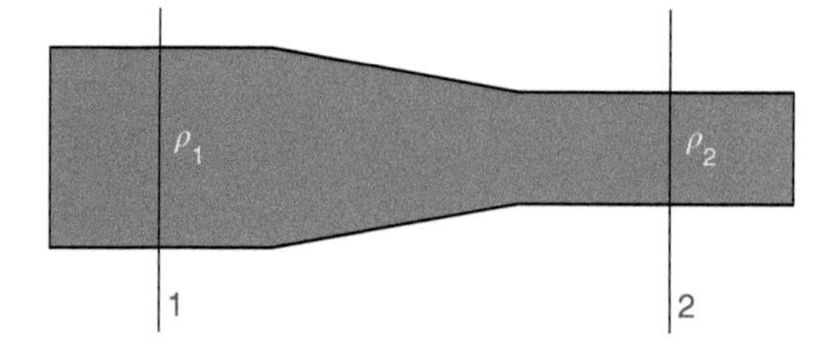

Figure 1.2 Flow.

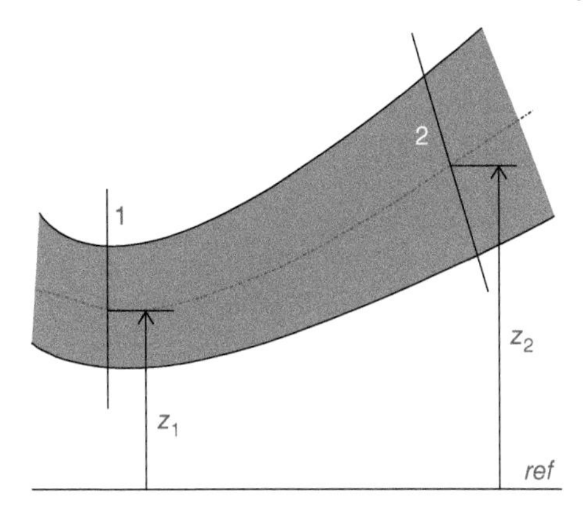

Figure 1.3 Law of Bernoulli.

1.3 Law of Bernoulli

Bernoulli expresses the conservation of energy for liquids. A fluid possesses pressure energy (dynamic pressure) and potential energy (Figure 1.3):

$$p_1 + \rho \cdot g \cdot z_1 + \rho \cdot \frac{c_1^2}{2} = p_2 + \rho \cdot g \cdot z_2 + \rho \cdot \frac{c_2^2}{2}$$

where z is the height coordinate.

1.4 Static and Dynamic Pressure

Consider a horizontal pipe. The law of Bernoulli applied to points 1 and 2 (Figure 1.4):

$$p_1 + \rho \cdot \frac{c_1^2}{2} = p_2 + \rho \cdot \frac{c_2^2}{2}$$

Point 2 is a *stagnation point*:

$$c_2 = 0$$

So:

$$p_2 = p_1 + \rho \cdot \frac{c_1^2}{2}$$

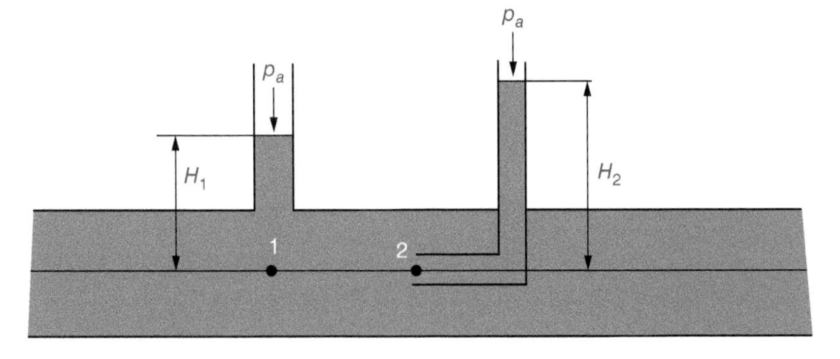

Figure 1.4 Static and dynamic pressure.

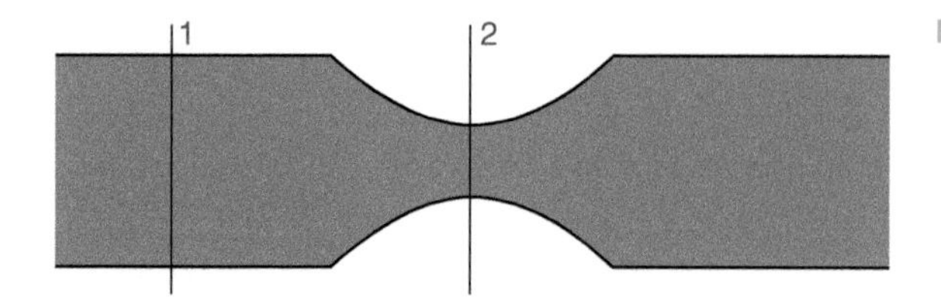

Figure 1.5 Venturi.

From which: $p_2 > p_1$
In vertical sense there is no movement and one can apply hydrostatics:

$$p_1 = p_a + \rho \cdot g \cdot H_1$$

$$p_2 = p_a + \rho \cdot g \cdot H_2$$

From which

$$p_2 = p_1 + \rho \cdot g \cdot (H_2 - H_1)$$

Look now at a tube with a narrowed passage (a venturi) (Figure 1.5):

$$Q_v = c_1 \cdot A_1 = c_2 \cdot A_2 \text{ with: } A_1 > A_2$$

$$c_1 < c_2$$

In the throat the liquid velocity will be greater than elsewhere.
The dynamic pressure in the throat is greater than elsewhere.
Application of Bernoulli leads to:

$$p_1 + \rho \cdot \frac{c_1^2}{2} = p_2 + \rho \cdot \frac{c_2^2}{2}$$

So that

$$p_1 > p_2$$

One finds that static pressure can be converted into dynamic pressure, and vice versa. That's why in many considerations in applied mechanics one often speaks of the *total pressure* of a fluid being the sum of static and dynamic pressure. The *total pressure* then expresses the "total energy content" of a liquid.

1.5 Viscosity

This paragraph is valid for liquids as well as for gasses, so we use the generic word "fluid."

Consider a fluid in an open channel (Figure 1.6). With the help of a plate a horizontal force *F* is applied in order to move the fluid. Let's imagine that the fluid consists of horizontal layers. At the top of the fluid the layer "adheres" by cohesion forces to the plate.

This layer thus moves with the velocity of the plate. At the bottom, however, the layer does not move at all because there it is bounded by cohesion forces to the bottom of the channel. It then is clear that every layer will possess its own velocity *c*, evolving from the highest velocity at the top to velocity zero at the bottom.

Every layer will exercise a resistance on the adjacent layer (a shearing stress): the layer at the top will have a braking action on the second layer, the third layer will be braked by the second one, and so on... One speaks of "viscous friction" between the layers.

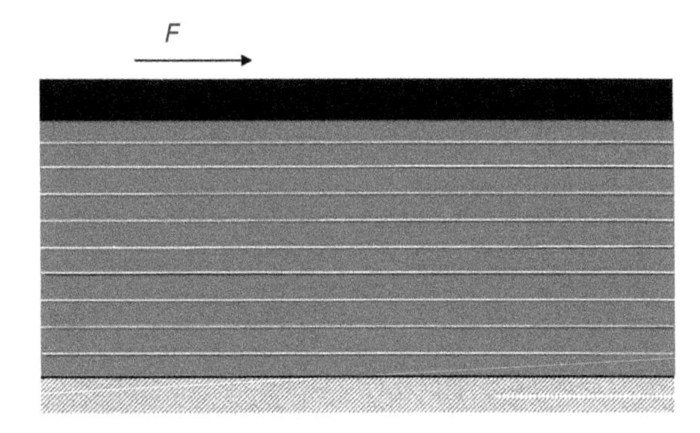

Figure 1.6 Viscosity.

If one wants to maintain a velocity of a moving fluid then a force F is necessary. Newton states that in order to keep an object at constant speed no force is necessary (of course this holds true only if there are no disturbing forces, like here). This force overcomes the internal friction of the fluid, i.e. the friction that the layers exercise on each other.

For so-called *Newtonian* liquids and gasses, the next law applies:

$$F = \eta \cdot A \cdot \frac{dc}{dz}$$

where:

A [m^2]: the section of which the force F is applied

c [m/s]: the local velocity of the fluid

z [m]: the position on the vertical axis

η [Pa · s]: the dynamic viscosity (old unit, non-SI: 1 Poise = 0.1 [Pa · s])

The unit Poiseuille is a shortcut for 1 [Pa · s].

The preceding case of an open channel can easily be extended to the case of a tube wherein a fluid moves: it suffices to mirror the case of Figure 1.6 around the plate (Figure 1.7).

Then, too, a force F is necessary to guide a fluid through a pipe, or in other words a pressure difference over the pipe is needed to compensate for the internal friction of the fluid.

The order of magnitude of the dynamic viscosity for gasses at room temperature is $10 \cdot 10^{-6}$[Pa·s].

In the range of low pressures, this is from 0.1 to 10 [bar], for ideal and real gasses η is independent of the pressure.

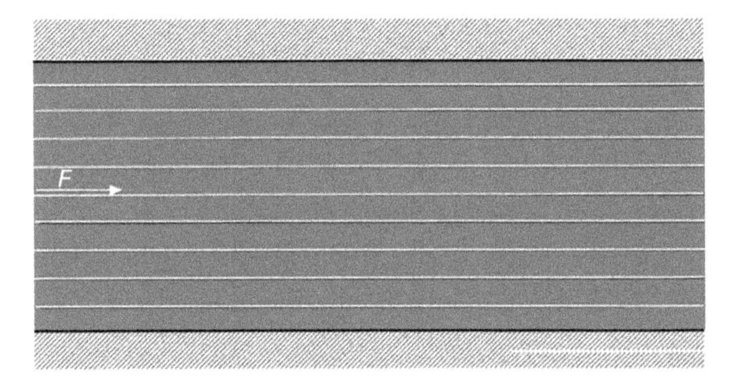

Figure 1.7 Channel.

The dynamic viscosity for liquids varies between large ranges, so typical values cannot be given. With increasing pressure, the dynamic viscosity of most liquids increases nearly proportionally with the pressure.

For gasses, the viscous effects come about by exchanges of impulse (mass multiplied by velocity) of the molecules: when a layer of gas is brought into movement the molecules in that layer will lose kinetic energy because of collisions with other molecules. For liquids, viscosity is caused by the intramolecular cohesion forces that brake the shift of the layers. That's why the viscosity of liquids decreases with increasing temperature there where for gasses the increasing motion of the molecules promotes the exchange of impulse so that the viscosity of gasses increases with temperature.

The dynamic viscosity of liquids at moderate pressures decreases exponentially with temperature. But, for gasses, by approximation:

$$\eta \div \sqrt{T}$$

The force F varies in every layer. So, we should write:

$$F = \eta \cdot A \cdot \frac{\partial c}{\partial z}$$

where F is the local force and $\frac{\partial c}{\partial z}$ is the local velocity gradient, or *rate of shear deformation.*

Dividing F by A leads to the local shear stress τ (Figure 1.8 for *a laminar flow* – see later).

$$\tau = \frac{F}{A} = \eta \cdot \frac{\partial c}{\partial z}$$

See Appendix A for a calculation of the velocity profile and the mean velocity c_m.

Not all liquids behave as Newtonian fluids (like water); in the pump industry the following liquids are common:

- *Pseudo plastic liquids*: the viscosity will decrease with increasing shear stress, e.g. paint and shampoos.
- The inverse includes *dilatant liquids*: the viscosity will increase as the shear stress increases, e.g. honey and quicksand.

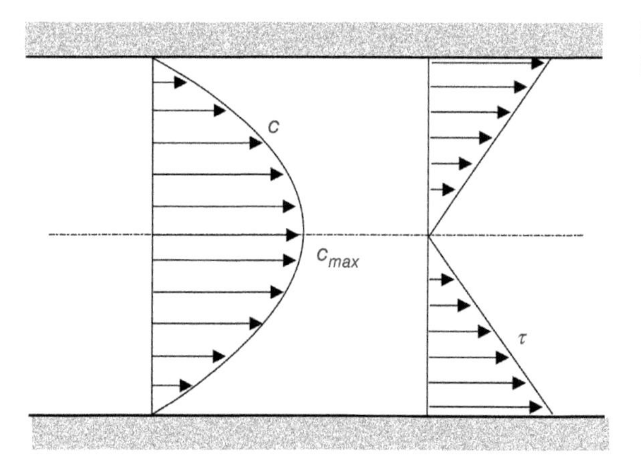

Figure 1.8 Velocity and shear stress profile in laminar flow.

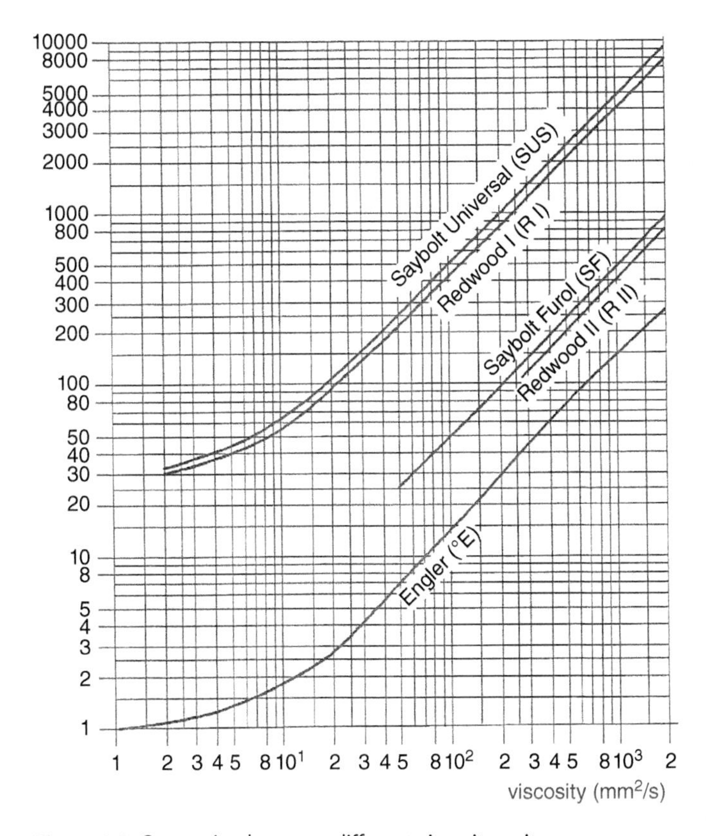

Figure 1.9 Conversion between different viscosity units.

Figure 1.10 Heating of heavy oils.

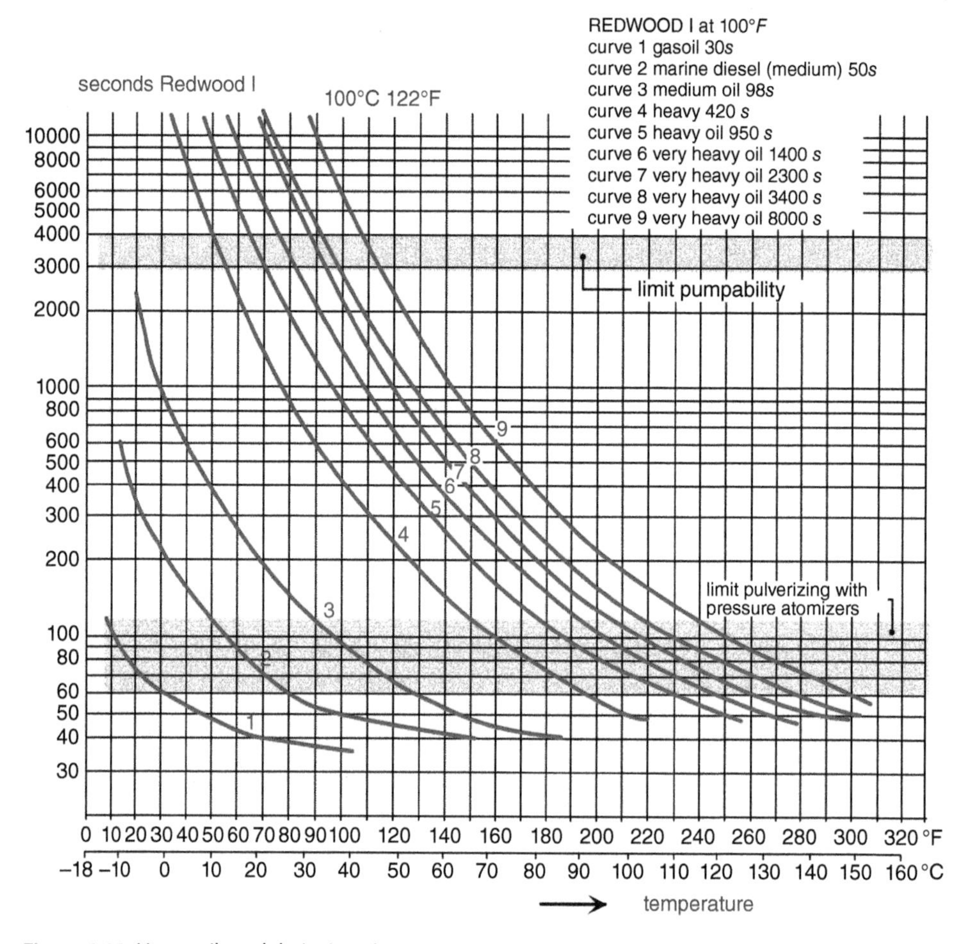

Figure 1.11 Heavy oils and their viscosity.

- A *plastic liquid* behaves very strangely: applying a certain shear stress leads to a Newtonian behavior, but with increasing shear stress it becomes pseudoplastic and then dilatant, e.g. ketchup and lubricating grease.
- *Thixotropic liquids* will, when the shear stress disappears, show an increasing viscosity with evolving time, e.g. yogurt, print ink, and sludge. When they are in a pump and the pump works they become thin. But if the pump stops, they become thick. After a while the pump is not able to pump them anymore.

In practice it turns out that the expression $\frac{\eta}{\rho}$ frequently appears in equations. That's why one defines the kinematic viscosity ν:

$$\nu \equiv \frac{\eta}{\rho}$$

The units are: $[\mathrm{m^2/s}]$

Sometimes one uses an old non-SI unit, the Stokes. 1 [St] = 10^{-4} $[\mathrm{m^2/s}]$. Other so-called technical units are Redwood I, Redwood II, Saybolt universal, Saybolt FUROL and Engler degrees.

Table 1.1 Guide values for fluids.

Medium		m/s
Oil	In pipelines	1–3
Water	In long pipelines	0.5–1
	Behind piston pumps	1–2
	Behind turbopumps	1.5–3
	For turbines	2–7
Gas	Low pressure	5–30
	Middle pressure	5–20
	High pressure	3–6
Compressed air	In pipelines	2–4
Vapor	1–10 [bar]	15–20
	10–40 [bar]	20–40
	40–125 [bar]	30–60

An oil with a viscosity of 3.6 [mm^2/s] at 20 °C will need twice as much efflux time than water with a viscosity of 1.8 [mm^2/s] at 20 °C.

The conversion between the different viscosity units is represented in Figure 1.9.

As pointed out earlier, the viscosity of a liquid decreases with temperature. That may imply that some liquids have to be heated before transport and to be able to form drops in an atomizing burner (Figures 1.10 and 1.11).

Guide values for fluids are given in Table 1.1.

1.6 Extension of Bernoulli's Law

Consider a pipeline with friction loss (Figure 1.12).

The classic law of Bernoulli:

$$p_1 + \rho \cdot \frac{c_1^2}{2} = p_2 + \rho \cdot \frac{c_2^2}{2}$$

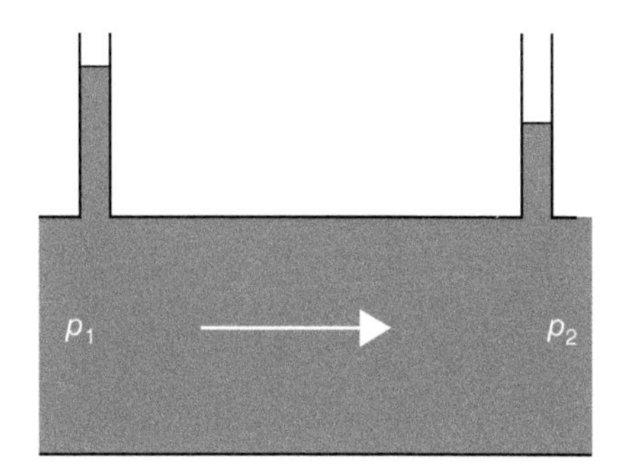

Figure 1.12 Friction loss.

is not valid anymore because:

$$p_1 + \rho \cdot \frac{c_1^2}{2} > p_2 + \rho \cdot \frac{c_2^2}{2}$$

Note the pressure loss by friction p_f, then:

$$p_1 + \rho \cdot \frac{c_1^2}{2} = p_2 + \rho \cdot \frac{c_2^2}{2} + p_f$$

In general this becomes:

$$p_1 + \rho \cdot \frac{c_1^2}{2} + \rho \cdot g \cdot z_1 = p_2 + \rho \cdot \frac{c_2^2}{2} + \rho \cdot g \cdot z_2 + p_f$$

Dividing by $\rho \cdot g$ leads to the law of Bernoulli being seen in the unit meter:

$$\frac{p_1}{\rho \cdot g} + \frac{c_1^2}{2 \cdot g} + z_1 = \frac{p_2}{\rho \cdot g} + \frac{c_2^2}{2 \cdot g} + z_2 + H_f$$

With the friction loss in meters:

$$H_f = \frac{p_f}{\rho \cdot g} \text{ [m]}$$

1.7 Laminar and Turbulent Flow

Consider a vessel where a pipeline is connected to (Figure 1.13). On the pipeline a valve is connected so that the flow can be regulated and thus the velocity of the liquid also. Color is then added via an opening. When the velocity is not too high the stream is very regular: the colored parts run in parallel layers. This is a *laminar* flow. At higher velocities the color particles form an irregular path and begin to mix with the liquid; the layers exchange energy with each other. This is a *turbulent* flow.

The difference between laminar and turbulent flow is made by the number of Reynolds (*Re*), defined by:

$$\text{Re} \equiv \frac{c_m \cdot D}{\nu}$$

where:

c_m: the mean velocity of the liquid [m/s]
D: the diameter of the pipeline [m]
ν: the kinematic viscosity [m^2/s]

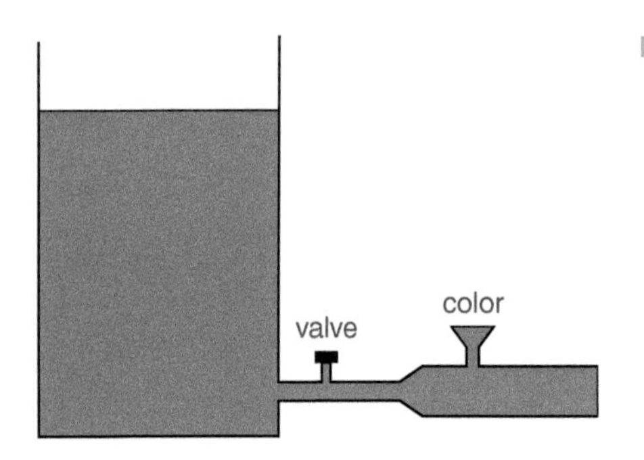

Figure 1.13 Types of flow.

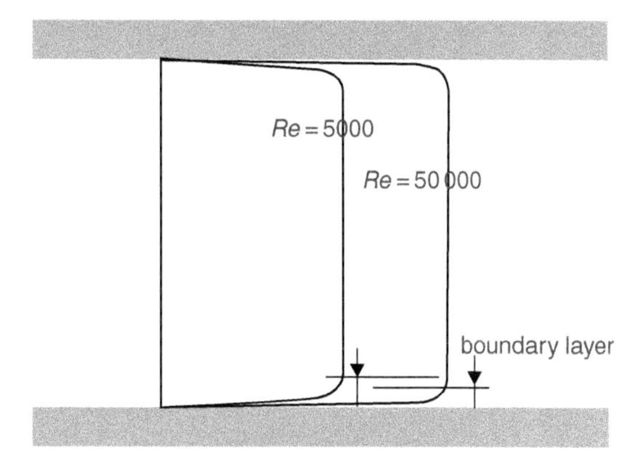

Figure 1.14 Turbulent flow: velocity profile.

It is easy to check that the number of Reynolds is, indeed, dimensionless.

When for a certain flow in a round tube Reynolds is lower than 2300 the flow is laminar; otherwise, it is turbulent.

In a laminar flow the friction losses p_f are proportional to the liquid velocity c, but in a turbulent flow they are quadratically proportional to this velocity.

The question arises if in practice one deals with laminar or turbulent flows. Take the case of water at 10 °C, flowing through a pipe with a diameter of 0.1 [m]. In order for the flow to be laminar the velocity may not be higher than:

$$c_m = \frac{2300 \cdot 1,3 \cdot 10^{-6}}{0,1} = 0.03\ [\mathrm{m/s}]$$

This velocity is very low. Practically one deals mostly with a turbulent flow. Only with very viscous oils or substances is the flow laminar. But, in a turbulent flow, the friction losses increase very quickly with the velocity. This last one will be limited as much as possible. For water in pipelines: ca. 1.5 [m/s].

In a turbulent flow the velocity profile is quite different from that of a laminar flow (Figure 1.14). Apart from a boundary layer, where the flow is laminar because of shear stresses, the velocity is constant. This is because the particles of the liquid are mixed and in this way exchange impulse with each other: so no difference in velocity can exist; if it did, *it would be destroyed by the interaction of the particles.*

Sometimes the liquid is so viscous that it has to be heated before transportation or pulverization in a nozzle from a burner.

1.8 Laminar Flow

1.8.1 Hydraulic Resistance

For a laminar or turbulent flow the following expression is valid:

$$p_f = \lambda \cdot \frac{L}{D} \cdot \frac{\rho \cdot c_m^2}{2}$$

It is possible to prove that for a laminar flow (see Appendix A):

$$\lambda = \frac{64}{Re}$$

For a noncircular section, still for a laminar flow, one introduces a correction factor:

$$\lambda = \zeta \cdot \frac{64}{Re}$$

Example 1.1 ***Concentric eccentric profile***
The value of ζ can be found in tables or diagrams (Figure 1.15).

Example 1.2 ***Rectangular profile***

a/b	1	2	3	4	6	8	∞
ζ	0.98	0.97	1.07	1.14	1.23	1.29	1.5

Example 1.3 ***Ellipse***

a/b	1	2	4	8	16
ζ	1	1.05	1.14	1.20	1.22

Example 1.4 ***Isosceles triangle***

θ	10°	30°	60°	90°	120°
ζ	1.79	0.82	0.83	0.82	0.80

1.8.2 Hydraulic Diameter

Consider a noncylindrical pipeline (Figure 1.16) of length L. In general that can also be an *open* channel. The cross-section where the fluid flows through is labelled A_w, the wet perimeter P_w. The aim is now to find an equivalent cylindrical tube for such cases. The diameter of that pipeline is called the *hydraulic diameter* D_H.

In order to let the noncylindrical pipeline behave like a circular line of diameter D and length L the viscous shear stresses in the fluid should lead to the same pressure drop p_f.

Isolate in both cases a piece of liquid mass between section 1 and 2.

On this mass the following forces are acting:

- The liquid on the left exercises a force $p \cdot A$, with:

$$A = \frac{\pi \cdot D_H^2}{4}$$

- The liquid on the right exercises a force $(p - p_f) \cdot A$

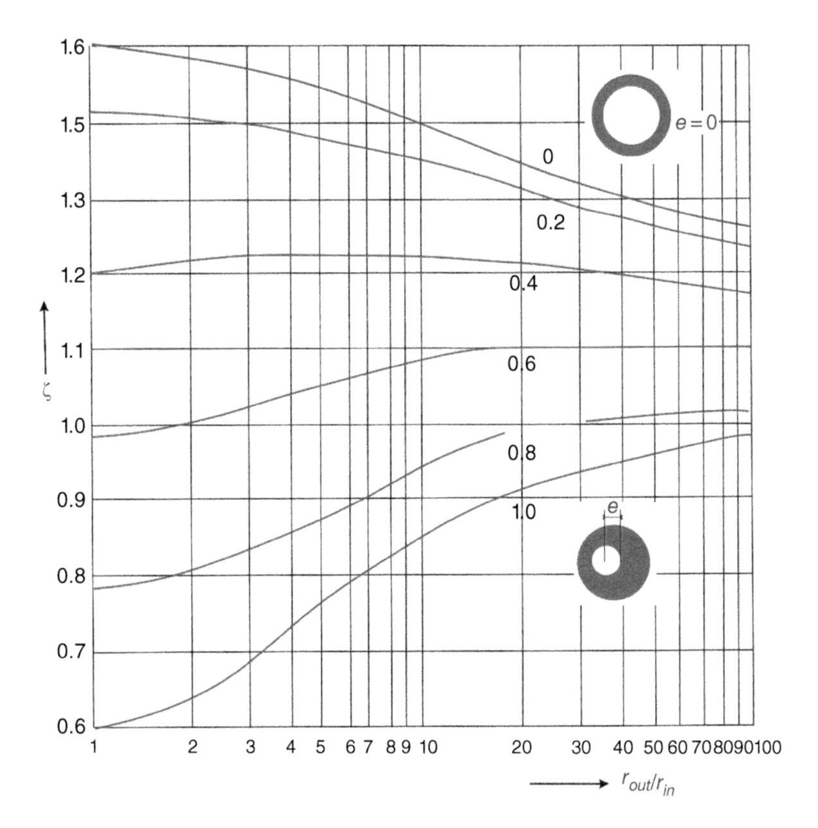

Figure 1.15 Concentric eccentric profile: (a) rectangular profile; (b) ellipse; (c) isosceles triangle.

Figure 1.16 Hydraulic diameter.

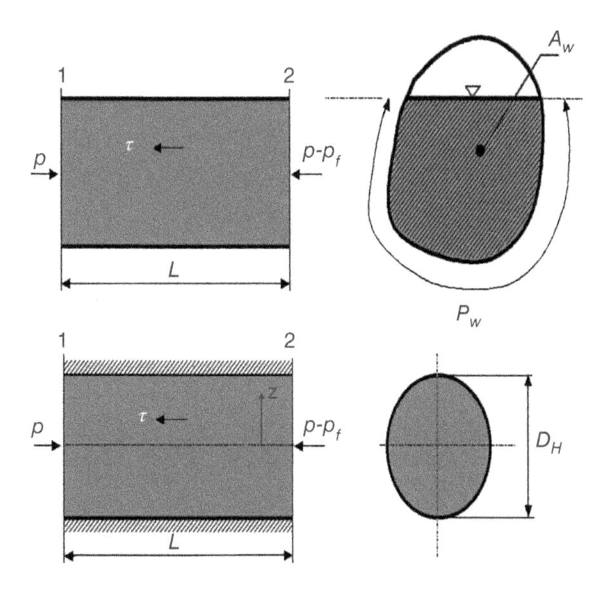

- At the cylinder wall shear stresses are exercised according to:

$$\tau = \eta \cdot \left(\frac{dc}{dz}\right)_{\frac{D_H}{2}}$$

The total shear force is then:

$$\tau \cdot (\pi \cdot D_H) \cdot L$$

where $\pi \cdot D_H$ is the surface on which the shear stress is active.

If the liquid mass in the pipeline or open channel moves with a constant mean velocity, the sum of all forces is zero, according to Newton's law:

$$\frac{\pi \cdot D_H^2}{4} \cdot p_f - \tau \cdot (\pi \cdot D_H) \cdot L$$

From where:

$$p_f = \frac{4 \cdot \tau \cdot L}{D_H}$$

Isolate now the liquid mass that flows between sections 1 and 2 of the noncylindrical channel.

On this mass the following forces are acting:

- The force $p \cdot A_w$
- The force $(p - p_f) \cdot A_w$
- The shear force $\tau \cdot P_w \cdot L$

where she shear stresses are active on the *wet* surface.

If the liquid mass is moving at a constant mean speed then the sum of all these forces is zero:

$$A_w \cdot p_f - \tau \cdot P_w \cdot L = 0$$

From where the pressure drop:

$$p_f = \frac{\tau \cdot P_w \cdot L}{A_w}$$

Identification of the two expressions for p_f:

$$D_H = \frac{4 \cdot A_w}{P_w}$$

where the number of Reynolds is calculated with the hydraulic diameter as character dimension:

$$Re = \frac{c \cdot D_H}{\nu}$$

Example 1.5 ***Square tube***

$$D_H = \frac{4 \cdot A_w}{P_w} = \frac{4 \cdot a^2}{4 \cdot a} = a$$

Example 1.6 ***Rectangular tube***

$$D_H = \frac{4 \cdot A_w}{P_w} = \frac{4 \cdot a \cdot b}{2 \cdot (a+b)} = \frac{2 \cdot a \cdot b}{a+b}$$

Example 1.7 ***Concentric passage***

$$D_H = \frac{4 \cdot A_w}{P_w} = \frac{4 \cdot \pi \cdot (r_{out}^2 - r_{in}^2)}{2 \cdot \pi \cdot (r_{our} + r_{in})} = 2 \cdot (r_{out} - r_{in})$$

1.9 Turbulent Flow

For straight pipes the pressure (friction) loss p_f is given by the formulae of Darcy–Weisbach:

$$p_f = \lambda \cdot \frac{L}{D} \cdot \left(\rho \cdot \frac{c_m{}^2}{2}\right)$$

Herein:

- c_m: mean velocity in pipe [m/s]
- D: diameter pipe [m]
- L: length of pipe [m]
- λ: hydraulic resistance factor (dependent on many factors, such as fluid, dimensions, etc.), dimensionless

To determine the value of λ one dispenses with tables, graphs, and empirical formulas.

Example 1.8 ***For smooth pipes and*** $3 \cdot 10^3 < Re < 10^6$

Formula of Blasius: $\lambda = 0.316 \cdot \mathrm{Re}^{-0.25}$

In case the pipe is rough inside, apart from the internal friction in the fluid, friction may occur against the wall of the pipe. This leads to additional losses (see later literature).

Example 1.9 ***For water***

Formula of Lang: $\lambda = 0.02 + \frac{0{,}0018}{\sqrt{c \cdot D}}$

(often the approximation: $\lambda = 0.03$ is made)

On the other hand, pressure losses occur in all sorts of "obstacles" like bends, elbows, widenings or restrictions.

For these cases one uses the general formula:

$$p_f = \xi \cdot \left(\rho \cdot \frac{c_m{}^2}{2}\right)$$

The values of ξ also are extracted from tables, graphs, and empirical formulas. Turbulent flows are characterized by the independence of *Re*.

Table 1.2 Perpendicular bend.

D/r	0.4	0.6	0.8	1	1.2	1.4	1.6	1.8
φ	0.13	0.18	0.25	0.4	0.64	1	1.55	2.17

Example 1.10 *Perpendicular bend (Table 1.2)*

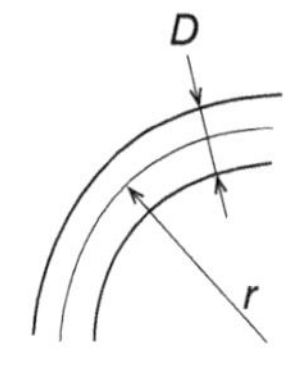

Example 1.11 *Slide shut-off valve*

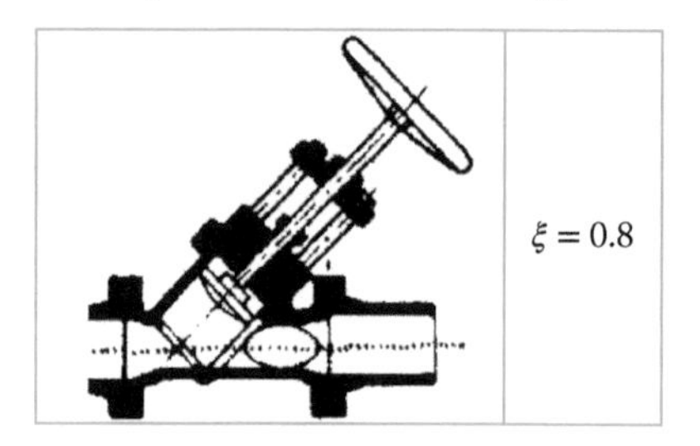

$\xi = 0.8$

Example 1.12 *Plug valve*

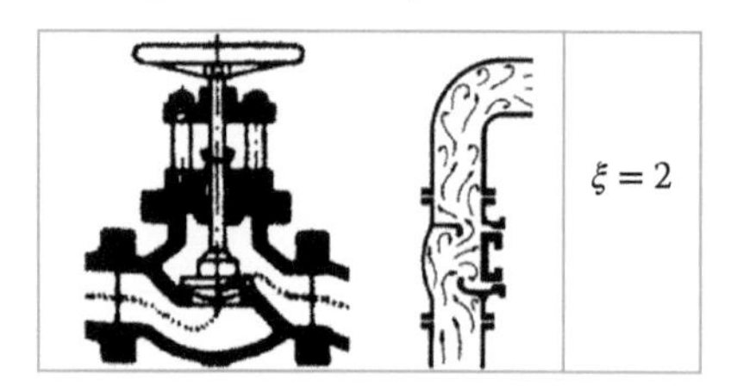

$\xi = 2$

Example 1.13 *Suction strainer*

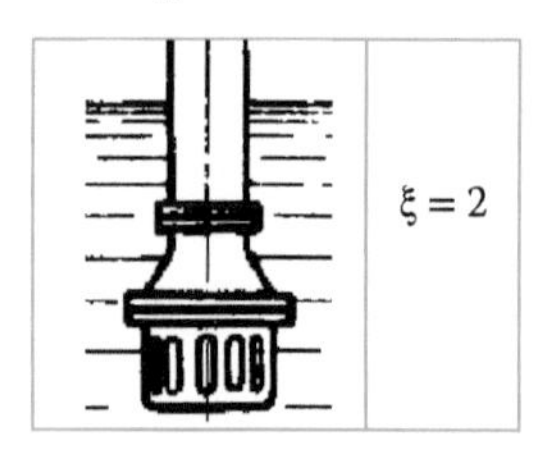

$\xi = 2$

Example 1.14 ***Diffusor***

When one looks at the value of ξ in the case of widenings (diffusors), it is clear that it increases very fast with the top angle of the cone ($2 \cdot \varphi$). For that reason, with regard to minimizing the friction losses, one will limit this top angle to 10°.

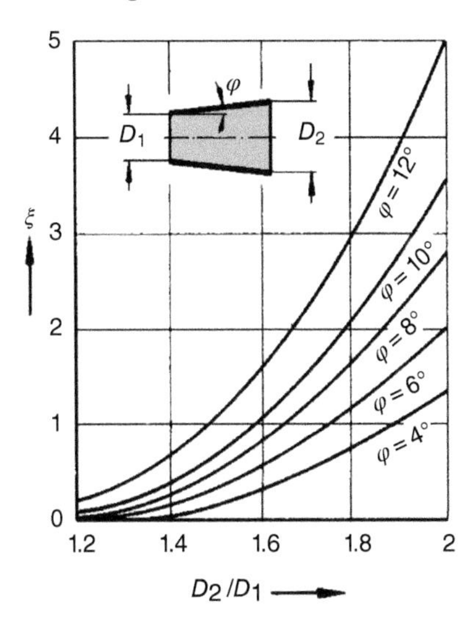

Example 1.15 ***Converging passage***

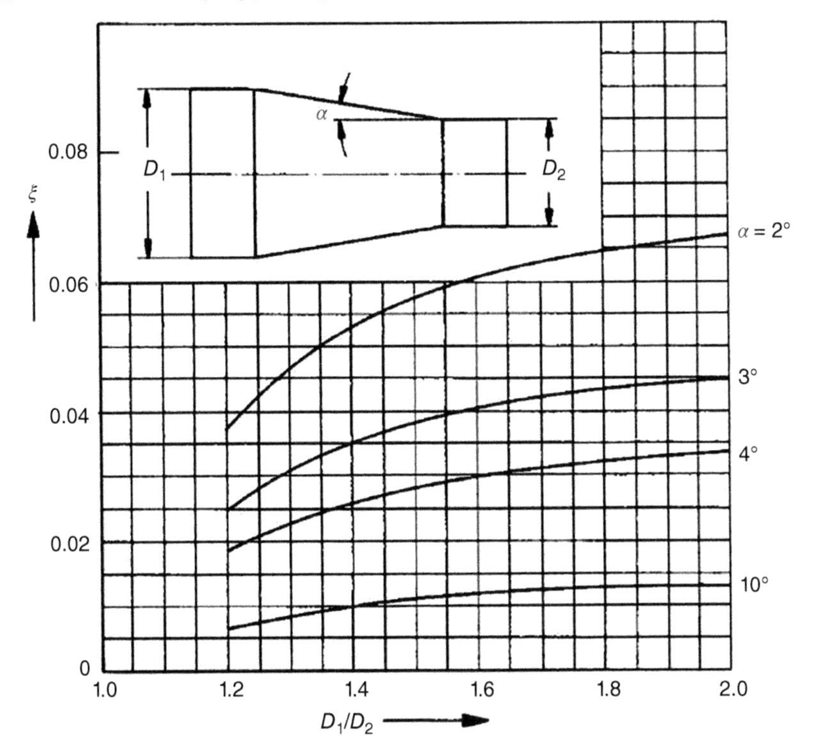

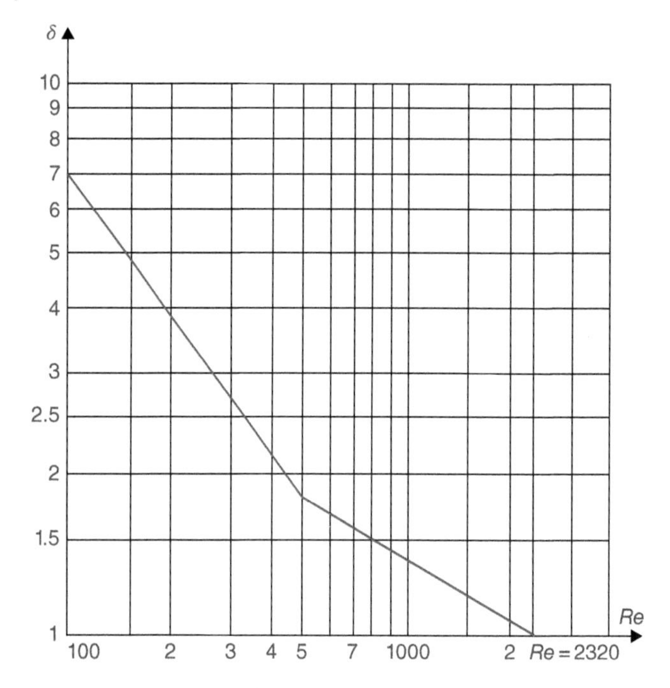

Figure 1.17 Conversion factor δ.

For a laminar flow the resistance factor ξ can be found by first calculating ξ for a turbulent flow and correcting with a conversion factor δ:

$$\xi_{lam} = \delta \cdot \xi_{turb}$$

where δ can be found from the diagram in Figure 1.17.

1.10 Moody's Diagram

Moody's diagram is valid for circular tubes in laminar flow and for all tubes with hydraulic diameter D_H in turbulent flow (Figure 1.18).

The surface condition of the inner side of the pipe is given by its relative roughness $\epsilon = \frac{k}{D}$, where D is the diameter of the tube and ϵ the roughness of the inner surface. The absolute roughness is not identical to the technical or natural roughness but is an equivalent, artificial roughness that is defined as *sand roughness*. It is given by sand grains with diameter k on smooth tubes that reproduce the natural roughness of sand (Figure 1.19).

In general: $\lambda = \varphi(Re, \varepsilon)$ (Table 1.3).

The friction factor λ increases in old pipes because of corrosion and sediments. According to the engineer who did the experiments the resistance must be multiplied with an aging factor(see Table 1.4).

Example 1.16 Consider a pumping installation with a water flow $Q_V = 360$ [m^3/h]. The geodetic suction head is 2.8 [m] and the geodetic pressure head 22 [m]. The suction pipe is 6 [m] long and is provided with a suction strainer with check valve ($\xi = 4$)

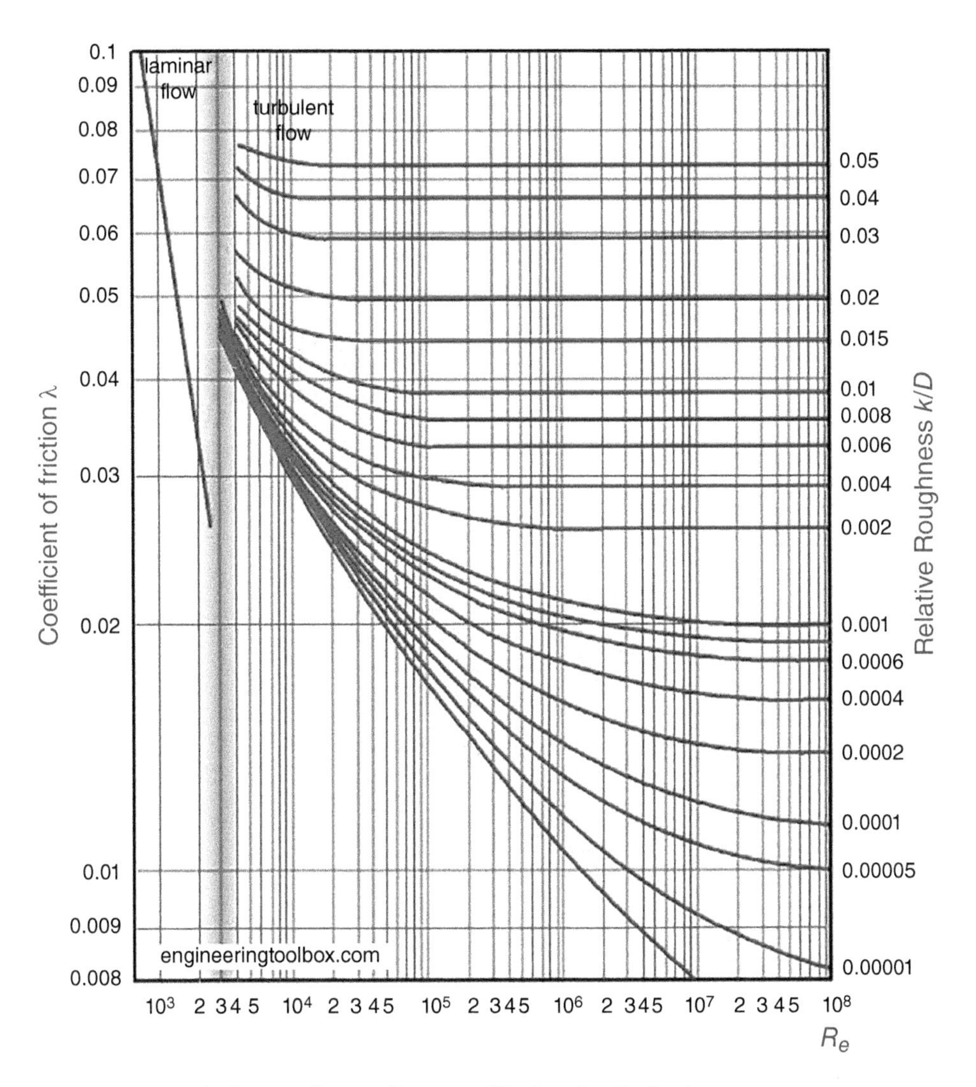

Figure 1.18 Moody diagram. (Source: Courtesy of Engineering Toolbox).

and a bend with radius 0.5 [m]. The press pipe is 150 [m] long and made up of five perpendicular bends (radius 0.5 [m]), five check valves ($\xi = 1$), and two sliding shut-off valves.

Determine the pressure that the pump should deliver (Figure 1.20).

At first, we determine the diameter D of the pipelines. We start by assuming a maximum water velocity of 1.5 [m/s]:

$$Q_V = \frac{\pi \cdot D^2}{4} \text{ from what: } D = 0.29 \text{ [m]}$$

We choose $D = 30$ [cm]. Backwards calculation of the diameter with this velocity leads to $c = 1.415$ [m/s].

From the various losses the ϕ values are calculated separately:

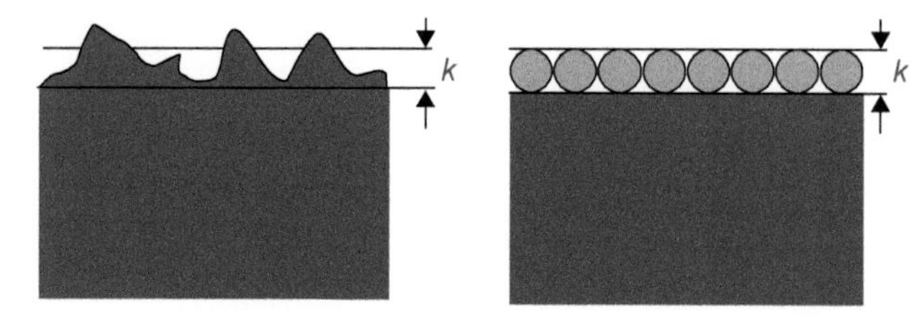

Figure 1.19 Absolute roughness k.

Table 1.3 Absolute roughness.

Sort	State	k (mm)
Drawn tubes in glass, lead, copper, brass		0.0015
PVC, polyethylene		0.05
Drawn steel tube	New	0.05
	Moderate rusted	0.4
	Heavy rusted	3
Welded steel tubes	New	0.05
	Moderate rusted	0.3
	Heavy rusted	4
Galvanized tubes		0.15
Cast iron	New	0.25
	Moderate rusted	1
	Heavy rusted	4
Concrete tubes	Smooth	0.3
	Rough	3
Wood	Scraped	0.18
	Not scraped	0.9

Table 1.4 Aging factor.

Year	2	5	10	20	30	40	50
Factor	1.1	1.2	1.35	1.75	2.10	2.60	3

The straight pipeline:

$$\xi_1 = \lambda \cdot \frac{L}{D} \text{ with } \lambda \cong 0.03$$

$$\xi_1 = 0.03 \cdot \frac{156}{0.3} = 1.6$$

The "obstacles"

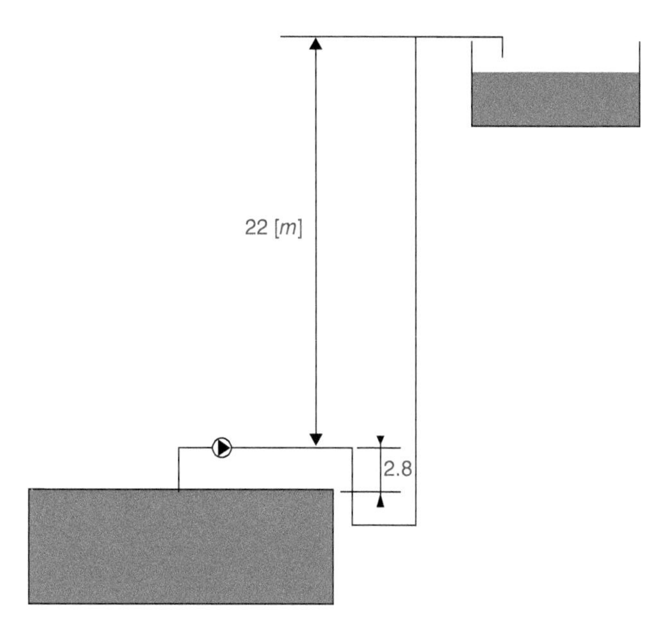

Figure 1.20 Example.

- Six bends with radius $r = 0.5$ [m], from which:

$$\frac{D}{r} = \frac{0.3}{0.5} = 0.6$$

According to the table: $\xi_2 = 6 \cdot 0.18 = 1.08$
- Two sliding vane valves:

$$\xi_3 = 2 \cdot 0.8 = 1.6$$

- Suction strainer with check valve:

$$\xi_4 = 4$$

- Five check valves:

$$\xi_5 = 5$$

The total pressure loss p_f caused by friction amounts to:

$$p_f = (15.6 + 1.08 + 1.6 + 4 + 5) \cdot \frac{10^3 \cdot 1.415^2}{2} = 27\,310 \text{ [Pa]}$$

Furthermore, a pressure p_{geo} is needed to overcome the geodetic height:

$$p_{geo} = \rho \cdot g \cdot H_{geo} = 10\,00 \cdot 9.81 \cdot 24.8 = 243\,288 \text{ [Pa]}$$

Finally, a dynamic pressure p_{dyn} is needed to give the water kinetic energy:

$$p_{dyn} = \rho \cdot \frac{c^2}{2} = 1001 \text{ [Pa]}$$

So, the pump must deliver a pressure p of:

$$p = p_f + p_{geo} + p_{dyn} = 271\,599 \text{ [Pa]} = 2.72 \text{ [bar]}$$

We find out that the dynamic pressure is so small it can be neglected.

1.11 Feed Pressure

1.11.1 Geodetic Feed Pressure

We introduce some terms (Figure 1.21):

H_p = difference in latitude between centerline pump and highest point of discharge = the *geodetic* press head
H_s = difference in latitude between centerline pump and liquid level suction reservoir = the *geodetic* suction head

The geodetic head is the head the pump has to deliver to overcome the suction and press height:

$$H_{geo} = H_s + H_p$$

The pressure that the pump must deliver to overcome this height between suction and pressure reservoir is called the *geodetic feed pressure* p_{geo}.

Remarks: some special cases

1) For the case of a negative geodetic suction head H_s, see Figure 1.22: the liquid flows toward the pump.
2) For the case of a closed circuit (house heating); see Figure 1.23. Here the geodetic press height is equal to the suction head, but this latter one is negative. The geodetic feed head is then zero: $H_{geo} = 0$. In the pipeline is a constant pressure $p_{geo} = \rho \cdot g \cdot H$ (not taken into account the friction losses).

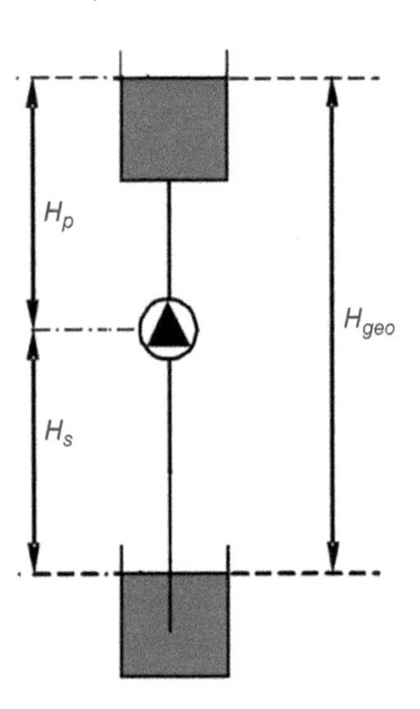

Figure 1.21 Geodetic feed pressure.

1.11.2 Static Feed Pressure

The static feed pressure p_s is defined as the pressure that the pump must deliver if the flow is zero. In the first instance, the pump must overcome the height. In the second instance, it must eventually overcome a pressure difference between the suction and the press reservoirs. Assume that the vapor pressure in the press reservoir is represented by

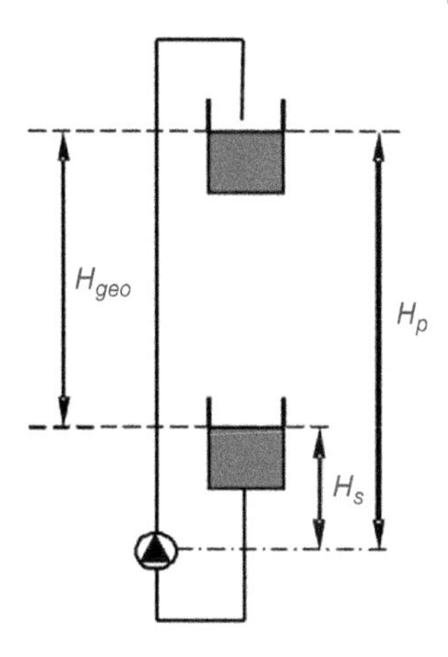

Figure 1.22 A negative geodetic suction head.

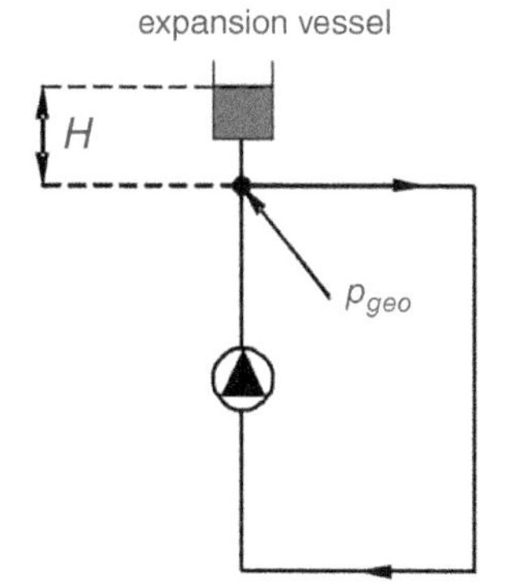

Figure 1.23 Closed circuit.

p_{vp} and the vapor pressure in the suction reservoir by p_{vs} then the static feed pressure amounts to:

$$p_s = p_{geo} + (p_{vp} - p_{vs})$$

In what follows it will always be assumed that the pressure above the two reservoirs is atmospheric pressure (Figure 1.24).

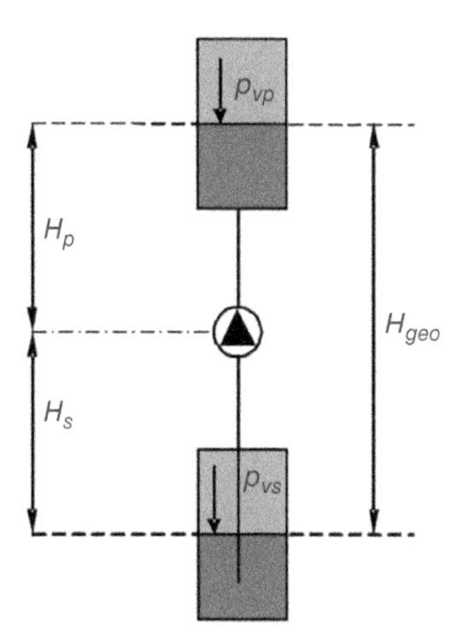

Figure 1.24 Static feed pressure.

1.11.3 Manometric Feed Pressure

Besides the just defined static feed pressure, necessary to overcome a height difference, the pump must deliver a pressure to surmount the friction losses in the pipes, bends, valves, etc. Furthermore, a pressure is necessary to permit liquid flow, that is the pump must give the liquid "velocity." If the liquid flows out of the press pipe with velocity c, then this corresponds, according to Bernoulli, to a dynamic pressure, with the value:

$$p_{dyn} = \frac{1}{2} \cdot \rho \cdot c^2$$

This velocity pressure, or dynamic pressure, will not be recuperated anymore: the liquid that flows into the press reservoir will lose its dynamic energy by friction.

Use these variable names for the resistance:

- p_{fs} the friction resistance in the suction pipe
- f_{fp} the friction resistance in the press pipe.

Then the total pressure to deliver by the pump is written as p_{man}.

$$p_{man} = p_{dyn} + p_{fz} + p_{fp} + p_s$$

This pressure is the *manometric feed pressure* (of the installation). It is the pressure that the pump must deliver to transport the liquid from the suction to the press reservoir. Often the dynamic pressure is small compared to the other pressures and consequently the term is neglected.

Let's look now at this manometric feed pressure from the standpoint of the pump. Consider the two places (1 and 2) before and behind the pump. When one works out the expressions for the total pressure at these two places, one finds that:

$$p_2 + \rho \cdot \frac{c_2^2}{2} > p_1 + \rho \cdot \frac{c_1^2}{2}$$

The classic law of Bernoulli is evidently not valid anymore in this case: the pump delivered energy to the liquid, so that the total pressure of the liquid at place 2 is greater than at place 1. Bernoulli's law should be corrected by adding the term p_{man}:

$$p_2 + \rho \cdot \frac{c_2^2}{2} = p_1 + \rho \cdot \frac{c_1^2}{2} + p_{man}$$

In other words, p_{man} is the *total pressure delivered by the pump.*

In another form:

$$p_{man} = (p_2 - p_1) + \frac{1}{2} \cdot \rho \cdot (c_2^2 - c_1^2)$$

It is self-evident that the manometric feed pressure delivered by the pump must exactly equal the manometric pressure demanded by the installation. We illustrate this as follows. In Figure 1.25 a pumping installation is represented. The law of Bernoulli is being applied twice, with the introduction of friction losses:

- Between points 0 and 1 on the figure with: $c_0 \ll c_1$:

$$p_a + 0 - p_{fs} = p_1 + \frac{1}{2} \cdot \rho \cdot c_1^2 + \rho \cdot g \cdot h_s \tag{1.1}$$

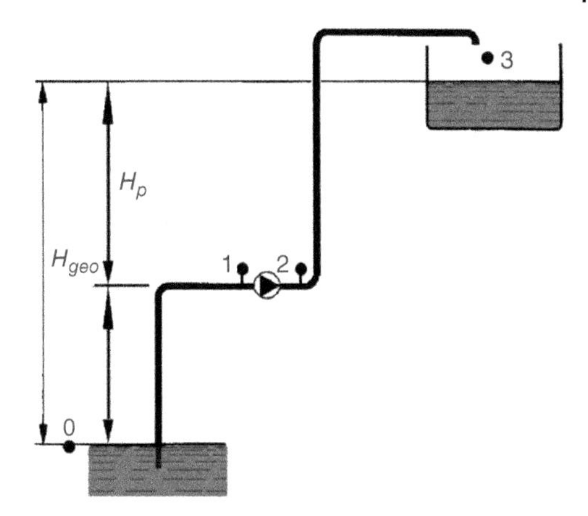

Figure 1.25 Example of a pumping installation.

- Between points 2 and 3 on the figure:

$$p_a + \frac{1}{2} \cdot \rho \cdot c_3^2 + \rho \cdot g \cdot h_p = p_2 + \frac{1}{2} \cdot \rho \cdot c_2^2 - p_{fp} \tag{1.2}$$

Equation (1.2) minus Equation (1.1) and rearranging the terms:

$$\rho \cdot g \cdot (h_p + h_s) + (p_{wp} + p_{wz}) + \frac{1}{2} \cdot \rho \cdot c_3^2 = p_{man} = (p_2 - p_1) + \frac{1}{2} \cdot \rho \cdot (c_2^2 - c_1^2)$$

So that *the pressure delivered by the pump is equal to the pressure demanded by the installation.* Logic.

1.11.4 Theoretic Feed Pressure

At the flanges of the pump, the manometric feed pressure is available. But *in* the pump itself, it will be necessary to apply a somewhat higher pressure on the liquid (by the driving motor). Indeed, when the liquid flows through the pump, inevitable friction losses will occur. The pressure that intrinsically must be present in the pump is the *theoretical feed pressure* p_{th}.

Interpret this as: it is the pressure that theoretically is available, but of which already a part is lost in the pump itself. The pressure loss in the pump is noted as p_{lp}:

$$p_{th} = p_{man} + p_{lp}$$

1.12 Law of Bernoulli in Moving Reference Frames

Consider a channel through which a liquid flows between two observer points 1 and 2 (Figure 1.26). Suppose that in the channel there is a permanent stream, there is no friction, and no external work is being applied to the liquid. The reference frame S where the study is being done is an *inertial frame.* This is a frame where Newton's law applies, and thus also the law of Bernoulli (because this is a variant of Newton's law applied to liquids). In other words:

$$p_2 + \rho \cdot \frac{c_2^2}{2} = p_1 + \rho \cdot \frac{c_1^2}{2}$$

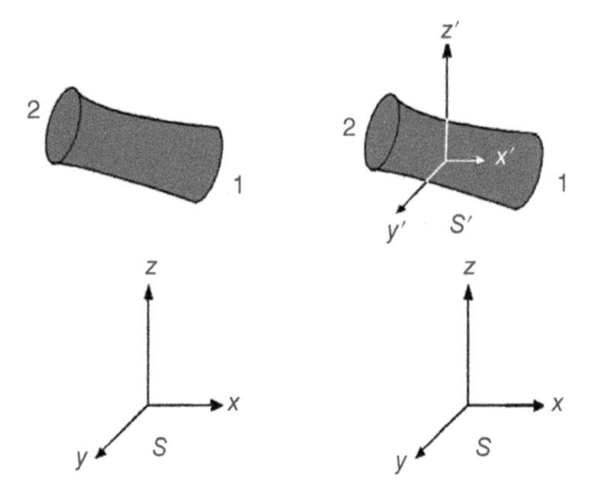

Figure 1.26 Bernoulli in moving frames.

Let this channel now move with a velocity u in reference to the "absolute" observer in S. In the new moving frame S′ the liquid possesses a relative velocity w. The observer in S, however, notes another velocity c. According to the laws of mechanics, the vector law applies:

$$\overline{c} = \overline{u} + \overline{w}$$

In other words, the *absolute vector velocity* $\overline{c}$ is equal to the *relative velocity* $\overline{w}$ plus the tow *velocity* $\overline{u}$. The tow velocity is the velocity with which the frame S′ moves in reference to the reference frame S (where the laws of Newton and Bernoulli are valid).

On the condition that the system S′ performs a *straight line uniform movement* with reference to S the law of Bernoulli *in the frame S′:*

$$p_2 + \rho \cdot \frac{w_2^2}{2} = p_1 + \rho \cdot \frac{w_1^2}{2}$$

But in the frame S′ Bernoulli's law is not valid! So, don't write:

$$p_2 + \rho \cdot \frac{c_2^2}{2} = p_1 + \rho \cdot \frac{c_1^2}{2} \text{ ERROR!}$$

So this formula may not be applied in this case.

If the movement of S′ is not uniform, but accelerated, for instance when observer S′ performs a rotation around S, Bernoulli may not be applied in the above-mentioned form. A correction term must be added, which is often called a *centrifugal term.* On the occasion of the study of centrifugal pumps this new form of Bernoulli's law will be reedited.

1.13 Water Hammer (Hydraulic Shock)

Warhammer is a shockwave that travels through a fluid in a pipeline. This occurs when a moving fluid is obliged to stop abruptly. The impulse the fluid is carrying creates a significant force in the closed system, on the pipe walls, a valve in the line, on the pump itself that drove the fluid, flowmeters, and pressure sensors. The abrupt stop can be caused by a:

Figure 1.27 Water hammer distractors. Source: Courtesy of Watts.

- Sudden start or stop of the pump.
- Sudden opening or closing of valves.
- Special case: flash gas. This is vapor that arises, for instance, in steam condensate that collects in a steam pipeline. The liquid water will then sometimes be transformed in steam, caused by a pressure decline. Because of this a volume increase of maybe 500 times takes place.

The pressure wave travels forwards and backwards in the closed pipe but will be damped in a short time. This short time can, however, be long enough to cause serious damage.

There are several solutions for this:

- At regular distances check valves are placed in the pipeline. The fluid will move forwards and backwards between two check valves. This will also give a bump, but the amount of fluid is smaller, so the damage to the check valves will be less. It is only a limited solution: most check valves close rapidly.
- Placing of distractors (Figure 1.27). These are vertical (or horizontal) tubes, placed on the pipeline that discharge in an air chamber, placed close to where the cause of the water hammer is situated. The pressure wave runs in the tube and presses against the air in the air chamber, compressing the air, and so dampens its energy.
- Solenoid activated valves: these are valves that open or close by activating the coil. The electric current in the coil is gradually increased, so that the valve closes or opens slowly (Figure 1.28).
- With a frequency converter the speed of the asynchronous motor can be increased gradually. This way the pump's speed increases gradually.

Figure 1.28 Slowly closing valve. Source: Courtesy of Parker.

1.14 Flow Mechanics

1.14.1 Hydrofoils

In what follows liquids and gasses with a constant specific mass, or one that does not change very much, are discussed. In this way, Bernoulli's law can be applied.

In order to understand the operation of a hydrofoil we start with a simple model: a fluid flows in a cylindric tube. This fluid (Figure 1.29) exercises a uniform pressure on the walls of the tube, as long as the section is constant.

If, however, we have a flexible tube, we can increase the velocity of the fluid at a certain location by squeezing the tube (Figure 1.30). In the throat the speed increases and the static pressure decreases.

The same effect can be obtained by placing an object, an obstacle, in the flow (Figure 1.31). On the upper side of the obstacle the velocity of the fluid will become

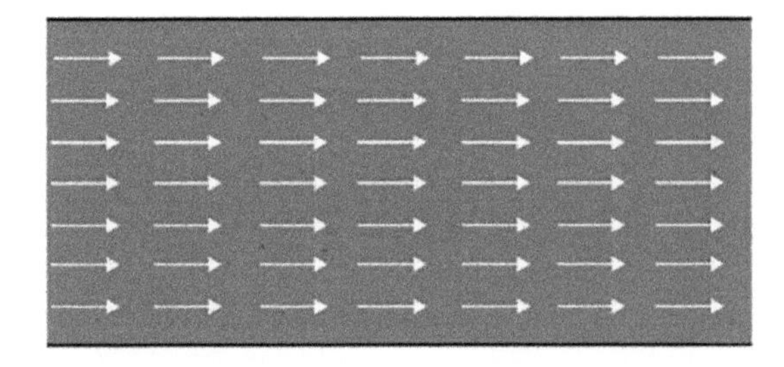

Figure 1.29 Tube.

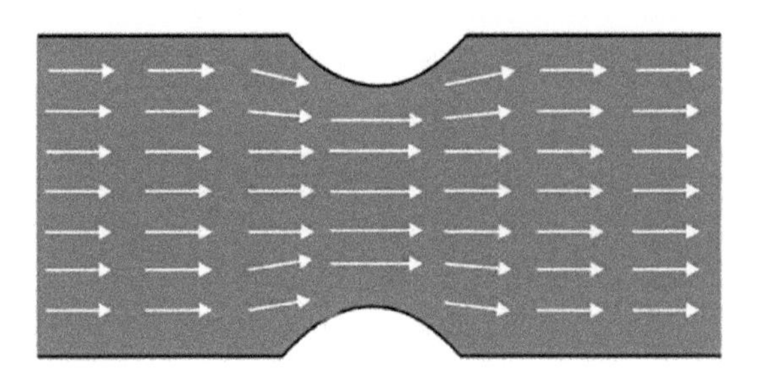

Figure 1.30 Flexible tube.

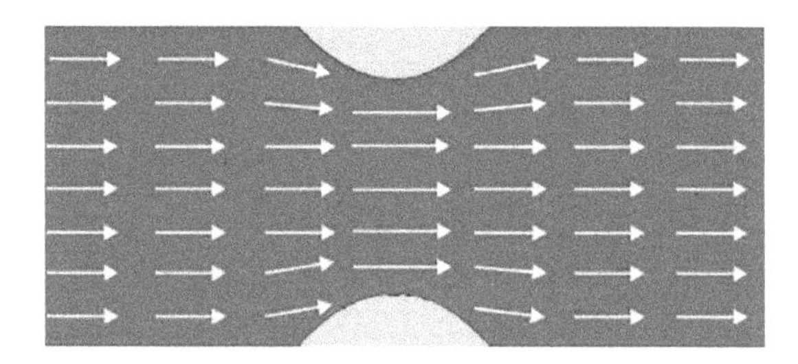

Figure 1.31 Obstacle.

greater and the static pressure lower. Because Figure 1.31 is symmetrical one can conclude by the same reasoning on one half of the tube (Figure 1.32).

The following jargon is used (Figure 1.33):

- 1: attack board
- 2: trailing edge
- 3: upper surface
- 4: lower surface
- δ: attack angle

Research on the streamlines of such a profile in a flow leads to the image in Figure 1.34. Upstream the static pressure and velocity are uniform with values p and c. The streamlines that are initially equally spaced become so distorted by the hydrofoil that on the upper side they come closer together, while on the lower side they disperse. Along a

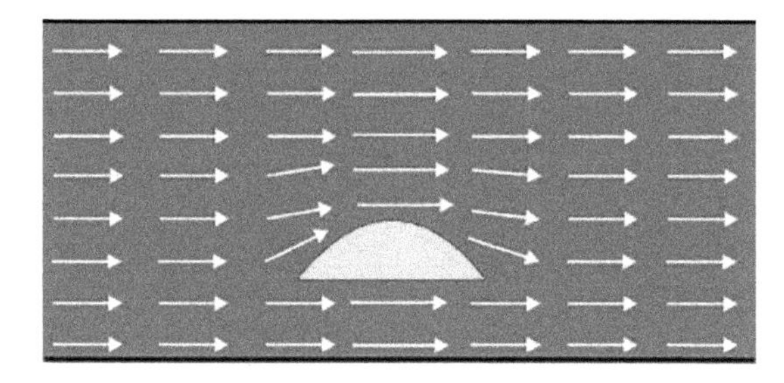

Figure 1.32 Flow over a hydrofoil.

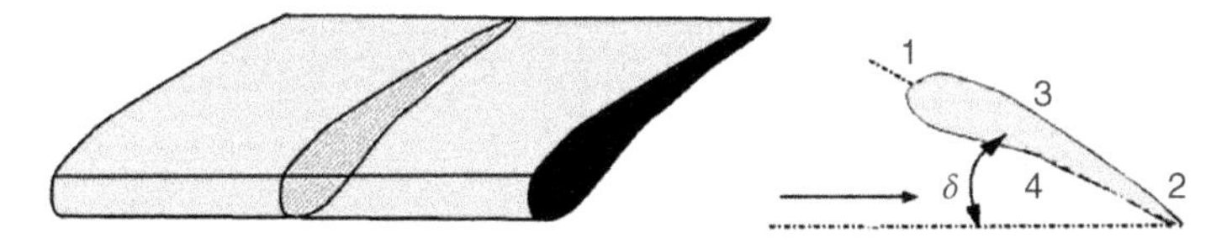

Figure 1.33 Hydrofoil.

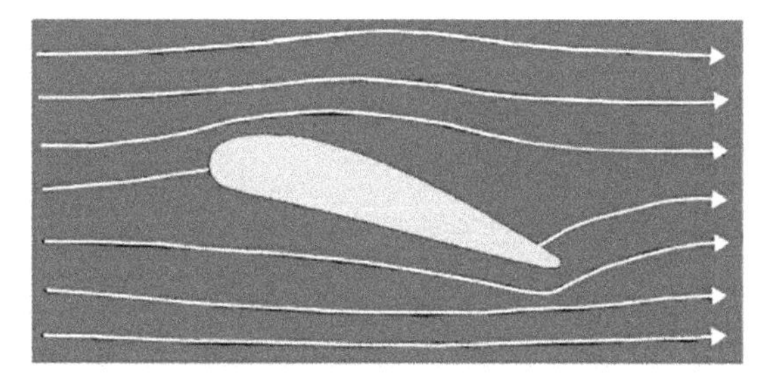

Figure 1.34 Hydrofoil in flow.

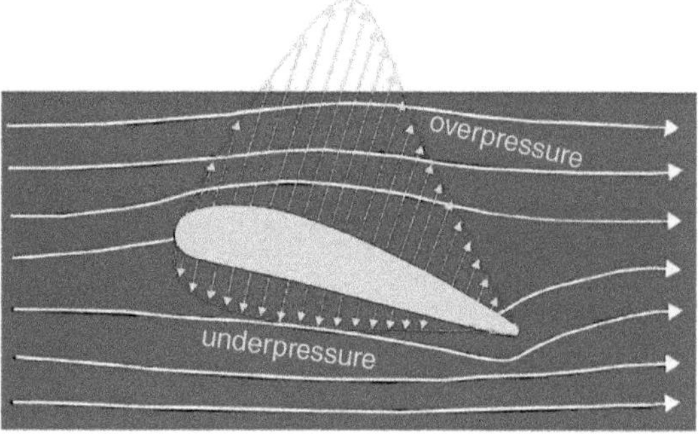

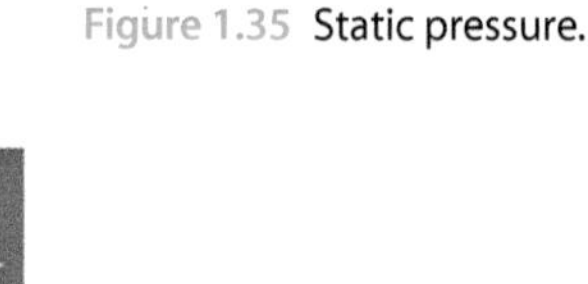

Figure 1.35 Static pressure.

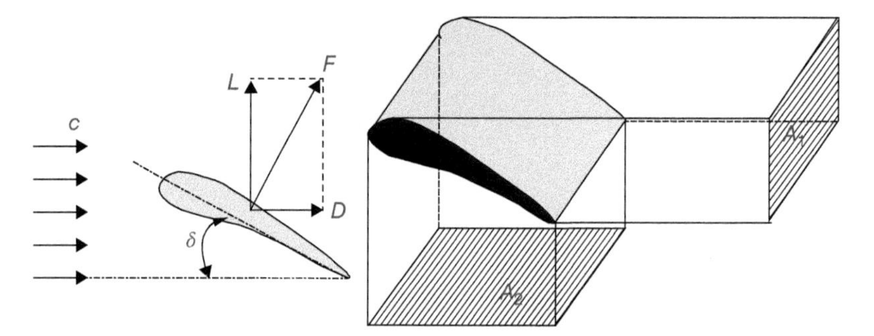

Figure 1.36 Projected surfaces.

streamline the *total* pressure is constant:

$$p + \rho \cdot \frac{c^2}{2} = p_l + \rho \cdot \frac{c_u^2}{2} = p_u + \rho \cdot \frac{c_l^2}{2}$$

where index l points to the "lower surface" and the index u to the "upper surface."

The local density of the streamlines is a measurement of the local velocity. c_u will be smaller than c and c_l greater than c. On the upper side there will be a static underpressure and on lower side an upper pressure.

The static pressure distribution is shown in Figure 1.35.

Integrating the pressure distribution over the whole profile leads to the dynamics of the profile; represented in Figure 1.36.

The total force F that the fluid exercises on the profile has two components: the lift L and the drag D, respectively perpendicular and along the velocity c.

Analogous to the law on friction resistance, one can state:

$$L = C_z \cdot A_2 \cdot \rho \cdot \frac{c^2}{2}$$

$$= C_x \cdot A_1 \cdot \rho \cdot \frac{c^2}{2}$$

The surfaces A are projected surfaces (Figure 1.36). The factors C_z and C_x are dependent on the form of the profile. Laboratory experiments show how the lift L and drag D change with the attack angle. In Figure 1.37 the results are given for a certain profile. The attack angle may change between certain limited values, but above a critical value (here 18°) the flow on the reverse becomes very chaotic: eddies (or vortexes) are formed. The

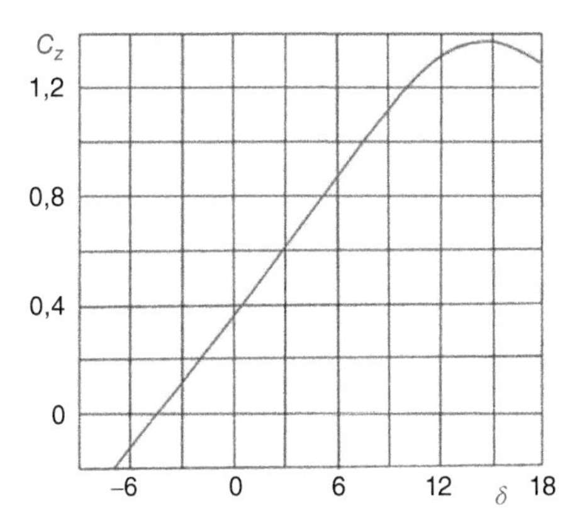

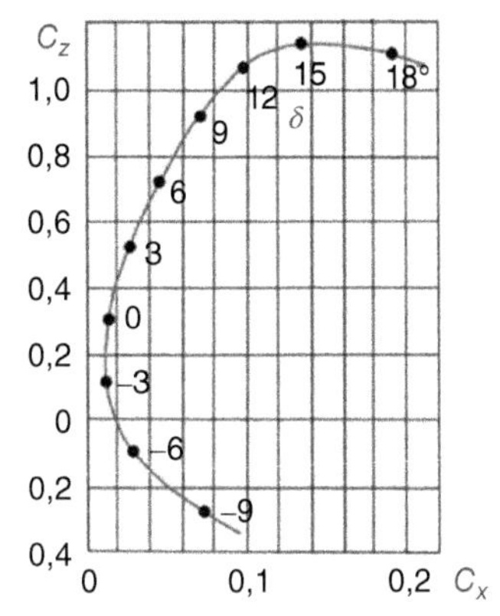

Figure 1.37 Behavior of a hydrofoil.

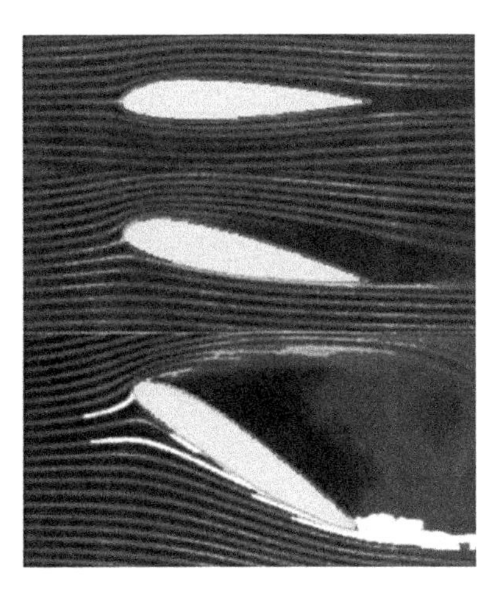

Figure 1.38 Stall.

upward stream may be laminar or turbulent. Turbulent does not mean that vortexes are formed. The pressure behind the profile becomes so low that there is no longer lift. The flow moves through a very narrow passage and stalls (Figure 1.38).

1.14.2 Applications

When objects move through a fluid, or inversely when a fluid flows around an obstacle, it is generally desirable that the drag D be as small as possible and the lift as high as

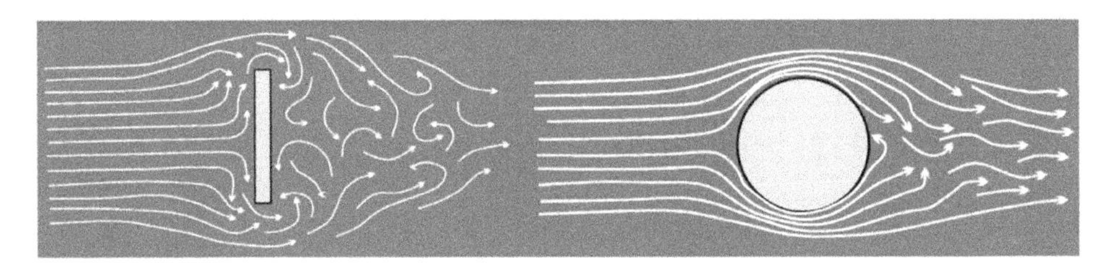

Figure 1.39 Flow around an obstacle.

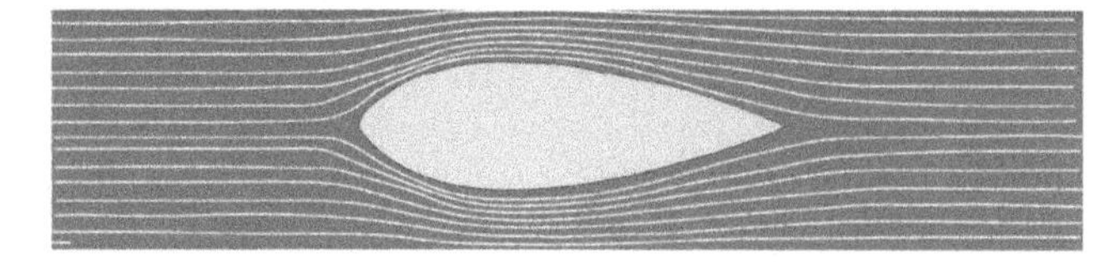

Figure 1.40 Streamlined object (aerodynamic profile).

possible. When the object is coarse (Figure 1.39) a wake will develop behind the object. Figure 1.40 shows an *aerodynamic profile where the drag is minimal*. This is also the case with profiles for airplane wings and helicopter screws (see Figure 1.41).

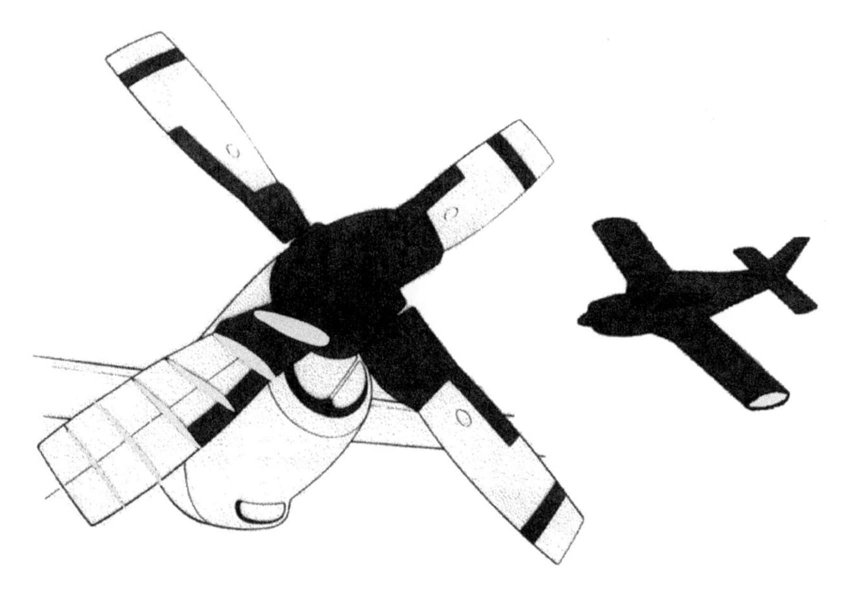

Figure 1.41 Helicopter screws and airplane wings.

2

Positive Displacement Pumps

CHAPTER MENU

2.1 Reciprocating Pumps

2.1.1 Operation

In Figure 2.1 a pump installation with a single acting piston pump is represented. The piston pump works according to the classic crank-connecting rod mechanism (Figure 2.2).

Let's assume that the pipeline is totally filled with water and that the piston is on the extreme left. If the piston moves to the right, an airless space grows in the cylinder. The atmospheric pressure above the suction reservoir presses the water from the suction pipe toward the suction valve. This one opens and the water flows into the cylinder. When the piston is in the extreme right-hand side, the pump cylinder is fully filled with water: the suction stroke ends. After that the piston will move to the left. Because of this a pressure occurs in the water that closes the suction valve and the press valve opens. The water can now flow in the press pipe and fill the press reservoir.

2.1.2 Flow

Note:

- A: the section of the piston
- s: the stroke length of the crank-connecting rod (twice the crank radius)
- N: the number of revolutions per minute

Then the quantity of transported fluid per back and forth stroke is $A \cdot s$.
The volumetric flow is then:

$$Q_V = A \cdot s \cdot \frac{N}{60}$$

Pumps and Compressors, First Edition. Marc Borremans.

Companion website: www.wiley.com/go/borremans/pumps

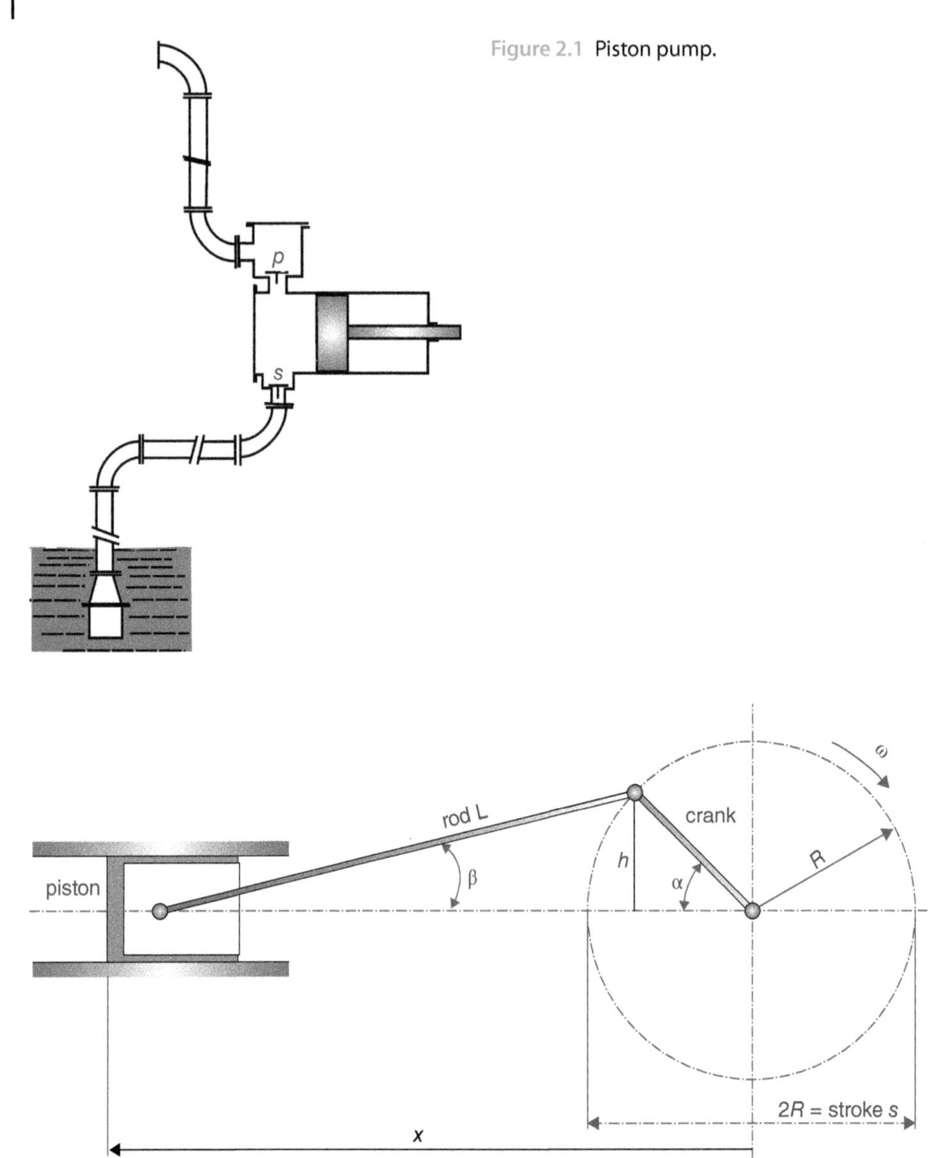

Figure 2.1 Piston pump.

Figure 2.2 Crank-connecting rod mechanism.

This is only true theoretically. In practice the flow will be lower. Indeed, the valves don't close abruptly at the moment the piston is in the dead position. For example, at the beginning of the suction stroke, the press valve will not be completely closed. Eventually, caused by wear, there can be leaks along the valves, so that during the press stroke liquid from the cylinder can pass via the suction valve, back to the suction pipe (and analogous during the suction stroke fluid will flow out of the press pipe back toward the cylinder).

The real *effective* volumetric flow Q_{Ve}:

$$Q_{Ve} = \eta_V \cdot Q_V$$

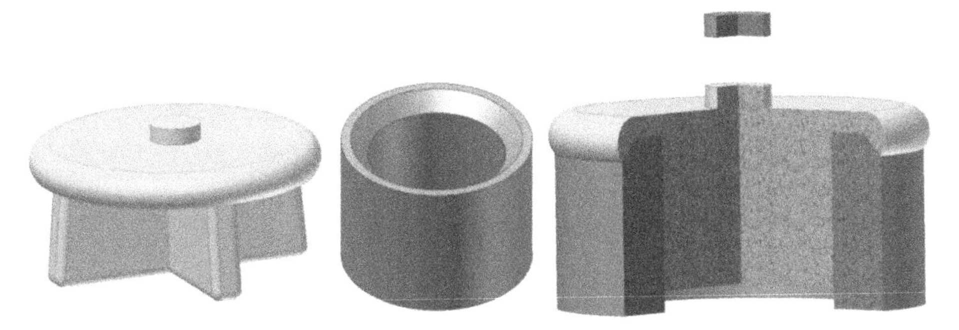

Figure 2.3 Disc valve: disc, seat.

where η_V is the *volumetric efficiency of the pump*. Dependent upon the pump, and its age, η_V amounts to 0.85–0.95.

Regulation of the flow can be done by:

- Regulation of the number of rotations N (this is the most economical but needs a motor).
- Using a bypass that brings back the surplus of pressed liquid back to the suction pipe or reservoir.
- A mechanism that keeps the suction valve open during a part of the press stroke.
- Use of multiple cylinders whereby one or more cylinders are temporarily switched off.

As far as usual rotations per minute is concerned:

- *Small fast running pumps*: 300 [rpm] or more.
- *Big slow running pumps*: 50 [rpm].

2.1.3 Valves

The valves are automatically opened and closed under the influence of the pressure difference in the pump. Several kinds of valves are available. So, for instance, in Figure 2.3 a disc valve is depicted and in Figure 2.4 a ball valve. Eventually the valves are spring loaded, so that they are called back to their seat by the pressure difference that opened them falls away.

2.1.4 Piston Sealing

The sealing between piston and cylinder can take place by means of leather cuff rings (Figures 2.5 and 2.6) or by piston rings made of plastic (Figures 2.7 and 2.8).

2.1.5 Plunger Pumps

In plunger pumps the piston is implemented another way. The piston does not rub *in* the pump housing (Figure 2.9). The sealing is provided by a gasket in the cylinder. The piston is now called a *plunger*. Plunger pumps are often applied for heavy work. The construction is more robust because the rod connects to a *cross* head.

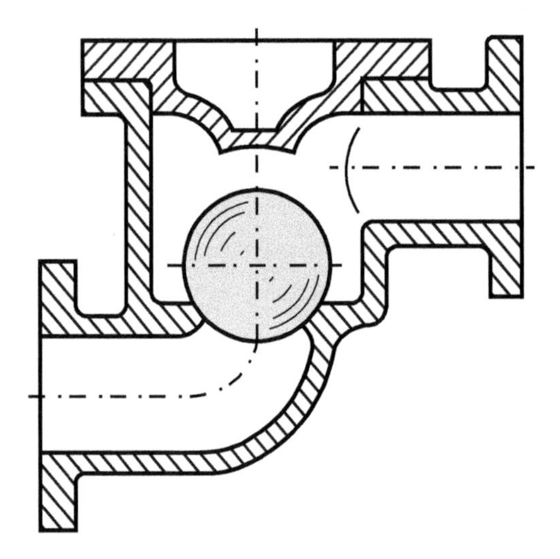

Figure 2.4 Ball valve.

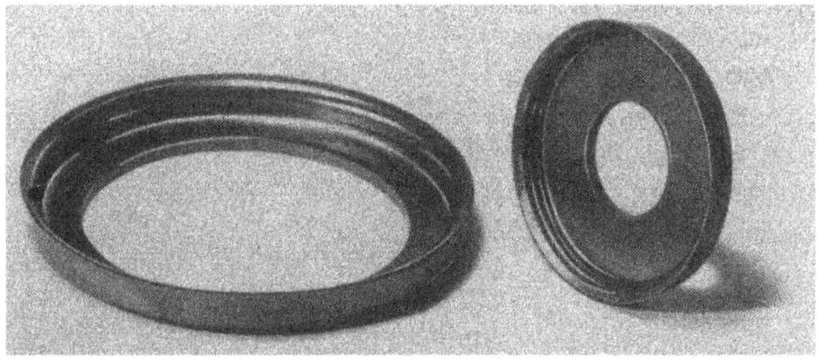

Figure 2.5 Cuff rings. Source: Eriks.

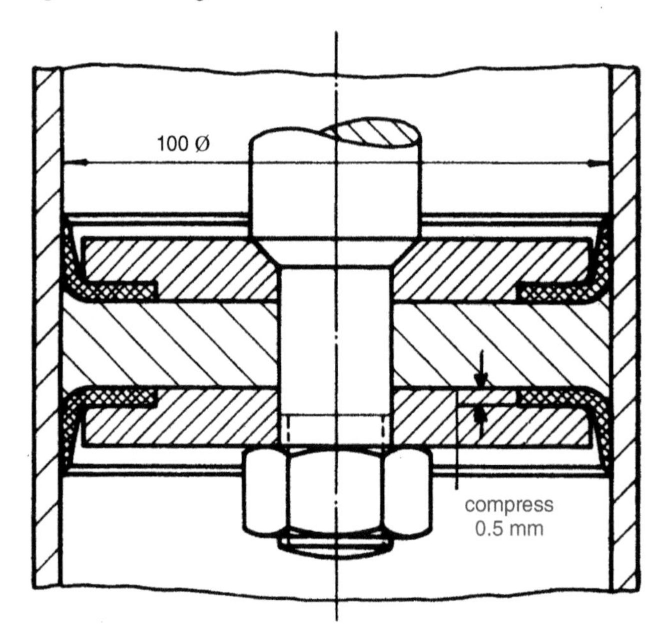

Figure 2.6 Cuff rings double acting pump. Source: Eriks.

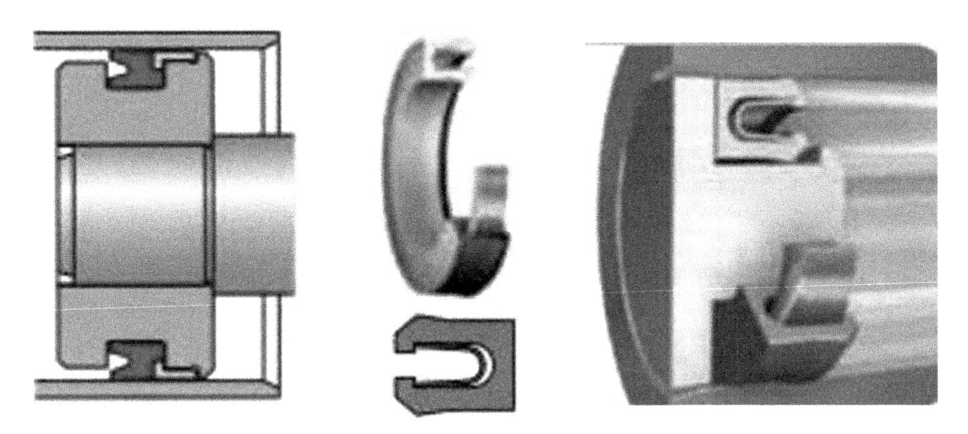

Figure 2.7 Piston rings. Source: SKF.

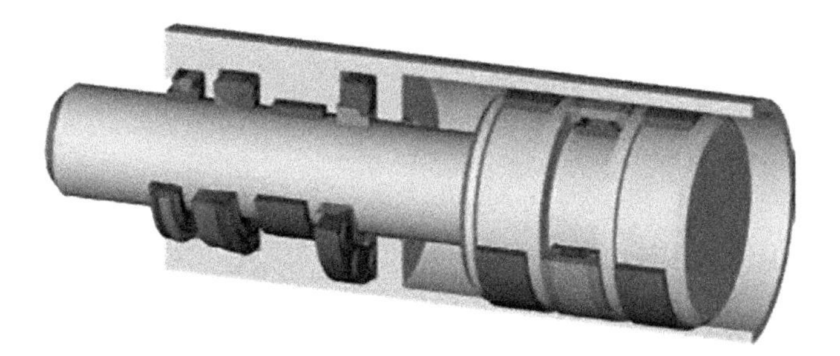

Figure 2.8 Piston rings double acting pimp. Source: SKF.

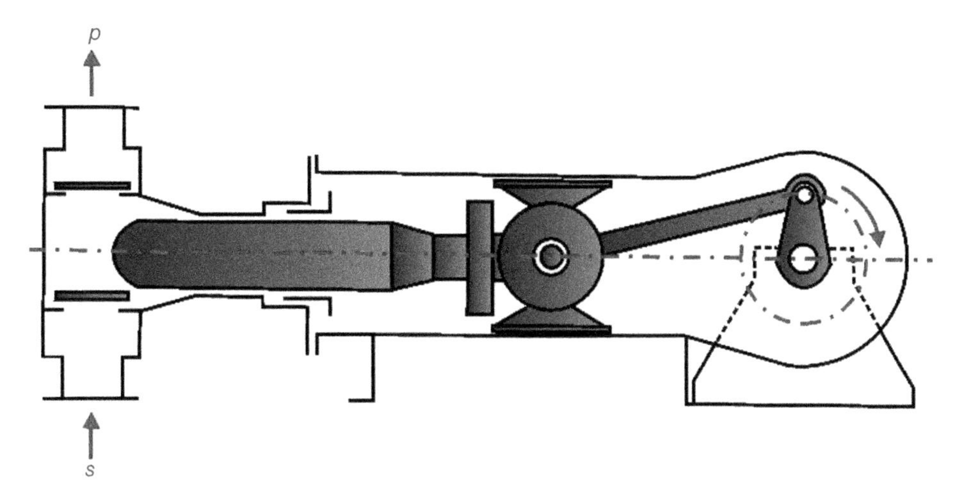

Figure 2.9 Plunger pumps.

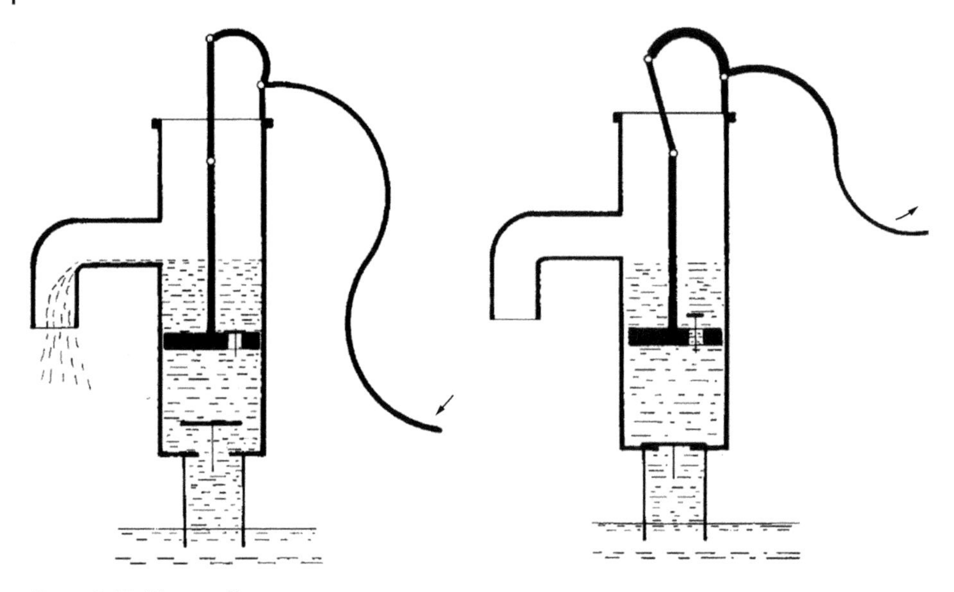

Figure 2.10 Water well pump.

Figure 2.11 Water well pump. Source: Clasal.

2.1.6 Hand Pump

In a hand pump (Figures 2.10 and 2.11) the press valve is built in the piston. When the handle moves downwards the piston goes upwards causing a suction action. Then the press valve closes, so that the water on the top of the piston is pushed upwards and can flow away to the press pipe.

This principle is also encountered in hydraulic manual-driven oil hand pumps.

2.1.7 Double Acting Pump

In order to get more capacity (flow) in *double acting pumps* both sides of the piston or plunger are utilized at the same time. The suction stroke of one side coincides with the press stroke of the other side (Figure 2.12).

This gives a higher flow and is more regular. A is, as always, the symbol for the surface of the piston. Let A^* be the surface of the piston rod:

$$Q_V = (2 \cdot s \cdot A - A^*)\frac{N}{60}$$

The operation of the pressure vessels is discussed later.

Figures 2.13 and 2.14 represent double acting pumps. Notice the transmission of motor to crank, lowering the speed of the pump versus that of the motor. Suction and press pipes are connected perpendicular to the frontal view of Figure 2.12.

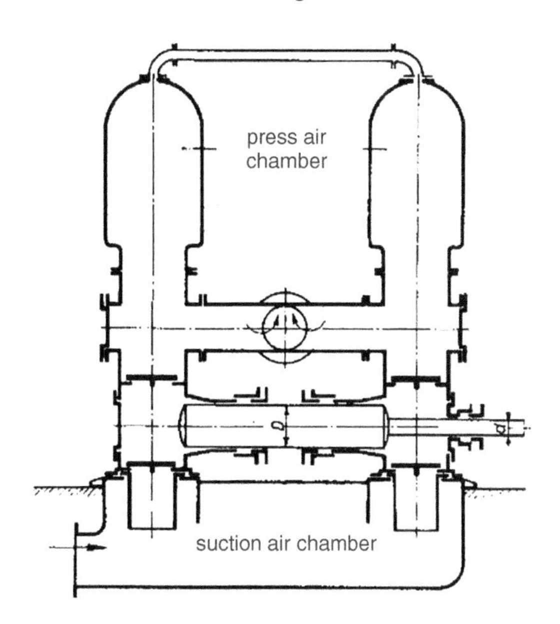

Figure 2.12 Double acting pump.

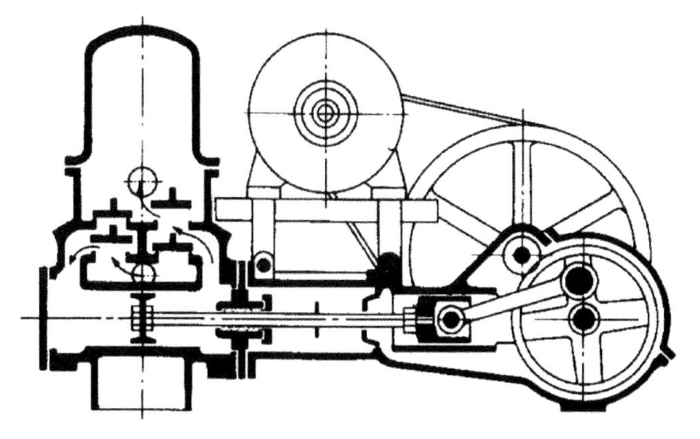

Figure 2.13 Double acting pump. Source: Stork.

Figure 2.14 Double acting pump.

2.1.8 Membrane Pumps

The membrane is moved manually or automatically up and down. This produces a suction and press action (Figures 2.15 and 2.16). In principle membrane pumps do not leak, because the liquid cannot penetrate the seal. This sort of pump is used where leaks are to be avoided. Such a pump is called a *hermetic pump*.

This pump is used where one has to deal with explosive liquids (gasoline), or radioactive, poisoned, or corrosive substances. In the latter case, only the pump housing should be protected.

Because of the lack of problems with a shaft seal, products such as glue, paint, waste liquids, sludge, and lubricating oil from garages can be pumped.

Nowadays, there are high tech rubbers (silicones) that can withstand temperatures of 400 °C.

The use of compressed air has a disadvantage: noise and a bad energy efficiency of the pump.

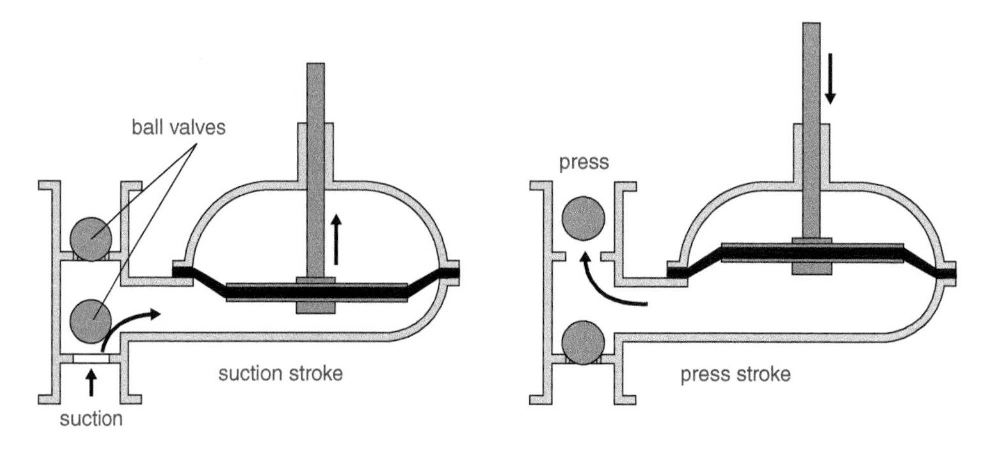

Figure 2.15 Operation principle membrane pump.

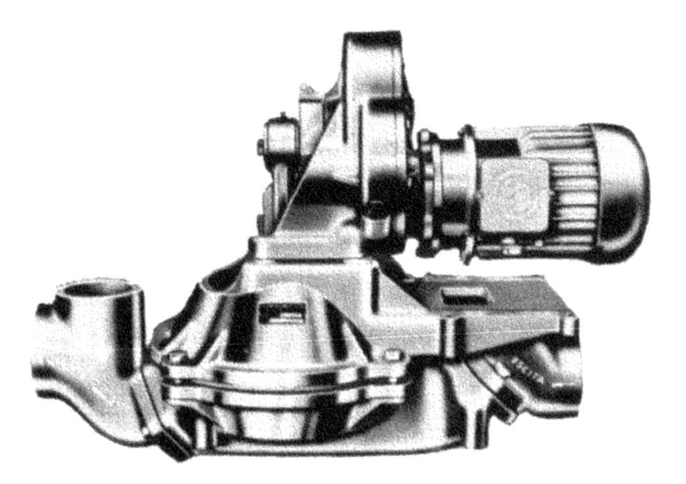

Figure 2.16 Electrically driven membrane pump.

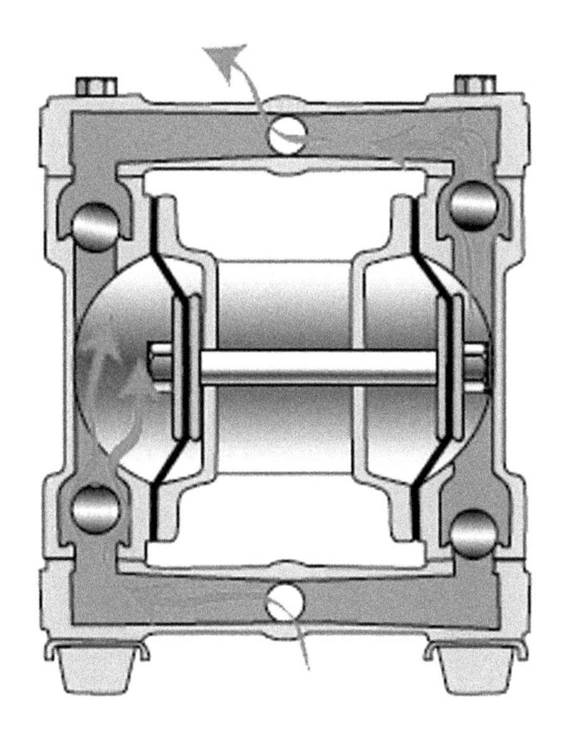

Figure 2.17 Principle compressed air drive membrane pump. Source: Johnson Pumps.

Another type of a double acting pump is the one in which two membrane pumps are set up back to back, being alternately driven by compressed air.

In the pump two membranes (yellow) are set, they are connected by a central axis. The membranes function as a separation between the compressed air and the liquid. In the middle block sits the air valve (not visible) (Figures 2.17–2.19). The pressure of the air is mostly no more than 7 [bar_g]. So even with an obstruction in the press pipe the delivered pressure of the liquid cannot be higher than 7 [bar_g].

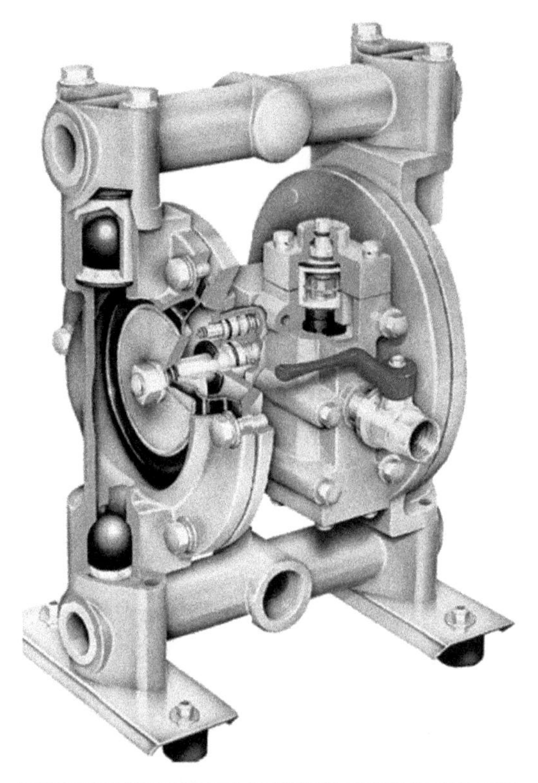

Figure 2.18 Section membrane pump. Source: Johnson Pumps.

Figure 2.19 Membrane pump. Source: Verderair.

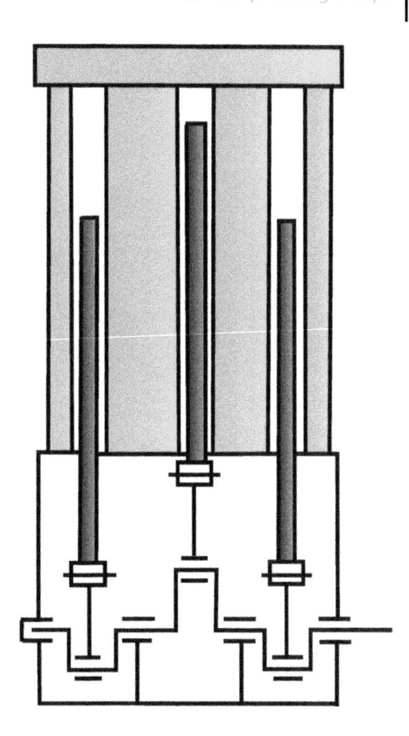

Figure 2.20 Triplex plunger pump.

The membranes are executed in plastics and elastomers as Santoprene rubber (trade mark of Advanced Elastomer Systems, a combination of polypropene and rubber), Teflon (PTFE), neoprene, nitrile rubber, depending on the medium. The lifespan is about 10–15 million cycles. Because they use rubbers, these pumps are often not adequate for high temperature applications.

2.1.9 Triplex Pumps

Triplex pumps are constituted of three pumps whereby the crank axis of one pump is shifted 120° relative to the next one (Figures 2.20–2.22). The advantage is that the volumetric flow is more regular; with a single acting pump there is only a discharge for half of the time, with a double acting pump the whole time, with a triplex pump the flow is even more regular (Figure 2.23).

2.1.10 Hydrophore

A *hydrophore installation* (Figures 2.24 and 2.25) is composed of an electric-driven pump that pumps the liquid in a nearby pressure vessel that functions as a reservoir. In addition, it consists of a regulation mechanism that turns the motor off or on when the reservoir is full or empty. When the pressure vessel is filled with liquid the air of the air chamber will be compressed. When it reaches a predefined maximum value, for instance 3 [bar], a pressure regulator will cause the motor to stop.

The vessel is now filled and can, dependent on the need of the consumers, deliver liquid underpressure of the air cushion.

Figure 2.21 Triplex plunger pump. Source: Speck pompen.

Figure 2.22 Triplex scouring-plunger pump, 60 [bar], 3700 [l/h]. Source: DUBA.

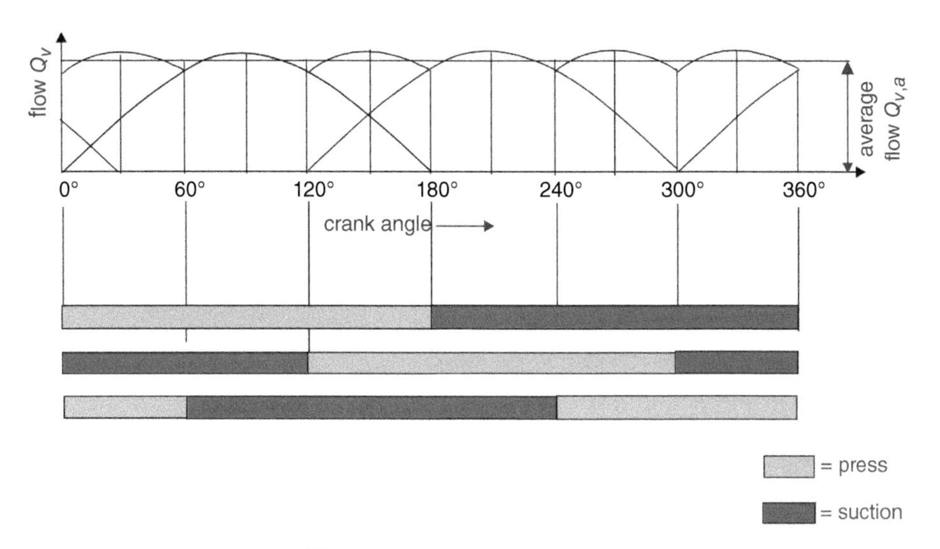

Figure 2.23 Triplex progress of flow.

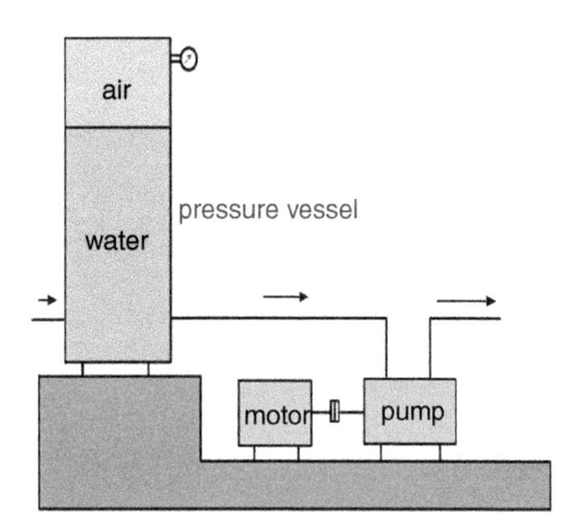

Figure 2.24 Hydrophore. Source: Stork.

Figure 2.25 Hydrophore group with piston.

Figure 2.26 "Booster pump." Source: KSB.

When liquid is tapped from the vessel, the air pressure will decrease. When the pressure reaches a predefined minimum value, for instance 1.5 [bar], the motor will automatically be restarted and the vessel replenished. The pump itself can be a piston pump of a radial pump (see later).

The vessel is made from plastic, sheet steel, or stainless steel. In the case of a plastic vessel it is implemented in a composite material. This leads to high durability and gives it a rustproof property. Any small tear in a composite material is, however, very difficult to repair. A vessel in steel is applied in heavy circumstances with high pressure and high temperatures.

Hydrophore installations are applied where the pipe pressure is insufficient (high buildings, buildings located in high places, etc.).

The same principle can be applied to *booster pumps*. A booster is a device that gives the incoming liquid, on a pressure higher than the atmospheric pressure, a boost, thus an increase in pressure. In Figure 2.26 a booster pump is pictured: it is an implementation with three radial pumps in parallel. The pumps are easily connected to the water network.

2.2 Maximum Suction Head

2.2.1 Theoretical

In this section we will check what the *maximum static (geodetic) suction head* H_{smax} is.

Consider the suction side of the pump (Figure 2.27). The pressure that the liquid column exercises the liquid column in the suction pipe caused by its weight must be compensated by the pressure p_a of the atmospheric air. Apply the law of Pascal at point Q and suppose that the piston ensures a perfect vacuum:

$$p_a = \rho \cdot g \cdot H_{smax}$$

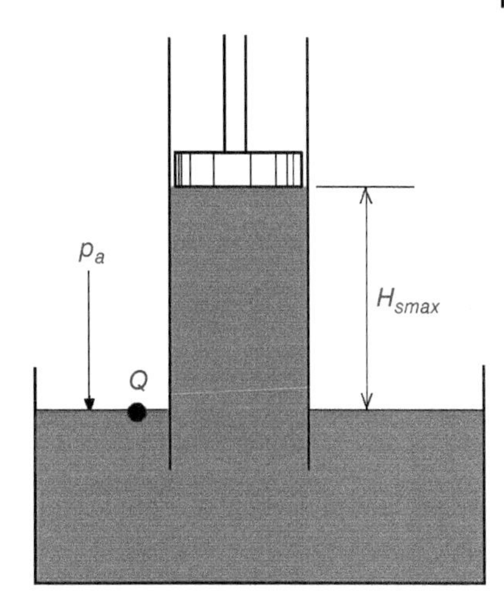

Figure 2.27 Suction head theoretically.

From which:

$$H_{smax} = \frac{p_a}{\rho \cdot g}$$

For water:

$$H_{smax} = \frac{p_a}{\rho \cdot g} = \frac{1{,}013 \cdot 10^5}{10^3 \cdot 9.81} = 10.33$$

For water the atmospheric pressure can only carry a water column of about 10 [m]. The pump may in no way be placed above the suction reservoir higher than this value.

2.2.2 Vapor Pressure

Beneath the piston the pressure is lower than the atmospheric pressure. This causes some of the liquid to turn into vapor near the piston (Figure 2.28). There arises an *equilibrium* between liquid and vapor with a certain pressure, called *vapor pressure* p_v.

The value of this vapor pressure can be read in vapor tables.

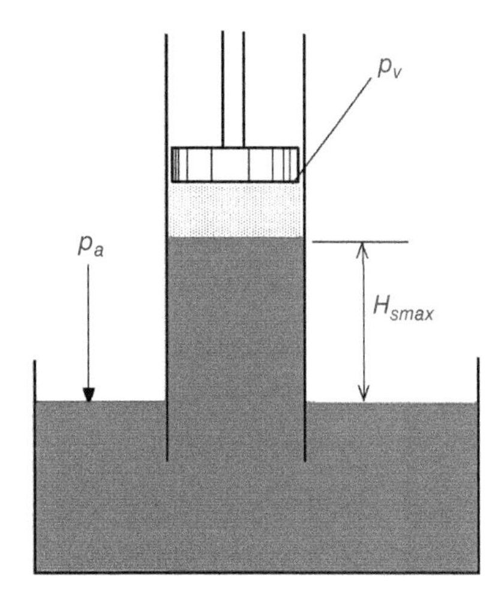

Figure 2.28 Influence vapor.

Table 2.1 Loss suction head caused by vapor pressure (water).

t° [°C]	0	10	20	30	40	60	80	100
p_v [bar]	0.0061	0.0123	0.0234	0.0424	0.0738	0.1992	0.4736	1.0132
$\frac{p_v}{\rho \cdot g}$ [m]	0.06	0.12	0.24	0.43	0.72	2.03	4.82	10.33

Applying the law of Pascal:

$$p_a = \rho \cdot g \cdot H_{smax} + p_v$$

$$H_{smax} = \frac{p_a}{\rho \cdot g} - \frac{p_v}{\rho \cdot g}$$

Let's treat water as an example once again and search for the values of p_v in the steam tables (Table 2.1): Example: with standard atmospheric pressure (1013 [hPa]) at 20 °C a column of water can be equilibrated:

$$(10.33–0.24)\,\text{m} = 10.09\,[\text{m}]$$

2.2.3 Velocity

Atmospheric pressure is necessary to push the fluid mass through the suction pipe (Figure 2.29):

Note:

- c' is the velocity of the water in the reservoir upon a point on the fluid level.
- c is the velocity of the fluid in the cylinder, thus the velocity of the piston (in a simple representation). Application of the Law of Bernoulli between a pointy on the liquid level and a point just beneath the piston:

$$\rho \cdot \frac{c_1^2}{2} + p_a = \rho \cdot \frac{c^2}{2} + p_v + \rho \cdot g \cdot H_{smax}$$

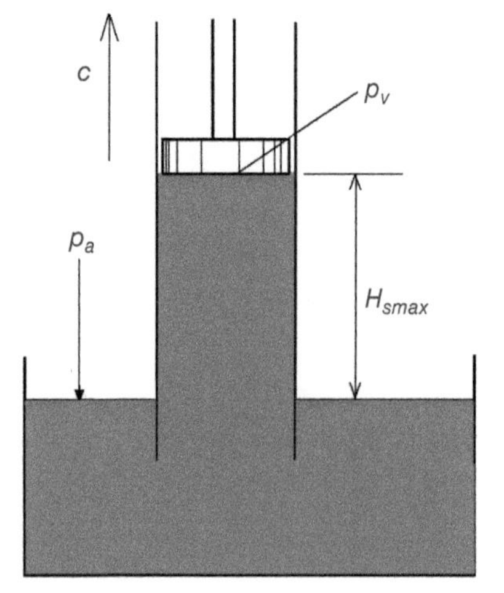

Figure 2.29 Influence velocity.

Because the total liquid surface delivers a contribution to the flow in the suction pipe and this surface is much greater than the section of the pipe, it can be assumed that:

$$c_1^2 \ll c^2$$

So:

$$H_{smax} = \frac{p_a}{\rho \cdot g} - \frac{p_v}{\rho \cdot g} - \frac{c^2}{2 \cdot g}$$

Example: Mostly the piston speed is 1–2 [m/s]. With that the loss in maximum suction head is about 0.05–0.2 [m]. A piston speed of 5 [m/s] would produce 1.27 [m] loss! And even then, with too high velocities the liquid will, when the piston moves backwards, no longer be able to follow the piston. A big vacuum is created under the piston and the small amount of water will evaporate. This will mean it is no longer possible to pump a significant amount of volume!

2.2.4 Barometer

The value of the atmospheric pressure p_a is dependent on the height above the sea level (Tables 2.2 and 2.3):

This you have to take into account when you place your pump in Mexico City on a level of 2000 [m].

2.2.5 Friction

To overcome the friction resistances that occur in the suction pipe (straight pipes, bends, valves, strainer) there must be a pressure drop available. This one can only be delivered by the atmospheric pressure and so there is less pressure available to hold the column liquid in the suction pipe. This resistance is noted as p_{fs}.

So, the expression for the maximum suction head is:

$$H_{smax} = \frac{p_a}{\rho \cdot g} - \frac{p_v}{\rho \cdot g} - \frac{c^2}{2 \cdot g} - \frac{p_{fs}}{\rho \cdot g}$$

Furthermore, the suction valve must be lifted and kept open. For this, too, pressure is necessary.

Table 2.2 Standard atmospheric pressures.

height	−500	0	100	200	500	1000	2000
p_a	1.08	1.013	1.00129	0.98945	0.95460	0.89873	0.79492

Table 2.3 Maximum suction head.

height	−500	0	100	200	500	1000	2000
$\frac{p_a}{\rho \cdot g}$	11	10.33	10.2	10.1	9.7	9.2	8.1

2.2.6 Acceleration

2.2.6.1 Kinematics

During the suction period the fluid must follow the piston. At the beginning of the suction stroke the piston accelerates and at the end it decelerates until it stops. Therefore, the aspirated liquid must also make an accelerated movement and afterwards a decelerated movement. For the acceleration a force is necessary (Newton), and only the atmosphere can fulfil this task.

A part of the atmospheric pressure will thus serve to accelerate the liquid in the suction pipe and the cylinder. This part cannot be used to hold the weight of the column of liquid.

First of all, let's see what is the order of magnitude of this acceleration. For that let's study the Kinematics of the crank-rod mechanism (Figure 2.30).

R is the radius of the crank, L the length of the rod, and ω the angular velocity of the crank. The position x of the piston is given by:

$$x = R \cdot \cos\alpha + L \cdot \cos\beta$$

with:

$$\alpha = \omega \cdot t$$

$$h = R \cdot \sin\alpha \quad h = L \cdot \sin\beta$$

$$\sin\beta = \frac{R}{L} \cdot \sin\alpha$$

$$\cos\beta = \sqrt{1 - \sin^2\beta} = \sqrt{1 - \frac{R^2}{L^2} \cdot \sin^2\alpha}$$

Normally, the ratio of $R : L$ is more or less $1 : 5$ so that:

$$\left(\frac{R}{L}\right)^2 \ll 1$$

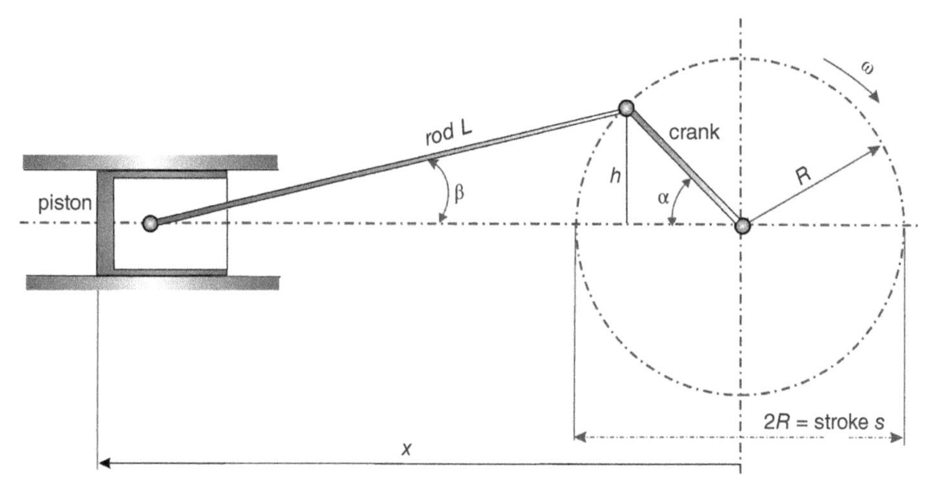

Figure 2.30 Crank-rod mechanism.

In that case the second last expression can on the basis of the Taylor approximation be written as:

$$\cos\beta = 1 - \frac{R^2}{2\cdot L^2}\cdot \sin^2\alpha$$

So that the expression for x becomes:

$$x = R\cdot\cos\alpha + L\cdot\left(1-\frac{R^2}{2\cdot L^2}\cdot\sin^2\alpha\right)$$

Or:

$$x = R\cdot\cos(\omega\cdot t) + L\cdot\left(1-\frac{R^2}{2\cdot L^2}\cdot\sin^2(\omega\cdot t)\right)$$

See the position, velocity, and acceleration of the piston in Figure 2.31 for the case:

$$\omega\cdot t\ \mathit{in\ radials}, \omega = 1, R = 1, L = 5$$

Deriving the expression to the time t leads to the piston velocity c:

$$c = \frac{dx}{dt} = -\omega\cdot R\cdot\sin(\omega\cdot t) - \frac{R^2\cdot\omega}{2\cdot L}\cdot\sin(2\cdot\omega\cdot t)$$

See position, velocity and acceleration of the piston ($\omega\cdot t$ *in radials,* $\omega = 1$, $R = 1$, $L = 5$ *in figure* 2.31, *done with gnuplot freeware*)

The acceleration a of the piston can be found by deriving a second time:

$$a = \frac{dc}{dt} = -\omega^2\cdot R\cdot\left(\cos(\omega\cdot t) + \frac{R}{L}\cdot\cos(2\cdot\omega\cdot t)\right)$$

Let's take as a calculation example:

$$\alpha = 0 \quad \frac{R}{L} = 0.2$$

Then:

$$a = \frac{6}{5}\cdot\omega^2\cdot R$$

2.2.6.2 Dynamics

Let's name A the piston surface, A' the transverse surface of the suction pipe, c' and a' resp. the velocity and acceleration of the liquid in the suction pipe, and L' the length of the suction pipe (Figure 2.32).

For the liquid to follow the piston it must be accelerated in the suction pipe. Here a force F is needed, according to Newton:

$$F = m\cdot a'$$

where m represents the mass of liquid in the suction line.

The force F is delivered by the atmosphere.

Because the volumetric flow is constant:

$$A'\cdot c' = A\cdot c$$

$$A'\cdot\frac{dc'}{dt} = A\cdot\frac{dc}{dt}$$

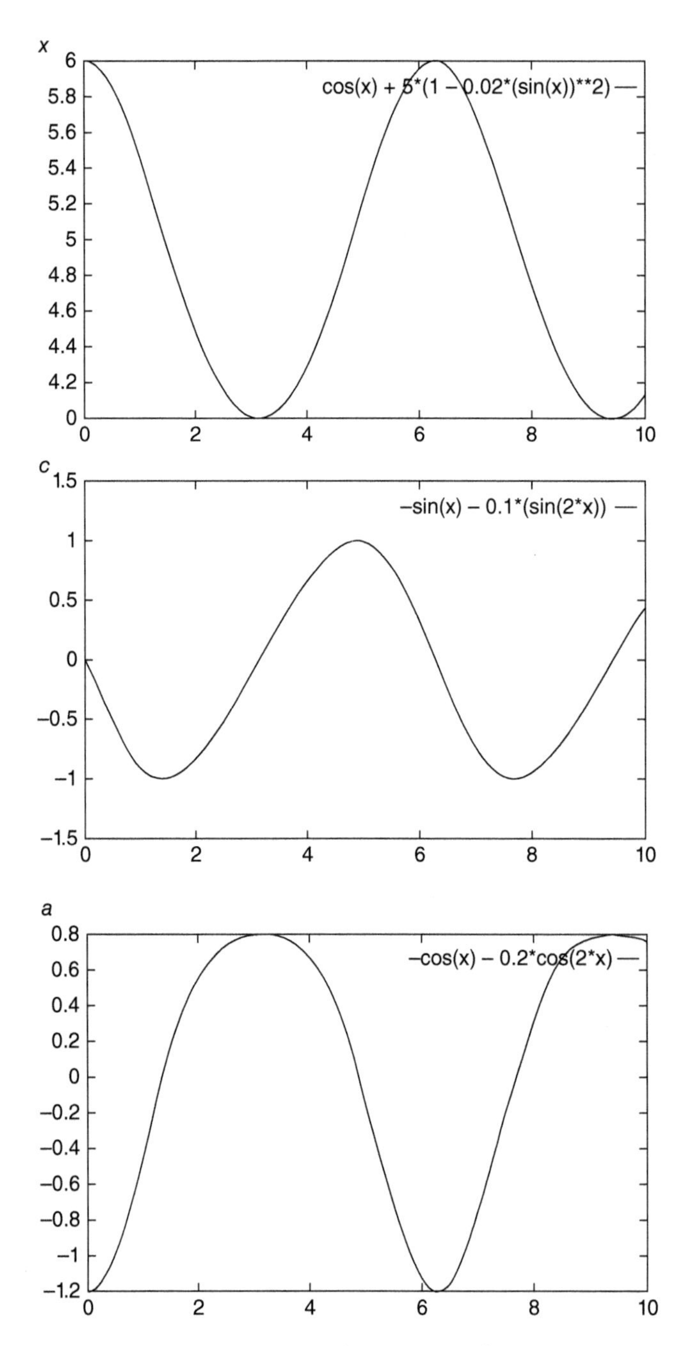

Figure 2.31 Position, velocity, and acceleration of the piston.

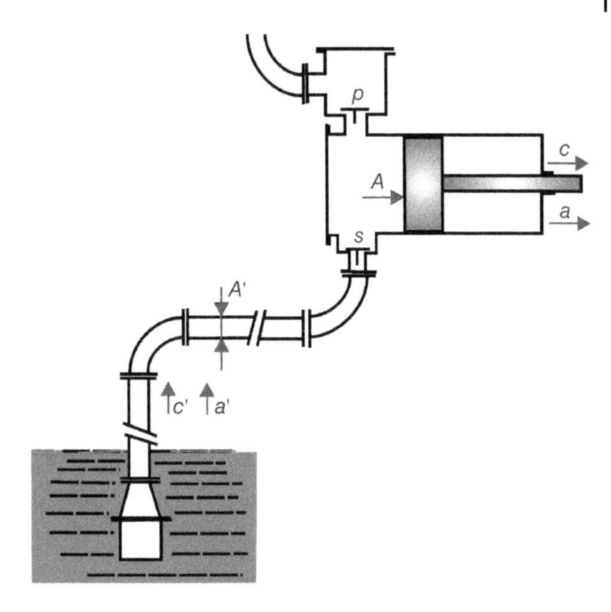

Figure 2.32 Acceleration loss.

$$A' \cdot a' = A \cdot a$$

$$a' = \frac{A}{A'} \cdot a$$

The portion of mass in the suction line (we neglect the mass in the cylinder):

$$m = \rho \cdot A' \cdot L'$$

So that:

$$F = \rho \cdot L' \cdot A \cdot a$$

From which the loss in pressure Δp:

$$\Delta p = \frac{F}{A'} = \rho \cdot L' \cdot A \cdot \frac{a}{A'}$$

And the according loss in suction head ΔH_{smax}:

$$\Delta H_{smax} = \frac{\Delta p}{\rho \cdot g} = \frac{L' \cdot A \cdot a}{g \cdot A'}$$

Example: filling in the value of a:

$$\Delta H_{smax} = \frac{6 \cdot L' \cdot A \cdot \omega^2 \cdot R}{5 \cdot g \cdot A'}$$

The diameter of the piston is 15 [cm], the diameter of the suction pipe 20 [cm], the 90 [rpm], the stroke length 30 [cm], and the length of the suction pipe 12 [m] (a piece can be horizontal). Filling in these values in the expression gives: $\Delta H_{smax} = 11$ [m]!

The acceleration of the discussed mass has yet another drawback. At the beginning of the suction stroke the piston accelerates and the liquid mass sets in motion. Nearly halfway through the stroke the piston reaches its maximum velocity. On the next part of the stroke, the piston velocity will decrease. But the liquid mass lags a little bit so that

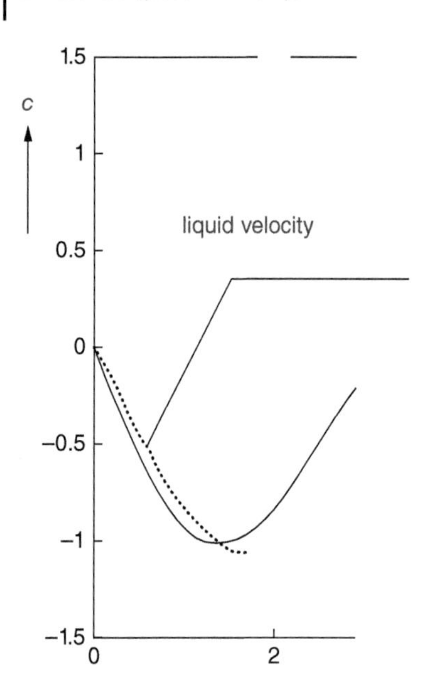

Figure 2.33 Liquid velocity.

a collision will occur between the liquid and the piston (Figure 2.33). This is the water hammer. The drive mechanism will be heavily charged. The more liquid there is, the more the damage may be.

2.2.7 Air Chambers

2.2.7.1 Suction Side

The previous section established that the loss in maximum suction head caused by the acceleration of the liquid mass in the suction pipe was so great that the pump could not suck anymore.

This can be remedied by the use of an air chamber underpressure. Consider now only the suction side (Figure 2.34). The liquid that is sucked out of the suction reservoir first comes in a space, the air chamber. The air chamber is placed as near as possible to the cylinder. In the chamber there is a spacious stock of liquid. Above the liquid there is a bellows where air is locked in. The air is in underpressure (vacuum).

Let's assume that the pressure in the chamber is constant. The chamber is so big that variations of the content during suction and press operation will be small. This means that the pressure will be constant in the bellows.

When the suction stroke starts, a vacuum is created in the cylinder, yet deeper than that in the bellows. So, liquid will flow to the cylinder. This time it is the air chamber that sucks the liquid in the cylinder out of the chamber, not the atmosphere.

The liquid between the chamber and the cylinder will accelerate. This is the first circuit. The chain is air chamber → cylinder. In Figure 2.35 the flow Q_V that flows here is set in function of the time.

A second circuit is formed by atmosphere → suction reservoir → air chamber. The atmospheric pressure is still higher than the pressure in the chamber, and during the press operation. In this way, liquid will flow all the time from the suction reservoir to the chamber. Because the pressure of the atmosphere and that in the chamber is *constant*

Figure 2.34 Suction side.

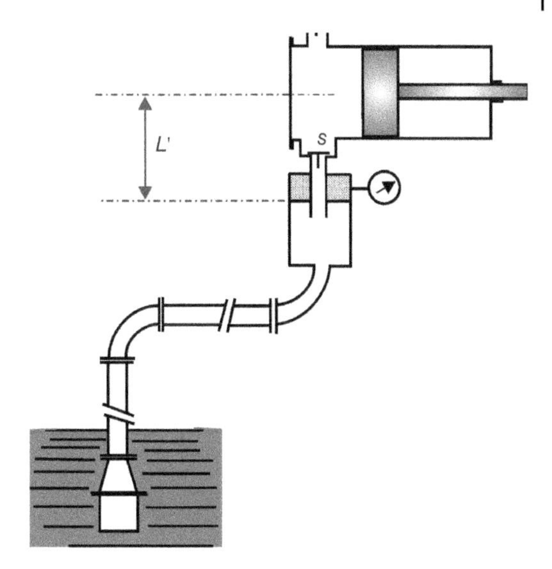

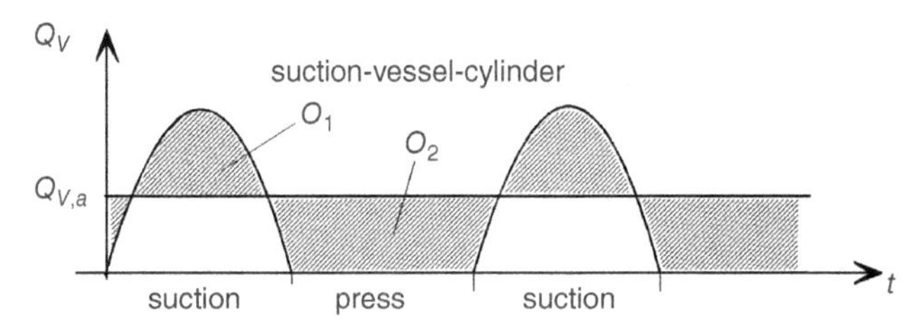

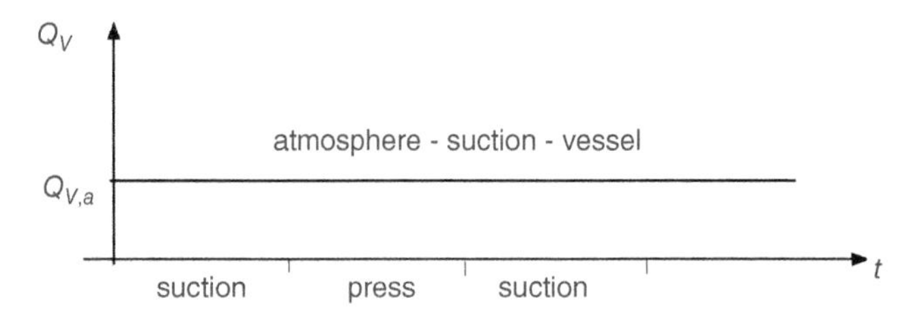

Figure 2.35 Flows in the pump installation.

the flow will also be constant! Figure 2.35 represents this flow Q_{va}. This all comes down to the fact that the average of flow Q_v must be equal to $Q_{V,a}$. In Figure 2.35 this means that the surfaces O_1 and O_2 must be equal.

It is only in the small piece of pipe L' in Figure 2.34 from the chamber to the cylinder that the pump must accelerate the liquid.

2.2.7.2 Press Side

Look now at the press side (Figure 2.36). The liquid mass in the press pipe is being pushed upwards by the piston. Because the piston accelerates and decelerates the liquid mass will follow this. If it concerns a very long press pipe, the force to apply to accelerate this liquid mass will be very high.

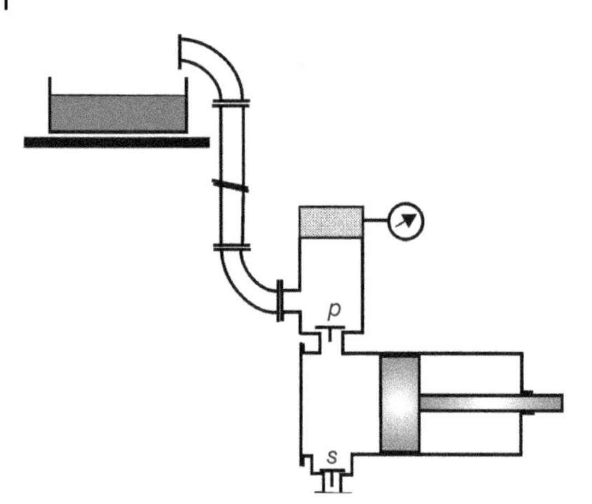

Figure 2.36 On press side.

In order to reduce this quantity an analogous solution will be implemented, as in the suction side. In Figure 2.34 an air chamber is placed on the press side. The vessel is in overpressure and if it is big enough one can assume that the pressure is constant.

Also during the press, at the suction stroke there will be a constant flow from the vessel to the press reservoir, because the pressure in the vessel is constant.

Only in the little piece cylinder → vessel the liquid is accelerated during the press stroke.

In Figure 2.37 an implementation of a piston pump with air chambers is represented.

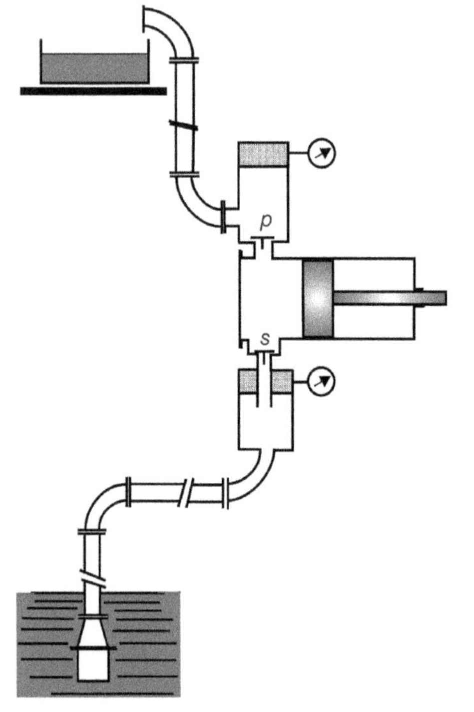

Figure 2.37 Implementation of a piston pump with air chambers.

2.3 Characteristic Values

2.3.1 Manometric Feed Pressure

Consider a piston pump with air chambers on which two manometers are connected (Figure 2.36).

Apply the law of Bernoulli between two points. The first point, where the velocity of the liquid is c', is positioned just beneath the air cushion in the pressure vessel. The second point is situated on the outlet of the press pipe where the velocity of the liquid is c. Because the liquid of the first point comes from a very big surface, its velocity c' is much smaller than the velocity c and can be neglected.

The indication of the *absolute* pressure $p_{r,p}$, at the press reservoir:

$$p_{r,p} = p_a + \rho \cdot g \cdot H_1 + p_{fp} + \rho \cdot \frac{c^2}{2}$$

Now apply the law of Bernoulli between a point that lies on the surface of the liquid of the suction reservoir and a second point that lies just beneath the surface of the air chamber. The liquid that comes from the first point comes from a large surface, so its velocity can be neglected. The same goes for the second point because the suction vessel is supposed to be big.

Taking this into consideration the *absolute pressure* $p_{r,s}$ in the suction vessel is:

$$p_{r,s} = p_a - \rho \cdot g \cdot H_2 - p_{fs}$$

The difference between the indications:

$$p_{r,p} - p_{r,s} = \rho \cdot g \cdot (H_2 + H_1) + (p_{fs} + p_{fp}) + \rho \cdot \frac{c^2}{2}$$

On the other hand, the manometric feed pressure was defined earlier as:

$$p_{man} = \rho \cdot g \cdot (H_2 + H_1 + H_3) + (p_{fs} + p_{fp}) + \rho \cdot \frac{c^2}{2}$$

This means that, taking apart the small term $(\rho \cdot g \cdot H_3)$ – the image is not to scale – the manometric feed pressure can be determined by the difference of the indications on the two manometers. In practice, however, manometers measure *overpressures* or *underpressures,* so to determine the manometric feed pressure one has to consider the sum of the indications (Figure 2.38).

The *manometric head* H_{man} is defined as:

$$H_{man} = \frac{p_{man}}{\rho \cdot g}$$

2.3.2 Theoretical Pressure

Outwards the pump delivers the manometric feed pressure p_{man} but it also has to overcome its own losses $p_{l,p}$ (friction in the pump housing, valve resistance, acceleration loss, etc.). The *theoretical feed pressure* p_{th} is the pressure that the pump could deliver if these losses did not exist. In Figure 2.39 this concept is illustrated.

$$p_{th} = p_{man} + p_{l,p}$$

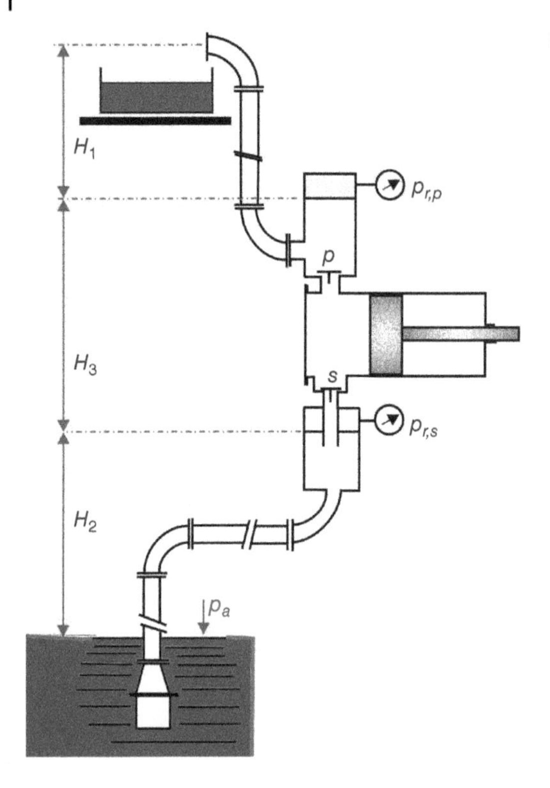

Figure 2.38 Manometric head.

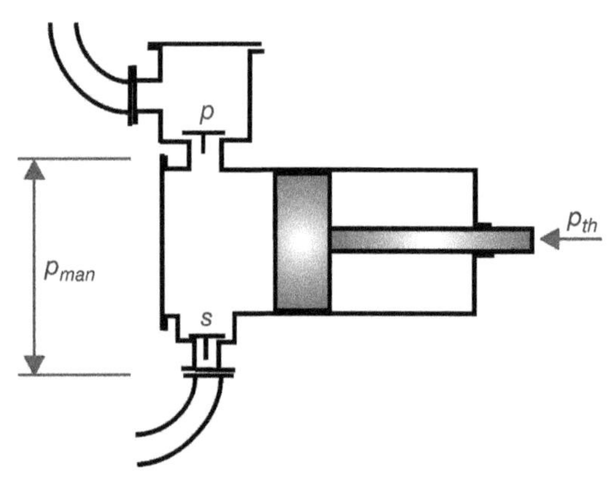

Figure 2.39 Theoretical pressure.

The *theoretical (pressure) head* is defined as:

$$H_{th} = \frac{p_{th}}{\rho \cdot g}$$

2.3.3 Power and Efficiency

Let's calculate the work to perform on the liquid during the suction stroke. Atmospheric pressure p always acts against the right-hand side of the piston (Figure 2.40). On the left-hand side there is an underpressure. Name the (absolute) pressure in the cylinder

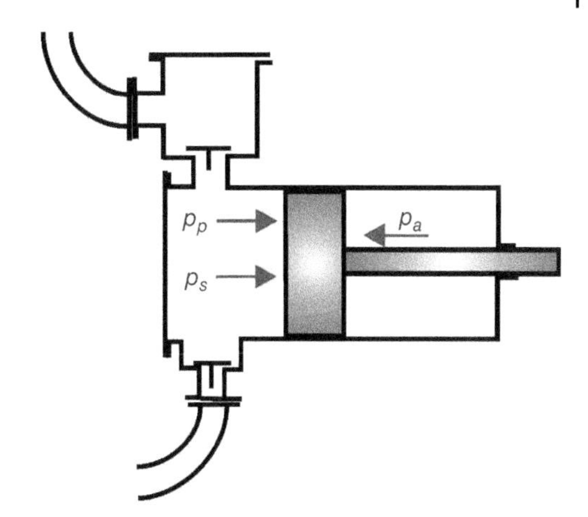

Figure 2.40 Pressure in pump.

during the suction stroke p_s, then the work (force × displacement) during the suction stroke:

$$(p_a - p_s) \cdot s \cdot \frac{\pi \cdot D^2}{4}$$

where s is the stroke length and D the diameter of the piston.

Analogues for the work during the press cycle:

$$(p_p - p_a) \cdot s \cdot \frac{\pi \cdot D^2}{4}$$

where p_p represents the pressure in the cylinder during the press stroke.

The work during one revolution of the crankshaft:

$$(p_a - p_s) \cdot s \cdot \frac{\pi \cdot D^2}{4}$$

On the other hand, we can set up an expression for p_p and p_s as follows (Figure 2.41), whereby the kinetic terms are neglected:

$$p_z = p_a - \rho \cdot g \cdot H_z - p_{fs} - p_{f,ps}$$

$$p_p = p_a + \rho \cdot g \cdot H_p + p_{fp} + p_{f,pp}$$

where $p_{f,ps}$ and $p_{f,pp}$ are respectively the losses in the cylinder during the suction and press stroke.

where from:

$$p_p - p_s = \rho \cdot g \cdot (H_s + H_p) + (p_{fs} + p_{fp}) + (p_{f,ps} + p_{f,pp})$$

Or, with $p_{lp} = p_{f,ps} + p_{f,pp}$

$$p_p - p_s = p_{man} + p_{l,p}$$

$$p_p - p_s = p_{th}$$

Substitution of previous expressions leads to the work in one revolution:

$$p_{th} \cdot s \cdot \frac{\pi \cdot D^2}{4}$$

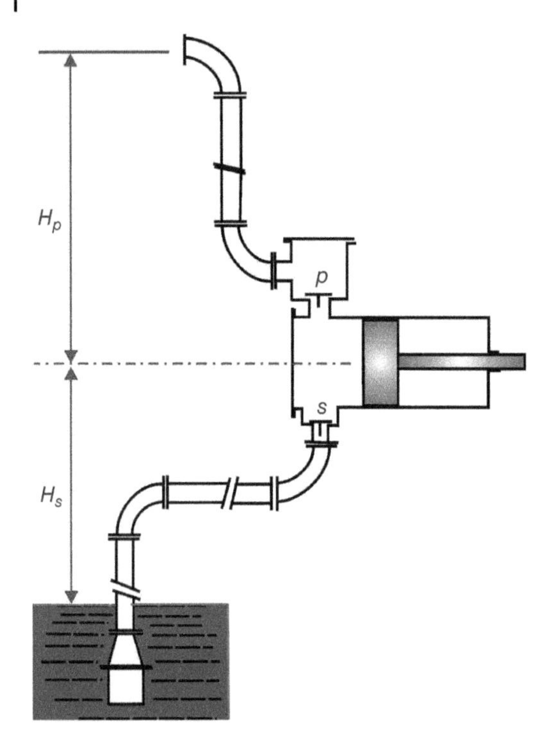

Figure 2.41 Pump installation.

1 second counts N revolutions.
From which the power P on the axis of the pump

$$P = p_{th} \cdot s \cdot \frac{\pi \cdot D^2}{4} \cdot \frac{N}{60}$$

One can prove that this expression is valid for all kinds of pumps.
The hydraulic efficiency of the pump is then defined as:

$$\eta_p = \frac{p_{man}}{p_{th}}$$

and the efficiency of the installation as:

$$\eta_i = \frac{p_s}{p_{th}}$$

2.3.4 Example

A pump installation (Figure 2.41) with air chambers (so acceleration loss is negligible) has a geodetic press head of 20 [m] and a geodetic suction head of 3 [m]. The pipe resistance of the suction pipe (straight pieces, bends, etc.) is 60 [hPa] [mbar], the one of the press pipe is 170 [hPa] [mbar]. The barometer indicates 1010 [hPa]. The temperature of the water is 60 °C.

The pump has the following characteristics:

- rpm: $N = 108$ [tpm].
- Stroke length: $s = 0.32$ [m].

- Piston diameter: $D = 200$ [mm].
- Distance between the liquid level of the suction vessel and the heart of the plunger 0.2 [m]. Analog for the press vessel: 0.2 [m].
- The valve resistances are both estimated to be 80 [hPa] [mbar].
- The friction in the pump housing is estimated to be 20 [hPa] [mbar] during the suction stroke. Which is the same as for the press stroke.
- The volumetric efficiency η_{ve} of the pump is 0.93.
- In this example we neglect the velocity loss (dynamic pressure) and acceleration losses.

Wanted:

a. The volumetric flows.
b. The indication of the vacuum meter on the suction vessel.
c. The pressure in the cylinder during the suction stroke.
d. Is the installation able to pump the water?
e. The indication of the manometer during the press stroke.
f. The pressure in the cylinder during the suction stroke.
g. The manometric feed pressure p_{man}.
h. The theoretical pressure p_{th}.
i. The power P needed to drive the pump.
j. The hydraulic efficiencies.

Solution:

a. *The theoretical (nominal) and effective volumetric flows:*

$$Q_V = A \cdot s \cdot \frac{N}{60} = \frac{\pi \cdot D^2}{4} \cdot 0.32 \cdot \frac{108}{60} = 0.00181\ [\mathrm{m^3/s}]$$

$$Q_{Ve} = \eta_v \cdot Q_V = 0.93 \cdot 0.00181 = 0.0168\ [\mathrm{m^3/s}]$$

b. *The indication of the vacuum meter on the suction vessel:*
The pressure losses are:
   - Height: $2.8 \cdot 10^3 \cdot 9.81$ [Pa] = 275 [hPa]
   - Friction $p_{f,s} = 60$ [hPa]

   Total = 335 [hPa]
   The indication is: 1010 – 335 = 675 [hPa]

c. *The pressure in the cylinder during the suction stroke:*
The supplementary losses are:
   - Height: $0.2 \cdot 10^3 \cdot 9.81$ [Pa] = 20 [hPa]
   - Friction = 20 [hPa]
   - Valve = 80 [hPa]

   Total = 120 [hPa]
   The (absolute) pressure is: 675 – 120 = 555 [hPa]

d. *Is the installation able to pump the water at 60 °C?*
$\frac{p_v}{\rho \cdot g}$ (at 60 °C) = 2.03 from where $p_v = 10 \cdot 9.81$ [Pa] = 199 [hPa]
This is less than 555 [hPa]: the pump is able to do the job.

e. *The indication of the manometer during the press stroke:*
Pressure losses:
- Height: $1.8 \cdot 10^3 \cdot 9.81$ [Pa] = 1942 [hPa]
- Friction $p_{f,p} = 170$ [hPa]
  Total: = 2112 [hPa]
  The indication is 1010 + 2112 = 3122 [hPa]

f. *The pressure in the cylinder during the press stroke:*
Supplementary losses:
- Height: $0.2 \cdot 103 \cdot 9.81$ [Pa] = 20 [hPa]
- Friction = 20 [hPa]
- Valve = 80 [hPa]
  Total: = 120 [hPa]
  The pressure is: 3122 + 120 = 3242 [hPa]

g. *The manometric feed pressure p_{man}:*
Seen as the sum of all “resistances” that the pump encounters in the installation:
Height $p_{geo} = 19.8 \cdot 10^3 \cdot 9.81$ [Pa] = 1942 [hPa]
Friction: $pfp + pfs = 170 + 60 = 230$

h. *The hydraulic efficiency of the pump:*

$$\eta_p = \frac{p_{man}}{p_{th}} = \frac{2486}{2686} = 0.93$$

i. *The hydraulic efficiency of the installation*

$$\eta_i = \frac{p_s}{p_{th}} = \frac{2256}{2686} = 0.84$$

j. Total p_{man}: = 2486 [hPa]

2.3.5 Characteristic Curve of the Pump

2.3.5.1 Characteristic of the Pipe Line

As discussed earlier, the manometric pressure p_{man} in an installation is needed to overcome the height, a dynamic pressure to give the liquid a velocity, and a pressure to overcome the friction resistance. In Figure 2.42 a graph is represented showing this manometric pressure against the effective volumetric flow Q_{ve}.The geodetic pressure is not dependent on the flow. This is shown as a horizontal line on the graph. The dynamic and friction pressures are quadratic dependent on the velocity, and thus quadratic on the flow. This is represented by a parabola, and in the graph this parabola is added to the straight line. The total is p_{man}.

2.3.5.2 Characteristic of the Pump

In a piston pump, and more generally in a positive displacement pump, a cavity is first created, then the cavity is closed by pressing the liquid in this cavity away, whereby the delivered pressure is exactly that one asked by the installation (Figures 2.42 and 2.43). In principle, as long as the driving motor is capable, there is no limit!

If the motor has no current limit, it will burn, because the torque delivered by an electric motor is proportional to the electric current.

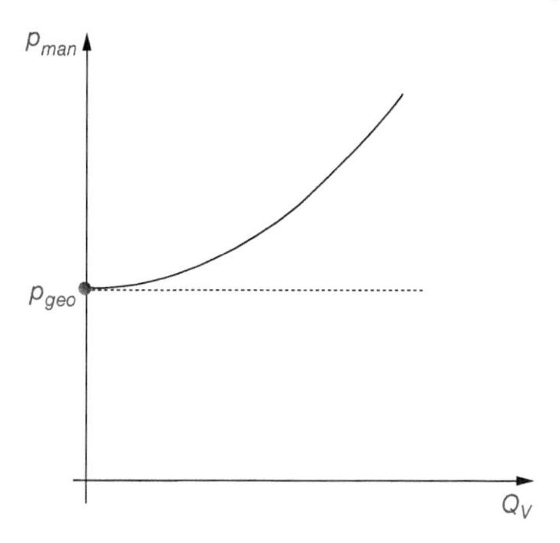

Figure 2.42 Characteristic of the installation (*system curve*).

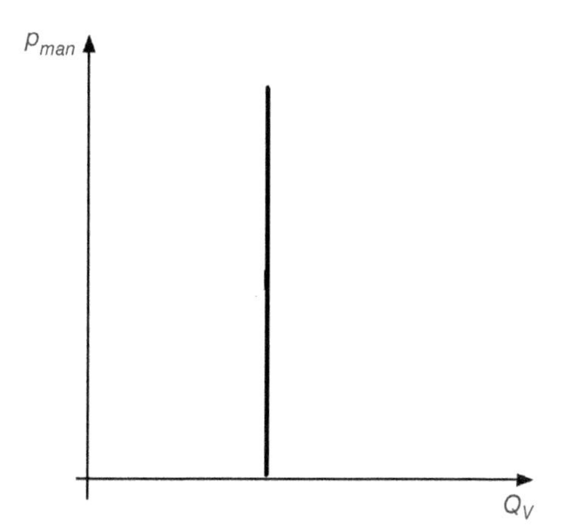

Figure 2.43 Characteristic of the pump.

The exaggerated pressure can be very dangerous. If a shut-off valve in the press pipe were to be closed, the pump would try to overcome this infinite pressure. Something would have to break. Never place a shut-off valve in the press pipe of a displacement pump. And even then, it is possible that the pump delivers the liquid to a pressure vessel. If there is no consumer of the liquid in this vessel, its valve is closed, the pressure in this vessel can increase to undesirable limits. It will explode.

Where the two curves meet each other, we find the duty point of the pump (Figure 2.44).

The pressure vessel will always be protected by a relief valve (safety valve, see Figures 2.45 and 2.46). Such a valve has a spring The tension of the spring is preset by a value that corresponds to the allowable pressure in the liquid acting on the valve. A maximum pressure is set. Above this maximum the spring will shrink so that it will release an opening in the valve, which drains the excess liquid away.

2.3.5.3 Regulation

How can one regulate the flow of a piston pump with a fixed displacement? The only obvious way is to change the rpm. If the rpm is increased, the pump characteristic will move to the right. A new work point is created (Figure 2.47).

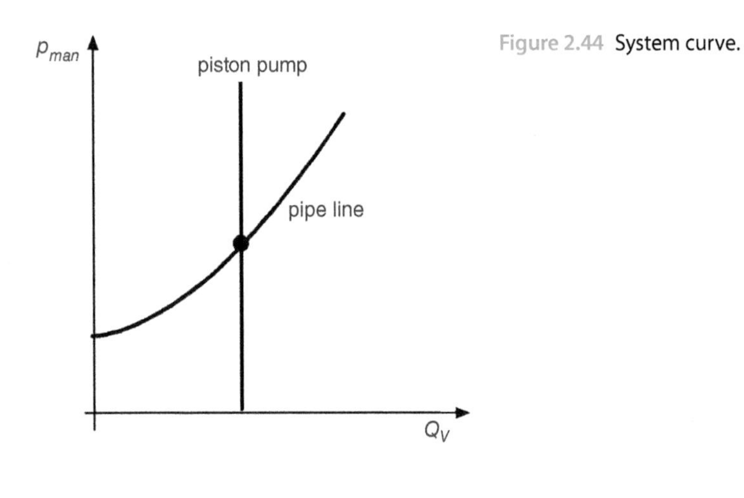

Figure 2.44 System curve.

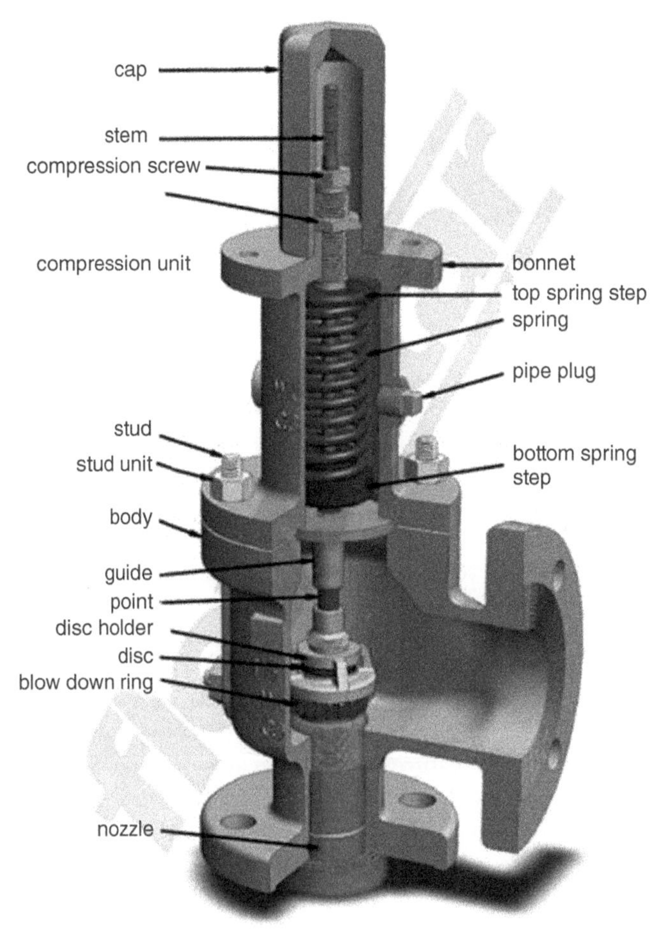

Figure 2.45 Pressure relief valve.

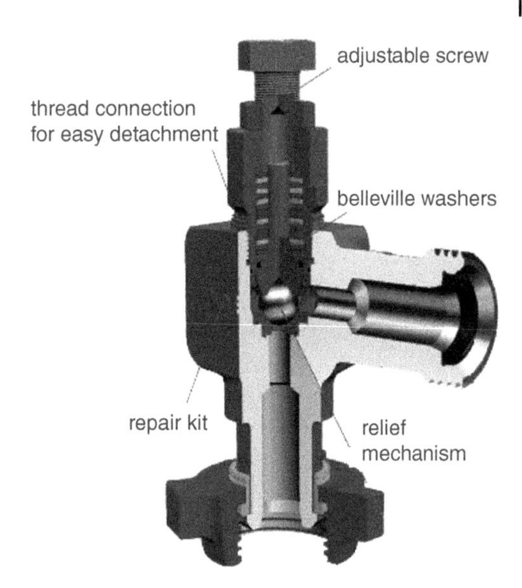

Figure 2.46 Ball relief valve. Source: Sepco.

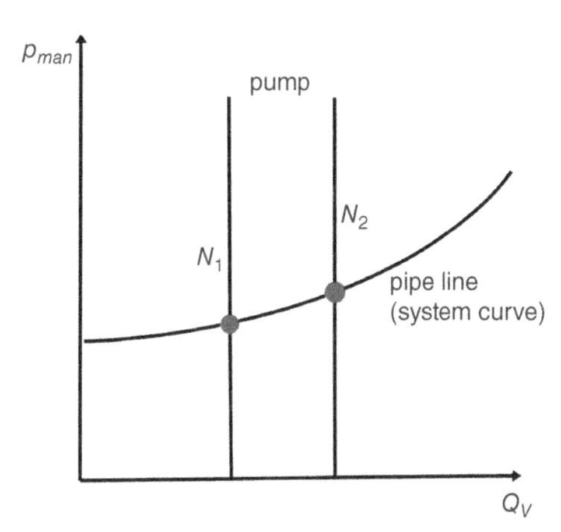

Figure 2.47 Changing duty point.

2.3.6 Conclusion

Piston pumps are very reliable devices that are used for applications where a high pressure is necessary. They are increasingly replaced by turbopumps, which are more flexible. Nowadays the principal applications are:

- Spring dewatering (Figure 2.48). Spring dewatering is a process whereby groundwater is pumped up. This is necessary to increase the stability of a dike or to increase slopes, or to maintain the water level in a well, to lower the water level in the bottom of a construction site so that it can be worked dry, among other things, to lay pipes and sewage pipes, or build cellars, underground car parks, etc.
- Concrete pumps (Figure 2.49) although they are often carried out as membrane pumps).
- High pressure hydraulic pumps, where they are unbeatable (see later).

Figure 2.48 Spring dewatering. Source: Wikipedia.

Figure 2.49 Concrete pump. Source: Wikipedia.

2.4 Hydraulic Pumps

2.4.1 Introduction

There are various designs of hydraulic pumps, but they have one thing in common: they are positive displacement pumps. Their aim is to produce pressure. This means that the delivered pressure is determined by the counterpressure in the discharge line. The volumetric flow is determined by the stroke volume of the pump, that is the moved volume per revolution and the rpm of the driving electromotor or diesel engine.

Some pumps allow the user to change the flow, and in particular the stroke volume.

Pumps with *constant volumetric flow* are pumps that, driven at a determined rpm, deliver all the time a constant flow. If the rpm changes, then the delivered flow will also change. Pumps with *variable volumetric flow* make it possible, with constant speed of the driving motor, to change the volumetric flow by changing the stroke volume.

In Figure 2.50 the various hydraulic pumps are classified.

2.4.2 Sliding Vane Pump

The operating principle of a sliding vane pump is represented in Figure 2.51 and an implementation in Figure 2.52. In the stator is a circular space. The circular rotor is eccentrically placed in the stator housing. The vanes are placed in the rotor. They can move radially. Rotor and vanes are locked up with little play between the cover plates. The rotor is fastened to the driving shaft. As soon as it moves a centrifugal force applies to the vanes. This force pushes the vanes against the wall of the casing. Chambers are formed between the vanes, the stator and rotor. On the bottom the chambers become bigger and a suction action arises. At the top the chambers become smaller and the liquid is discharged.

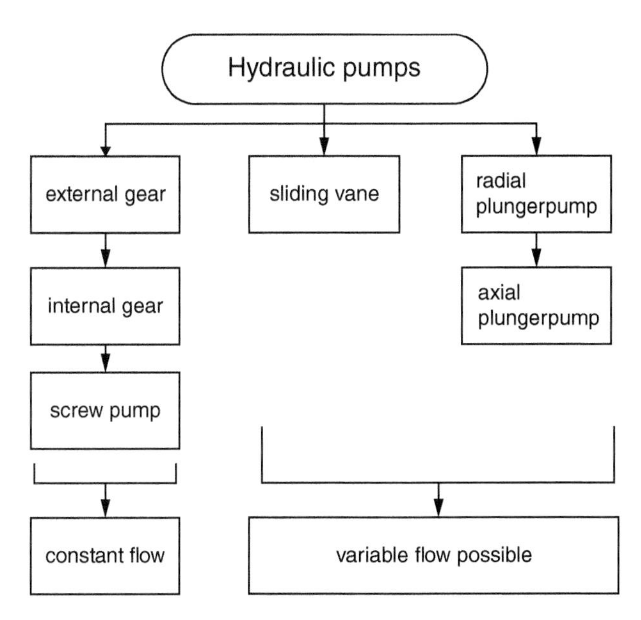

Figure 2.50 Classification hydraulic pumps.

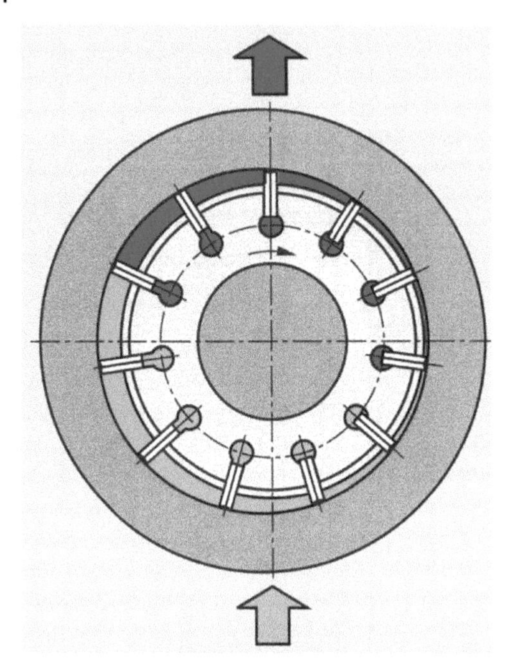

Figure 2.51 Sliding vane pump. Source: Bosch Rexroth AG.

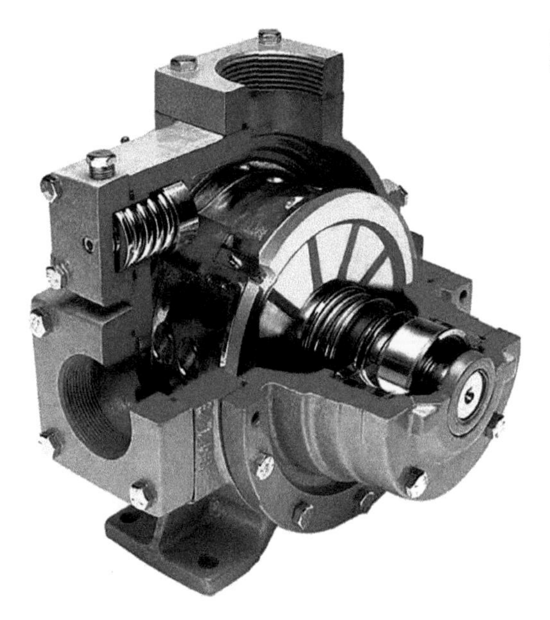

Figure 2.52 Sliding vane pump. Source: Cork.

The eccentricity – this is the perpendicular distance between the centerline of housing and rotor – determines the stroke volume. When the eccentricity is zero, the pump delivers no flow. By regulating the eccentricity (Figure 2.53) the volumetric flow can be decreased or increased. In principle, it can even be reverted, at least if the construction is provided for this.

Sliding vane pumps have rpms of 3000 and nominal pressures of up to 160 [bar].

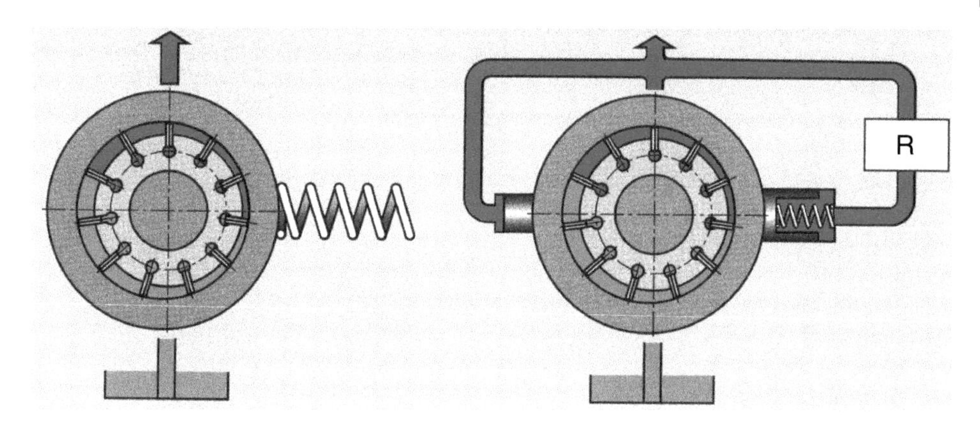

Figure 2.53 Sliding vane pump with variable flow (a) direct and (b) indirect piloted via pressure regulator R. Source: Bosch Rexroth AG.

2.4.3 Gear Pumps

2.4.3.1 External Toothing

When the viscosity of the liquid is too high, centrifugal pumps are not adequate because the friction in the vane's channel is too high. Other pumps are used to transport the liquid with low velocity. They are always positive displacement pumps.

The gear pump is constituted of two precisely machined gears that rotate in the pump housing (Figures 2.54 and 2.55). On the suction side the liquid fills a tooth cavity and on the discharge side the liquid is pressed away by repression.

One of the gears will be connected with the driving shaft, while the second one will be dragged by the first one. The mechanism of the operation happens as follows: when a tooth rotates out of a tooth cavity of the other gear an underpressure will be created in the tooth cavity causing liquid to be sucked away. This is the suction side. This liquid will be dragged with t the tooth cavity along the pump housing. On the other side of the pump a tooth cavity of the other gear will be filled with liquid, the cavity will then be reduced, and so the liquid will be discharged on the press side.

Properties:

- This pump can be driven directly.
- Be
  - N: rpm
  - Z: Number of teeth of one gear
  - V: volume tooth cavity.
- Then the volumetric flow is: $Q_V = 2 \cdot Z \cdot V \cdot \frac{N}{60}$

  So, the volumetric flow is proportional to the rotation speed, that runs from 500 to 6000.

  The quantity of pumped volume per rotation is the stroke volume and is indicated on the pump. It is 0.2 to 200 [cm^3].
- Delivered pressure: to 300 [bar].
- Viscosity to 80 000 [mPa s].
- Between the teeth of the gripping gears on one hand and between the gears and the housing on the other hand liquid can leak from discharge to suction side. The flow will be higher with higher discharge pressures.

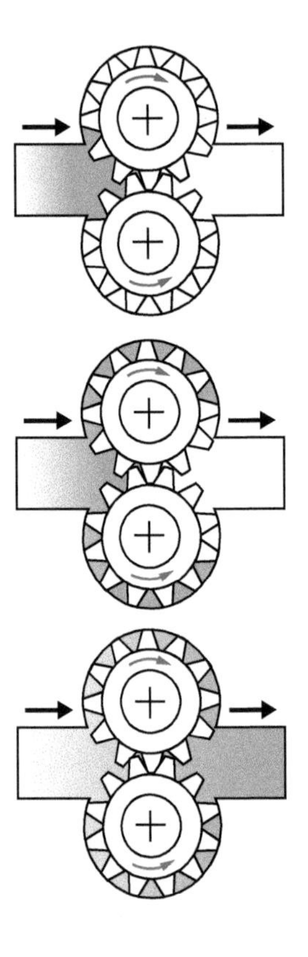

Figure 2.54 Gear pump.

Figure 2.55 Teeth on shaft.

- When a tooth, before it reaches the lowest depth, touches both sides of the tooth cavity on the other teeth, the oil can no longer escape. It will be compressed and very high forces arise. This causes extra wear. In order to let the oil escape, relief grooves are applied. These are cavities in the walls next to the gears so that the trapped liquid can escape to the discharge side, without the possibility of leaks. The minimal dimensions

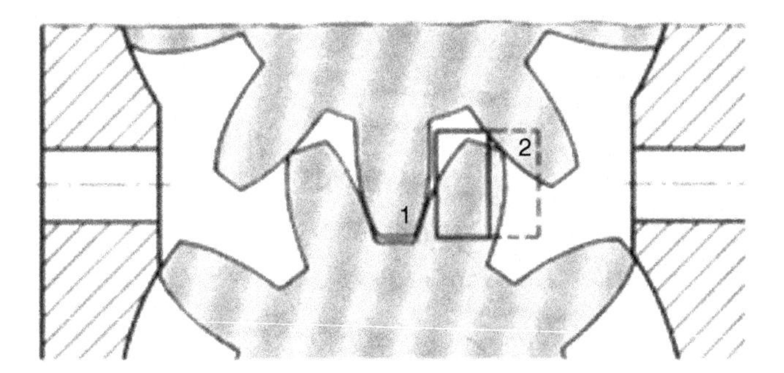

Figure 2.56 Relief grooves.

Figure 2.57 Axial leak compensation. Source: Bosch-Rexroth GmbH.

of these grooves are shown in Figure 2.56 with a full drawn rectangle. But for certainty they are made a little bit bigger; see the rectangle drawn with a dashed line.

Besides, an axial leak can originate along the flat sides of the gears (axial leak slits). A minimum leak split is necessary to allow movement. To have a constant volumetric flow in function of the discharge pressure and wear, slits are placed in the housing (Figure 2.57). This is the leak slit compensation. The operation of this compensation is explained in specialized literature.

Because the gear pump is a positive displacement pump, like the piston pump, there is, in principle, no limit to the delivered discharge pressure. In the case of a closed press valve or obstruction in the discharge line, one will still experience problems, because the mechanism of the pump means the liquid will be discharged anyway, no matter how great the press pressure is. That's why a safety valve is placed on the discharge side. It opens when the discharge pressure reaches the preset pressure of the spring (Figure 2.58) and the liquid returns to the suction side.

2.4.3.2 Internal Toothing

There are many variants of the gear pump. One type is a gear pump with internal toothing (see Figures 2.59–2.61). It works as follows: the rotor with internal toothing B is

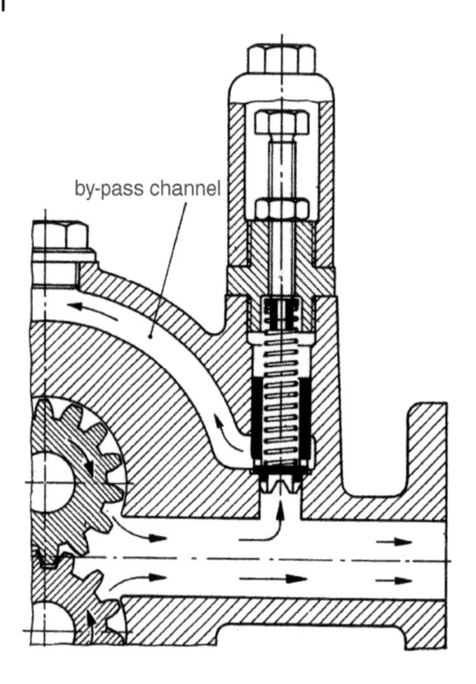

Figure 2.58 Bypass.

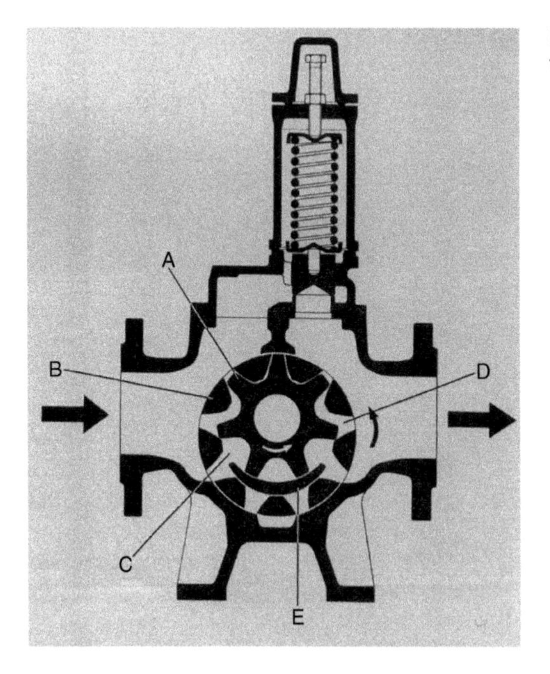

Figure 2.59 Gear pump with internal toothing. Source: Stork.

driven and drags with its gear A rotates in a closed housing. Gear A rotates eccentrically relative to gear B. While rotating in the indicated direction the teeth of the rotor slide in the linker half out of each other. The space C between the teeth is now filled with liquid. At D the teeth move out of each other and this causes liquid to move to the

Figure 2.60 Parts. Source: Johnson Pump.

Figure 2.61 Gear pump with internal toothing. Source: Stork.

discharge side. The separation between the press and suction side is made by the wall of the pump housing and the sickle-shaped separation piece E between rotor and gear.

Let's look at some properties. On the inlet side the offered space covers 120° (Figure 2.62). This causes the inlet chamber to not open abruptly but it will fill slowly. The consequence is that the pump works particularly regularly and has a very good suction property.

At the place of the filling piece the liquid is transported without any volume change.

The space behind the filling piece connects to the discharge side. This causes the volume between the tooth cavities to diminish and the liquid to be displaced.

Where the teeth grip one another, the displacement property of the pump becomes clear, because, other than in the gear pump with external toothing, nearly no dead zones are present.

The oil volume in the dead zones is compressed and causes pressure peaks and noise. Internal gear pumps exhibit nearly no pressure pulsation and have a noiseless operation.

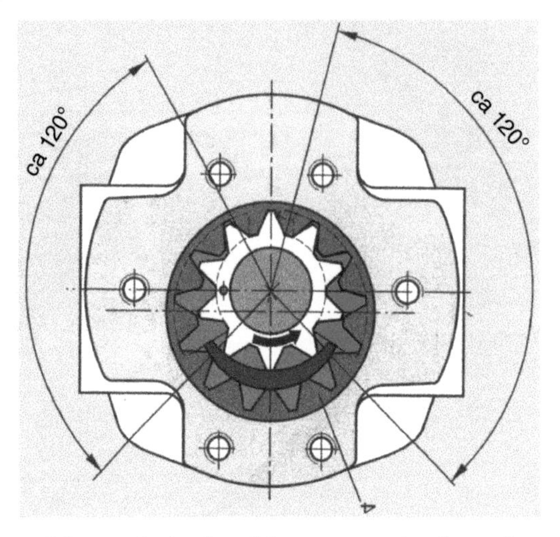

Figure 2.62 Gear pump with internal toothing. Source: Bosch Rexroth AG.

We conclude that this construction has a better volumetric efficiency than the classic gear pump, and so higher discharge pressures are reachable. The flow is pulsation free.

2.4.4 Screw Pumps

The screw pump in its simplest form is shown in Figure 2.63. It is composed of a shaft with a right-handed screw thread. When the shaft is rotated to the right it seems to be like the thread moves to the left and an imaginary nut would go to the right. That's why the shaft is rotated in the left sense, then the thread appears to go to the right and an imaginary nut would go to the left. Replace now the nut with liquid, then it will run from right to left in Figure 2.63. The rotating movement of the screw thus pushes the liquid in an axial direction. The liquid does not rotate but moves in a straight line axially.

One could say, therefore, that an axial turbopump has one continuous vane. So, a pulsation free transport is realized and this causes a noiseless action. One application would be in theaters.

The flows go as high as 1500 [m^3] per hour and the discharge pressures as much as 100 [bar], if not more.

Screw pumps are intended as hydraulic pumps, for liquids with high viscosity. The leak from discharge at the suction side is minimal. One possible application is pressing plastic granulates in extrusion molds.

Figure 2.64 shows a two-shaft screw pump. It consists of a right- and a left-handed screw. Both axes rotate in counter rotation so that their treads fit closely together. But

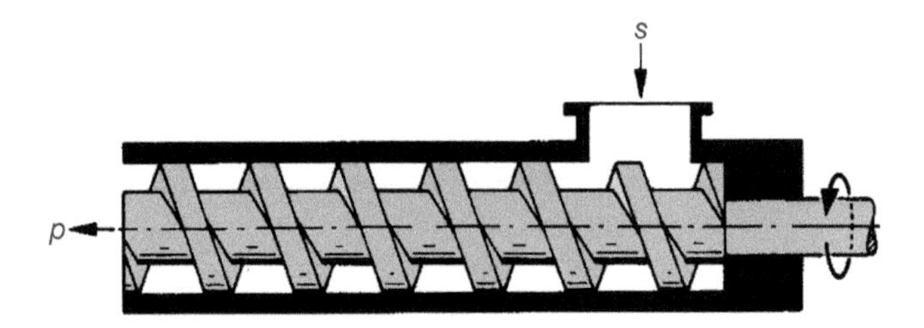

Figure 2.63 One-shaft screw pump.

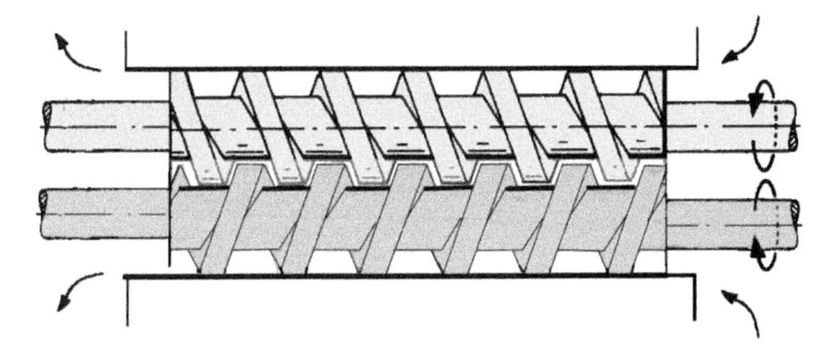

Figure 2.64 Two-axis screw pump.

there is no contact between the screws or relative to the housing. Executions of the pump screw are represented in Figures 2.65–2.77.

In principle screw pumps need an axial thrust bearing. Indeed, on one side (Figure 2.63) there is a pressure of, say, 50 [bar] and on the other side the suction pressure is 1 [bar]. This pressure difference, and thereby the axial pressure on the rotors, must be caught by an axial bearing. This can be by-passed by placing two screw pumps back to back (Figures 2.68–2.71).

There are also screw pumps with three screws (Figures 2.72 and 2.73). The liquid flows quasi pulsation free, thus it has a noiseless action.

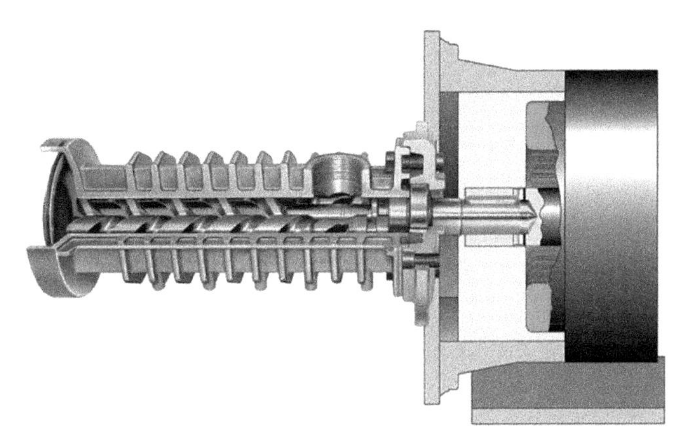

Figure 2.65 Two-axis screw pump. Source: Allweiler.

Figure 2.66 Screws. Source: DERA.

Figure 2.67 Screws. Source: Borneman-Gouldpumps.

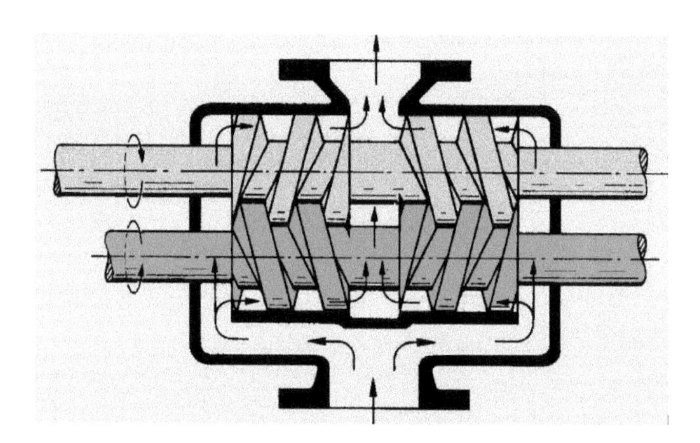

Figure 2.68 Axial equilibrium.

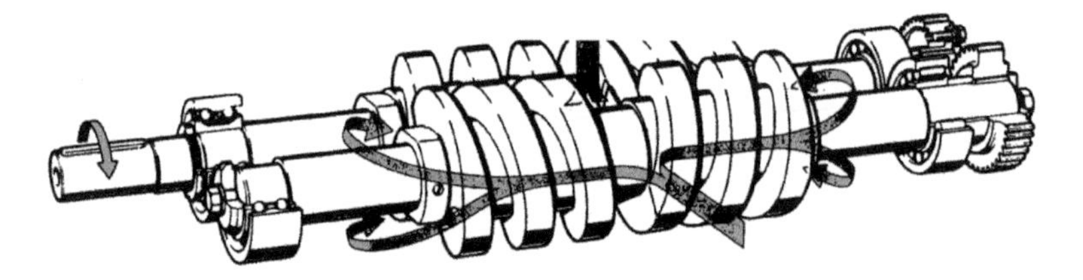

Figure 2.69 Operation. Source: Bornemann.

2.4.5 Radial Plunger Pumps

The pump with radially moving plungers consists of a disc rotor placed eccentrically in the housing (Figures 2.74 and 2.75). There are radial holes where the plungers can slide. At the end of the plungers are slide block or guide rollers. The rotor has two channels, *s* and *p*, that respectively are connected with suction and discharge sides. When the rotor

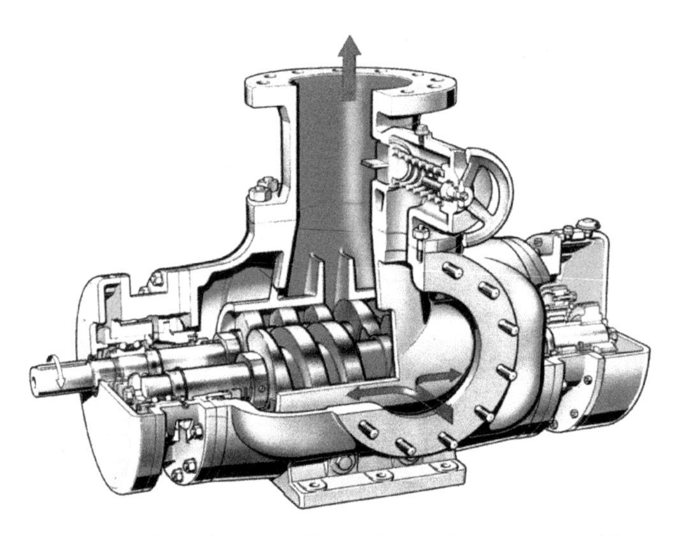

Figure 2.70 Three-dimensional view. Source: Bornemann-Gouldpumps.

Figure 2.71 Woman with screw pump.
Source: Bornemann.

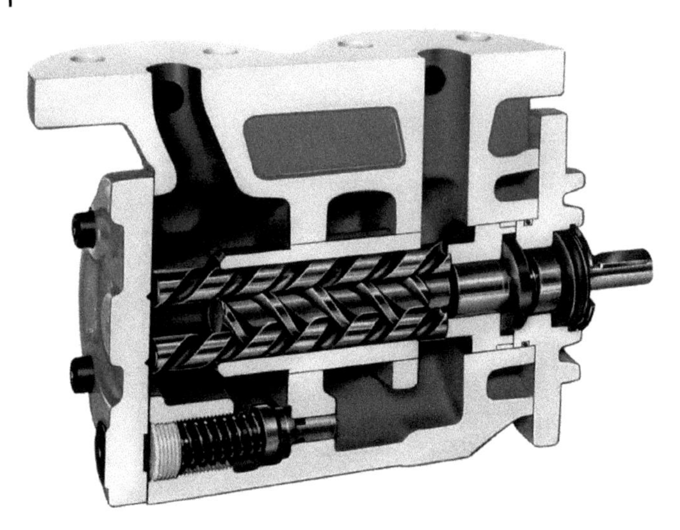

Figure 2.72 Three-axis screw pump. Source: Allweiler.

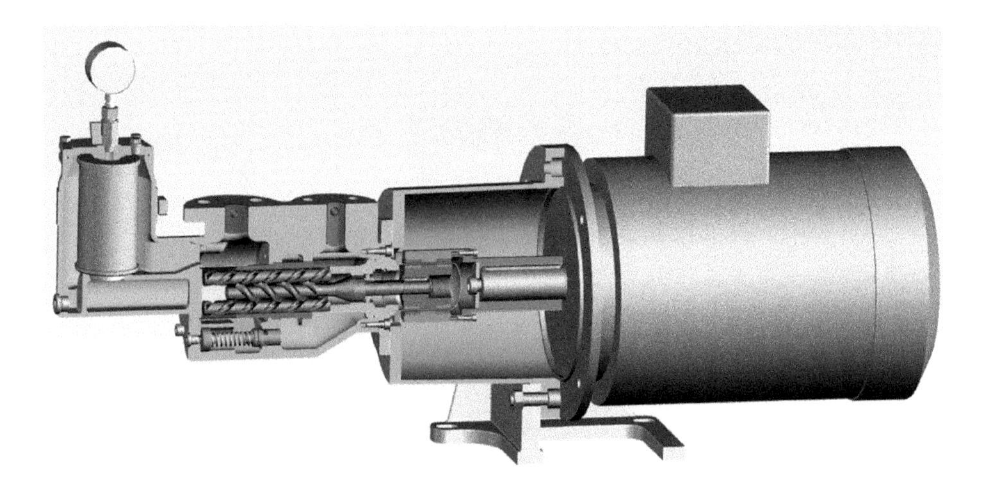

Figure 2.73 Three-axis screw pump with magnetic coupling. Source: Allweiler.

rotates the plungers are alternatively pushed in and out of the holes so that suction and discharge actions occur. The eccentricity determines the stroke length $s = 2.e$ of the plunger and so the stroke volume of the pump. Regulating apparatus can be applied in order to displace the location of the housing and so the eccentricity e and the stroke volume. In this way, without changing the rotation speed of the pump, the volumetric flow can be altered from zero to maximum. Furthermore, the eccentricity can be reversed: in this way the pump can deliver flow in two directions, and that with a constant direction of the drive rotation. Radial pumps have rpm's of as much as 300 with a nominal pressure of 320 [bar].

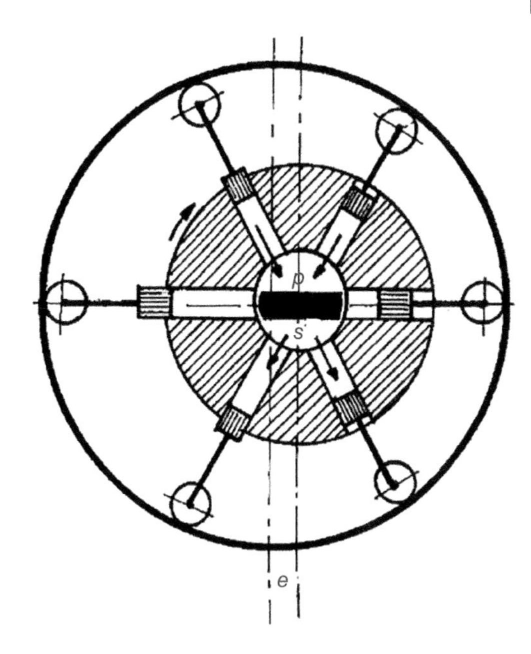

Figure 2.74 Radial plunger pump.

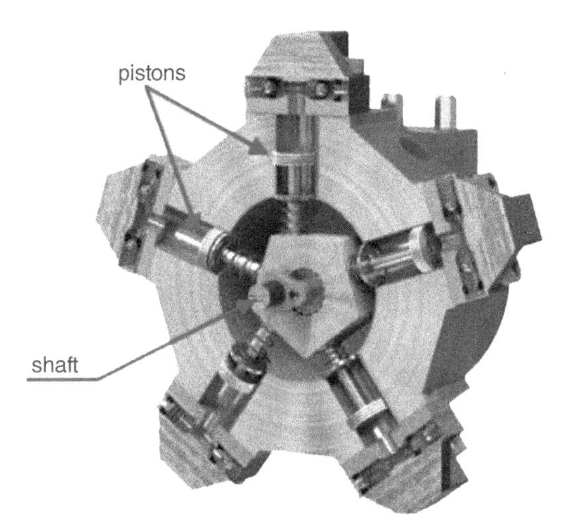

Figure 2.75 Radial pump.

2.4.6 Axial Plunger Pumps

There are various sorts of axial plunger pumps. One sort has a fixed plate, a "wobble plate", that is fitted in an oblique manner on the shaft. When the wobble plate is brought into rotation the plungers will alternatively be pushed into axial holes so that suction and press action occur (Figure 2.76).

The suction action takes place by spring force (Figure 2.77). The pump has press valves but no suction valves: at every press stroke the suction port is covered by the plungers; at every suction stroke the suction port is freed by the plungers.

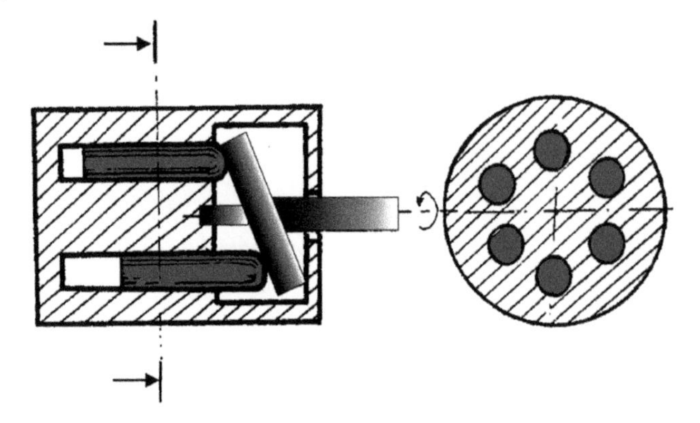

Figure 2.76 Plunger pump with rotating disc.

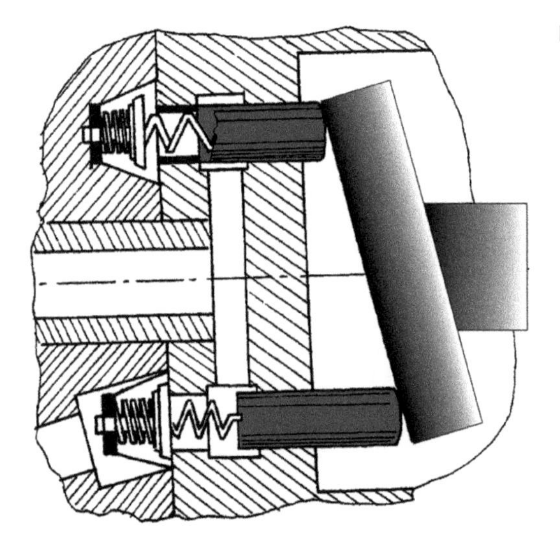

Figure 2.77 Suction and press action.

The volumetric flow is determined by the rotation speed of the driving motor, the stroke length of a plunger, and the number of plungers. These last two are constant for a determined construction, so only the rotation speed can realize a variation of the volumetric flow. This is called a *hydraulic pump with constant volumetric flow*.

A second kind of axial plunger pump has an oblique disc that stands still in relation to the now rotating cylinder block (Figure 2.78). Once every rotation of the cylinder block the plungers slide in and out so that a suction and press action occurs. When the plunger is at the top (as in Figure 2.79), there is suction action; when the plunger is at the bottom, there is press action. The press and suction openings are situated respectively under and at the top in the stationary cylinder block. Internal springs are pushed against the swash plate. By changing the angle of the swash plate, the volumetric flow can be regulated, and this at a constant rotation speed of the motor. See Figure 2.80 for an implementation.

At last there is an axial pump where the cylinder block is erected obliquely. At each rotation of the driving shaft, every plunger is pushed in and out and a suction and press action occur. The supply and discharge of liquid (see s and p in Figure 2.81) is done via openings in a stationary plate (called the *mirror plate*). Two circular grooves in the

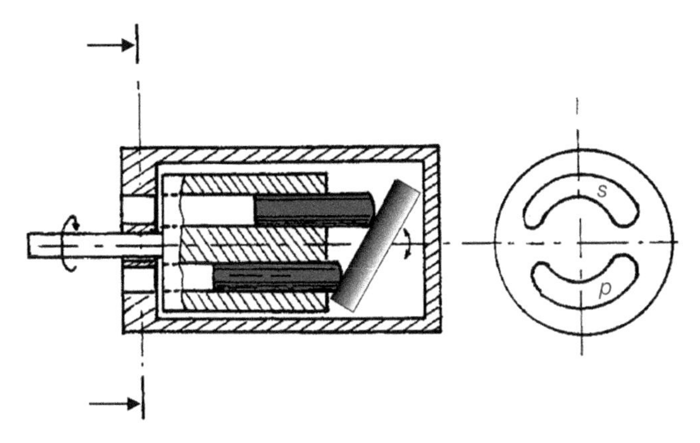

Figure 2.78 Axial plunger pump with rotating cylinder block (swash plate pump).

Figure 2.79 Swash plate pump. Source: Bosch Rexroth AG.

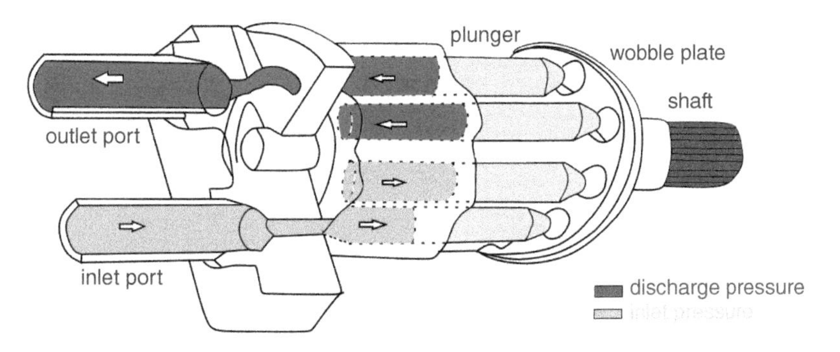

Figure 2.80 Three-dimensional view swash plate pump. Source: Free-ed net.

mirror plate connect the pump spaces with the suction reservoir during the suction and with the press reservoir during the pressing (Figures 2.81–2.84). By changing the angle of the pump housing, the volumetric flow can be regulated. In order to do this, the plunger rods are hung up with ball and socket joints.

Axial plunger pumps have rpm's of as much as 3000 and pressures of up to 1000 [bar].

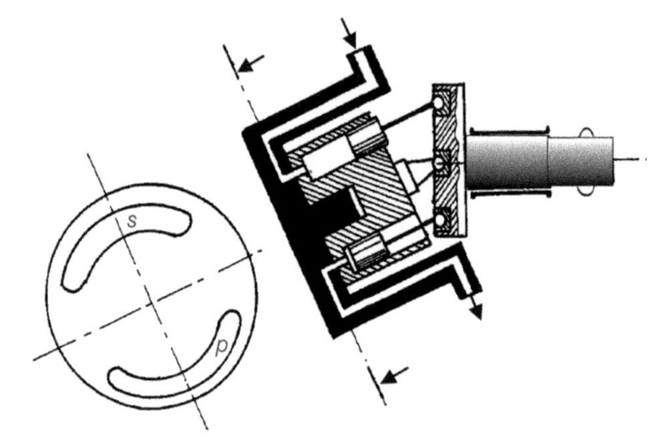

Figure 2.81 Oblique cylinder block.

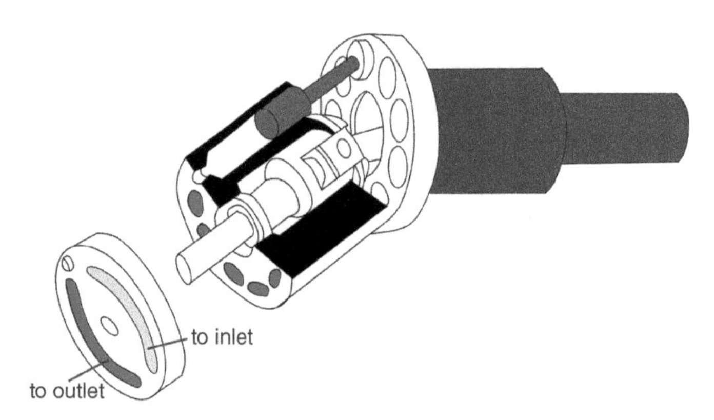

Figure 2.82 Three-dimensional view knee pump. Source: Free-ed net.

Figure 2.83 Knee pump. Source: Bosch Rexroth AG.

Figure 2.84 Parts.

Figure 2.85 Two-lobe pump. Source: Viking Pumps.

2.5 Other Displacement Pumps

2.5.1 Lobe Pump

This pump is like a gear pump but has only two or three teeth (Figures 2.85–2.88). The name of the pump is lobe pump or roots pump. But the limit of the number of teeth causes a very big passage for the liquid. The aim is then, too, to pump contaminated liquids, or liquids with solid parts (yogurt from fruit, for instance). The mode of operation is the same as that of gear pumps (Figure 2.89).

The rotors are connected with gears but don't touch each other. Therefore, the rotors must be driven with a synchronous drive (Figure 2.90). The clearance between the rotors has to be small in order to have self-priming properties. The thicker the liquid, the better.

Because the rotors do touch each other these pumps may run dry, i.e. no lubrication is necessary.

The lack of dead corners makes the pump very hygienic. It can be used in the alimentation, chemical, and pharmaceutical industries. One speaks of *CIP* (cleaning in place), i.e. self-cleaning.

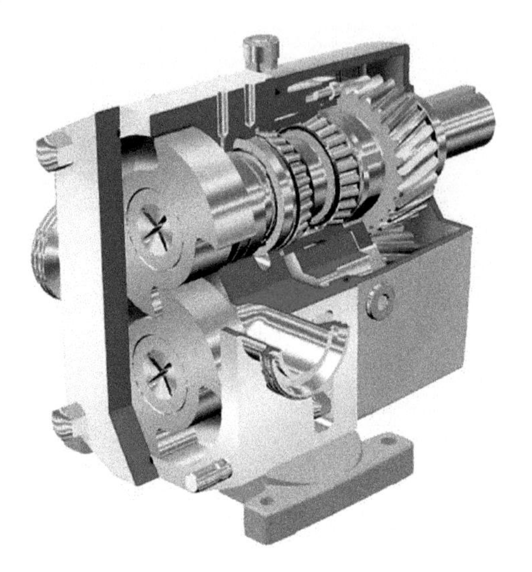

Figure 2.86 Lobe pump. Source: Allweiler.

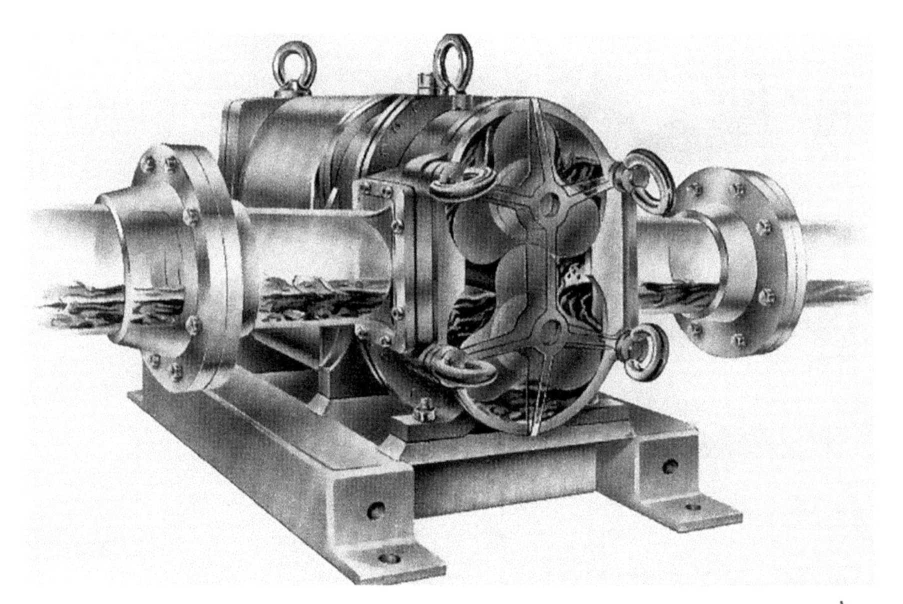

Figure 2.87 Roots pump. Source: Börger GmbH.

Some other properties include:

- Speed between 50 and 500 [rpm].
- Flows: 5–600 [m^3/h].
- Pressures up to 14 [bar].
- Viscosity up to 16 500 [mm^2/s].

Depending on the intended purpose, different rotor forms and materials are used: the rotor coating can be of rubber, plastic, or metal (Figure 2.91).

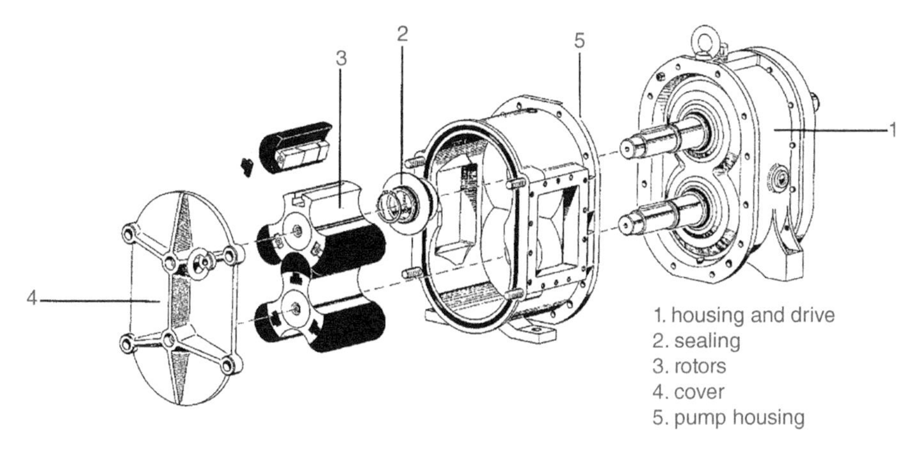

Figure 2.88 Roots pump. Source: Börger GmbH.

Figure 2.89 Operation principle.

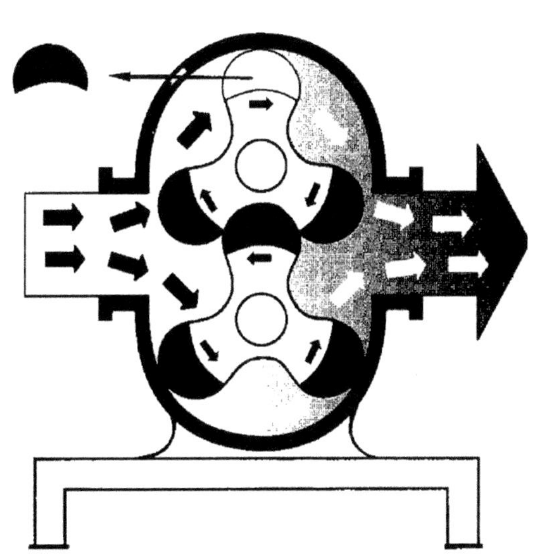

Figure 2.90 Lobe pump. Source: Allweiller.

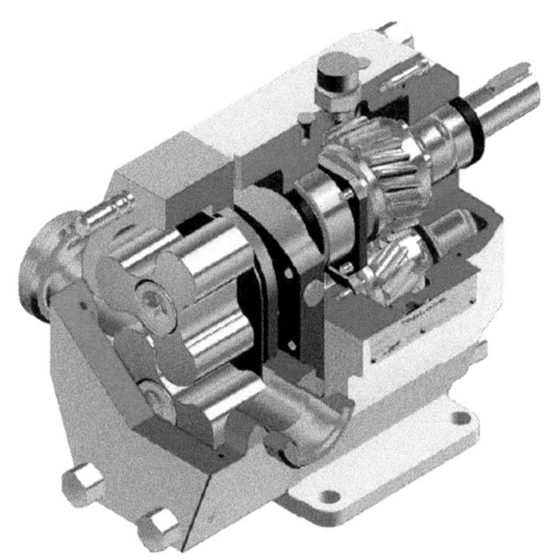

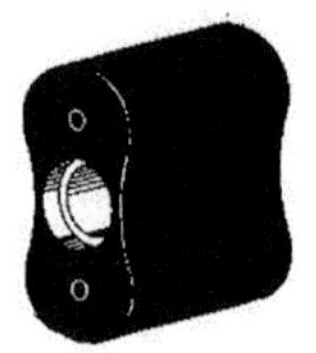

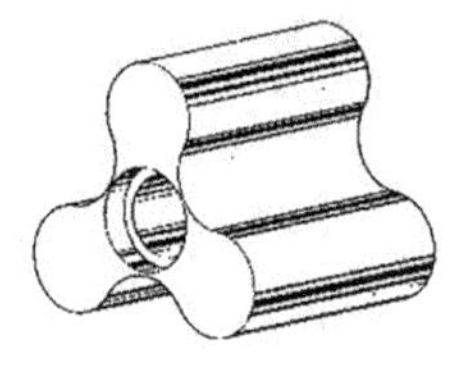

Figure 2.91 Types of rotors. Source: Börger GmbH.

Figure 2.92 Five-lobe pump. Source: Vikings Pump.

A variant of a roots pump with three teeth is a five-lobe pumps (Figure 2.92). It is mainly used for shampoos, cream, vaccinations, drinks, etc. The advantage is that the flow is cut into five pieces, more than the two of a classical two-lobe pump. This way the flow is more or less pulsation free. As every fifth of a rotation a constant amount of liquid is delivered, the pump can be used as a dosing pump.

There are on the market variants of gear pumps that work as cutters (Figures 2.93–2.96). They are only designed to cut, not to deliver pressure, and they only process the material for a sucking lobe pump.

2.5.2 Peristaltic Pump

The peristaltic movement is well known from the throat. When eating or drinking the food is sent to the stomach by a pulsating movement of the throat. If one hangs upside down and drinks a glass of water the liquid will travel to the stomach against gravitational force. This is much like how a snake propels a rabbit through its body. It is from this that we get the name *snake pump*.

Figure 2.93 Lobe pump. Source: Johnson.

Figure 2.94 Cutter knives. Source: Börger GmbH.

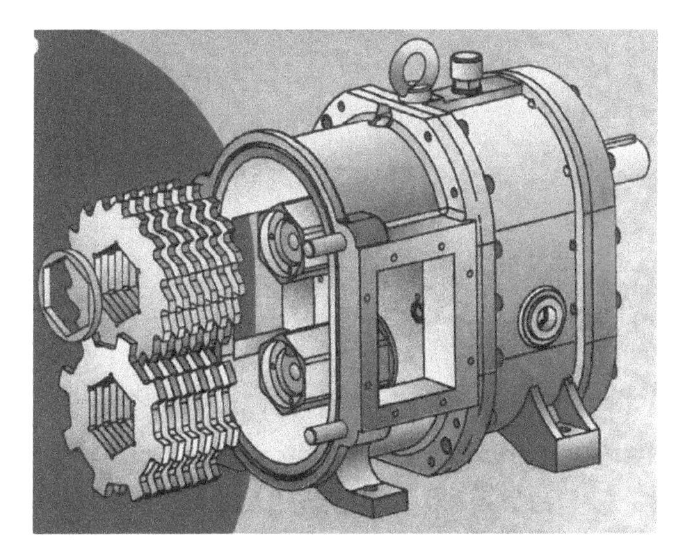

Figure 2.95 Cutter. Source: Börger GmbH.

A hose is being pushed with pressure rolls: in the hose a closed space is created that moves by the rotating movement of the rotor (Figures 2.97–2.99). The liquid or gas is transported from the suction to the discharge side. The hose makes a U-turn of 180°.

This pump works as a membrane pump but has the advantage that it delivers a continuous flow and has no valves, nor a shaft sealing.

2.5.2.1 Properties

Like the membrane pump this pump is leakless; applications are the pumping of sterile liquids like blood in a heart/lung machine, corrosive, abrasive, highly viscous substances,

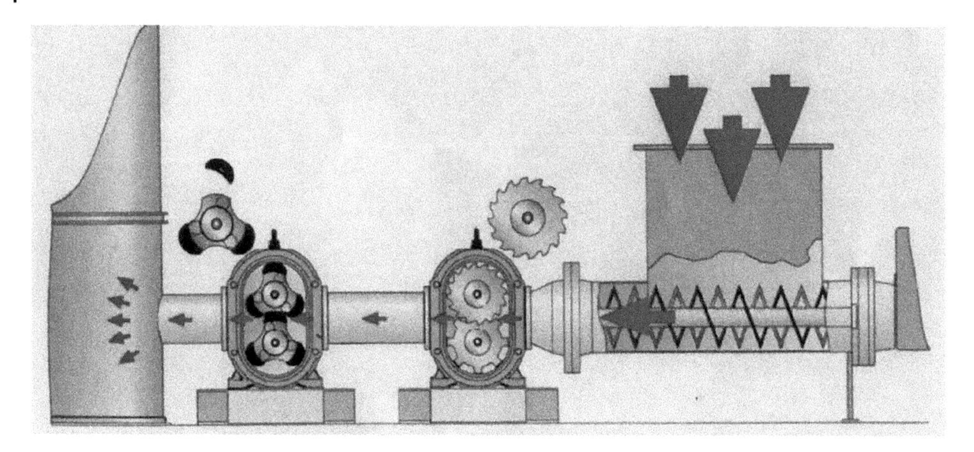

Figure 2.96 The pressure delivering roots pump is preceded by a cutter and a transport screw. Source: Börger GmbH.

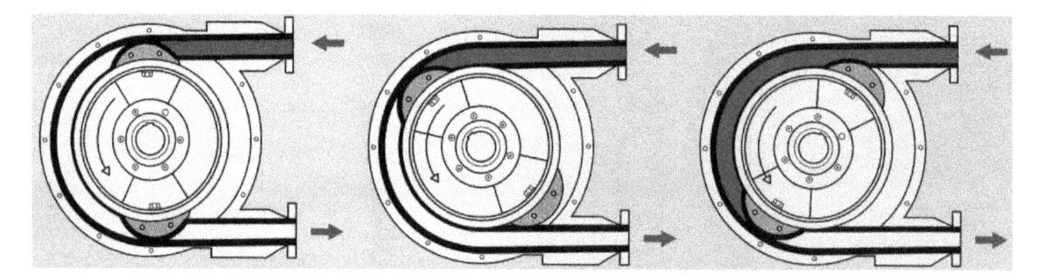

Figure 2.97 Principle operation peristaltic pump. Source: Mouvex Company.

Figure 2.98 Implementations. Source: Allweiler.

acids and bases, paints, ink, coatings, glue, sludge, waste water and radioactive substances, and in general in the chemistry industry.

The flexible hose, made of plastic, silicones, or rubber, is heavily subject to wear and must be regularly replaced. To withstand high pressures (up to 16 [bar]) a hose with armoring is used.

The capacity: as much as 100 [m^3/h].

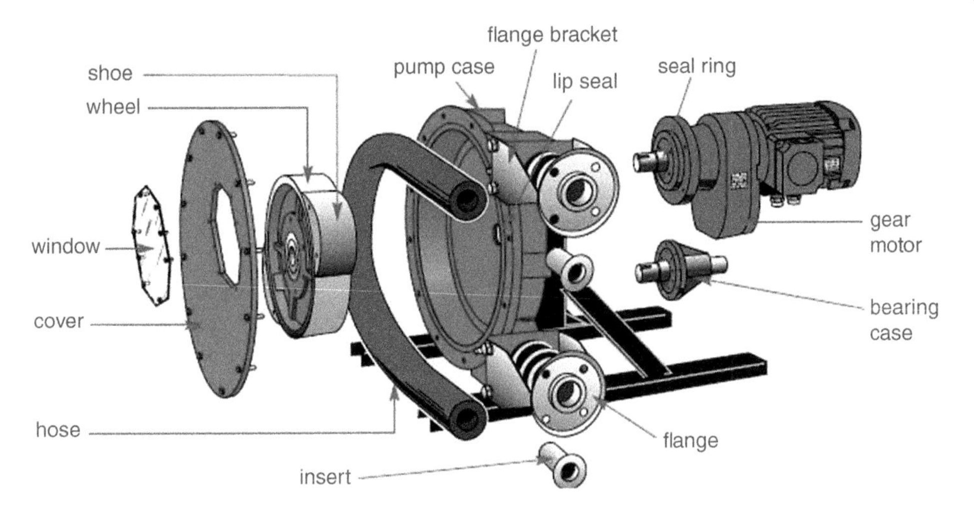

Figure 2.99 Peristaltic pump. Source: Picture from Mouvex Company.

2.5.2.2 Applications

- *Small implementations as dosing pumps*
  These pumps work mostly with pressure rollers and runs without lubricant over the hose. Near the press and suction connections, the hose is clamped and runs in one piece from the suction reservoir to the discharge reservoir.
  Peristaltic pumps are often used as dosing pumps because the volumetric flow is proportional to the rotation speed and independent of the discharge pressure. Applications include for blood circulation, where the pulsation fits perfectly well with that of the heart.
  As dosing pumps, these small pumps can be found on the market as multi-blocks, where 2–20 pumps are shoved over one driving shaft. The various hoses can then simultaneously dose various products, pigments, etc.
- *Big pumps for industrial applications*
  These pumps have mainly two pressure shoes that rotate in a lubricant (Figure 2.100). This lessens the friction and reduces the friction heat. By adding removing shims the pressure on the hose is increased or decreased and the pressure regulated.
  It is easy to clean the hose and causes minimal damage. This is necessary, for instance, in the food industry for the pumping of live fish, etc.
  Further, a snake pump is pre-eminently appropriate for abrasive media with big solid particles and highly viscous properties. Examples include: concrete, sewage sludge, and dredging sludge.

2.5.3 Mono Pump

This pump (Figure 2.101–2.104) is known by several names, like eccentric screw pump or Moineau pump (after the inventor). There are no valves.

The metal rotor has a screw form, with a circular section. Herein fits a rubber stator with an oval cross cut. To withstand high pressures (some bars) the stator is surrounded by a metal casing.

Figure 2.100 Big peristaltic pump. Source: Picture from Mouvex Company.

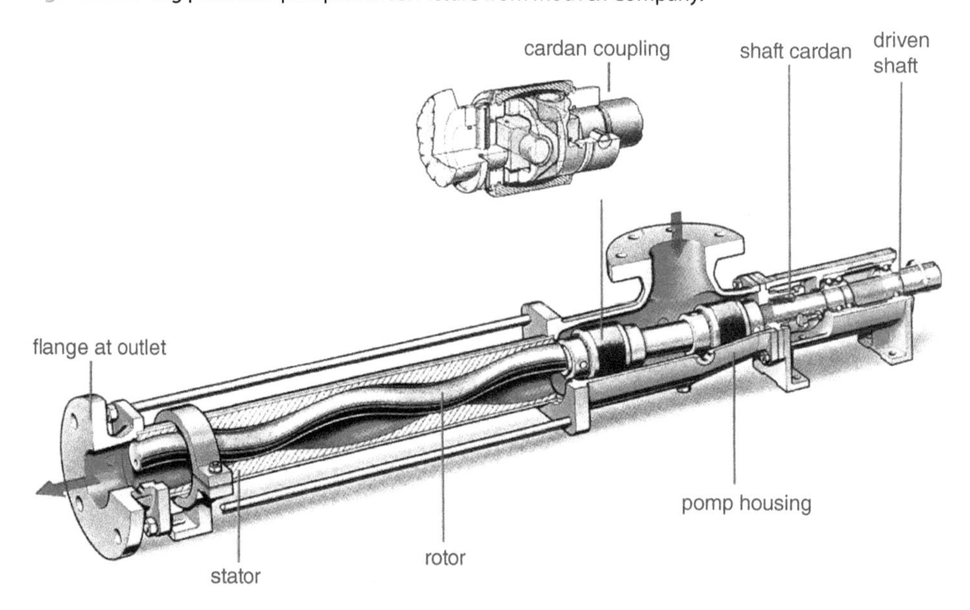

Figure 2.101 Mono pump. Source: Bornemann – Gould pumps.

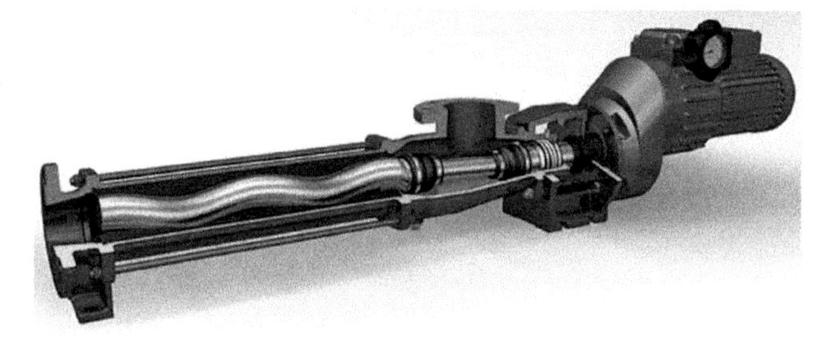

Figure 2.102 Three-dimensional view of a mono pump.

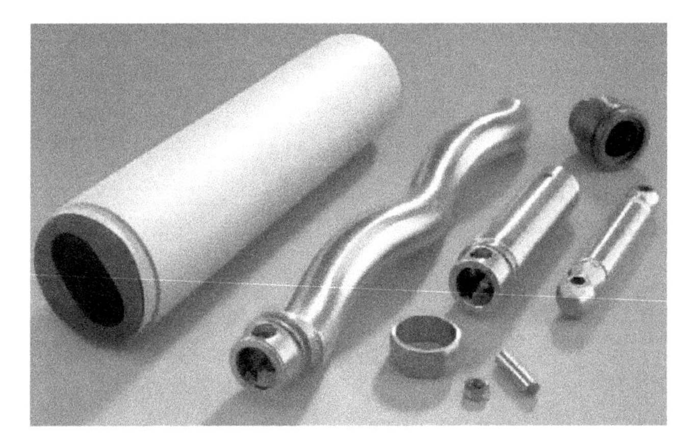

Figure 2.103 Mono pump parts.

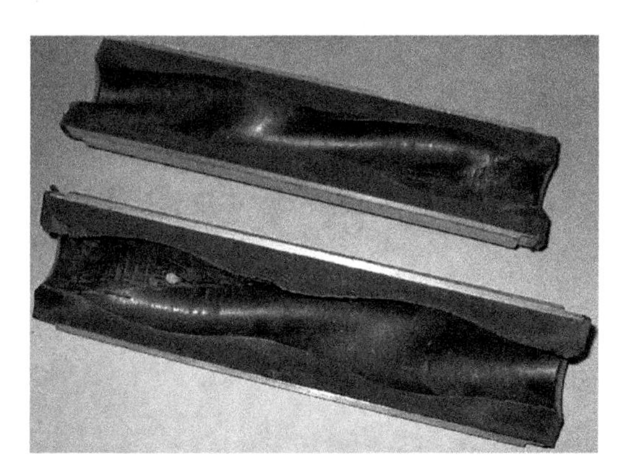

Figure 2.104 Mono pump stator. Source: Seepex.

The rotor is a kind of screw with a big pitch, with a very deep thread and closely fitted shaft. The stator has a pitch that is twice that of the rotor. This causes cavities between the stator and the eccentric rotating rotor. The cavities progress continuously from the suction to the press side.

The passage is very spacious, and so slurries can be pumped that cannot be transported with any other pump of comparable price. Hard particles that are stuck between the rotor and stator are pushed in the stator and are afterwards transferred to the liquid stream.

Because of the tight fitting of the rotor against the stator the pump exhibits very good self-priming properties.

Rotation of the rotor induces chambers behind each other, at standstill these chambers are closed. The number of chambers depends on the number of windings of the stator. When the rotor starts rotating the chambers move in a screw form in an axial direction. Be careful: depending on the rotation sense this can be forwards or backwards.

In Figure 2.105 a rotor and stator are represented. The liquid (the blue line) is completely locked up in the first image. After a turn of 90° (the second image), this chamber moves to the press side. The third image gives the situation when the rotation has been rotated 180°.

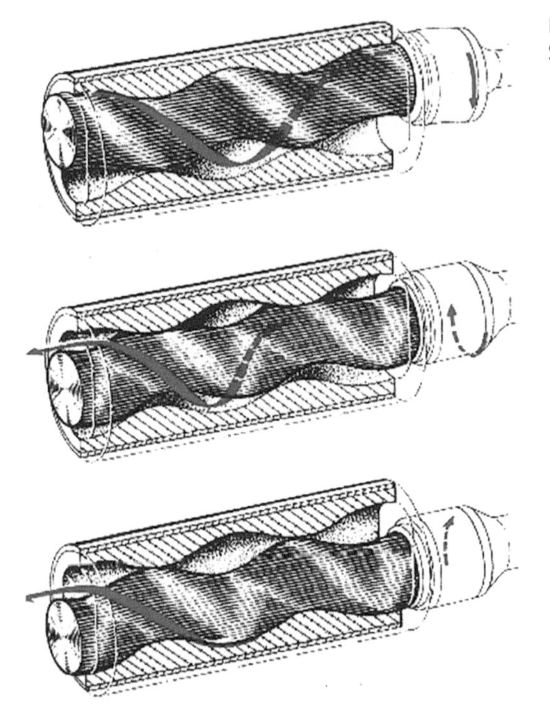

Figure 2.105 Operation mono pump. Source: Bornemann - Gouldpumps.

Eccentric screw pumps need to convert the rotation movement of the shaft in an eccentric movement of the rotor. Therefore, a cardan coupling is used (Figure 2.106).

DAUREX offers two types of coupling:

- The open coupling is used in the food industry because of its good rinsing capability. This coupling also proves its usefulness when the pumps are used for lubricating liquids, fats, and oils.
- The closed coupling is used for abrasive liquids. The coupling is filled with fat for an optimal lubrification of the coupling.

The pumped liquid forms a lubricating film, so the pump doesn't run dry. Some pumps therefore have a thermostat that measures the temperature of the stator and stops the pump functioning when a certain limit, for instance 100 °C, is reached.

Applications can be found for the pumping of viscous liquids with abrasive parts, sludge from domestic wastewater, synthetic resin, paste, liquid from chromate treatment, diluted sulfuric acid, fuels, cosmetic products, etc.

The liquid is barely damaged. That's why the pump is used for foods like yogurt, mustard, tomato concentrate, honey, sauces, chocolate. Sometimes the mono pump is preceded by a transport pump (Figure 2.107).

The flow rate is determined by the pitch of the rotor/stator, the diameter, and the eccentricity, as well as by the pumping rotation speed. The pressure capability depends on the number of stages, with the differential pressure being as much as 6 [bar] per stage.

At higher pressures the sliding seals between cavities will leak some fluid rather than pump it, so when pumping against high pressures a longer pump with more cavities is more effective, since each seal has only to deal with the pressure difference

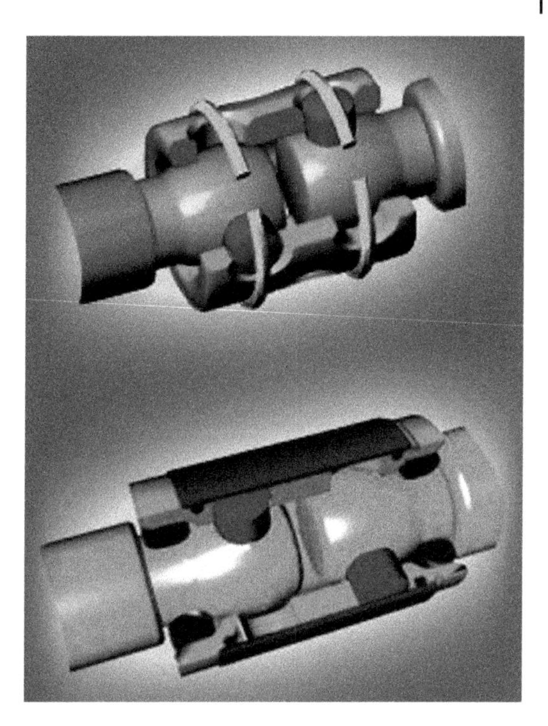

Figure 2.106 Cardan coupling. Source: DAUREX.

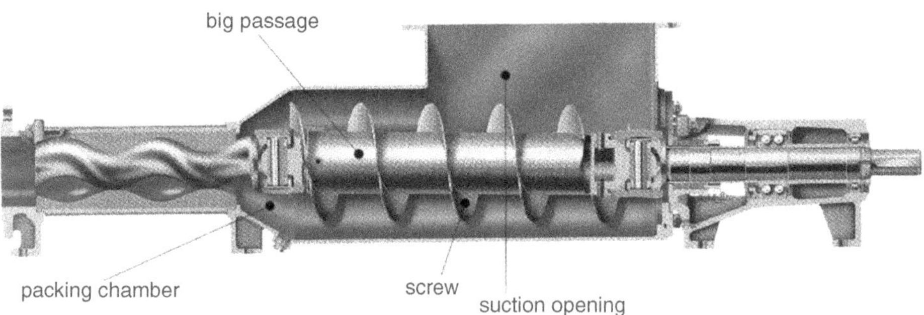

Figure 2.107 Mono pump preceded by transport screw. Source: Allweiler.

between adjacent cavities. Pump design begins with two (to three) cavities per stage. The number of stages (currently up to 25) is only limited by the ability to machine the tooling (Figure 2.108).

Mono pumps are found in designs of as much as 160 [bar] and capacities of 1 [m^3/h].

2.5.4 Flex Impeller Pump

The flex impeller pump consists of a rotor with flexible vanes, eccentrically positioned in the pump housing (Figures 2.109–2.111). In Figure 2.109 the operation is sketched. First the pump sucks air, then the liquid. The pump is self-priming.

Because of the eccentricity a partial vacuum is created when the liquid streams between the flexible vanes at the inlet port. The resulting sucking action soaks the liquid up.

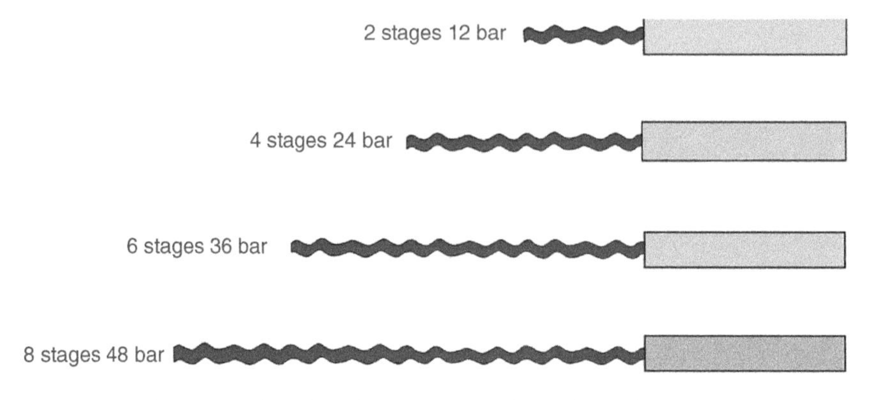

Figure 2.108 Pressure and number of stages. Source: Nemo, Netzsch.

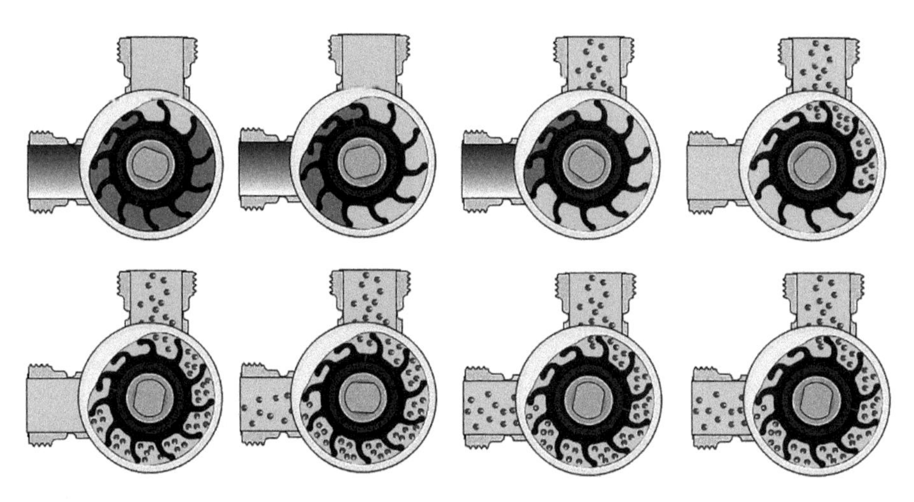

Figure 2.109 Operation. Source: Johnson Pumps.

Figure 2.110 Impeller. Source: Johnson Pumps.

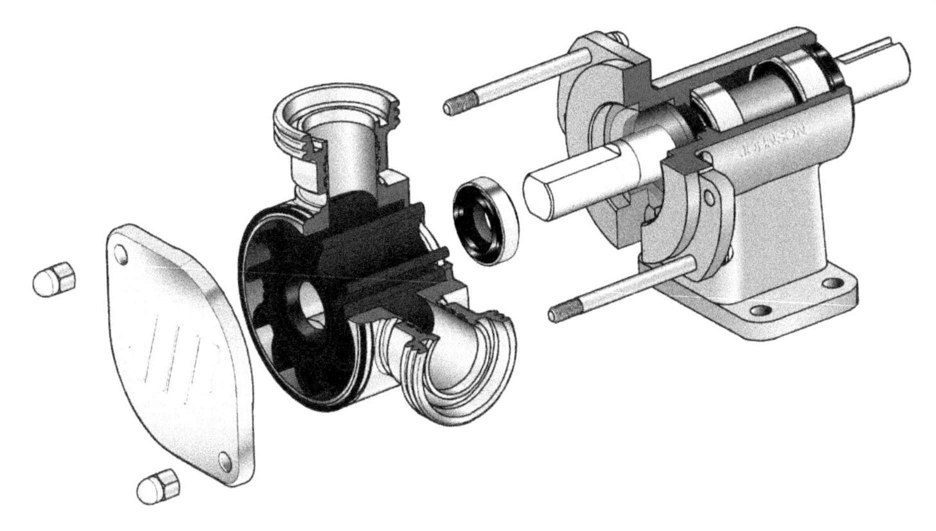

Figure 2.111 Flex impeller pump. Source: Johnson Pumps.

The rotating impeller transports the liquid from inlet to outlet port. During this period the quantity of liquid is constant. The distance between the vanes offers enough space to transport considerably big solid parts, without the liquid experiencing hinder from it.

The liquid leaves the pump in a continuous, uniform stream at the moment that the vanes curve

By inverting the rotation sense of the rotor, the liquid can be transported in the other sense.

The flex impeller pump can pump liquid with either high or low viscosities, liquid with solid parts, air, and gasses.

The impeller is a part that is subject to wear and loss of elasticity and this has consequences on the presentation of the pump. Volumetric flow and pressure decline as the pump's operation lengthens.

Aggressive liquids shorten the lifecycle of the impeller. A low rpm is recommended. Then a constant flow can be guaranteed for a longer time.

The following factors have an influence on the impeller lifecycle:

- The nominal pressure.
- The speed.
- Properties of the rotor material.
- Lubrification properties of the pumped liquid.

In Figure 2.112 the pump characteristic is represented. This gives an idea of common pressures and flows.

2.5.5 Side Channel Pump

This pump is mistakenly taken for a centrifugal pump because at first sight it has a resemblance with it and also because it has one or more impellers. But this pump has the same properties as any other rotating positive displacement pump, and so its energy consumption increases as its discharge pressure is increased. Every impeller rotates in a pump housing segment closely to a plane where a side channel is placed (Figures 2.113–2.117).

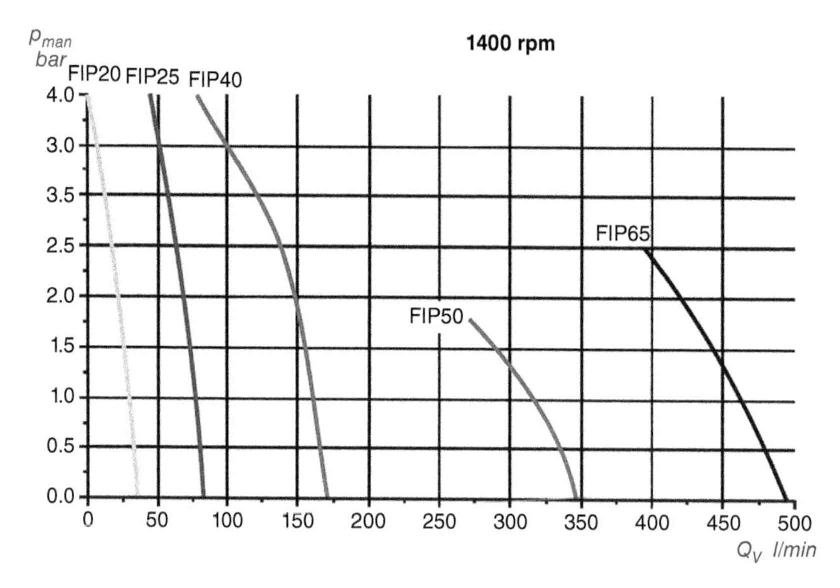

Figure 2.112 Characteristics. Source: Johnson Pumps.

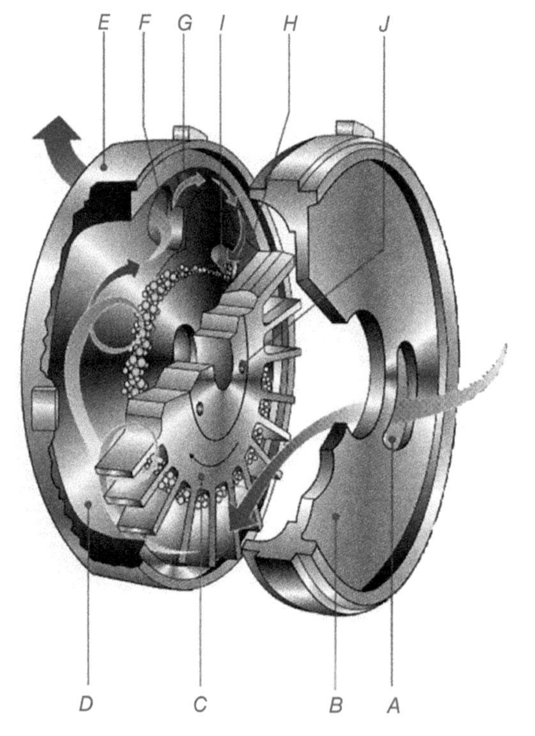

Figure 2.113 Operation side channel. Source: Sihi-Stirling.

Figure 2.114 Side Channel pump.

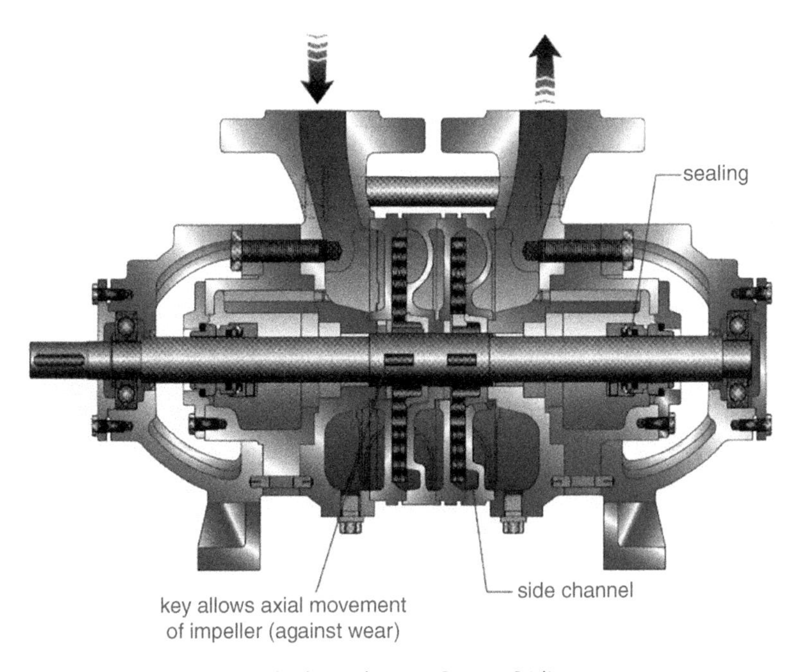

Figure 2.115 Two-stage side channel pump. Source: Stirling.

This side channel is only placed at a part of the circumference, is deepest in the middle, and becomes gradually less deep at the endings. When the impeller rotates, the space between the vanes will, because of the form of the side channel, become greater and smaller – and this gives a pumping action.

The design principle of the side channel pump is so that it can work at low NPSH (net positive suction head) (see later) and that trapped air or gas is eliminated (to 50% gas). If a wet gas (vapor) enters the impeller C, the centrifugal force will swing the heavier

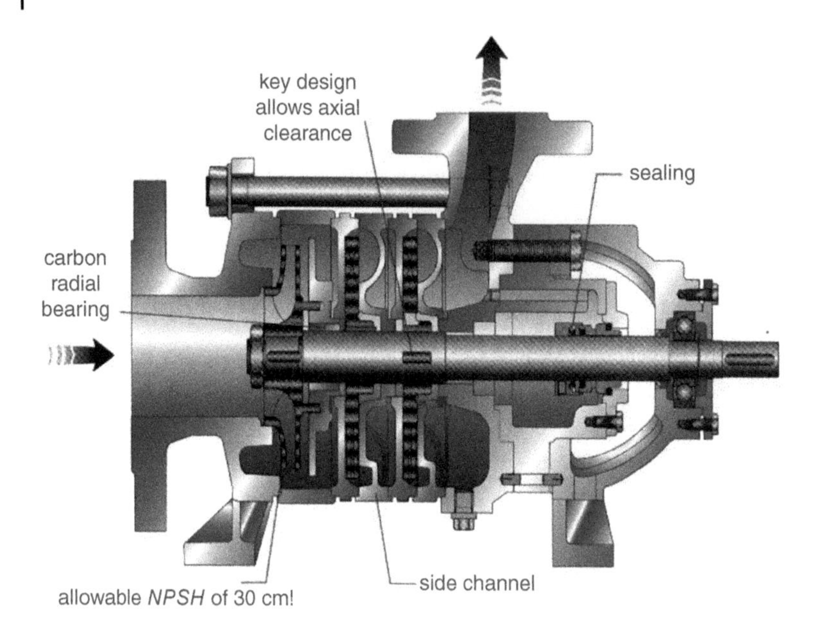

Figure 2.116 Side channel pump preceded by a radial impeller. Source: Stirling.

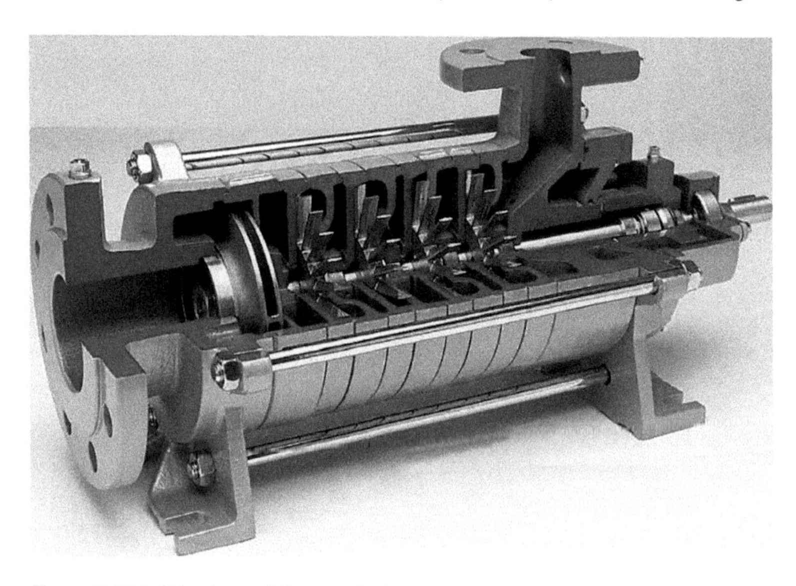

Figure 2.117 Side channel. Source: Corken.

parts to the outer circumference of the impeller, while the lighter parts will gather at the foot of the impeller. The centrifugal force will force the liquid to quit the impeller at the top and flow to the side channel D, which is situated in the outlet plate E. Then the liquid will flow back to the foot of the impeller and follow a spiral path. This spiral path causes the pump to deliver a higher pressure than a comparable centrifugal pump, with the same rpm and dimensions. This is shown in Figure 2.113. The liquid or vapor enters the pump via opening A in the suction plate B. This plate fits tightly to the impeller.

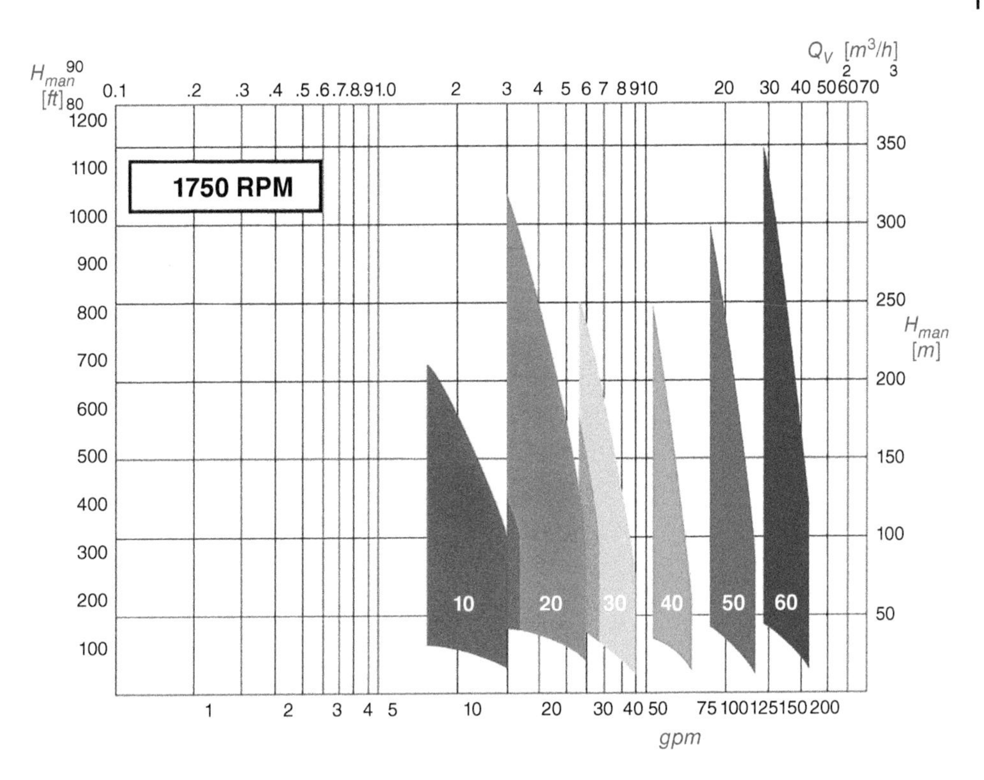

Figure 2.118 Characteristic side channel pump. Source: Corken.

Higher pressures can be reached by series connection. The impeller in every stage, though radially keyed, allows some axial movement. Equilibrium holes J in the impeller hub and the pump housing center the impeller hydraulically, and this eliminates the axial force on the bearings.

The greatest part of the liquid is delivered through the outlet port F; the rest is led to a mini channel G, which eventually comes to a stop at point H. The liquid is obliged to move to the foot of the vanes that compress eventually present vapor. If the liquid is evaporated for a part, then the vanes compress this vapor and the vapor becomes liquid. Is that a good thing? Yes, because a classic centrifugal pump, and this pump is a variant of it, cannot deliver vapor. The compressed gas and liquid is now pressed to the secondary outlet port I, where it adds with the liquid that was forced through port F.

Every time the liquid streams back into the impeller, it adds new energy there: a pressure rise arises that is five times greater than with a centrifugal pump with comparable impeller diameter and rpm.

The problem of continuous air or gas buildup is now overcome and this is repeated in the following stages.

Sometimes the side channel pump is preceded by a centrifugal impeller; the aim is to improve the NPSH (see later); see Figure 2.116.

Applications are the pumping of low to light viscous liquids and clean liquids with low flows but high pressures, like water, diesel oil, acid, and bases. Because of its self-priming property and the ability to treat vapors and gasses, the pump is also used for drainage by well points (gassy water), volatile liquids (solvents), gasoline, liquid gasses (LPG, CO_2, refrigerants), and sometimes as a fire pump to save time.

Drawbacks include: bad efficiency, noisy, the pump may not run dry (often suction and press pipe are mounted above the pump), very sensitive for contaminated liquids.

The characteristic of a side channel pump (Figure 2.118) is descending, and steeper than that of a centrifugal pump. This implies that the power characteristic is descending. This means the most appropriate flow regulation is not a throttle regulation but a by-pass regulation (the power increases with throttling). A throttle regulation is energy consuming, therefore a by-pass regulation is more appropriate. A speed regulation does not consume additional energy, but is expensive.

3

Dynamic Pumps

CHAPTER MENU

3.1 Radial Turbopumps (Centrifugal Pumps)

3.1.1 General

In Figure 3.1 (and Figure 3.2) the operation of a centrifugal pump is represented. An impeller with curved vanes rotates in the pump housing.

Suppose for the moment that the pump and the suction pipe are filled with liquid. When the impeller rotates there will be a centrifugal force on the liquid. This causes a pressure rise in the liquid toward the outer circumference and the liquid parts are dragged to the outside of the impeller where they are transported to the discharge pipe.

On the other hand, the moved liquid will create an underpressure at the suction side. This attracts new liquid. The liquid enters the pump axially and leaves the impeller radially.

Because the impeller delivers liquid on every point of the circumference, the quantity of liquid will increase with the increasing circumference. If one wants to limit the velocity of the liquid, and with that the friction losses, the pump housing should gradually increase. So, the pump housing will have the form of a spiral and its name is the *volute*.

Figures 3.3 and 3.4 show two implementations and Figure 3.5 a technical drawing.

3.1.2 Impeller Forms

In Figure 3.6 some forms of impellers are represented.

3.1.2.1 Closed Impeller

A closed impeller (Figure 3.7) consists of two discs, with the vanes in between. The sealing between the suction and press opening consists of a cylinder ring that fits with a small clearance in the replaceable *wear ring* (Figure 3.8) of the pump housing. This is

Pumps and Compressors, First Edition. Marc Borremans.

Companion website: www.wiley.com/go/borremans/pumps

Figure 3.1 Operation.

Figure 3.2 Operation: axial inlet.

important for minimizing internal leaks. This impeller has a good volumetric efficiency but is only appropriate for clean liquids, liquids with eventually small solid particles, without long fibers, and with little abrasive parts (Figure 3.9).

3.1.2.2 Half-Open Impeller

A half-open impeller consists of one disc with vanes on one side (Figure 3.10). There is a small gap between the impeller and the pump housing. This is necessary to ensure minimal leakage losses. A half-open impeller is appropriate for abrasive liquids, for instance water with sand. The wall of the pump housing consists often of wear-resistant materials (for instance stellite) or replaceable wear parts made of rubber. To compensate for wear the distance between impeller and pump housing is sometimes adjustable.

Figure 3.3 Implementation. Source: Begemann.

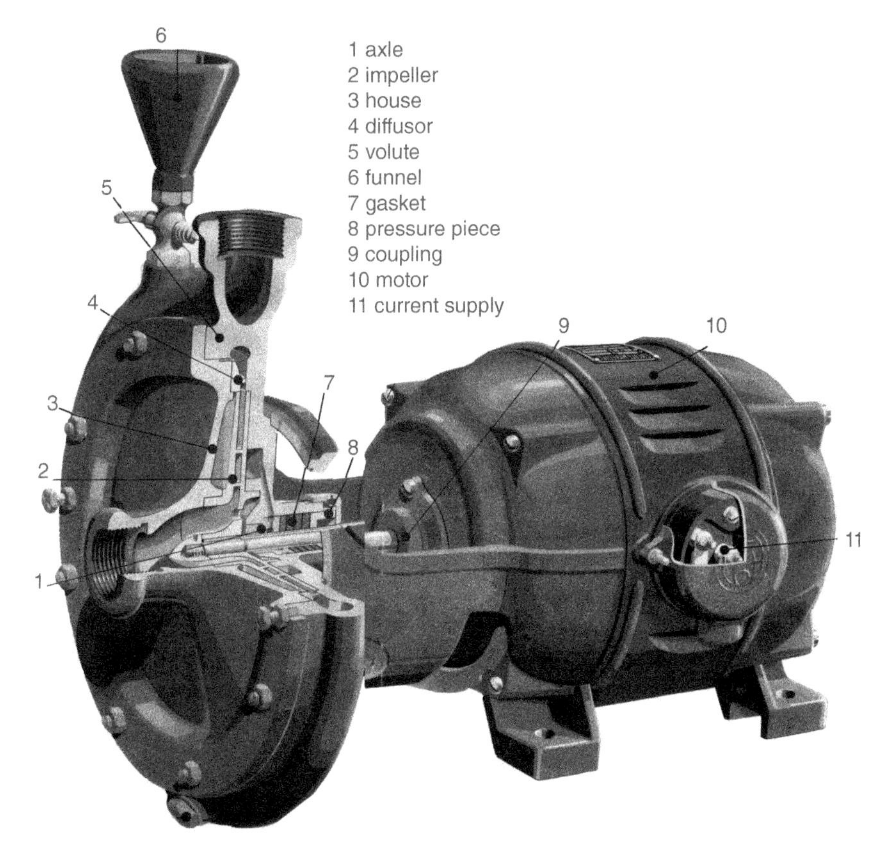

Figure 3.4 Section of centrifugal pump.

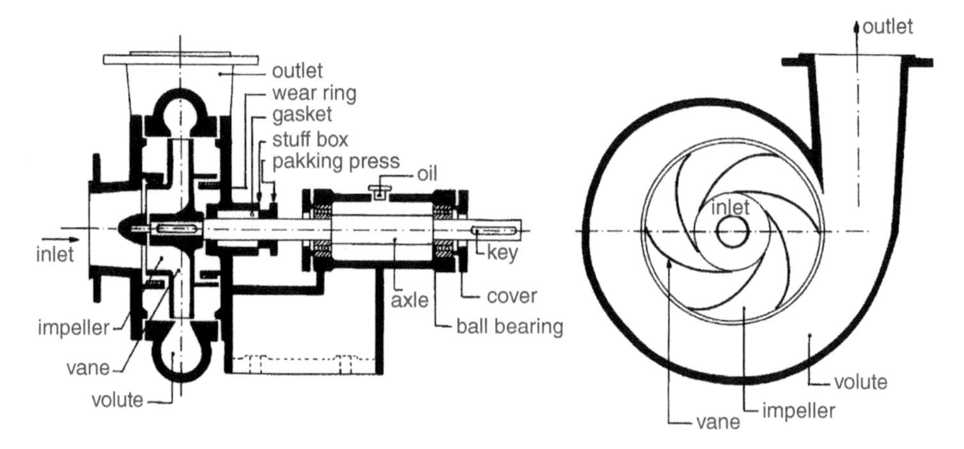

Figure 3.5 Technical drawing.

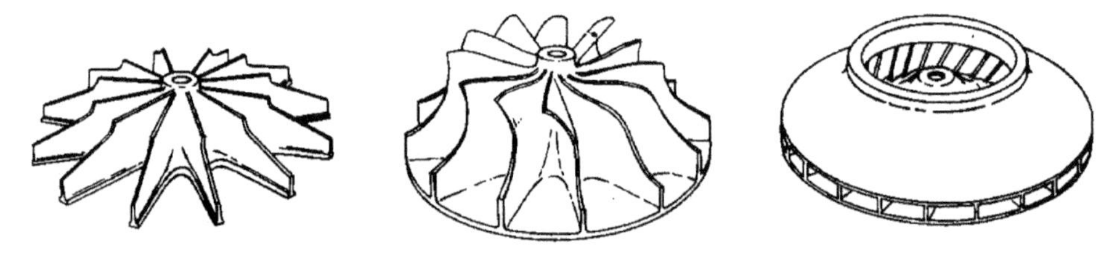

Figure 3.6 Impeller forms.

Figure 3.7 Closed impeller.

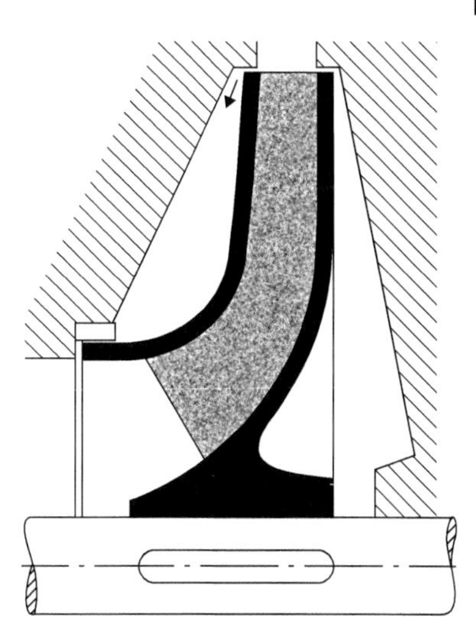

Figure 3.8 Wear ring.

Figure 3.9 Wear ring. Source: Gould Pumps.

Figure 3.10 Half-open impeller.

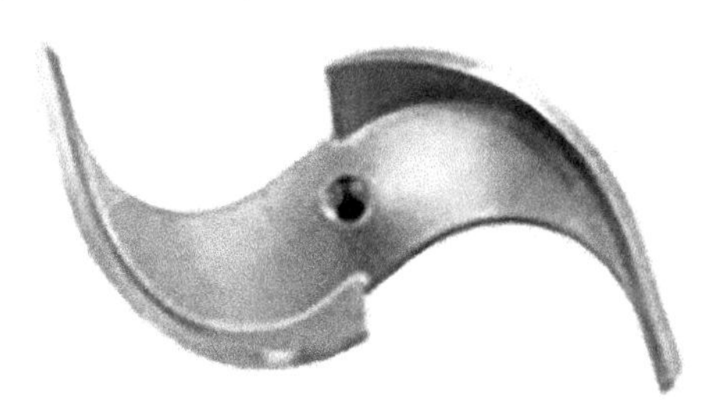

Figure 3.11 Open impeller.

3.1.2.3 Open Impeller

The disc discussed with the half-open impeller is now practically completely gone. Because the vanes have a relatively large clearance between the two covers in the pump housing, the volumetric efficiency is very low (leak losses). Open impellers are appropriate for liquids that contain gas (vapor) with little contamination, and eventually paper dust (Figure 3.11).

3.1.3 Velocity Triangles

Consider in Figure 3.12 an impeller. On inlet and outlet of a vane channel one can discern the following velocities:

c_1 : the absolute velocity at inlet of the impeller
c_2 : the absolute velocity at outlet
w_1 : the relative velocity at inlet
w_2 : the relative velocity at outlet
u_1 : the circumferential velocity at inlet
u_2 : the circumferential velocity at outlet

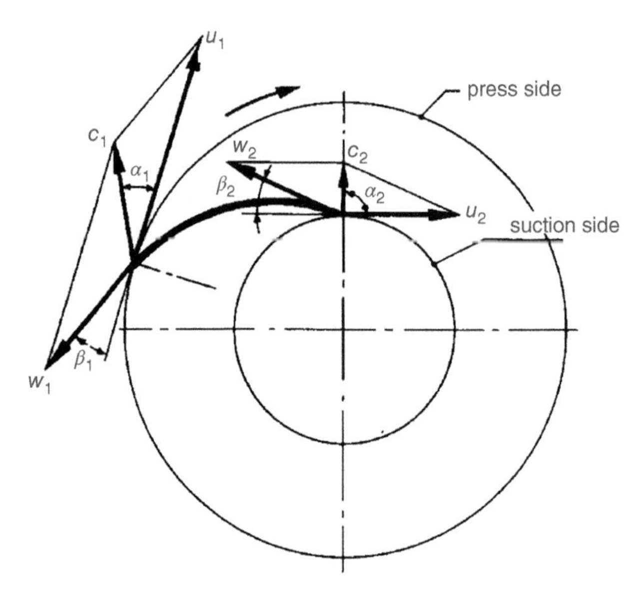

Figure 3.12 Velocity triangles.

It is always assumed that the liquid enters and quits the impeller bump free. This means that the relative velocities w are tangential to the vane profile. The vane profile is given by the vane angles ß. These vane angles are given by construction.

Vectoral:

$$\overline{c_1} = \overline{u_1} + \overline{w_1}$$

$$\overline{c_2} = \overline{u_2} + \overline{w_2}$$

The parallelogram rule leads to triangles, the velocity triangles (Figure 3.13). In these triangles all sorts of relations can be written, for instance the cosine rules:

$$w_1^2 = u_1^2 + c_1^2 - 2 \cdot u_1 \cdot c_1 \cdot \cos \alpha_1$$

$$w_2^2 = u_2^2 + c_2^2 - 2 \cdot u_2 \cdot c_2 \cdot \cos \alpha_2$$

$$c_1^2 = u_1^2 + w_1^2 - 2 \cdot u_1 \cdot w_1 \cdot \cos \beta_1$$

$$c_2^2 = u_2^2 + w_2^2 - 2 \cdot u_2 \cdot w_2 \cdot \cos \beta_2$$

3.1.4 Flow

3.1.4.1 Definition

Turbomachines need a review of the definition of volumetric flow. Suppose in Figure 3.14 that a liquid with vector velocity c streams in a tube. The problem now is to know how much volume of this fluid per unit of time flows through the surface A.

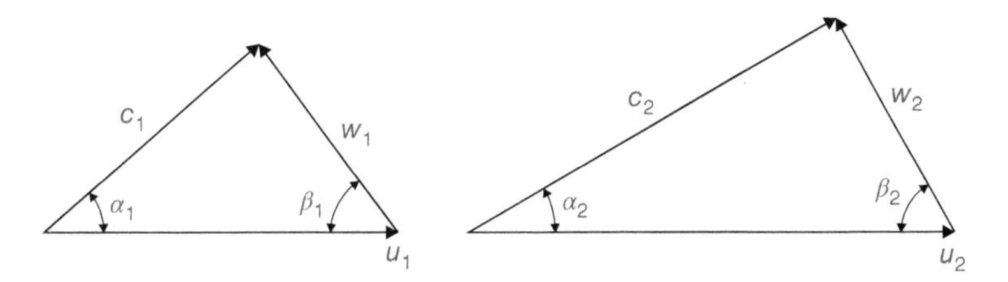

Figure 3.13 Velocity triangles.

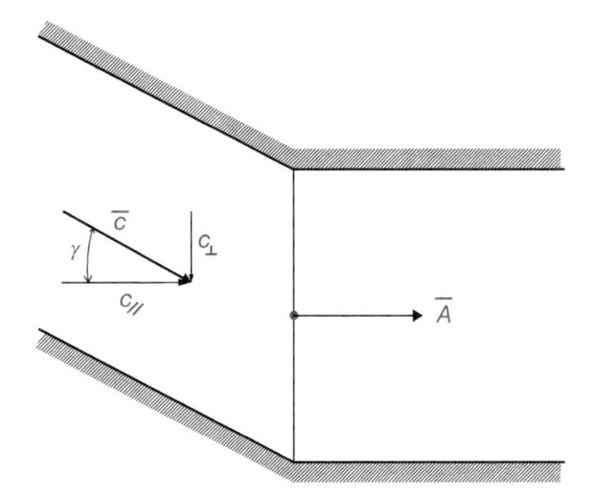

Figure 3.14 Definition volumetric flow.

Therefore, the velocity c is decomposed in two rectangular components, $c_\perp$ and $c_\parallel$ respectively across and along the surface A.

It is only the cross component that causes the transport of mass through section A, and that's why:

$$Q_V = c_\perp \cdot A$$

The surface vector A is introduced (perpendicular on A and in magnitude equal to A). The volumetric flow can be defined as the scalar product of the vectors $\overline{c}$and $\overline{A}$:

$$Q_V = \overline{c} \cdot \overline{A} = c \cdot \cos\gamma = c_\perp \cdot A$$

3.1.4.2 Flow Determining Component of the Velocity

The previous section stipulates that in some geometrical configurations only a determined component of the velocity is important for the value of the flow. Look now at the outlet of the impeller (Figure 3.15). The outlet surface of 1 vane channel be A_2.

Be z the number of vanes, this is also the number of channels:

$$A_2 = \frac{1}{z} \cdot 2 \cdot \pi \cdot r_2 \cdot h$$

where r is the radius of the impeller on outlet and h the depth, perpendicular on the page and supposed to be constant.

The volumetric flow Q_V for the whole pump:

$$Q_V = z \cdot \overline{c}_2 \cdot \overline{A_2} = z \cdot c_2 \cdot \sin\alpha_2 = z \cdot c_{2r} \cdot A_2$$

where:

$$c_{2r} = c_2 \cdot \sin\alpha_2$$

c_2 is the radial component of the absolute outlet velocity $\overline{c}$. This determines the flow.

3.1.4.3 The Relative Flow

One analogous method for the volumetric flow can be studied in a frame that moves with the rotating vane channel.

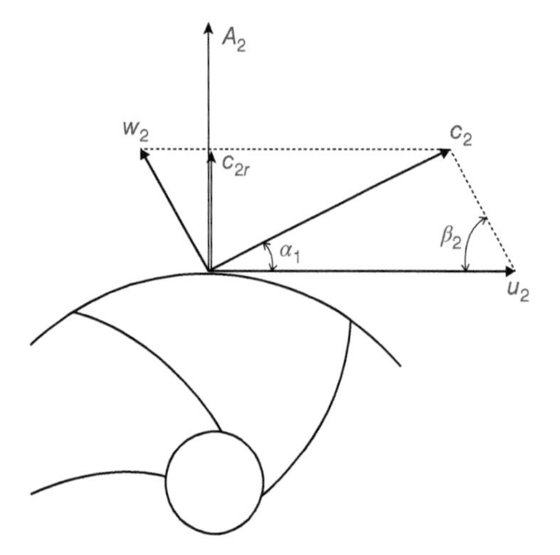

Figure 3.15 c_{2r} determines the flow.

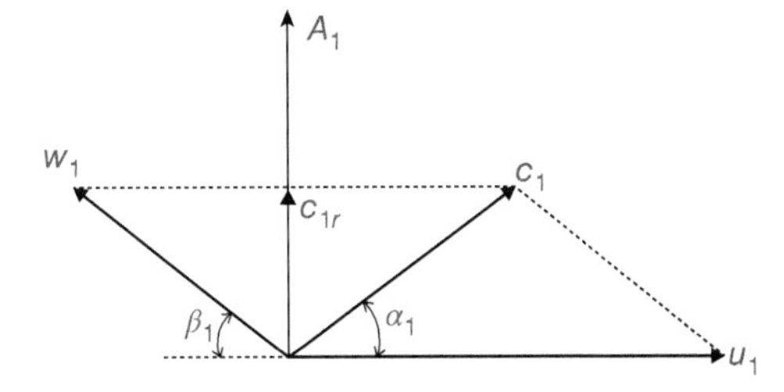

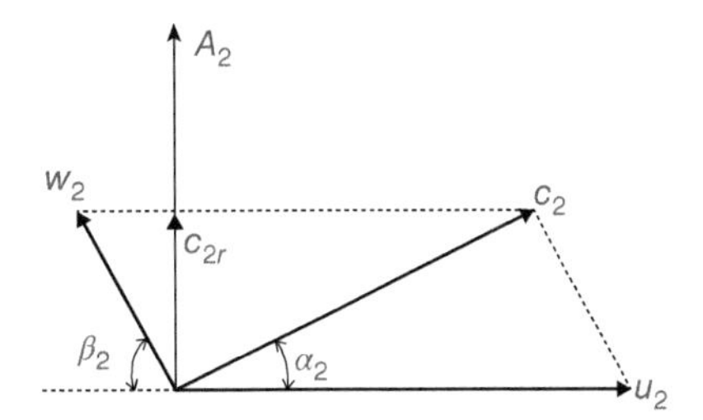

Figure 3.16 Velocity triangles.

By definition:

$$Q_V = z \cdot \overline{w_1} \cdot \overline{A_1} = z \cdot \overline{w_2} \cdot \overline{A_2}$$

Work out the scalar products:

$$Q_V = z \cdot w_1 \cdot A_1 \cdot \sin \beta_1 = z \cdot w_2 \cdot A_2 \cdot \sin \beta_2$$

From Figure 3.16:
$w_1 \cdot \sin \beta_1 = c_{1r}$ and $w_2 \cdot \sin \beta_2 = c_{2r}$

In other words: the absolute and relative volumetric flows are equal.

Examine now the flow in the relative frame. In Figure 3.17 the streamlines $\overline{w}$ are perpendicular on the surfaces A'. At the inlet this perpendicular surface is named A_1' and analogous A_2' at the outlet.

The relationship between the surfaces A and A' can be found in Figure 3.18.

$$A_2' = A_2 \cdot \cos \gamma = A_2 \cdot \sin \beta_2$$

And what is drawn at the outlet of the vane channel can, of course, be written for the entry and all places in the channel:

$$A' = A \cdot \sin \beta$$

This way the volumetric flow becomes:
$Q_V = z \cdot A' \cdot w = z \cdot A_1' \cdot w_1 = z \cdot A_2' \cdot w_2$
where w is the relative velocity on a place in the vane channel where the section, perpendicular on w, is A'.

Figure 3.17 Perpendicular surface.

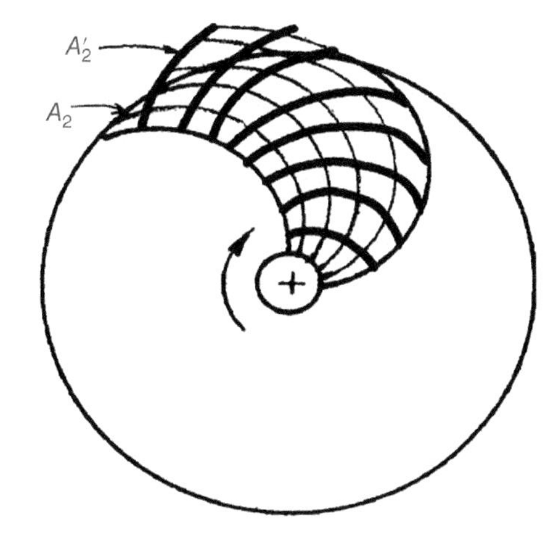

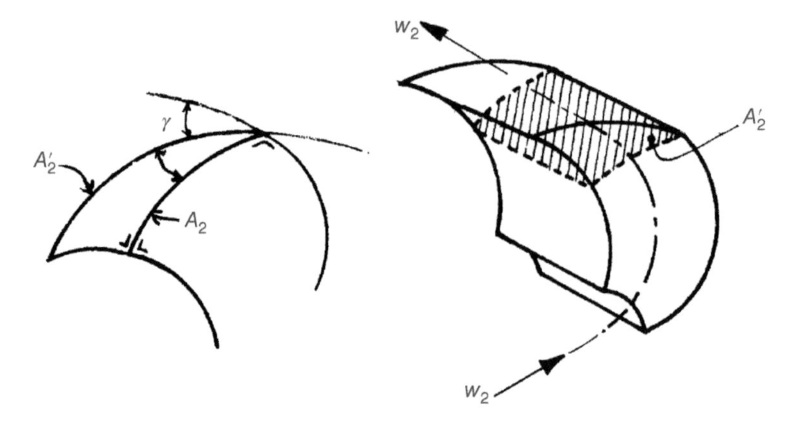

Figure 3.18 Perpendicular section A'.

3.1.5 Static Pressure in a Closed Pump

Consider the case where there is liquid in the impeller and that the pump is closed on inlet and outlet, so there can be no flow. When the impeller rotates and is placed in the rotating frame, the impeller and the liquid undergo a centrifugal force toward the outer circumference. The magnitude of this force increases with the radius and this leads to the fact that the liquid parts will exercise a pressure on one another that will increase toward the outer circumference. So, a pressure buildup will originate from inlet to outlet (Figure 3.19).

Mathematically, this is treated by considering an elementary part with mass dm and volume dV (Figure 3.19) on a radius r of the center of the impeller. Assume, for simplicity's sake, that the dimension of the channel perpendicular on the paper has a constant value h.

The impeller rotates with angle velocity ω.

$$-(p + dp) \cdot A_I + p \cdot A_{II} + 2 \cdot \left(p + \frac{dp}{2}\right) \cdot \sin\frac{\Delta\theta}{2} \cdot A_{III} + dF_c = 0 \tag{3.1}$$

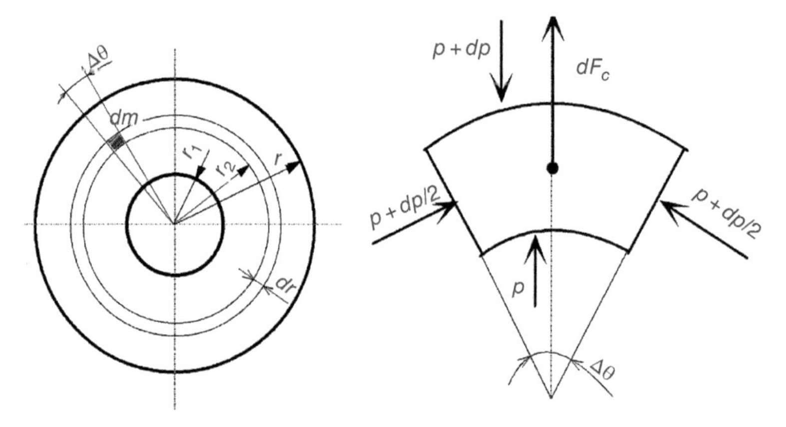

Figure 3.19 Static pressure rise in closed pump of part dm.

$$A_I = (r + dr) \cdot \Delta\theta \cdot h \tag{3.2}$$

$$A_{II} = r \cdot \Delta\theta \cdot h \tag{3.3}$$

$$A_{III} = dr \cdot h \tag{3.4}$$

The elementary centrifugal force, acting on dm:

$$dF_c = dm \cdot \omega^2 \cdot r \tag{3.5}$$

With:

$$dm = \rho \cdot dV \tag{3.6}$$

$$dV = h \cdot \frac{\Delta\theta}{2 \cdot \pi} . (\pi \cdot (r + dr)^2 - \pi \cdot r^2)$$

$$dV = h \cdot \frac{\Delta\theta}{2 \cdot \pi} . (2 \cdot \pi \cdot r \cdot dr + \pi \cdot dr^2)$$

Neglecting the second order term dr^2 :

$$dV = h \cdot \Delta\theta \cdot r \cdot dr \tag{3.7}$$

Equation (3.7) in Equation (3.6):

$$dm = \rho \cdot h \cdot \Delta\theta \cdot r \cdot dr \tag{3.8}$$

Equation (3.8) in Equation (3.5):

$$dF_c = \rho \cdot h \cdot \Delta\theta \cdot r^2 \cdot dr \cdot \omega^2 \tag{3.9}$$

Substitution of Equation (3.9), (3.2), (3.3), and (3.4) in (3.1), taking into account the first order approximation for sinuses of small angles:

$$\sin\frac{\Delta\theta}{2} \cong \frac{\Delta\theta}{2}$$

$$-(p + dp) \cdot (r + dr) \cdot \Delta\theta \cdot h + p \cdot r \cdot \Delta\theta \cdot h + 2 \cdot \left(p + \frac{dp}{2}\right) \cdot \sin\frac{\Delta\theta}{2} \cdot dr \cdot h$$
$$+ \rho \cdot h \cdot \Delta\theta \cdot r^2 \cdot dr \cdot \omega^2 = 0$$

Dividing by $h \cdot \Delta\theta$:

$$-(p + dp) \cdot (r + dr) + p \cdot r + \left(p + \frac{dp}{2}\right) \cdot dr + \rho \cdot r^2 \cdot dr \cdot \omega^2 = 0$$

Work out:

$$-p \cdot r - p \cdot dr - r \cdot dp - dp \cdot dr + p \cdot r + p \cdot dr + \frac{1}{2} \cdot dp \cdot dr + \rho \cdot r^2 \cdot dr \cdot \omega^2 = 0$$

Neglecting second order terms:

$$-p \cdot r - p \cdot dr - r \cdot dp + p \cdot r + p \cdot dr + \rho \cdot r^2 \cdot dr \cdot \omega^2 = 0$$

Simplification:

$$-r \cdot dp + \rho \cdot r^2 \cdot dr \cdot \omega^2 = 0$$

$$dp = \rho \cdot r \cdot dr \cdot \omega^2$$

The total pressure rise in radial direction from inlet to outlet of the impeller:

$$\int_1^2 dp = \int_{r_1}^{r_2} \rho \cdot r \cdot dr \cdot \omega^2$$

$$p_2 - p_1 = \rho \cdot \frac{\omega^2}{2} \cdot (r_2^2 - r_1^2)$$

With the circumferential velocity $u = \omega \cdot r$:

$$p_2 - p_1 = \frac{\rho}{2} \cdot (u_2^2 - u_1^2).$$

3.1.6 Theoretical Feed Pressure

3.1.6.1 Law of Bernoulli in Rotating Frame

When the pump is *open* and the fluid moves with relative velocity w through the vane channel, the liquid moves with a *total pressure*:

$$p + \frac{\rho}{2} \cdot w^2$$

The total pressure rise when moving through the impeller:

$$(p_2 - p_1) + \frac{1}{2} \cdot \rho \cdot (w_2^2 - w_1^2)$$

The previous section demonstrates that, in a closed pump ($w = 0$), the centrifugal force is responsible for the increase of the static pressure, with an amount of:

$$\frac{\rho}{2} \cdot (u_2^2 - u_1^2)$$

When the liquid streams through the impeller, the value of the centrifugal force does not change. So, it can be assumed that in the case of an open pump the quantity of *total pressure energy* added to the liquid is equal to that of a closed pump. In other words, in an open pump the centrifugal energy divides in static and dynamic pressure energy:

$$\frac{\rho}{2} \cdot (u_2^2 - u_1^2) = (p_2 - p_1) + \frac{1}{2} \cdot \rho \cdot (w_2^2 - w_1^2)$$

In a rotating frame the law of Bernoulli can then be written as:

$$p_1 + \frac{1}{2} \cdot \rho \cdot w_1^2 - \frac{1}{2} \cdot \rho \cdot u_1^2 = p_2 + \frac{1}{2} \cdot \rho \cdot w_2^2 - \frac{1}{2} \cdot \rho \cdot u_2^2 \tag{3.10}$$

3.1.6.2 Discussion

If the longitudinal section (the perpendicular section A') of the vane channel is drawn (Figure 3.20), one can see that:

$$A_2' > A_1'$$

In other words, the channel behaves as a *diffusor*.
Because:

$$Q_V = z \cdot w_1 \cdot A_1' = z \cdot w_2 \cdot A_2'$$

$$\rightarrow w_2 < w_1$$

In other words, the relative velocity in the impeller *decreases*.

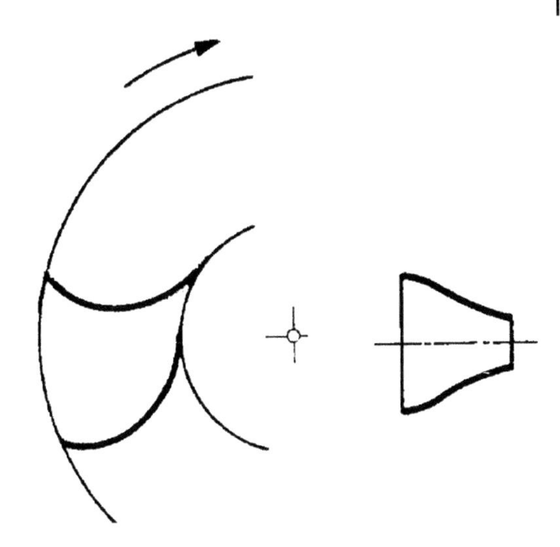

Figure 3.20 Vane channel as diffusor.

From Equation (3.10) it follows that the static pressure rise in the impeller is:

$$(p_2 - p_1) = \frac{\rho}{2} \cdot (u_2^2 - u_1^2) + \frac{\rho}{2} \cdot (w_1^2 - w_2^2) \tag{3.11}$$

So, the second term in Equation (3.11) is positive (pay attention on the inversion of the indices with w!).

$$\frac{\rho}{2} \cdot (w_1^2 - w_2^2) > 0$$

From this we get the following interpretation: the static pressure rise $(p_2 - p_1)$ in the pump is composed of:

- The influence of the centrifugal force: $\frac{\rho}{2} \cdot (u_2^2 - u_1^2)$.
- The diffusor action of the vane channel, where relative dynamic pressure is converted to static pressure $\frac{\rho}{2} \cdot (w_1^2 - w_2^2)$.

3.1.6.3 Theoretical Feed Pressure

The theoretical feed pressure is the total pressure that is executed by the impeller on the fluid when it runs through the vane channel. If there is no friction in the impeller, this total pressure rise is given by:

$$p_{th} = (p_2 - p_1) + \frac{\rho}{2} \cdot (c_2^2 - c_1^2) \tag{3.12}$$

If there is friction then the total pressure rise will be less, the manometric feed pressure, but that is discussed later.

Substitution of Equation (3.12) in (3.11):

$$p_{th} = \frac{\rho}{2} \cdot (u_2^2 - u_1^2) + \frac{\rho}{2} \cdot (w_1^2 - w_2^2) + \frac{\rho}{2} \cdot (c_2^2 - c_1^2)$$

Substitution of Equations (3.13) and (3.14) in the velocity triangles (Figure 3.21):

$$w_1^2 = u_1^2 + c_1^2 - 2 \cdot u_1 \cdot c_1 \cdot \cos\alpha_1 \tag{3.13}$$

$$w_2^2 = u_2^2 + c_2^2 - 2 \cdot u_2 \cdot c_2 \cdot \cos\alpha_2 \tag{3.14}$$

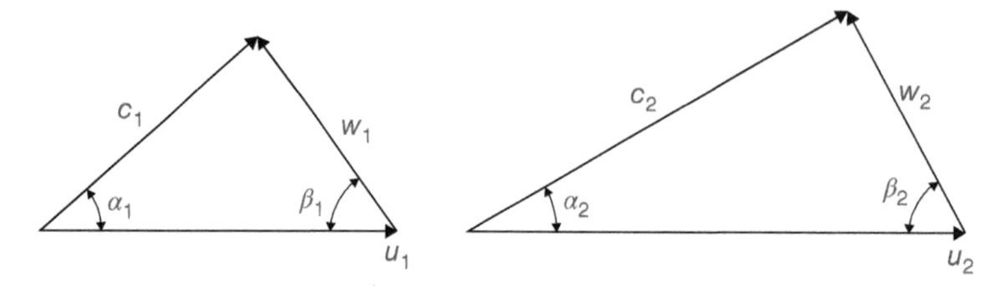

Figure 3.21 Velocity triangles.

Leads to:

$$p_{th} = \rho \cdot (u_2 \cdot c_2 \cdot \cos \alpha_2 - u_1 \cdot c_1 \cdot \cos \alpha_1)$$

Mostly it can be assumed that the liquid enters the impeller radially, so that $\alpha = 90°$ and that's why:

$$p_{th} = \rho \cdot (u_2 \cdot c_2 \cdot \cos \alpha_2).$$

Example 3.1 ***Velocity Triangles***
Be given a centrifugal pump (Figure 3.22) with:

$N = 1500$ [rpm]
$c_1 = c_{1_r} = c_{2_r} = 3$ [m/s]
$\beta_2 = 40°$
$u_2 = 20.5$ [m/s]
$u_1 = 10.8$ [m/s]

To pump: water.
Determine:

- The two velocity triangles.
- The static pressure rise $(p_2 - p_1)$.
- The theoretical feed pressure.

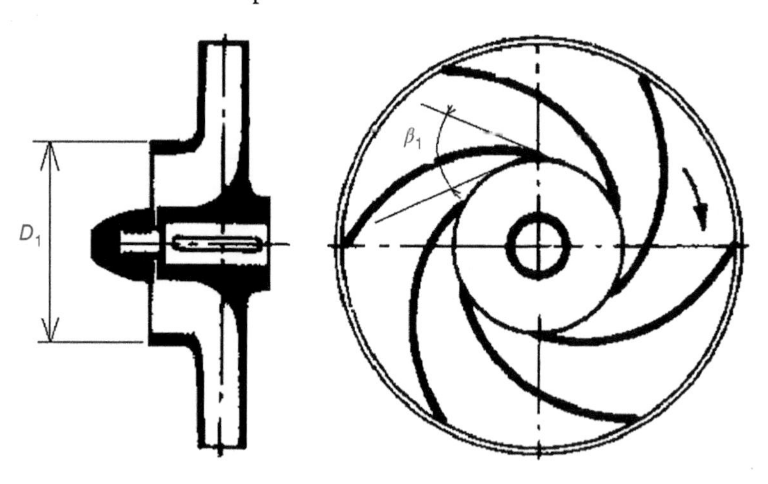

Figure 3.22 Example 3.1.

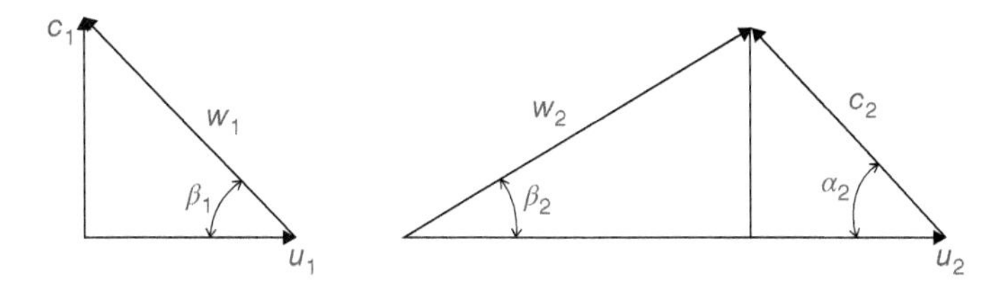

Figure 3.23 Velocity triangles, Example 3.1.

Solution

In the first rectangular velocity triangle (Figure 3.23):

$$c_1 = u_1 \cdot tg\,\beta_1$$

$$\beta_1 = arctg\,\frac{3}{10,8} = 16^\circ$$

$$w_1 = \frac{u_1}{\cos 16^\circ} = \frac{10,8}{\cos 16^\circ} = 1.2\,[\mathrm{m/s}]$$

In the second velocity triangle

$$c_{2r} = w_2 \cdot \sin\beta_2 = c_2 \cdot \sin\alpha_2 = 3\,[\mathrm{m/s}] \quad (3.15)$$

$$w_2 = \sin\beta_2 = \frac{3}{\sin 40^\circ} = 4.7\,[\mathrm{m/s}]$$

$$u_2 = c_2 \cdot \cos\alpha_2 + w_2 \cdot \cos\beta_2$$

$$c_2 \cdot \cos\alpha_2 = u_2 - w_2 \cdot \cos\beta_2 = 20.5 - 4.7 \cdot \cos 40^\circ = 1.9\,[\mathrm{m/s}] \quad (3.16)$$

(3.15)/(3.16):

$$\frac{c_2 \cdot \sin\alpha_2}{c_2 \cdot \cos\alpha_2} = tg\,\alpha_2 = \frac{3}{16.9}$$

where from: $\alpha_2 = 10^\circ$

Equation (3.15):

$$c_2 = \frac{3}{\sin\alpha_2} = \frac{3}{\sin 10^\circ} = 17.3\,[\mathrm{m/s}]$$

Calculation of the dynamic pressure differences:

$$\frac{\rho}{2} \cdot (u_2^2 - u_1^2) = \frac{10^3}{2} \cdot (20.5^2 - 10.8^2) = 1.52 \cdot 10^3\,[\mathrm{Pa}] = 1.52\,[\mathrm{bar}]$$

$$\frac{\rho}{2} \cdot (w_1^2 - w_2^2) = \frac{10^3}{2} \cdot (4,7^2 - 11,2^2) = -0.52 \cdot 10^3\,[\mathrm{Pa}] = -0.52\,[\mathrm{bar}]$$

$$\frac{\rho}{2} \cdot (c_1^2 - c_2^2) = \frac{10^3}{2} \cdot (17,3^2 - 3^2) = 1.45 \cdot 10^3\,[\mathrm{Pa}] = 1.45\,[\mathrm{bar}]$$

The static pressure difference:

$$p_2 - p_1 = \frac{\rho}{2} \cdot (u_2^2 - u_1^2) - \frac{\rho}{2} \cdot (w_2^2 - w_1^2) = (1.52 - -0.52) = 2.04\,[\mathrm{bar}]$$

The theoretical feed pressure:

$$p_{th} = (p_2 - p_1) + \frac{\rho}{2} \cdot (c_2^2 - c_1^2) = (1.45 + 2.04) = 3.46\,[\mathrm{bar}]$$

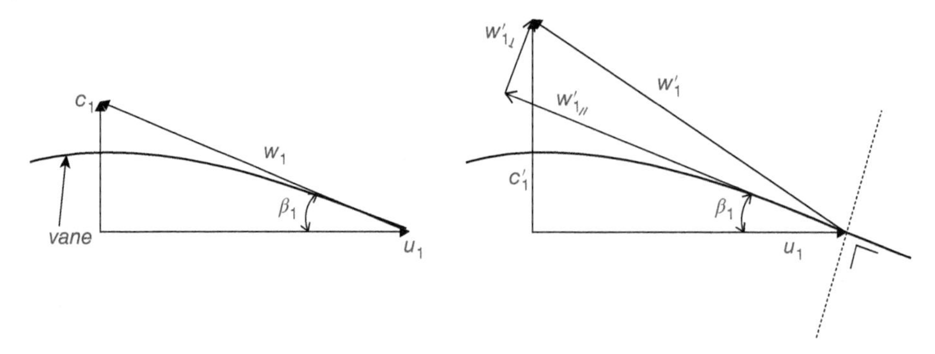

Figure 3.24 Bump free and non-bump free entry.

Example 3.2 ***Bump Free Inlet***
Given a centrifugal pump (backwards bended vanes) with diameter $D_i = 14$ [cm] (Figure 3.24).

Suppose that the pump has a speed of 1500 rpm and that the flow is equal to the nominal flow where the inlet velocity $c_1 = c_{1r} = 3$ [m/s]. What value should the constructer give at the vane angle β_1 in order that the liquid should enter the impeller bump free? What happens in non-nominal situations?

Determine first the peripheral velocity u_1:

$$u_1 = \omega \cdot r_1 = \frac{D_1}{2} \cdot \omega = \pi \cdot D_1 \cdot N = \pi \cdot 0.14 \cdot \frac{1500}{60} = 11\,[\text{m/s}]$$

$tg\,\beta_1 = \frac{c_1}{u_1} = \frac{3}{11}$ From where : $\beta_1 \cong 15°$

$$w_1 = \sqrt{u_1^2 + c_1^2} = \sqrt{11^2 + 3^2} = 11.4\,[\text{m/s}]$$

Suppose now that the manufacturer has given the vane angle β_1 the value 15° then will at 1500 rpm and $c_1 = 3$ [m/s] the relative velocity w_1 also make an angle of 15° with the circumferential velocity u_1. If now for one reason or another the volumetric flow increases, then c_1 $(= c_{1r})$ will increase (flow determining component) to a value c_1'. If the rotation speed of the drive does not change, then u_1 will keep its value.

The three vectors $\overline{u}$, $\overline{w}$, and $\overline{c}$ will still form a triangle and this will mean that the new velocity w_1' will not be tangential anymore to the constructed vane angle (see Figure 3.24). There will be a bump.

The velocity w_1' can be decomposed in two components $w_{1\perp}'$ and $w_{1\|}'''$ (Figure 3.24).

The bump component $w_{1\perp}'$ is perpendicular on the vane and the associated kinetic energy is lost in friction heat.

In an analogues way, one can see that the velocity triangles will be deformed when the rpms are different from the nominal one (by changing u_1). There also will be no bump free entry anymore.

The pump is thus designed for a determined rpm and flow where the stream is bump-free. Non-nominal circumstances give energy loss by bumps. This loss increases the more it deviates from the nominal circumstances.

3.1.7 Diffusor

The velocity c_2 with which the liquid flows out of the impeller is much too high to be sent in the press pipe. First a great part of the obtained dynamic pressure:

$$\frac{\rho}{2} \cdot (c_2^2 - c_1^2)$$

must be converted in static pressure. This can be done by widening the passage, thus in a *diffusor*. Such diffusor action takes place in a stationary *radial diffusor*, placed after the impeller (Figures 3.25 and 3.26). After this diffusor the liquid is caught in the volute. Eventually the volute can be widened to keep the velocity of the liquid constant. Finally, an outlet piece is made like a diffusor.

One has to be conscious that the conversion of dynamic to static pressure always experiences attendant friction loss. This loss can be limited by making the angle of the diffusor no bigger than 9°.

Eventually a guide wheel is placed in the diffusor (Figure 3.27) or partitions (Figure 3.28) to guide the liquid. But because these partitions only work with one operating point it is not common.

Till now only the impeller has been studied:

$$p_{th} = (p_2 - p_1) + \frac{\rho}{2} \cdot (c_2^2 - c_1^2)$$

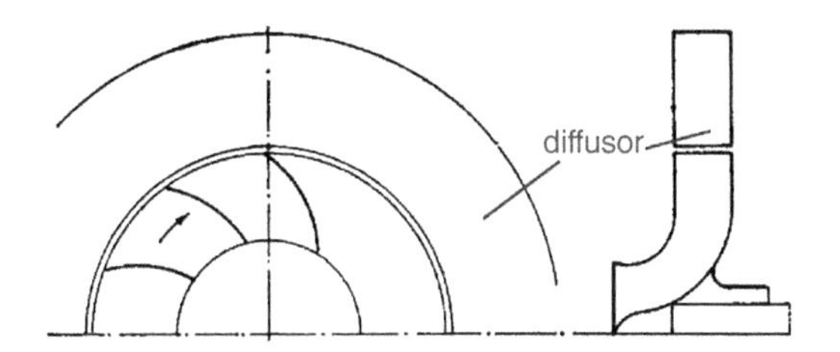

Figure 3.25 Diffusor.

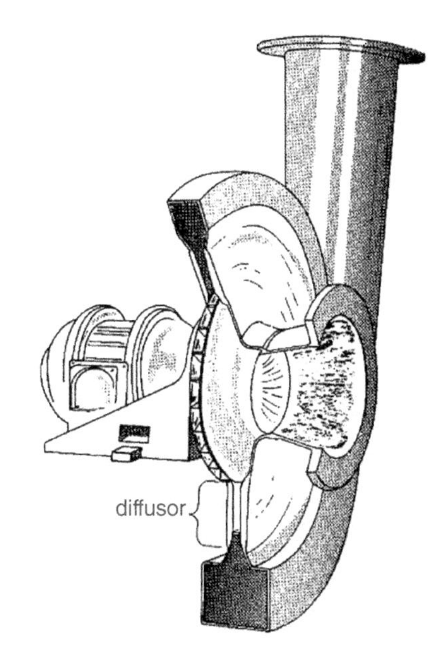

Figure 3.26 Diffusor.

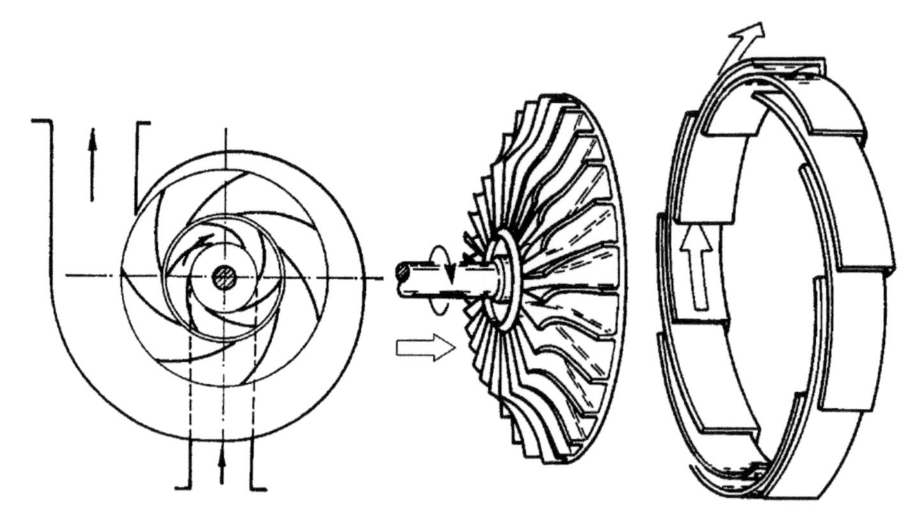

Figure 3.27 Guide wheel and diffusor.

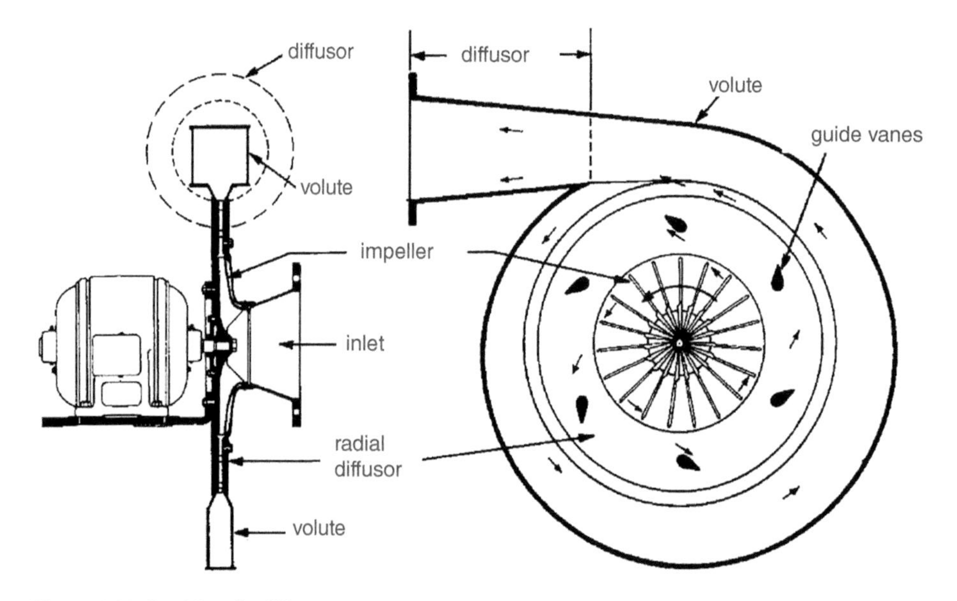

Figure 3.28 Partitions in diffusor.

But in the inlet piece, the volute, the diffusor, and the outlet piece there are transfers between dynamic and static pressure. On the *flanges* of the pump the velocities c_1' and c_2' are equal and the pressure difference is (Figure 3.29):

$$p_{th} = (p_2 - p_1) + \frac{\rho}{2} \cdot (c_2^2 - c_1^2) = p_2' - p_1'$$

One will remark that the inlet piece is always convergent. In a divergent piece the liquid cannot follow very well the presented path and turbulence occurs. This is not the case with a convergent piece. Here the liquid is "forced" to follow the path.

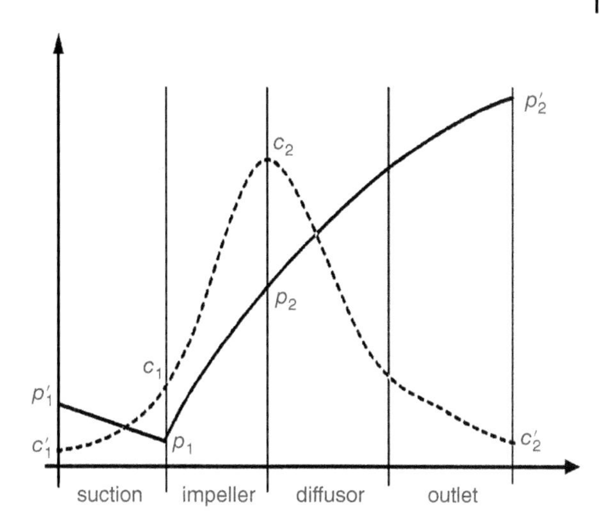

Figure 3.29 Pressure and velocity progress.

3.1.8 Influence of Vane Angle

3.1.8.1 Graphically

Three types of vane angles are compared: backwards bended, radial ending, and forward curved vanes (Figure 3.30). To compare the three types, they must be compared in the same conditions: a bump free inlet, the same dimensions, the same rpm thus same rotation speed, the same circumferential velocity, the same volumetric flow, and so the same radial component of the absolute velocity c_{2r}, and at last the same liquid (ρ).

Suppose radial inlet, then:

$$p_{th} = \rho \cdot u_2 \cdot c_2 \cdot \cos \alpha_2$$

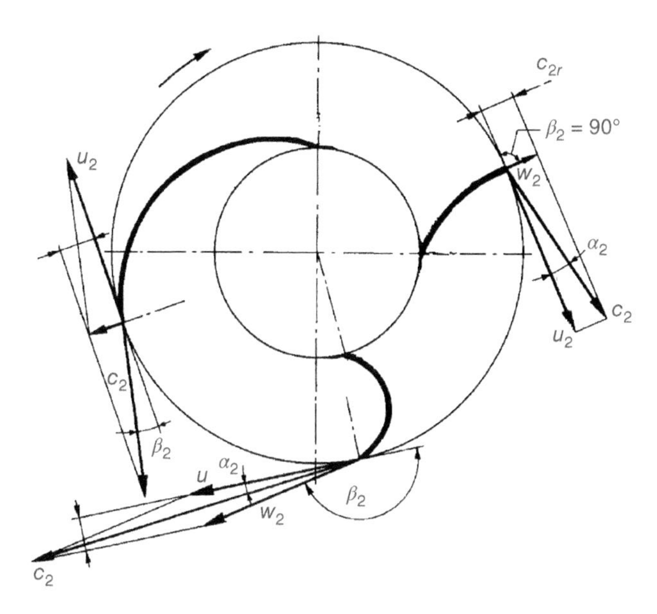

Figure 3.30 Influence of vane angle.

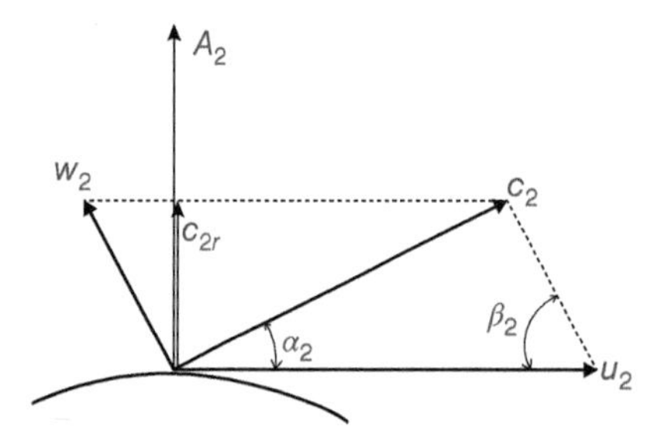

Figure 3.31 Velocity triangle.

$c_2 \cdot \cos \alpha_2$, this is the projection of $\overline{c}$ on $\overline{u}$. At constant value u_2 (and for the same fluid, this is same value of ρ) the feed pressure is proportional to $c_2 \cdot \cos \alpha_2$.

By comparing the various cases in Figure 3.30 it can be determined that how bigger the angle β_2, how bigger the feed pressure. As far as that is concerned one would prefer a pump with forward-bended vanes. But, on the other hand, c_2 itself is bigger as β_2 increases. This is not interesting: in the diffusor the velocity will have to be converted to static pressure and this will be with concomitant friction losses, which will reduce the pump's efficiency. That's why the value of c_2 is limited and one generally prefers backward-bended vanes.

3.1.8.2 Analytically

As always radial inlet is supposed. The problem now is to find an expression between p_{th} and Q_V.

The volumetric flow is given by:

$$Q_V = z \cdot A_2 \cdot c_{2r}$$

with

$$p_{th} = \rho \cdot u_2 \cdot c_2 \cdot \cos \alpha_2$$

In Figure 3.31:

$$c_2 \cdot \cos \alpha_2 = u_2 - w_2 \cdot \cos \beta_2$$

$$w_2 \cdot \sin \beta_2 = c_{2r}$$

$$c_2 \cdot \cos \alpha_2 = u_2 - c_{2r} \cdot \text{cotg } \beta_2$$

So,

$$p_{th} = \rho \cdot u_2 \cdot \left(u_2 - \frac{Q_V}{z \cdot A_2} \cdot \text{cotg } \beta_2 \right) \tag{3.17}$$

Equation (3.17) is graphically represented in Figure 3.32.

3.1.9 Pump Curve

3.1.9.1 System Curve

In general, a pump connected to an installation (the *system*) must deliver a pressure p_s to overcome the height, a dynamic pressure to give the liquid a velocity, and a pressure to overcome the friction in the pipe lines. The sum of all these pressures is called the *manometric feed pressure of the installation*. In Figure 3.33 this system curve is set out

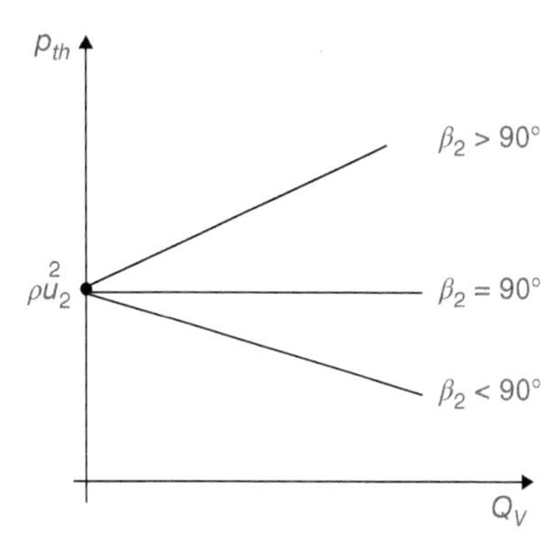

Figure 3.32 Feed pressure and flow.

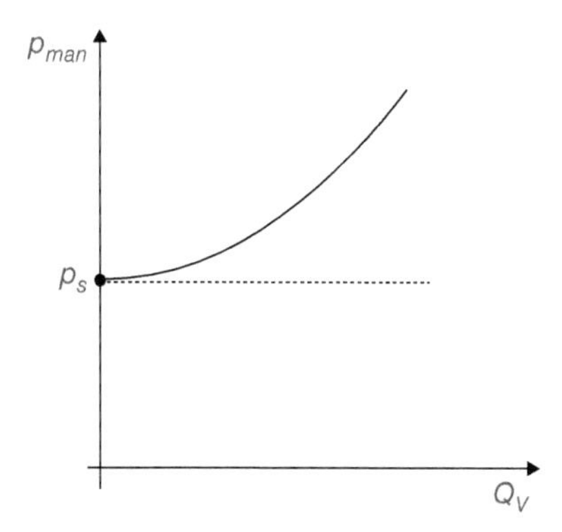

Figure 3.33 System curve.

in function of the volumetric flow Q_V. The static feed pressure $p_s = \rho \cdot g \cdot h_s$ does not depend on the flow. The dynamic pressure and friction pressure are quadratic uniform with the velocity, and thus with the flow. This gives a parabola in the graph.

3.1.9.2 Build Up Pump Curve

In the pump itself there are:

- The classic viscous friction losses that the fluid undergoes in the inlet and outlet piece, the bends, the volute, the curve vane channel; they are uniform with the quadratic of the volumetric flow (Figure 3.34).
- The shock losses: at a certain flow the liquid streams shock free into the pump, but at other flows there is a bump between liquid and entrance of the impeller; these losses increase as the flow increasingly deviates from the nominal flow.

This friction and shock loss form together the pressure loss p_{lp}. The manometric feed pressure is found by subtracting the losses from the theoretical pressure:

$$p_{man} = p_{th} - p_{lp}$$

Graphically this is represented in Figure 3.35.

In reality the theoretical pressure is estimated too high. In the vane channel a circular flow is induced and this means that the feed pressure is weakened. But because this book has no intention to design turbopumps this problem is not discussed here.

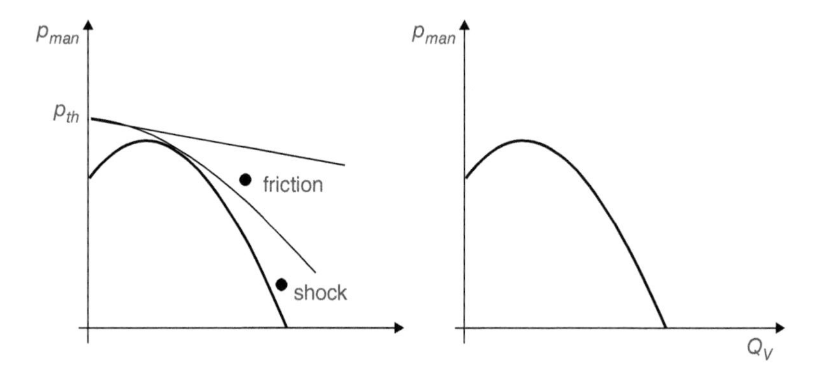

Figure 3.34 Friction losses and bump losses.

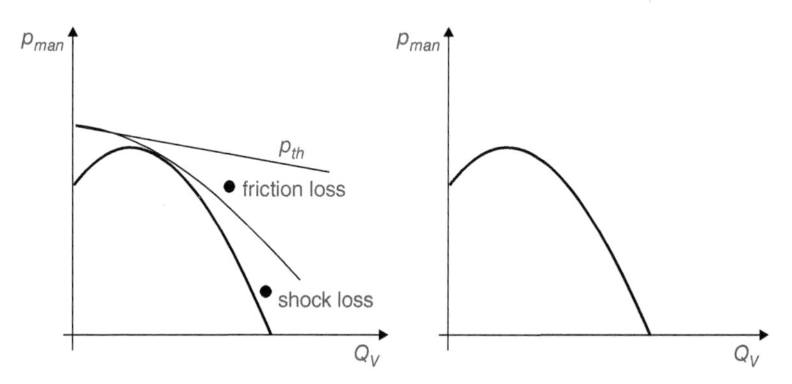

Figure 3.35 Manometric pressure.

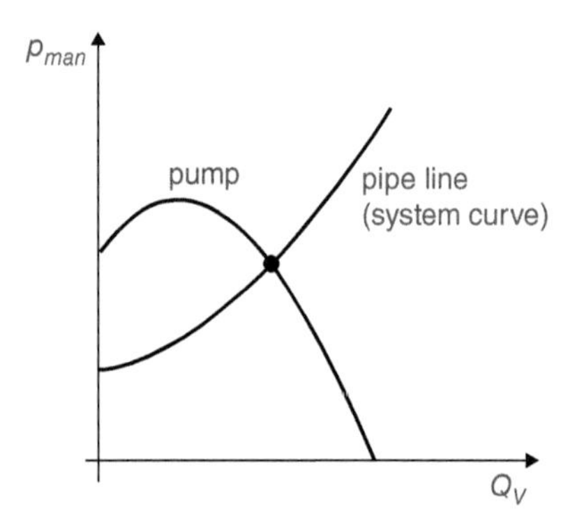

Figure 3.36 Operating point.

3.1.9.3 Operating Point

The *operating point* is the intersection between the system curve and the pump curve (Figure 3.36). Another name for an operating point is *duty point.*

3.1.10 Pump Efficiency

The hydraulic efficiency of the pump:

$$\eta_p \equiv \frac{p_{man}}{p_{th}}$$

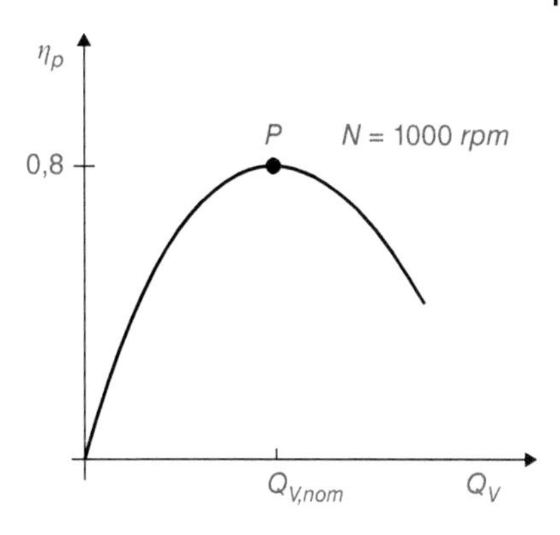

Figure 3.37 Efficiency. Point P is the BEP.

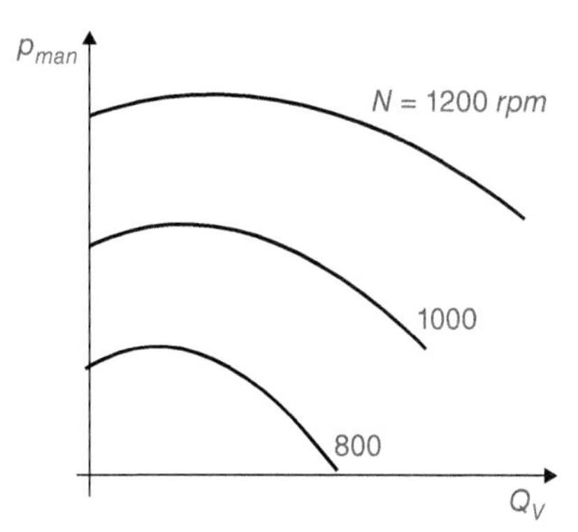

Figure 3.38 Influence rpm.

In Figure 3.37 η_p is set out in function of the volumetric flow Q_V with constant rpm. At the nominal flow $Q_{v,\,nom}$ the pump efficiency is maximum. The point on the curve with best efficiency, at a certain rpm, is the BEP (best efficiency point) or optimal operating point.

3.1.11 Influence RPM

According to Equation (3.17) one can pose qualitatively:

$$p_{th} = \frac{p_{man}}{\eta_p} \cong p_{man} \cong \rho \cdot u_2^2 \div N^2$$

The result is set out in Figure 3.38.

3.1.12 First Set of Affinity Laws

Consider a pump with two states, 1 and 2, with rpm's N_1 and N_2, same dimensions D (and r), and same pump efficiencies η_{p1} and η_{p2}, and the same liquid ρ. The two states work in the BEP (at different rpm's) (Figure 3.39).

$$\frac{Q_{V,1}}{Q_{V,2}} = \frac{c_{1,1}}{c_{1,2}} = \frac{u_{1,1} \cdot \sin\beta_1}{u_{1,2} \cdot \sin\beta_1} = \frac{u_{1,1}}{u_{1,2}} = \frac{N_1}{N_2} \cdot \frac{D_1}{D_2}$$

$$\frac{Q_{v,1}}{Q_{v,2}} = \frac{N_1}{N_2}$$

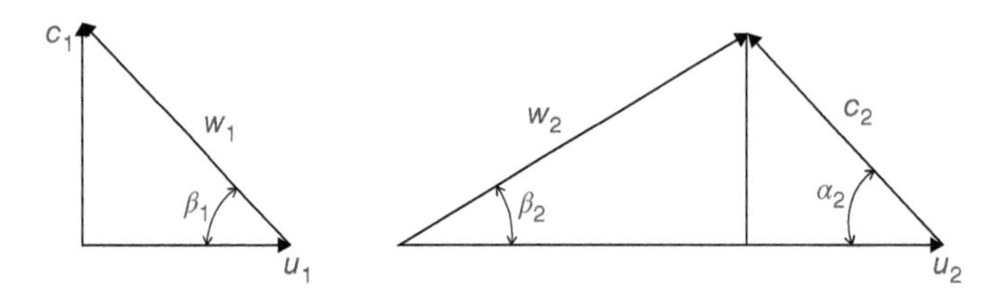

Figure 3.39 Velocity triangles.

$$\frac{p_{man,1}}{p_{man,2}} = \frac{p_{th,1}}{p_{th,2}} \cdot \frac{\eta_{p,1}}{\eta_{p,2}} = \frac{p_{th,1}}{p_{th,2}} = \frac{\varrho \cdot u_{2,1} \cdot c_{2,1} \cdot \cos\alpha_{2,1}}{\varrho \cdot u_{2,2} \cdot c_{2,2} \cdot \cos\alpha_{2,2}}$$

If the two states have the same pump efficiencies, this means that the second velocity triangles of the two states will be uniform, because the angles β will have to be the same value. Then the angles α will also be the same. So:

$$\frac{p_{man,1}}{p_{man,2}} = \frac{u_{2,1} \cdot c_{2,1}}{u_{2,2} \cdot c_{2,2}}$$

Because the velocity triangles are uniform:

$$\frac{c_{2,1}}{c_{2,2}} = \frac{u_{2,1}}{u_{2,2}}$$

And:

$$\frac{u_{2,1}}{u_{2,2}} = \frac{N_1}{N_2}$$

Finally:

$$\frac{p_{man,1}}{p_{man,2}} = \left(\frac{N_1}{N_2}\right)^2$$

$$\frac{p_{man,1}}{p_{man,2}} = \left(\frac{N_1}{N_2}\right)^2$$

Or:

$$\frac{H_{man,1}}{H_{man,2}} = \left(\frac{N_1}{N_2}\right)^2$$

At last:

$$P_t = Q_v \cdot p_{th} = Q_V \cdot \frac{p_{man}}{\eta_p}$$

$$\frac{P_{t,1}}{P_{t,2}} = \left(\frac{N_1}{N_2}\right)^3$$

In the supposition that: $\eta_{p,1} = \eta_{p,2}$!
Eliminating the rpm's:

$$\frac{H_{man,1}}{H_{man,2}} = \left(\frac{Q_{v,1}}{Q_{v,2}}\right)^2$$

The boxed expressions are the *affinity laws*. Be careful, this is only a way to calculate the pump curve to another rpm. It says nothing about the state of the pump that will really be worked in at changing speed. This is discussed later.

In the $H_{man} - Q_V$ diagram this last expression is a parabola through the origin. One can draw a bundle of parabola. On one parabola one works in the same circumstances at different rpm's. That is, all velocity triangles are uniform with each other. All points on such a parabola are *homologous points*. One parabola collects all points with the same pump efficiency $\eta_{p,\,opt}$. The other parabolas are a collection of points with the same operating conditions, with other series of homologous velocity triangles than the nominal velocity triangles.

Example 3.3

A centrifugal pump runs at $N_1 = 3500$ [rpm], discharges a volumetric flow $Q_{V,1} = 1$ [m^3/s], the manometric head is $H_{man,1} = 90$ [m] and the technical power is $P_{t,1} = 27$ [kW].

What will be the flow, the head, and the power when the rpm is changed to 2900 [rpm]?

Solution

$$Q_{v,2} = \frac{Q_{v,1}}{N_1} \cdot N_2 = \frac{1}{3500} \cdot 2900 = 0.83\,[\text{m}^3/\text{min}]$$

$$P_{t,2} = P_{t,1} \cdot \left(\frac{N_2}{N_1}\right)^3 = 27 \cdot \left(\frac{2900}{3500}\right)^3 = 15.4\,[\text{kW}]$$

$$H_{man,2} = H_{man,1} \cdot \left(\frac{Q_{v,2}}{Q_{v,1}}\right)^2 = 90 \cdot \left(\frac{0.83}{1}\right)^2 = 62\,[\text{m}]$$

3.1.13 Second Set of Affinity Laws

For the second set of affinity laws the rpm is constant, but the diameter D is variable. On the same base as the formulas of the previous section the following relationships can be found:

$$\frac{Q_{v,1}}{Q_{v,2}} = \frac{D_1}{D_2}$$

$$\frac{p_{man,1}}{p_{man,2}} = \left(\frac{D_1}{D_2}\right)^2$$

$$\frac{H_{man,1}}{H_{man,2}} = \left(\frac{D_1}{D_2}\right)^2$$

$$\frac{P_{t,1}}{P_{t,2}} = \left(\frac{D_1}{D_2}\right)^3$$

3.1.14 Surge

In Figure 3.40 a pump installation is represented. The press pipe discharges into a press reservoir. Operating point C corresponds to the level of the liquid in the reservoir in

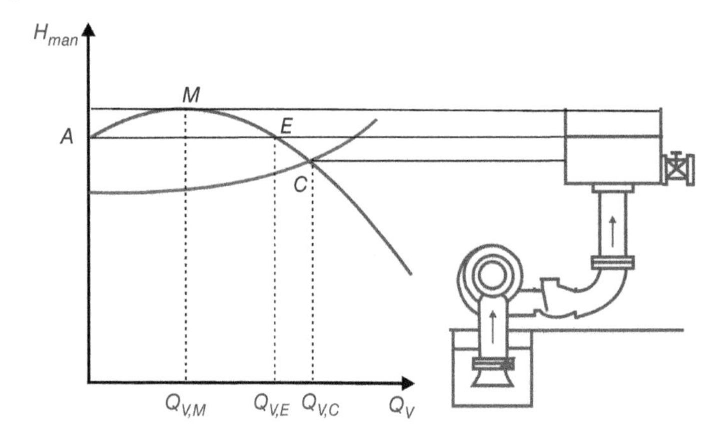

Figure 3.40 Surge.

Figure 3.40. Suppose that the demand for liquid by the consumer decreases. The valve is partially closed. But the pump delivers the same flow to the reservoir. The level in the reservoir will rise. The operating point shifts from C to E. If, however, the demand stays low, the operating points will further shift from point E to the left. Finally, it arrives at point M. As soon as this point M is reached the pump delivers its highest possible pressure. If the level in the reservoir still keeps increasing there is no operating point anymore and suddenly the operating point of the pump jumps from M to A. The pump delivers no flow anymore. But the pressure in the reservoir is higher than that of the pump. A possibility is that there could be a flow from the reservoir to the pump. This can be prevented by placing a check valve in the press pipe. The supply to the consumer will lower the level in the press reservoir and all of a sudden the pump will be capable of delivering liquid again. The operating point A jumps back to E, but this phenomenon can happen again. This leads to a pulsating action. This is *surge* (stall).

In Figure 3.41 a pump curve is shown where a special case of a system curve is happening. There are two intersection points. This would mean that there are two possible operating points. Strange.

What is happening? Suppose that the operating point is point 1. Let there be a small perturbation of the operating point. Suppose the flow increases a bit. On the pump curve one can see that the delivered pump pressure increases. This in turn leads to a rise in flow. This continues and finally we arrive at point 2.

But is point 2 a stable point? Suppose that the flow increases a bit. On the pump curve one can see that the delivered pump pressure decreases, which makes the operating point shift to the right. This causes the pump to deliver a lower flow. In other words, the operating point will shift back to the left. And so we return to point 2. Point 2 is a stable point.

One thing or another could lead to a forward and back swinging between points 1 and 2. That's why care must be taken to always stay in the right side of the pump curve.

3.1.15 Application Field

The pump characteristic $H_{man} = \varphi(Q_V)$ at various rotation speeds can be extended with ISO-efficiency curves (see Figure 3.42). N is the nominal rpm, $H_{man, nom}$ the manometric head at nominal rpm, and $Q_{V, nom}$ the nominal flow.

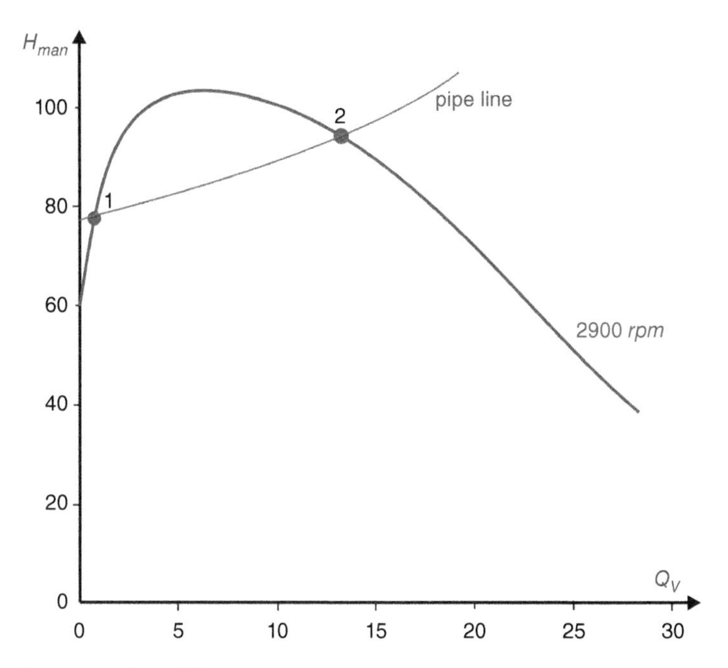

Figure 3.41 Instability.

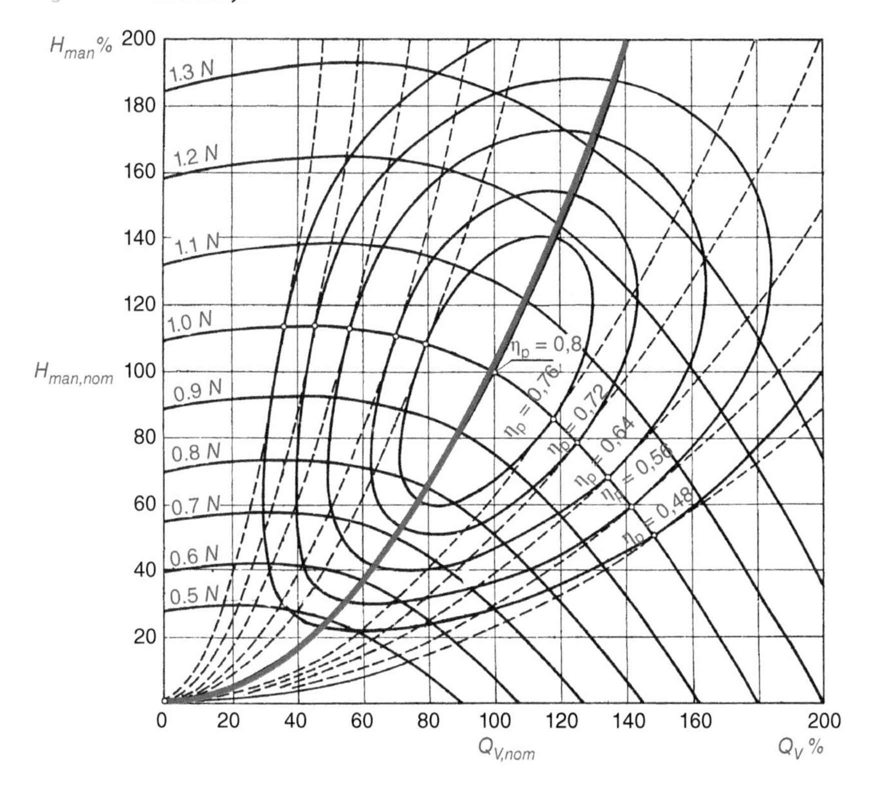

Figure 3.42 Pump curves.

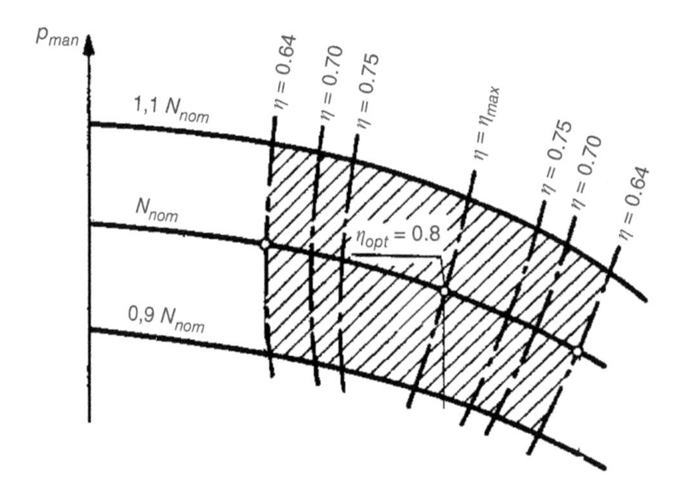

Figure 3.43 Application field.

The red parabola from the origin connects all homologous points at various rpm's with maximum efficiency, thus all BEP points. The dashed parabola are curves of equal operating points (lines of equal specific speed, see later).

Not all points in the graphic of Figure 3.43 are states that an engineer considers reasonable. A certain part of the graphic in the $H_{man} = \varphi(Q_V)$ quadrant limited by a maximum and minimum acceptable efficiency is considered as the operating area of the pump, called the *application field* (Figure 3.43).

The manufacturer of pumps is confronted with the issue of needing different pumps for different applications. When a manufacturer designs a series of pumps, they will take care that their application fields touch each other or overlap.

Figure 3.44 shows such a combined diagram of a normalized series of one-impeller pumps (ISO 2858). Vertically the manometric head H_{man} is set out.

3.1.16 Flow Regulation

Adjusting the operating point to changing circumstances can be done by:

- A change of the system curve: throttle and bypass regulation.
- A change of the pump curve: speed regulation.

3.1.16.1 Throttle Regulation

When the pump works at constant speed the pump curve is fixed.

A regulation of the volumetric flow occurs by working on the system curve. This can be done by placing a valve in the press pipe. By changing the passage of the valve more or less resistance is brought in the press pipe so that the "friction part" of it changes. The static part does not change. With throttling the system curve will increase and the operating point shifts to the left. The feed pressure increases and the volumetric flow decreases (see Figure 3.45).

The head over the throttle valve (Figure 3.46):

$$\Delta H_{man} = H_{man,2} - H_{man,1}$$

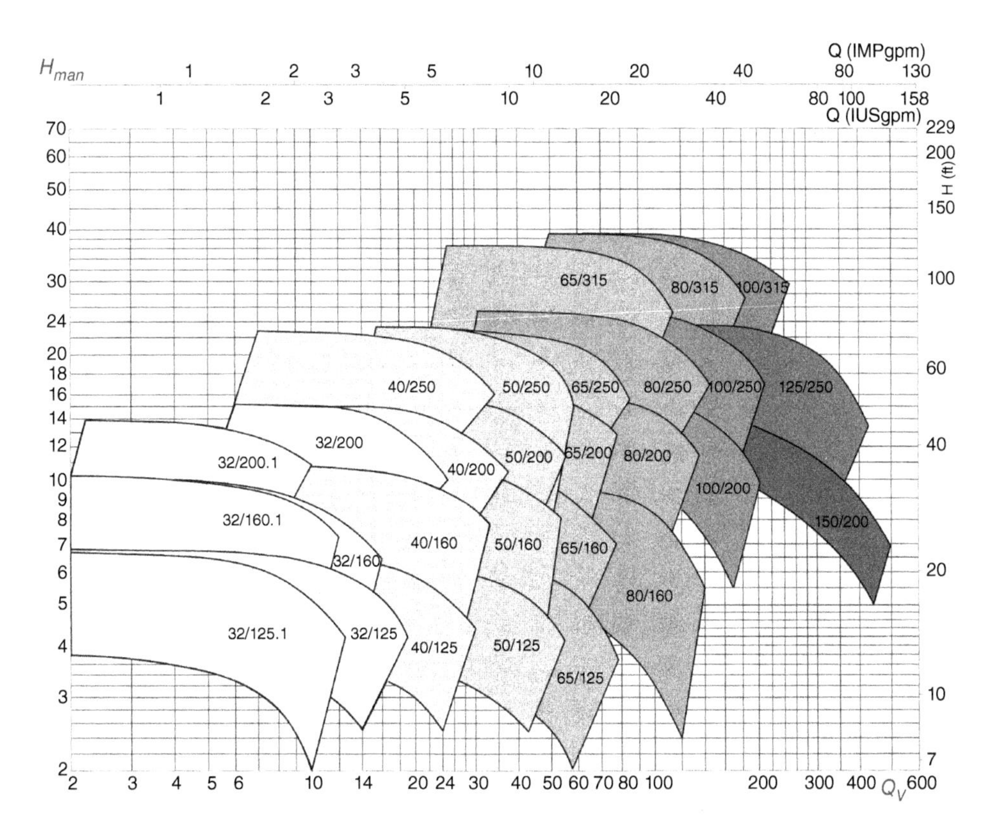

Figure 3.44 Homologous series of pumps. Source: DAB 50/160 means pump outlet diameter 40 [mm], impeller diameter 16 [mm].

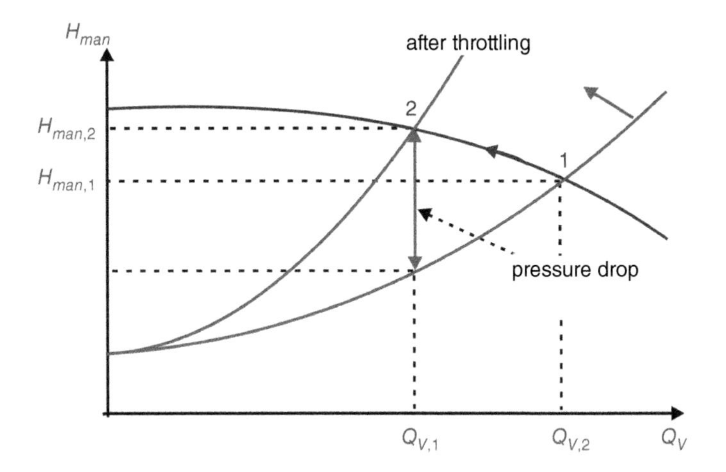

Figure 3.45 Throttling: evolution of characteristic.

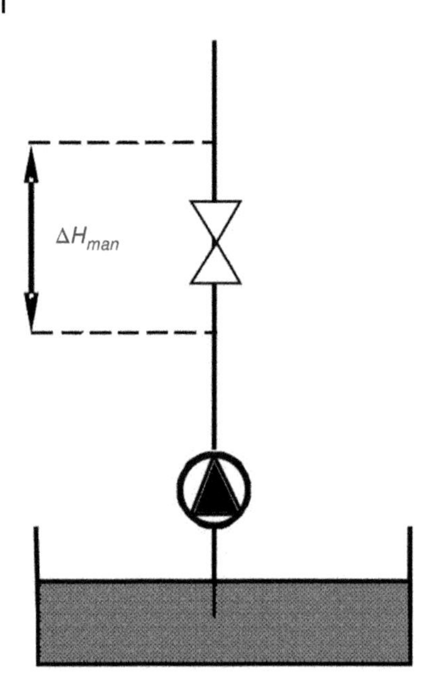

Figure 3.46 Throttling: pressure loss.

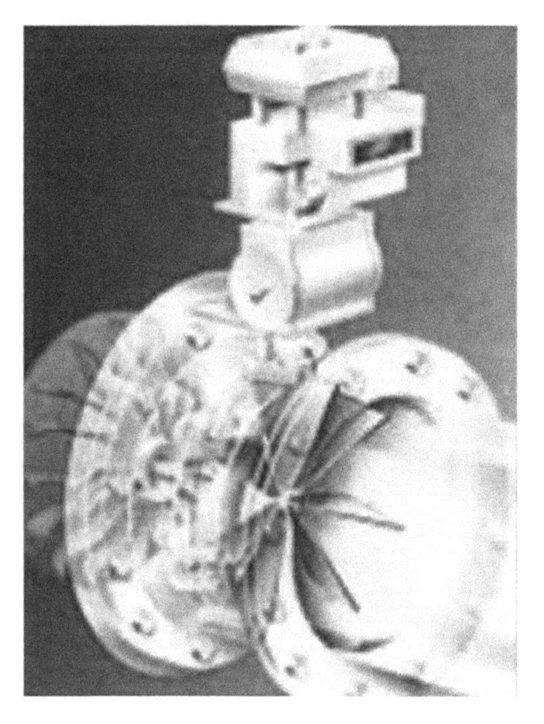

Figure 3.47 Throttle vanes (IGVs). Source: Atlas Copcoc.

Throttling must happen in the discharge line, not in the suction line, because of the danger of cavitation.

Throttling at inlet is a very dangerous business. The pressure at the inlet will decrease. The net positive suction head (NPSH) will deteriorate (see Section 3.3). The pressure at

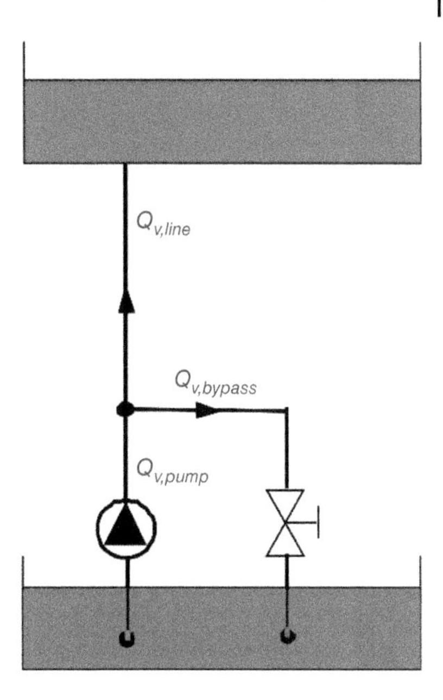

Figure 3.48 Bypass regulation.

the inlet will be so low that vapor will create *cavitation* (see later). It is preferable for throttling to be performed at the outlet.

An inlet guide vane (IGV) – see Section 3.3 – can be placed at the entrance. The vanes area is arranged as radial blades in the intake, which will cause the liquid to rotate (*peroration*) while the flow is throttling. This method offers a greater regulation range than classic throttling, where there is always a risk that *surge* could take place. Variable outlet vanes at the discharge side have a turn-down ratio down to 30% with maintained pressure.

In a very sophisticated implement inlet guide vanes (IGVs) can be turned according to the desired flow, and offer a bump-free inlet. The rotation of the IGV is done either by a servomotor or manually (Figure 3.47).

3.1.16.2 Bypass Regulation

Via a bypass line a part of the volumetric flow can flow back to the suction reservoir. Theoretically can the liquid also be flown back to the suction line (or the suction flange of the pump) but then the pump power ($Q_V \cdot p_{th}$) ends up in a small part of liquid, that will heat up.

Opening the bypass valve (Figure 3.48) will increase the flow and the flow through the discharge line will decrease. The difference in flow will flow back. On the pump curve the operating point shifts to the right; the pump pressure decreases. Opening the by-pass valve (Figure 3.48) will increase the flow $Q_{V,pump}$ delivered by the pump, but the flow $Q_{V,line}$ through the discharge line will decrease. The difference in flow, $Q_{v,bypass}$ will flow back. On the pump curve the operating point shifts to the right; the pump pressure decreases. The pump pressure – this is the pressure difference over the bypass valve – must decrease to a value on the system curve that meets this pump pressure.

The system curve of the bypass line runs through the origin because it is a closed loop (Figure 3.49). One common characteristic can be found as follows: for a determined feed

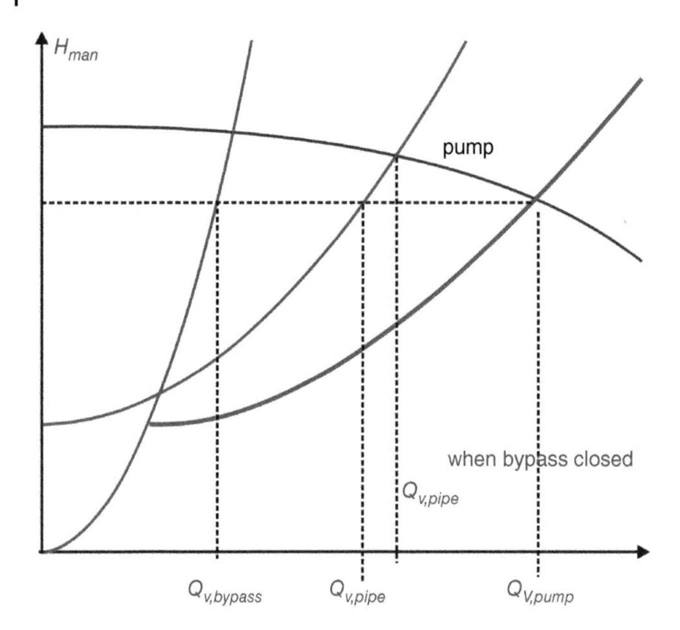

Figure 3.49 Characteristic behavior.

head (horizontal line) one counts the flows of both parallel lines together. The operating point is the intersection between the pump curve and the new system curve.

Regulating the flow using a bypass line consumes a great deal of energy (Figure 3.49).

3.1.16.3 Speed Regulation

Increasing the speed changes the pump curve, while the system curve stays the same. The operating point shifts (Figure 3.50).

3.1.16.4 Comparison

- With speed regulation (Figure 3.51) the mechanical load on the pump parts is lower (bearing and impeller, deflection of pump shaft, vibrations, and wear of shaft sealing)

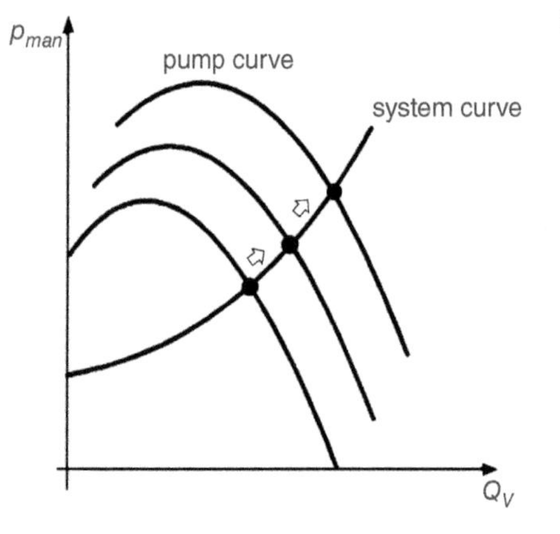

Figure 3.50 Regulation speed.

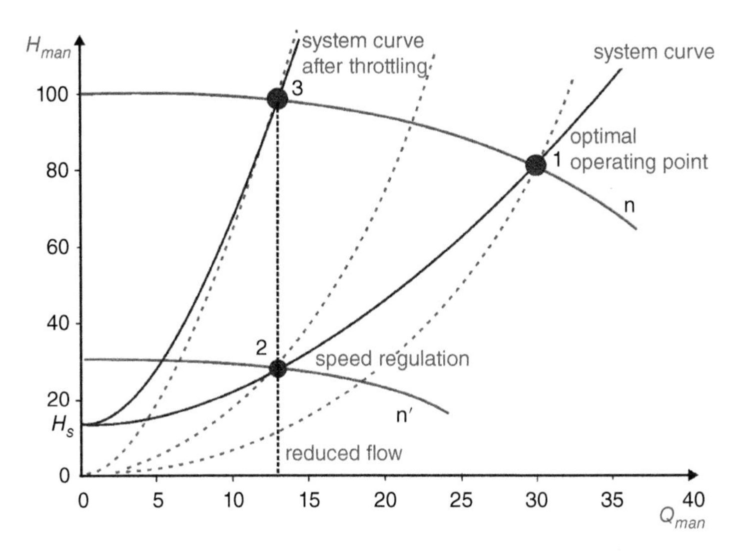

Figure 3.51 Comparison by-pass versus throttling regulation of flow.

because the pump does not run all the time on full load. This causes the MTBF (mean time between failure) to be greater and decreases the cost of maintenance.

- Because of the lower load of the shaft sealing and the lack of a throttle valve, there is less loss in product (every leasing leaks). The product loss through the axis of the valve of a regulation valve is much greater than through the mechanical seal of the pump (see later).
- Regulation of the speed is excellent: it is an energy-saving regulation.
- For the same flow decreasing the power using a speed regulator is more energy-efficient than using throttling. This becomes apparent in the formula:

$$P_t = \frac{Q_V \cdot p_{man}}{\eta_p}$$

where:

- the feed pressure of the pump with speed regulation is lower than with throttling.
- the pump efficiency is higher (Figure 3.51).

3.1.17 Start Up of the Pump

When a centrifugal pump (and in general every turbopump) does not work for a time this is mostly because all of the liquid has gone and pump space and suction line is empty and filled with air. When the pump starts up the air should be evacuated, but that is a problem, when one considers the formula of the theoretical pressure:

$$p_{th} = \rho \cdot (u_2 \cdot c_2 \cdot \cos \alpha_2)$$

It appears that the feed pressure is uniform with the specific mass of the fluid that is in the pump. Take a centrifugal pump that normally delivers 3 [bar] with water (ρ =1000 [kg/m]^3). When the pump housing is filled with air ($\rho \cong 1$ [kg/m^3]) it would

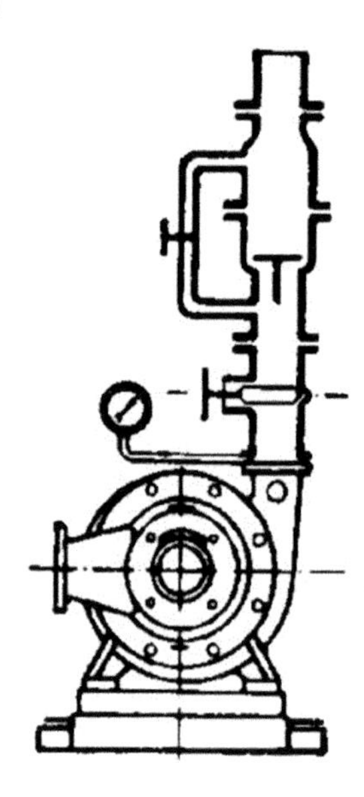

Figure 3.52 Starting up the pump.

deliver only 3 [mbar (hPa)]. This is insufficient to evacuate the air out of the pump housing. The atmospheric pressure is not capable of pushing water out of the suction reservoir to the impeller against the pressed air in the suction line and the impeller.

That's why first of all the air must be evacuated from the pump. For this purpose, on the highest point of the pump is a filling opening and a purging valve foreseen. At the bottom of the suction line is a valve so that the liquid that is filled via the opening does not leak in the suction reservoir.

When one has to do with an installation as in Figure 3.52 whereby the press line and press reservoir are above the pump the discharge line is foreseen with a press valve. The purging is done by opening the press valve and letting liquid flow back to the pump and the suction pipe. Big pumps are foreseen with a bypass pipe line.

The pump can then be started up.

But in fact, we forgot something. The feed pressure is also uniform with the speed of the pump. When the driving motor starts up, its speed is zero. It will take some time before the feed pressure of the pump attains its nominal value, and meanwhile there is not enough pressure to hold the column of liquid in the press pipe.

The liquid in the press line will flow down and the pump will work as turbine, with all problems that that gives.

That's why the valve in the press line will gradually be opened, the pump will start rotating and when the manometer on the press pipe indicates enough pressure the valve in the press pipe can be opened. Al this can be overcome by placing a check valve in the press line.

From the above mentioned it is clear that the startup of a turbopump is cumbersome. That's why on the market there are self-priming pumps. These pumps need only to fill the pump housing with liquid.

There are various systems. We give two examples.

Self-priming pumps are mostly made so that the suction pipe line is placed above the pump housing. At the first the first commissioning the pump housing is filled with liquid (Figure 3.53a). When the pump becomes operative some air is sucked from the suction

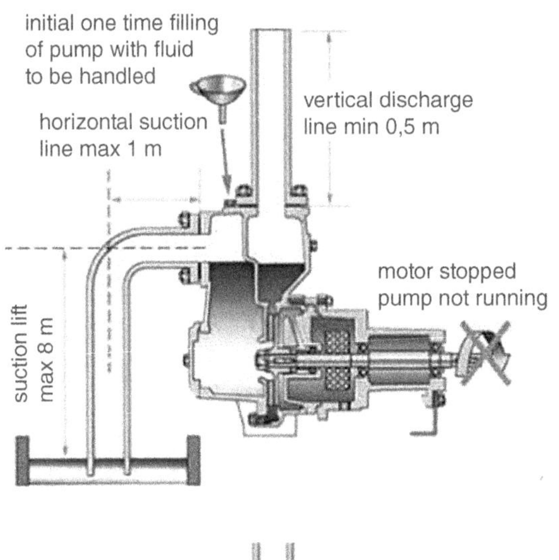

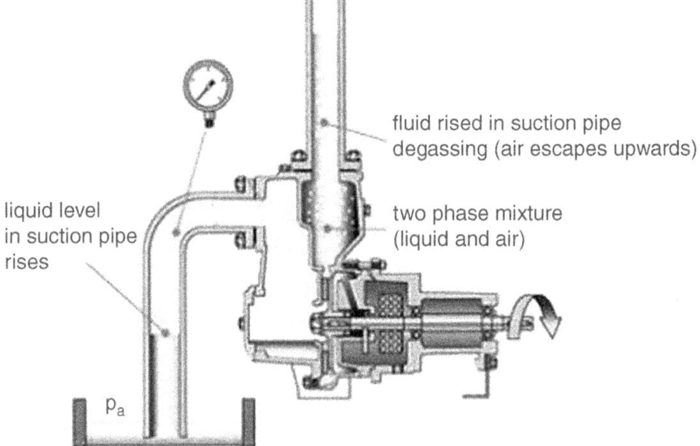

Figure 3.53 Self-priming centrifugal pump. Source: KSB.

line, but this air mixes with the liquid and is pressed out of the pump housing. But this mix is recirculated to the pump itself (Figure 3.53b); this way the pump keeps a sufficient pump pressure. Until all the air is evacuated. Then the pump sucks liquid and delivers sufficient feed pressure (Figure 3.54).

Figure 3.55 is a section drawing of what is called a jet pump. It is a self-priming pump that in fact is a double stage pump consisting of a jet pump (an ejector, see later) and a centrifugal pump as second stage. The centrifugal pump takes care for the driving force and the pressure. The jet pump consists of a narrowing channel (1) and a widening channel (2).

When the pump is started up, the liquid, that is previously present in the impeller, is tossed away and led to the diffusor, the velocity will be converted to static pressure. A part of the liquid leaves the diffusor to the outlet line, but the biggest part is being led to the ejector (1). This part works as a driving flow and drags a water–air mix from suction. The velocity of this mix lies between that of the driving flow and the sucked liquid. This

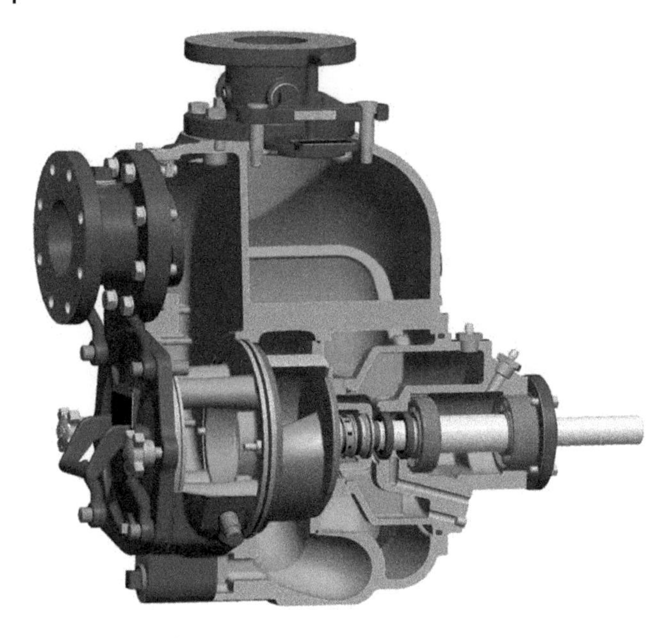

Figure 3.54 Self-priming pump. Source: Wemco.

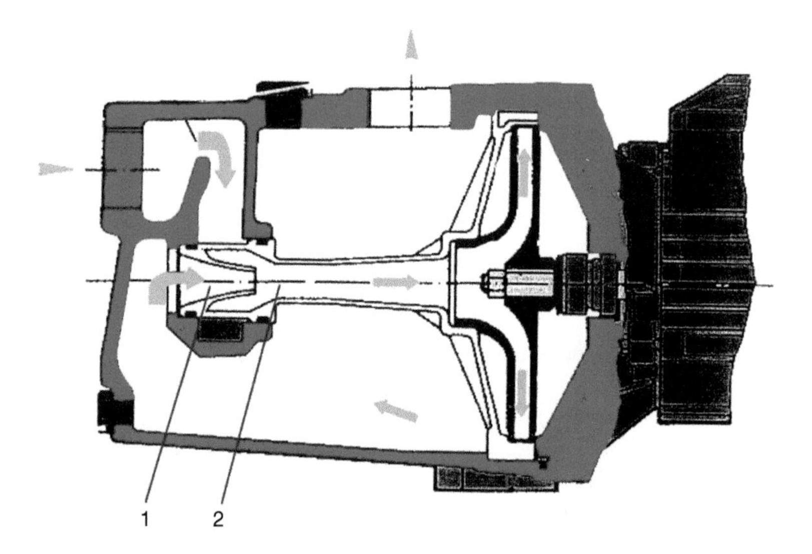

Figure 3.55 Self-priming pump with ejector.

mix flows through the ejector diffusor (venturi, see Figure 3.55), where the velocity is converted to static pressure. The static pressure of this mix that leaves the first stage of the pump lies between that of the two liquid sources.

The air is being evacuated in a deaerator space and transported via the press opening.

The double stage construction will cause the delivered pressure to be higher than normally, while by the recirculation of the liquid in the pump the delivered flow will be less than normal.

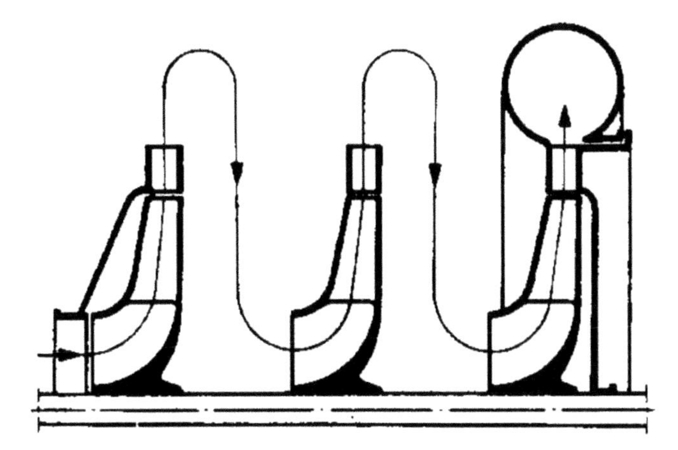

Figure 3.56 More impellers.

To avoid having to refill the pump every time, a foot valve is placed in the suction line. This prevents the pump and/or the suction pipe line from running dry.

3.1.18 High Pressure Pumps

The maximum feed pressure of a single impeller centrifugal pump is limited. Take the case of water, a maximum circumferential velocity u_2 of 35 [m/s] (cast iron) and a maximum velocity c_2 of 20 [m/s] (otherwise the friction losses become too great) then the feed pressure becomes:

$$p_{th} = \rho \cdot u_2 \cdot c_2 \cdot \cos \alpha_2 \cong 10^3 \cdot 35 \cdot 20 \cdot 1 \text{ [Pa]} \cong 7 \cdot 10^5 \text{ [Pa]} \cong 7 \text{ [bar]}$$

If higher pressures are wanted multistage pumps should be used: a shaft with many impellers (Figure 3.56). When the liquid leaves the first impeller it is led through a diffusor and then to a connection channel to the inlet of the second impeller.

If the first stage delivers 3 [bar] and there are five stages, every stage delivers 3 [bar] extra so that the whole pump delivers 15 [bar]. In reality the increase per stage is not so high because every impeller delivers a counterpressure, so that some less than 15 [bar] will be the result. It is even so that after a lot of stages a supplementary stage will not add much pressure.

Figure 3.57 represents a high pressure centrifugal pump. Figures 3.58 and 3.59 show clearly that these multistage pumps consist of many one-impeller cells.

3.1.19 Roto-jet Pump

For applications with low volumetric flow and high pressure a centrifugal pump is not adequate. As an alternative to multistage pumps and displacement pumps there is the roto-jet pump. The roto-jet pump looks like a centrifugal pump (Figure 3.60).

The operating principle is simple: a rotor brings the liquid in rotation, a stationary channel (the pickup tube) catches the liquid and converts dynamic pressure to static pressure. Feed pressures as high as 160 [bar] can be attained.

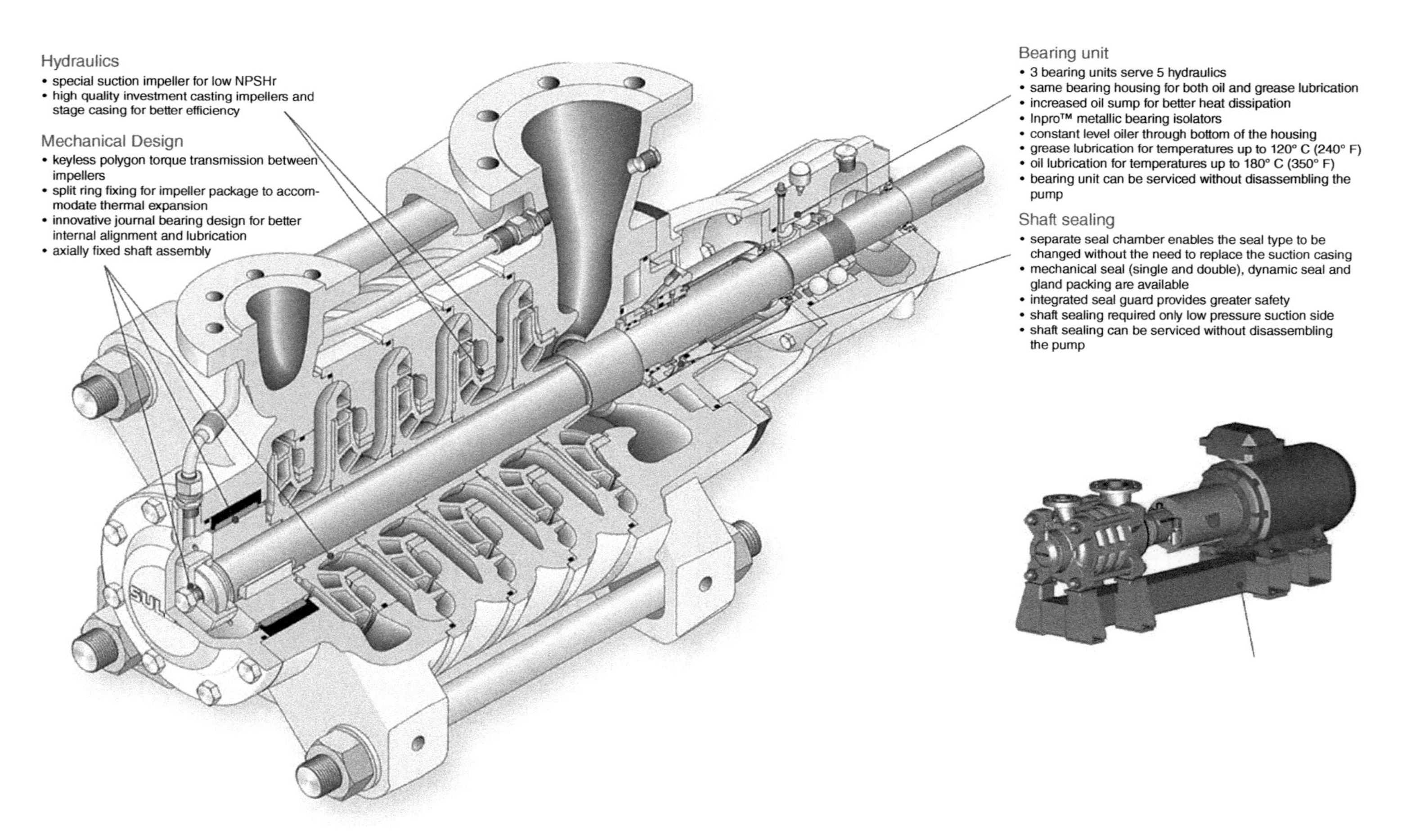

Hydraulics
• special suction impeller for low NPSHr
• high quality investment casting impellers and stage casing for better efficiency
Mechanical Design
• keyless polygon torque transmission between impellers
• split ring fixing for impeller package to accommodate thermal expansion
• innovative journal bearing design for better internal alignment and lubrication
• axially fixed shaft assembly
Bearing unit
• 3 bearing units serve 5 hydraulics
• same bearing housing for both oil and grease lubrication
• increased oil sump for better heat dissipation
• Inpro™ metallic bearing isolators
• constant level oiler through bottom of the housing
• grease lubrication for temperatures up to 120° C (240° F)
• oil lubrication for temperatures up to 180° C (350° F)
• bearing unit can be serviced without disassembling the pump
Shaft sealing
• separate seal chamber enables the seal type to be changed without the need to replace the suction casing
• mechanical seal (single and double), dynamic seal and gland packing are available
• integrated seal guard provides greater safety
• shaft sealing required only low pressure suction side
• shaft sealing can be serviced without disassembling the pump
SUL

Figure 3.57 High pressure centrifugal pump.

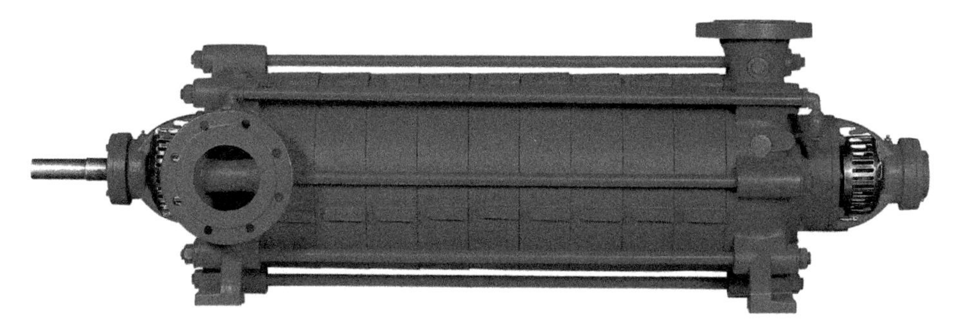

Figure 3.58 Implementation high pressure pump. Source: Hydrotec.

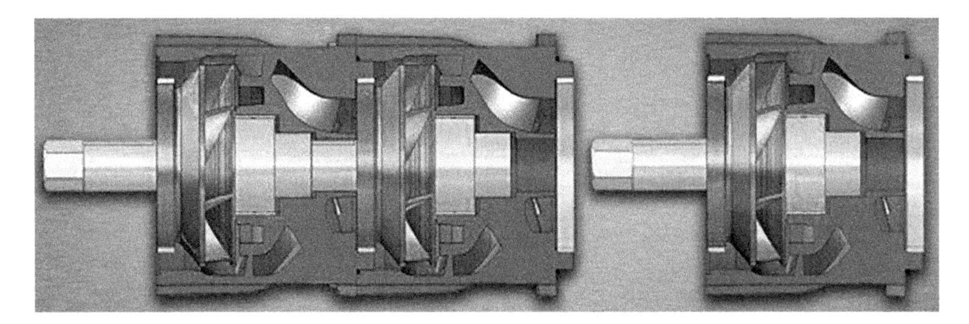

Figure 3.59 Multi-cell pump, modular design pump. Source: Ritz.

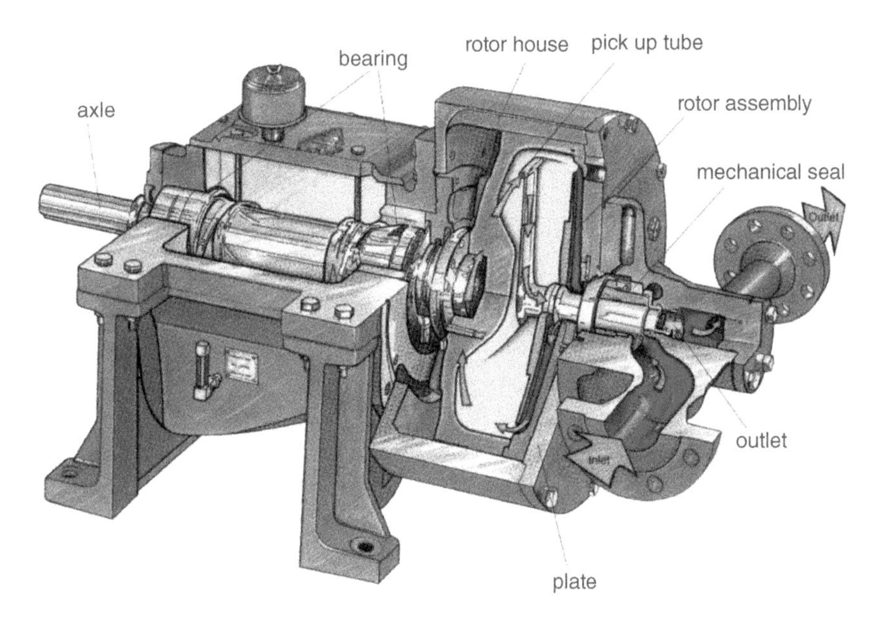

Figure 3.60 Roto-jet pump. Source: Weir, Roto-Jet.

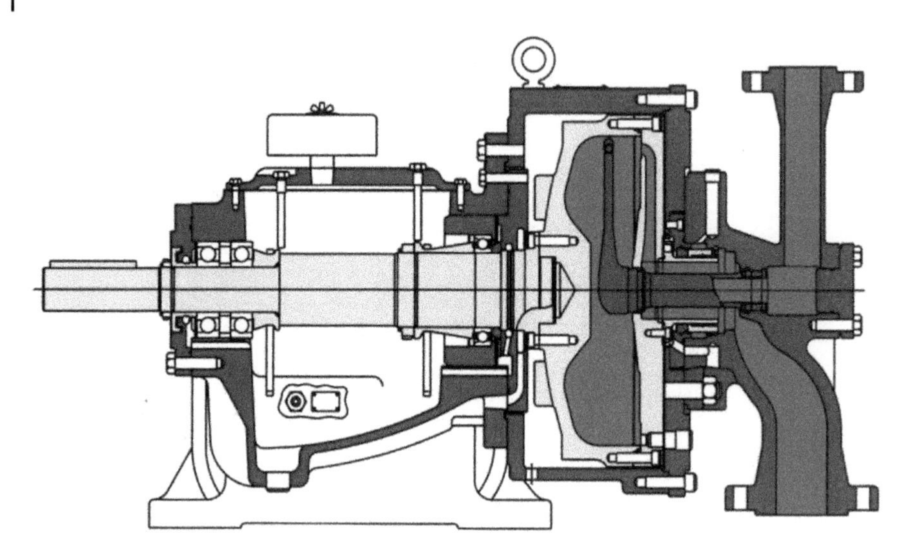

Figure 3.61 Section roto-jet pump. Source: Weir.

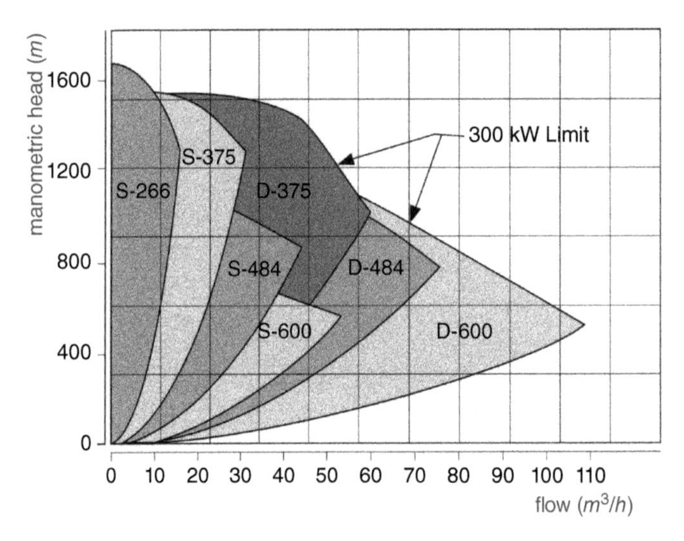

Figure 3.62 Characteristics. Source: Weir.

Because the pressure build-up takes place in a stationary channel, the rotating and wet part of the pump is on suction pressure and that is a big advantage for the shaft sealing. The speed is relatively low, so is the noise level. The pump is easy to maintain.

These pumps are used in all industries. Don't underestimate them: they can produce as much as 300 [kW] and flows of 100 [m^3/h] (Figures 3.61 and 3.62).

3.1.20 Vortex Pumps

The principle of a 100% withdrawn vortex impeller pump is simple. It is a kind of centrifugal pump where the impeller is mounted in the back of the pump housing, rotating

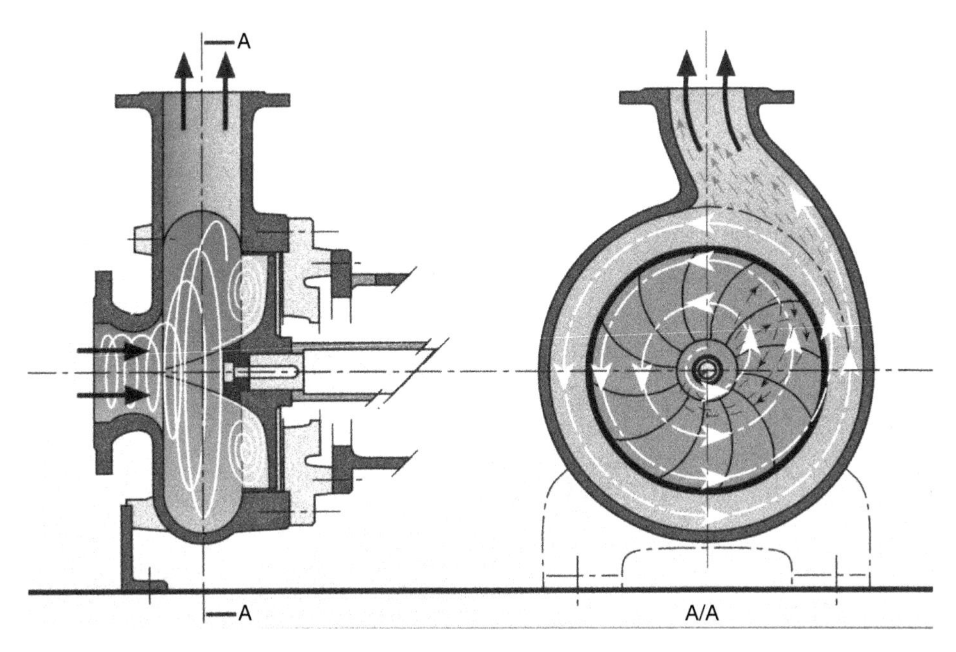

Figure 3.63 Operating principle vortex pump. Source: Egger.

Figure 3.64 Implementation. Source: Weir.

in a space that is completely outside the path of the liquid. The rotation creates a vortex similar to that found in a lavabo when it runs dry (or in a tornado). Because the impeller is not in the path of the liquid, the pump is ideally fitted for the treatment of liquids that contain long fibers, solid parts with big diameter, mud, or a combination of these without any obstruction taking place (Figure 3.63).

The flow in a vortex pump is not caused by the impeller but by the rotating vortex that is created by the impeller. Solid parts don't come into contact with the impeller. The impeller is resistant to abrasive solid parts. The housing is hardened.

Another application is the pumping of solid parts that may not be damaged (Figures 3.64 and 3.65).

Figure 3.65 Implementation vortex pump. Source: Egger.

In Figure 3.66 an implementation is given of a submersible pump with vortex impeller.

Vortex pumps provide a solution to difficult pumping problems such as sludges and slurries with large particle solids, materials with entrained air, and stringy and fibrous materials (Figure 3.67). The efficiency of a vortex pump is lower than that of a classic centrifugal pump.

3.2 Axial Turbopumps

3.2.1 Operation

Consider a flat plate consisting of oblique vanes with regard to the movement u of the plate (Figure 3.68). When the plate moves to the right, liquid in the channel will be pushed outwards. The remaining vacuum cavity of the vane channel will suck in new liquid. Consider now a circular plate and the same thing happens. The flow, globally seen, is axial (Figures 3.69 and 3.70).

The study of axial pumps is done in plan view (Figure 3.71). The liquid is sucked axially from the inlet pipe and moves tangentially; there is no radial component. The housing round the impeller inhibits that. Later we will find out that straight vanes don't work: only curved vanes give pressure rise.

3.2.2 Volumetric Flow

3.2.2.1 Axial Velocity v

All turbomachines can be described by velocity triangle (Figure 3.72). Axial pumps have the same peripherical velocity u at inlet and outlet of the channel. Also, the surfaces A_1 and A_2 are equal (Figures 3.73 and 3.74).

The volumetric flow Q_V on inlet:

$$Q_V = z \cdot \overline{A_1} \cdot \overline{c}_1 = z \cdot A_1 \cdot c_1 \cdot \sin \alpha_1$$

Figure 3.66 Vortex Submersible pump. Source: Wemco CES.

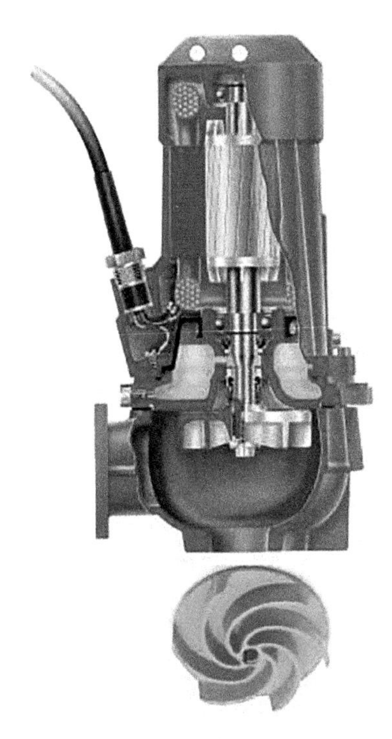

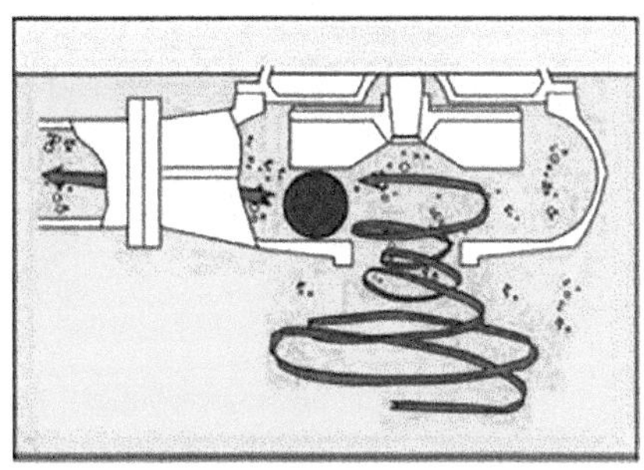

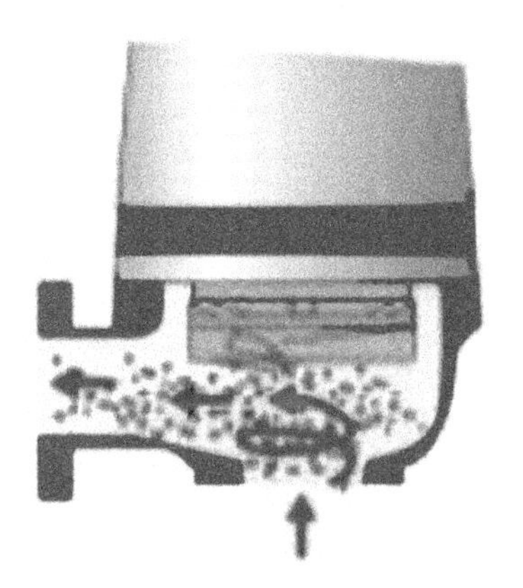

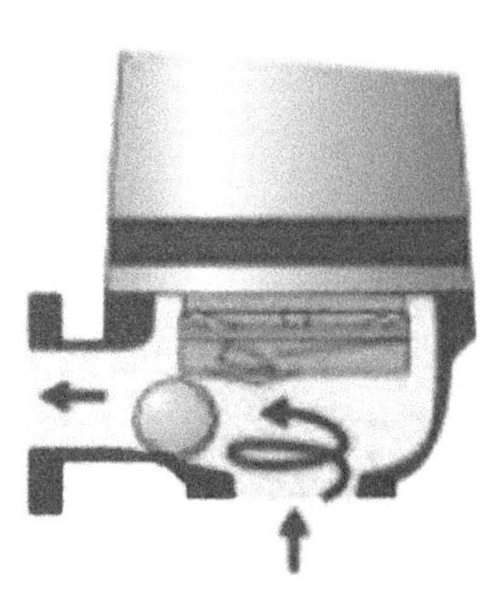

Figure 3.67 Avoidance of clogging. Source: Weir.

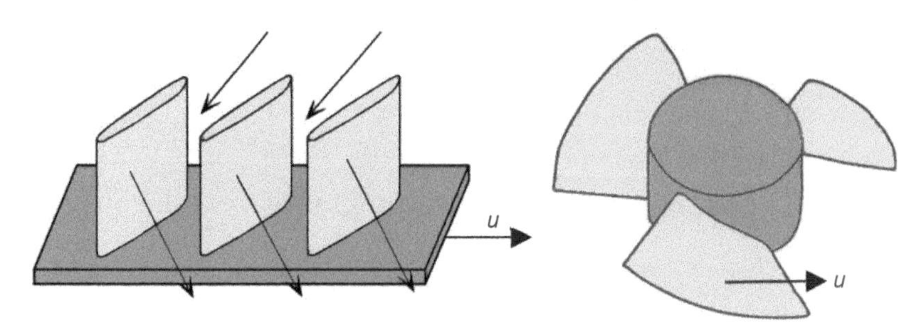

Figure 3.68 Principle of an axial pump.

Figure 3.69 Axial pump. Source: Sulzer – Ensival.

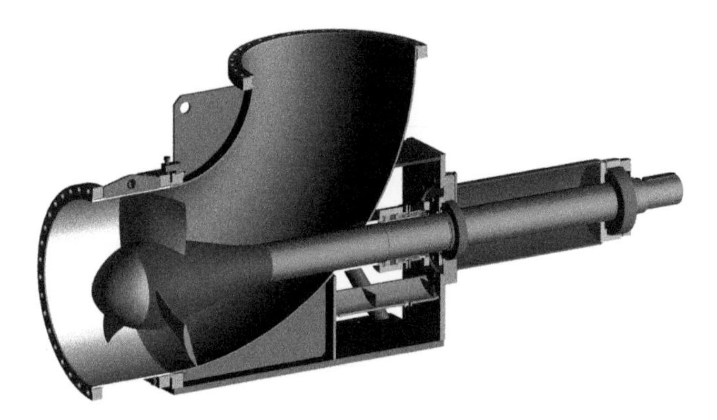

Figure 3.70 Axial pump. Source: Allweiler.

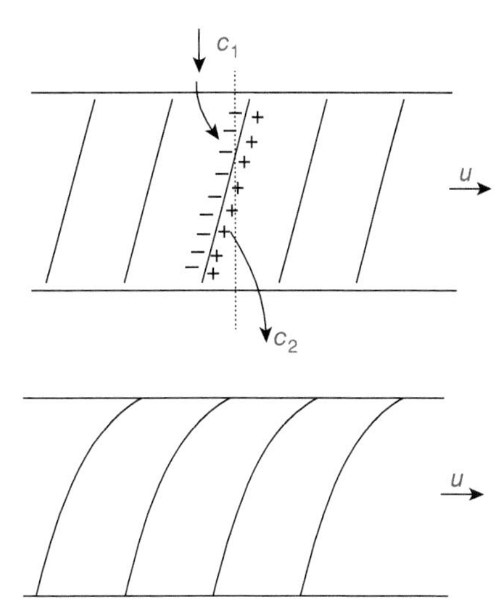

Figure 3.71 Plan view.

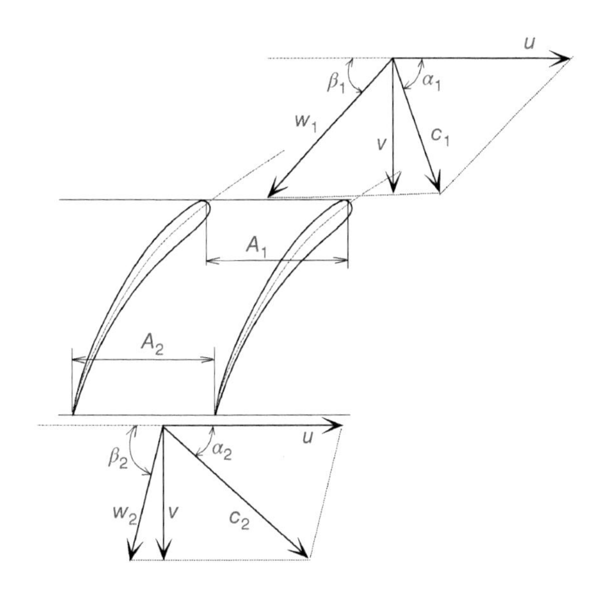

Figure 3.72 Velocity triangles.

The axial component v of the absolute velocity c_1 is:

$$v = c_1 \cdot \sin \alpha_1$$

$$Q_V = z \cdot A_1 \cdot v$$

On outlet:

$$Q_V = z \cdot \overline{A_2} \cdot \overline{c_2} = z \cdot A_2 \cdot c_2 \cdot \sin \alpha_2$$

Because $A_1 = A_2 = A$, scalar-wise as vectorially, and the volumetric flow is the same at inlet and outlet, so for the passage velocity v (Figure 3.72):

$$v = c_2 \cdot \sin \alpha_2$$

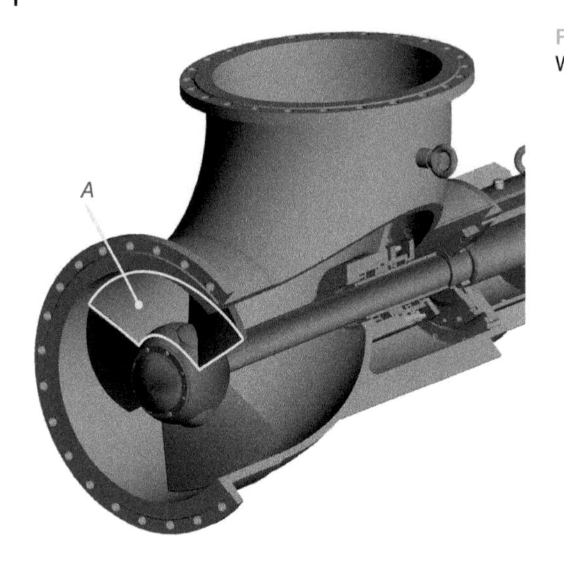

Figure 3.73 Passage area A. Source: Warman.

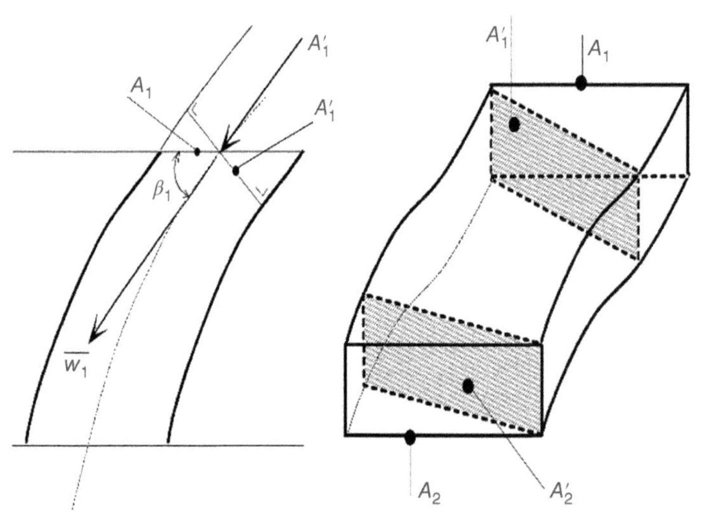

Figure 3.74 Perpendicular surface *A*′.

In other words, the altitude lines v in the velocity triangles are equal; v is called the *axial velocity*.

3.2.2.2 Perpendicular Surface *A*′

Take the standpoint of a relative observer that travels with the channel (Figure 3.74). Consider the surfaces A_1' and A_2' along with the streamlines of $\overline{w}$, met:

$$\overline{w_1}//\overline{A_1'} \text{ and } \overline{w_2}//\overline{A_2'}$$

In Figure 3.74 it is easy to prove that:

$$A_1' = A_1 \cdot \sin \beta_1 \text{ and } A_2' = A_2 \cdot \sin \beta_2$$

Figure 3.75 Channel acting as a diffuser.

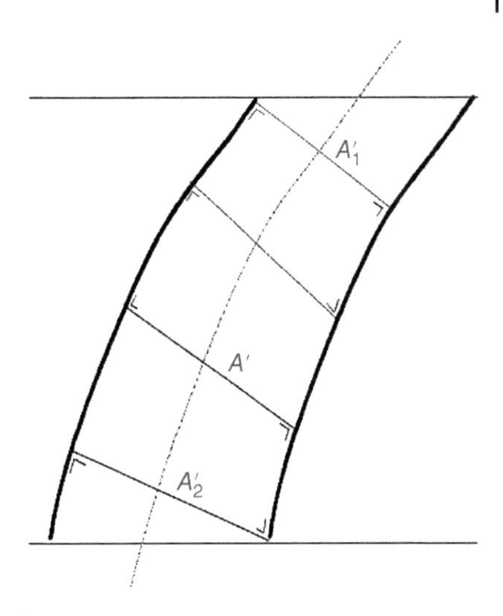

The relative volumetric flow in the vane channel:

$$Q_V = z \cdot \overline{w_1} \cdot \overline{A_1} = z \cdot \overline{w_2} \cdot \overline{A_2}$$

can be written as:

$$Q_V = z \cdot w_1 \cdot A_1 \cdot \sin\beta_1 = z \cdot w_2 \cdot A_2 \cdot \sin\beta_2$$

$$Q_V = z \cdot w_1 \cdot A_1' = z \cdot w_2 \cdot A_2'$$

In Figure 3.75 the surface A' is presented. The form of the vane channel is such that the surface A' gets bigger as the flow reaches the outlet, in other words the channel behaves as a *diffusor*.

$$A_1' < A_2' \text{ and } w_1 > w_2$$

3.2.3 Theoretical Feed Pressure

According to the definition of the theoretical feed pressure:

$$p_{th} = (p_2 - p_1) + \frac{1}{2} \cdot \rho \cdot (c_2^2 - c_1^2)$$

Application of Bernoulli in the moving vane channel:

$$p_2 - p_1 = \frac{1}{2} \cdot \rho \cdot (w_1^2 - w_2^2)$$

Both expressions lead to:

$$p_{th} = \frac{1}{2} \cdot \rho \cdot (w_1^2 - w_2^2) + \frac{1}{2} \cdot \rho \cdot (c_2^2 - c_1^2) \tag{3.18}$$

The cosine rules in the velocity triangle:

$$c_1^2 = w_1^2 + u^2 - 2 \cdot u \cdot w_1 \cdot \cos\beta_1$$

$$c_2^2 = w_2^2 + u^2 - 2 \cdot u \cdot w_2 \cdot \cos\beta_2$$

$$c_2^2 - c_1^2 = w_2^2 - w_1^2 - 2 \cdot u \cdot (w_2 \cdot \cos\beta_2 - w_1 \cdot \cos\beta_1) \tag{3.19}$$

Figure 3.76 Velocity triangles.

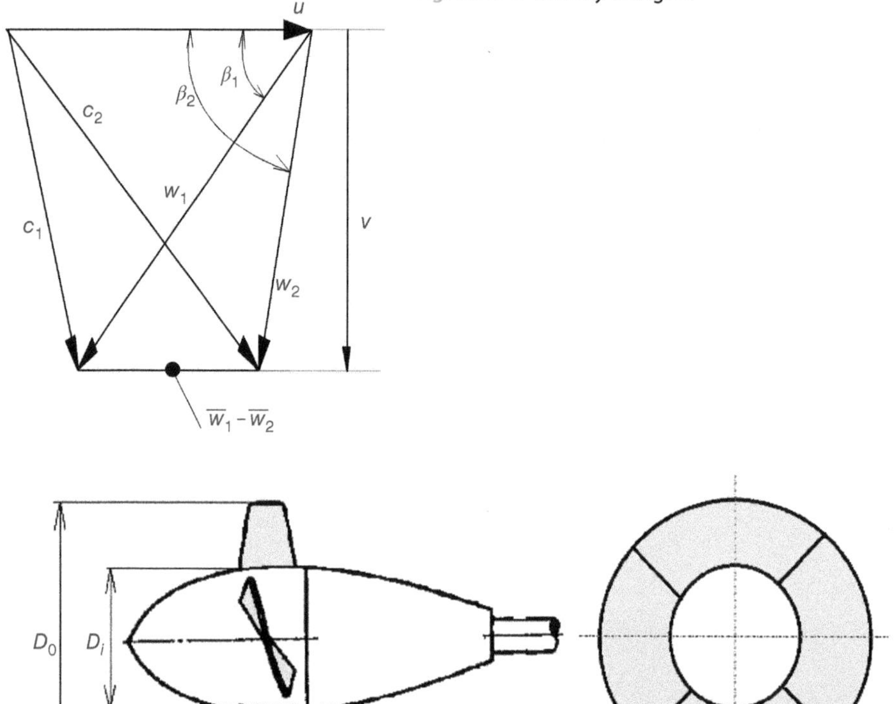

Figure 3.77 Example 3.4.

Equation (3.19) in (3.18):

$$p_{th} = \rho \cdot u \cdot (w_1 \cdot \cos\beta_1 - w_2 \cdot \cos\beta_2) \tag{3.20}$$

Suppose now that the angles β_1 and β_2 are equal. In that case $\overline{w_1}$ and $\overline{w_2}$ (Figure 3.76) are equal. Equation (3.20) teaches us that in that case there is no value for p_{th} (Figure 3.76), in other words a pump with straight vanes wouldn't work. The vanes have to be curved.

From Equation (3.20) and Figure 3.76 it can be derived that the theoretical pressure is uniform with the difference of the vectors:

$$p_{th} \div |\overline{w_1} - \overline{w_2}|$$

Example 3.4

Given an axial pump with vane angles $\beta_1 = 50°$ and $\beta_2 = 70°$. The peripherical velocity u is 10 [m/s]. Water is pumped. Suppose that the flow enters the impeller axially. Dimensions (Figure 3.77): $D_0 = 30$ [cm], $D_i = 15$ [cm].

Determine:

a) The theoretical feed pressure
b) The power needed to drive the pump

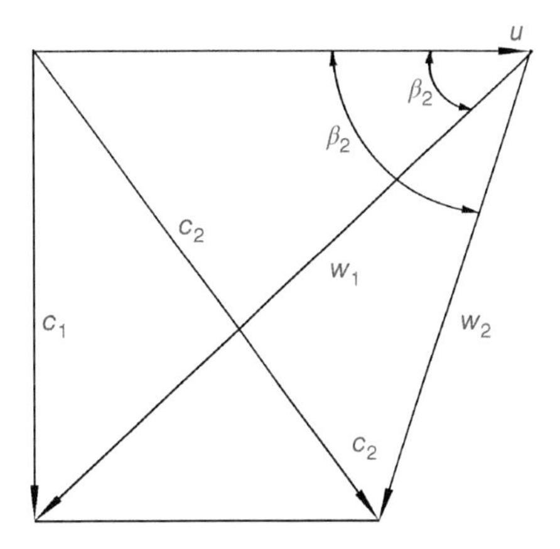

Figure 3.78 Example 3.4.

Solution

From the velocity triangles (Figure 3.78):

$$c_1 = u \cdot \tan \beta_1 = 10 \cdot \tan 50^\circ = 11.92\,[\mathrm{m/s}]$$

$$w_1 = \frac{u}{\cos 50^\circ} = \frac{10}{\cos 50^\circ} = 15.56\,[\mathrm{m/s}]$$

$$w_2 = \frac{c_1}{\sin \beta_2} = \frac{11.92}{\sin 70^\circ} = 12.68\,[\mathrm{m/s}]$$

$$c_2 = \sqrt{w_2^2 + 10^2 - 2 \cdot u \cdot w_2 \cdot \cos \beta_2}$$
$$= \sqrt{12.68^2 + 10^2 - 2 \cdot 10 \cdot 12.68 \cdot \cos 70^\circ} = 13.19\,[\mathrm{m/s}]$$

The static pressure rise in the pump:

$$p_2 - p_1 = \frac{1}{2} \cdot \rho \cdot (w_1^2 - w_2^2) = \frac{1}{2} \cdot 10^3 \cdot (15.56^2 - 12.68^2)$$
$$= 4.067 \cdot 10^4\,[\mathrm{Pa}] = 406.7\,[\mathrm{hPa}]$$

The dynamical pressure rise:

$$\frac{1}{2} \cdot \rho \cdot (c_2^2 - c_1^2) = \frac{1}{2} \cdot 10^3 \cdot (13.19^2 - 11.92^2) = 1.594 \cdot 10^4\,[\mathrm{Pa}] = 159.4\,[\mathrm{hPa}]$$

The theoretical feed pressure:

$$p_{th} = p_2 - p_1 + \frac{1}{2} \cdot \rho \cdot (c_2^2 - c_1^2) = 566.1\,[\mathrm{hPa}]$$

The volumetric flow:

$$Q_V = z \cdot A \cdot v = z \cdot A \cdot c_1 = \frac{\pi}{4} \cdot (D_u^2 - D_i^2) \cdot c_1$$
$$= \frac{\pi}{4} \cdot (0.3^2 - 0.15^2) \cdot 0.92 = 0.63\,[\mathrm{m^3/s}]$$

The technical power on the shaft:

$$P = Q_V \cdot p_{th} = 0.63 \cdot 566.1 \cdot 10^2 = 35.7 \cdot 10^3\,[\mathrm{W}] = 35.7\,[\mathrm{kW}]$$

Looking at the numbers of this example and comparing them with the example in the part on radial turbopumps, it is striking that:

- The radial pump has greater static pressures. This is caused by the centrifugal effect, which is not available with axial pumps.
- The velocities in an axial pump are greater. These pumps are suitable for big volumetric flows.
- The absolute velocities c_1 and c_2 in the centrifugal pump are resp. 3 and 17 [m/s] and are now 12 and 13 [m/s]. This means that the dynamic pressure in a centrifugal pump must be converted to static pressure in a diffusor. This leads to friction losses. The hydraulic efficiency of an axial pump will then be better.

3.2.4 Diffusors

Like with centrifugal pumps a static pressure occurs in the rotor, but also a (small) increase for the velocity c. Therefore the rotor must be followed by a diffusor (Figure 3.79) where dynamic pressure will be converted to static pressure.

Because the velocity c_2 does not leave the rotor axially, it is accompanied with friction losses in the diffusor. This can be avoided by placing a guide wheel (which does not rotate!) that reorients the absolute velocity in an axial direction. As represented in Figures 3.80 and 3.81, it is clear that this guide wheel behaves as a diffusor.

3.2.5 Vane Profile

Axial pumps used to be designed with a lot of vanes, where the vanes lead the flow in its trajectory. A pure geometrical description is applied to the flow.

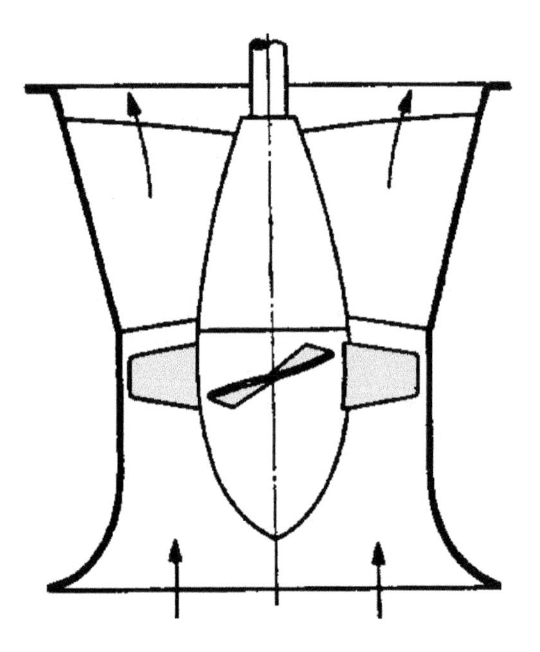

Figure 3.79 Diffusor.

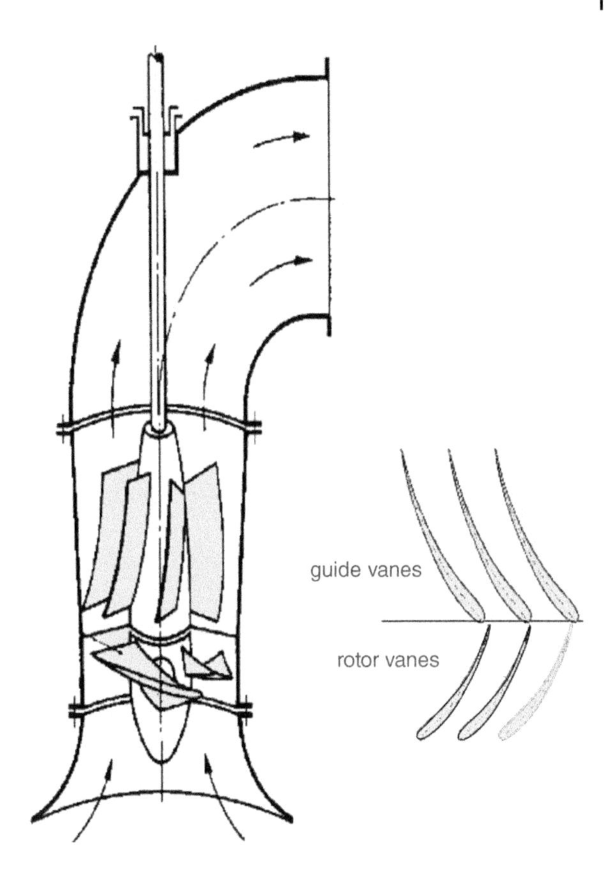

Figure 3.80 Guide wheel.

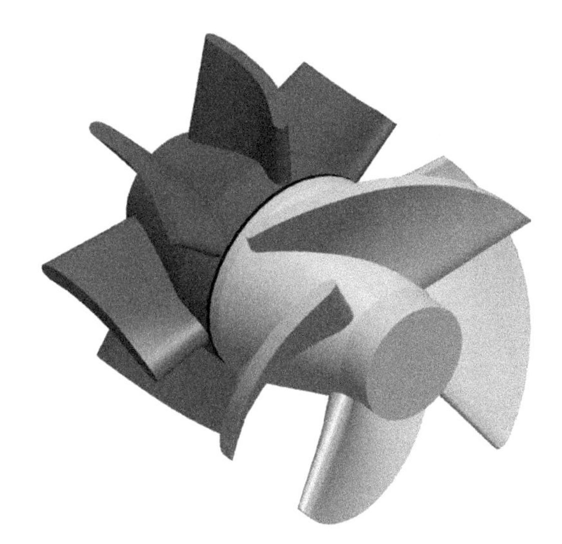

Figure 3.81 Guide wheel. Source: Wikipedia.

Nowadays one encounters axial pumps with only a few vanes (4–6). The flow of the liquid gets the character of a flow round a hydro foil, and an aerodynamic concept can be applied. The vanes are designed aerodynamically in order to produce as little friction as possible.

On the other hand, the foot of the vane is bigger than the top, for considerations of strength.

If one takes into account that the peripheral velocities at the top and the foot of the vane are different the vane angles β on foot and top should be different in order to avoid bump (Figures 3.82–3.84).

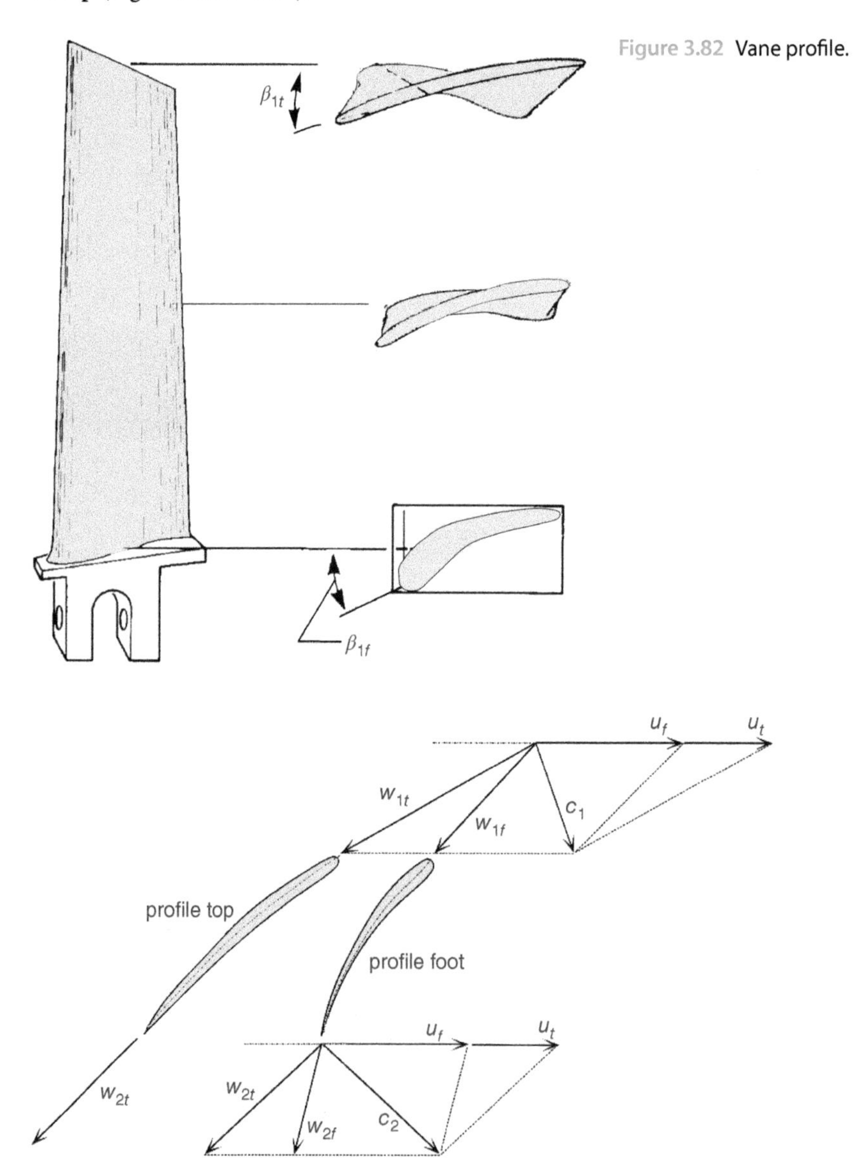

Figure 3.82 Vane profile.

Figure 3.83 Velocity triangles on top and foot of vane.

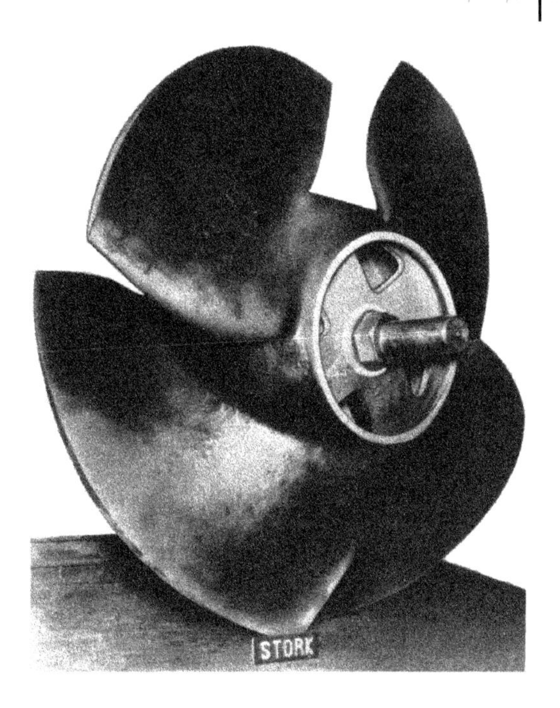

Figure 3.84 Axial pump. Source: Stork, now Johnson Pumps.

3.2.6 Half-Axial Turbopumps

3.2.6.1 Motivation

In axial pumps the centrifugal force plays no role in increasing the static pressure in the rotor. Only diffusor operation leads to a static pressure rise. In this way, the static pressure rise in an axial pump will be lower than that of a comparable centrifugal pump. But, on the other hand, the vanes of an axial pump are much smaller, which means that higher peripherical velocities can be allowed. This leads to higher volumetric flows.

The vane channels of axial pumps are much simpler than those of radial turbopumps. This means that the friction losses in the pump are less and that axial pumps have a better hydraulic efficiency.

Centrifugal pumps can be used for moderate volumetric flows and high pressures. Axial pumps are more adequate for great volumetric flows with moderate pressures (transport of liquids). Sometimes pumps are required that exhibit both properties. These pumps are semi-axial pumps.

3.2.6.2 Francis Vane Pump

The Francis vane (screw) turbopump has vanes that start with a screw form and end with a more radial form (Figure 3.85).

The liquid enters the pump axially and leaves radially. There is a volute. This pump has primarily the properties of a centrifugal pump, so has a reasonable static pressure rise. The simpler form of the vane channel makes the pump work with a higher efficiency than that of a centrifugal pump.

This pump is sometimes referred to as a *screw centrifugal pump*.

Another variant of the axial pump is presented in Figure 3.86. It consists of a screw form impeller that sucks the liquid, then the liquid is thrown away by centrifugal force,

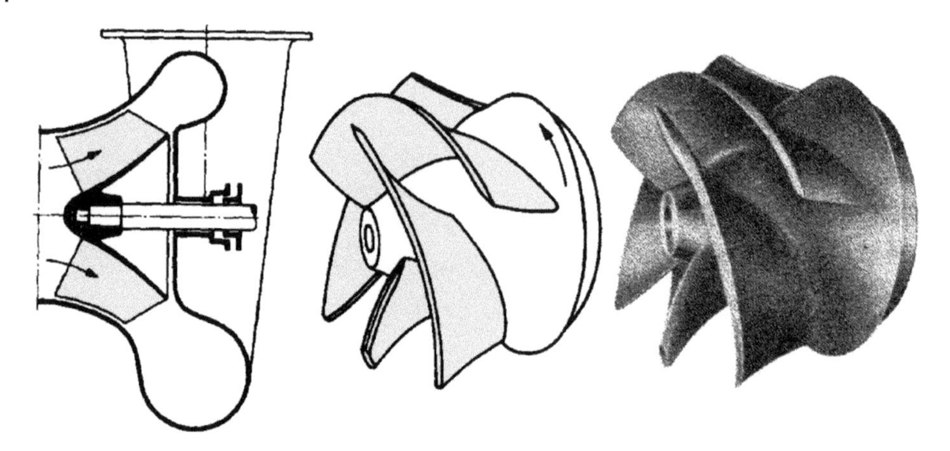

Figure 3.85 Francis vane (Screw) pump. Source: *Les Turbopompes*, by Troskolanski.

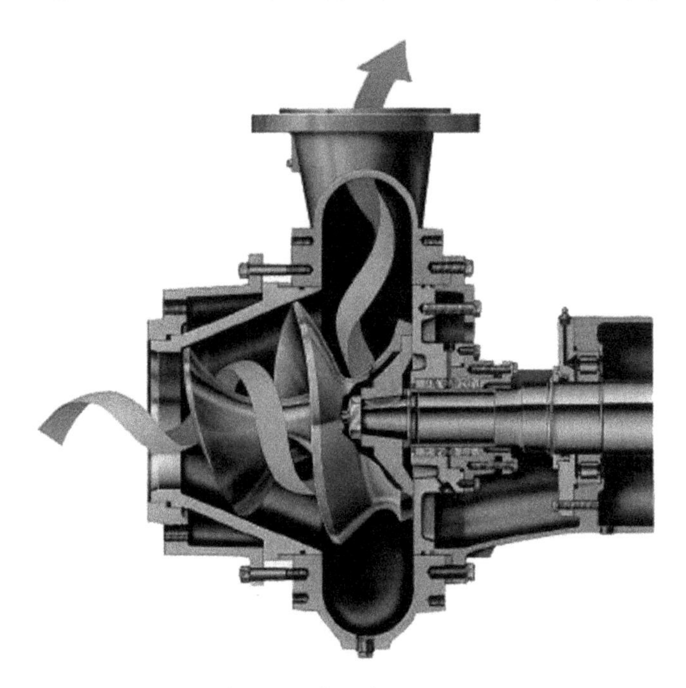

Figure 3.86 Variant. Source: Hidrostal.

thereafter the flow leaves the pump through a volute. A big advantage is the large passage. This pump is suitable for contaminated liquids (Figures 3.87–3.90).

3.2.6.3 Mixed Flow Pump

In a mixed flow pump the flow is predominantly axial, but a little radial component is allowed. This is performed by an increase of the rotor diameter toward the outlet (Figure 3.91).

The liquid enters the pump axially and is led to a guide wheel that functions as a diffuser (Figures 3.92 and 3.93).

Figure 3.87 Francis vane pump. Source: Stork, now Johnson Pumps.

Figure 3.88 Francis vane pump. Source: Hus.

Figure 3.89 Implementation Source: Hidrostal.

Figure 3.90 Implementation. Source: Hus.

In Figure 3.94 the profiles of the different turbopumps are represented: from left to the right there is an evolution of a pure axial flow, the diagonal or mixed-flow pump, the screw of francis-vane pump and at last the centrifugal pump. From the left to the right the volumetric flow and the hydraulic efficiency decrease while the static pressure rises.

3.2.6.4 Characteristics of Turbopumps

The characteristic pump curve of an axial pump is much steeper than that of a radial pump. The power curve decreases with increasing flow, the opposite of a radial pump. Mixed flow pumps have a character that lies behind those two extremes (Figures 3.95–3.97).

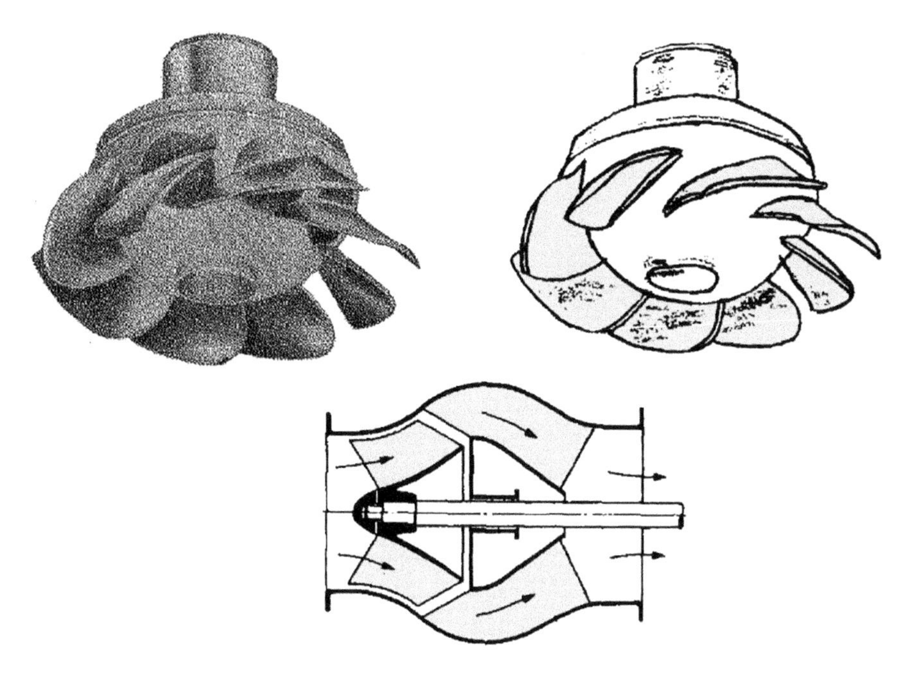

Figure 3.91 Mixed flow pump.

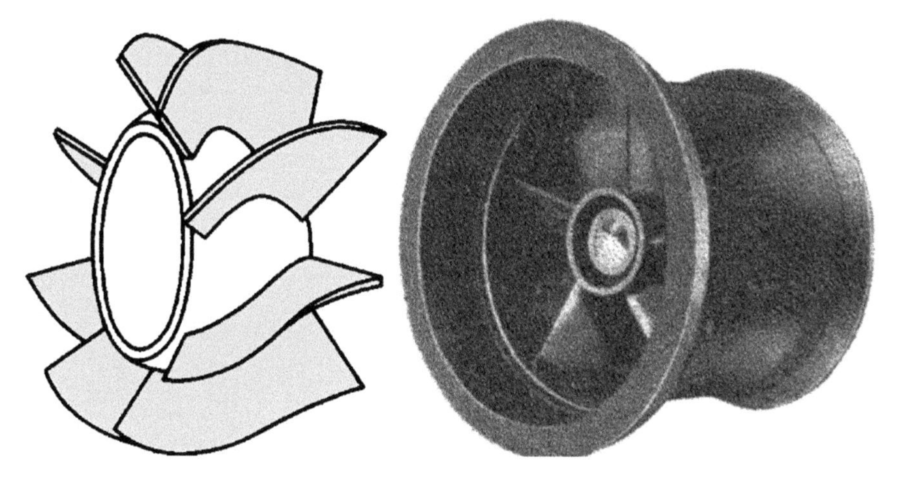

Figure 3.92 Guide wheel.

3.2.7 Archimedes Screw

The Archimedes screw pump consists of an Archimedes screw. This screw is, in fact, a long axial vane rotor.

The liquid moves axially.

These pumps find their application in transporting water, in particular the transport of contaminated water in water purification stations. Because of the big passage that the channel offers the liquid, it is possible to transport solid parts without risk of obstruction.

Screw pumps have a very good hydraulic pump efficiency.

Figure 3.93 Mixed flow pump. Source: Stork, now Johnson Pumps.

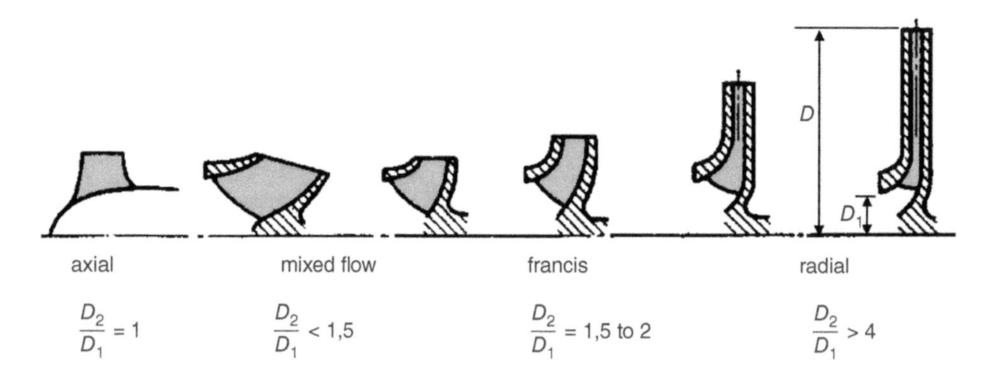

Figure 3.94 Turbopumps.

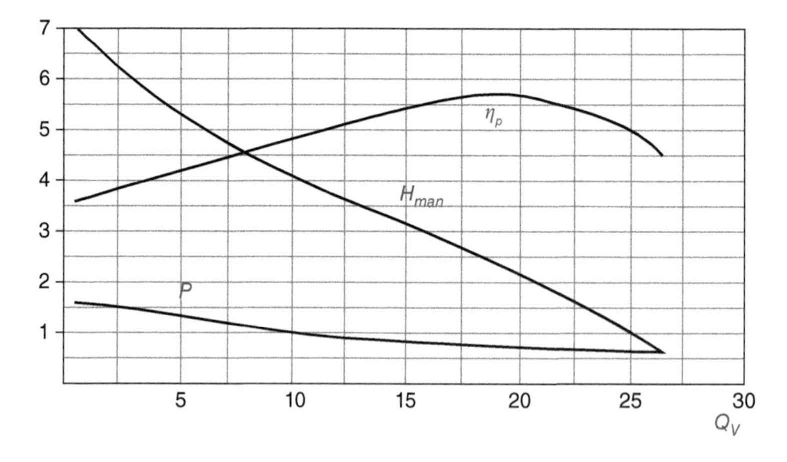

Figure 3.95 Characteristic curves of an axial pump.

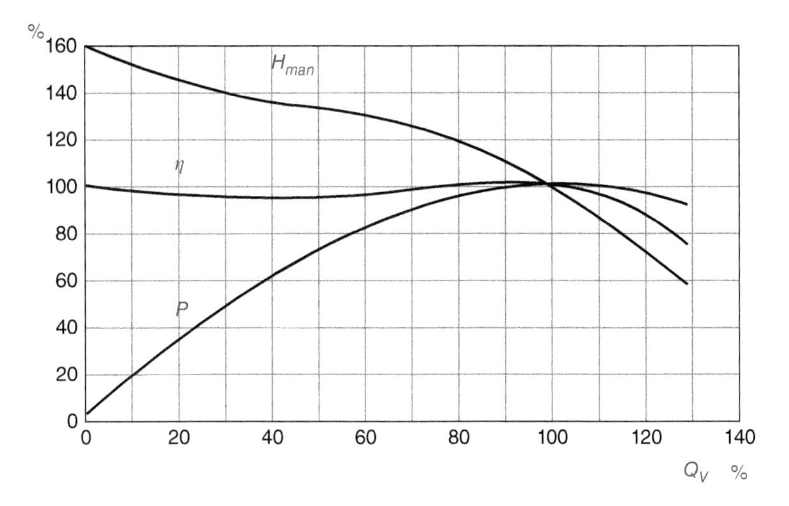

Figure 3.96 Characteristics of a mixed flow pump.

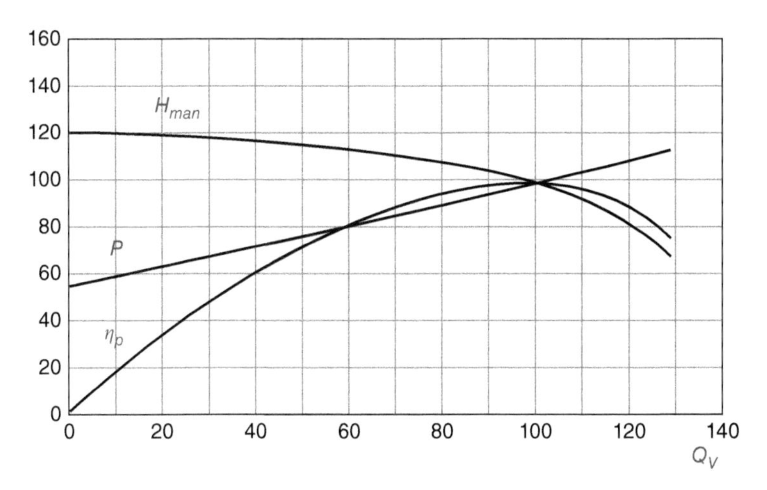

Figure 3.97 Characteristic curves of a radial pump.

In order to keep the wear caused by the solid parts to a minimum, they are driven at speeds from 10 to 100 [rpm].

Flow rates can attain 3000 [l/sec] for an impeller diameter of 3 [m], at manometric feed heads of 10 [m] (Figures 3.98–3.102).

3.3 Turbopumps Advanced

3.3.1 1st Number of Rateau

Consider centrifugal pumps. In Figure 3.103 the velocity triangles are shown.

The circumferential velocity u_2 is uniform with the diameter D_2 of the impeller and with the rotation speed N:

$$u_2 \div D_2 \cdot N$$

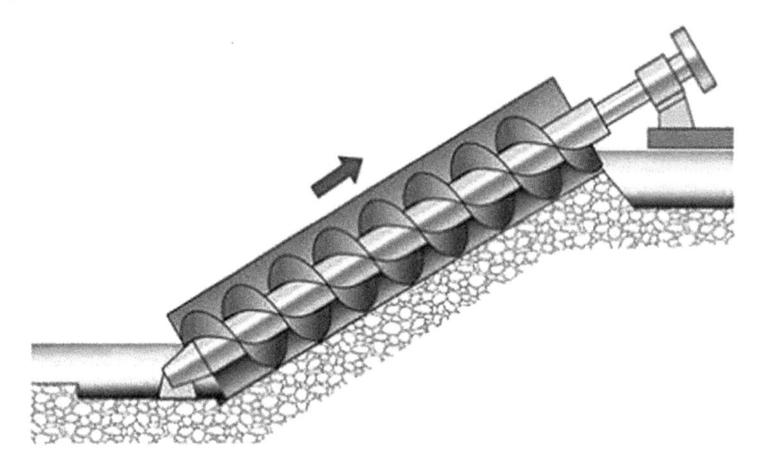

Figure 3.98 Archimedes pump. Source: KSB.

Figure 3.99 Archimedes screw. Source: Dewekom Engineering.

If the rpm of the machine is increased, the circumferential velocity u_2 will increase in proportion. If now the machine – this is a determined design with known angles β_1 and β_2 – is made bigger (thus let D_2 increase) then also u_2 will increase proportionally.

On the other hand, if one way or another u_2 changes then the velocity triangle will change. The velocities w_2 and c_2 will become greater if u_2 increases, and vice versa. But the way in which they become greater (or smaller) depends on the design (β_1 and β_2) and on the duty point of the pump.

That's why:

$c_2 \cdot \sin\alpha_2 \div u_2 \cdot \varphi_1$ (duty point, type pump) where φ_1 means *function of*.

Expressed another way:

$c_2 \cdot \sin\alpha_2 \div D_2 \cdot N \cdot \varphi_1$ (duty point, type pump)

Figure 3.100 Implementation. Source: Dewekom Engineering.

Figure 3.101 Implementation. Source: Dewekom Engineering.

The volumetric flow of the pump is given by:

$$Q_V = z \cdot A_2 \cdot c_2 \cdot \sin \alpha_2$$

Or:

$Q_V = D^3 \cdot N \cdot \varphi_1$ (duty point, type pump)

This function φ_1 is the first number of Rateau:

$\Pi_1 = \frac{Q_V}{D^3 \cdot N} = \varphi_1$ (duty point, type pump)

It is easy to control that Π_1 is indeed a number: it has no dimensions.

3.3.2 2nd Number of Rateau

Analogues the manometric feed pressure can be described:

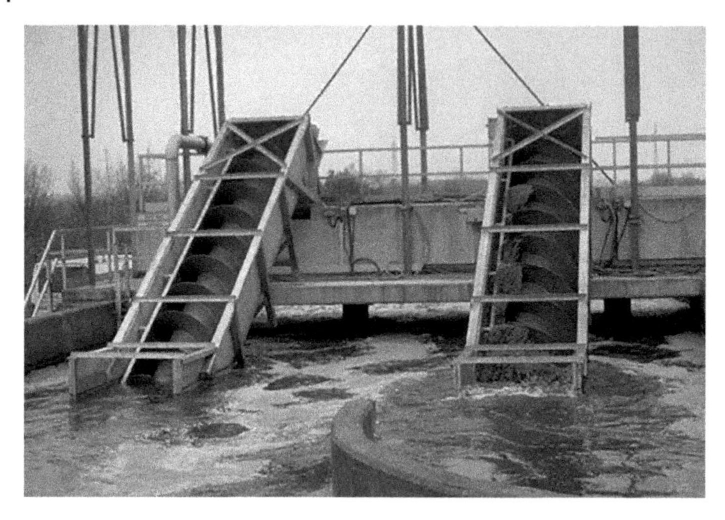

Figure 3.102 Archimedes screw on a site. Source: Dewekom Engineering.

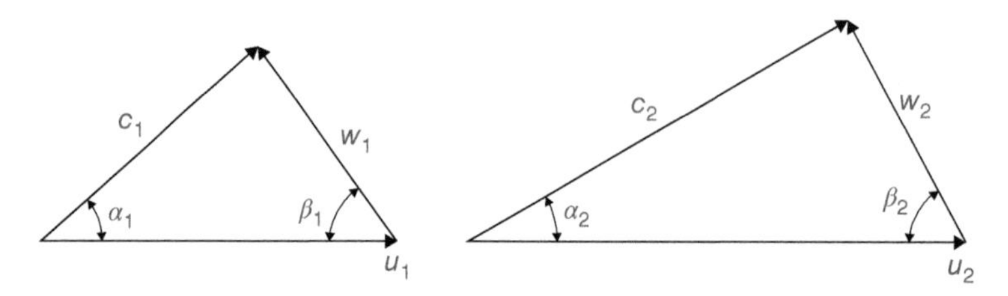

Figure 3.103 Velocity triangles.

$p_{man} = \rho \cdot u_2 \cdot c_2 \cdot \cos\alpha_2 \div \rho \cdot D_2 \cdot N \cdot D_2 \cdot N = \varphi_2$ (duty point, type pump)
Better than p_{man} a value is used that is independent of the used liquid:

$$p_{man} = \rho \cdot g \cdot H_{man}$$

where H_{man} is the manometric head:

$H_{man} \div D_2^2 \cdot N^2 \cdot \varphi_2$ (duty point, type pump)

The function φ_2 is the 2nd number of Rateau:

$\Pi_2 = \frac{H_{man}}{D^2 \cdot N^2} = \varphi_2$ (duty point, type pump)

The dimensions of Π_2:

The dimensions in SI, MKSA are (m stands for meter and s for second)

$[\Pi_2] = \text{m}^{-1}\,\text{s}^2$

So, the name "number" is disturbing.

3.3.3 Homologous Series

Consider a series homologous pumps. These are pumps that can differ qua dimensions (D_2, N) but are all of the same design, thus they have the same vane profile.

Consider first one pump out of this set and draw its characteristic (Figure 3.104).

Of course, if we note in general φ as *function of*:

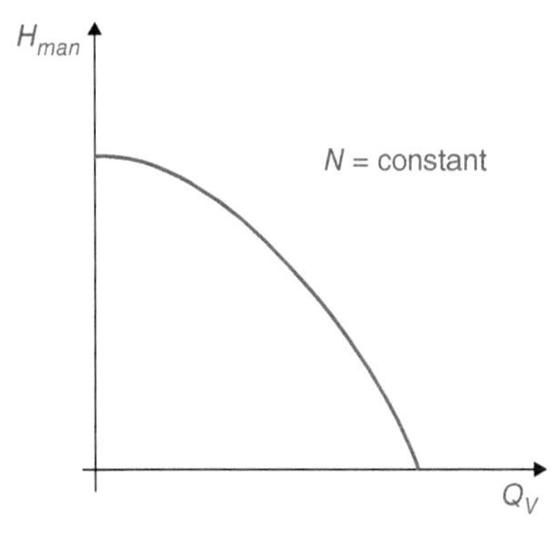

Figure 3.104 Characteristic.

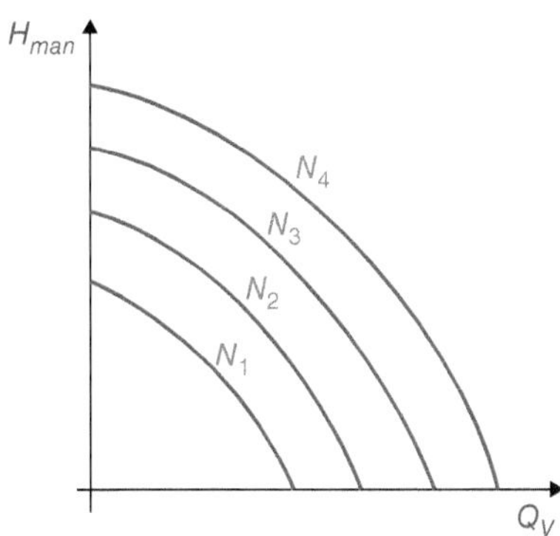

Figure 3.105 Bundle of characteristics.

$H_{man} = \varphi$ (duty point) and $Q_V = \varphi$ (duty point)

As known, we a get bundle characteristics as in Figure 3.105 when the speed is changed with this pump.

We can, however, make one curve of this bundle if the quantities on the axes are differently dimensioned:

According to the above:

$$\frac{H_{man}}{N^2} = \varphi(\text{duty point})$$

$$\frac{Q_V}{N} = \varphi\ (\text{duty point})$$

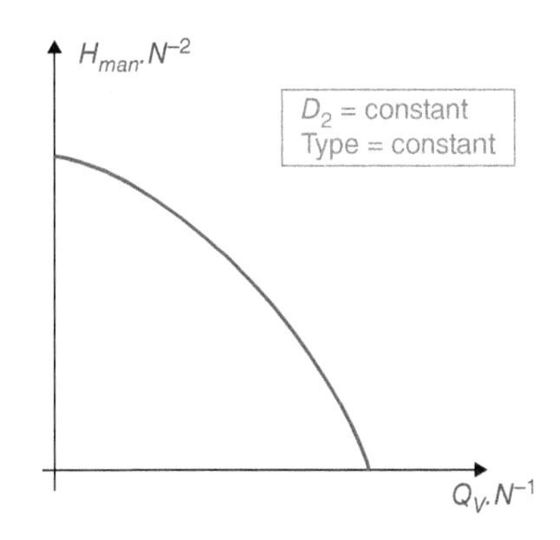

Figure 3.106 Characteristics bundled together.

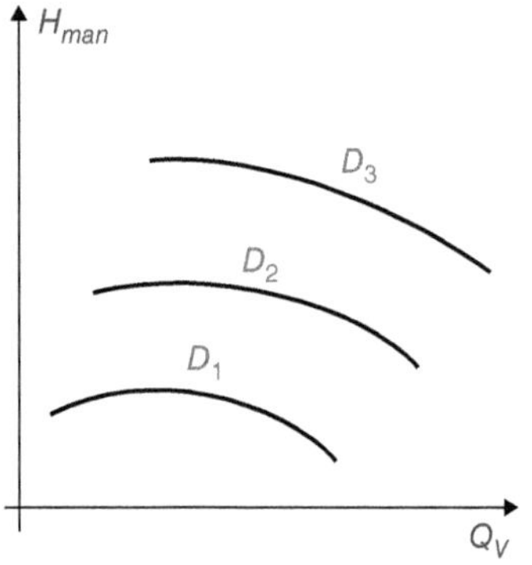

Figure 3.107 Characteristics at various impeller diameters.

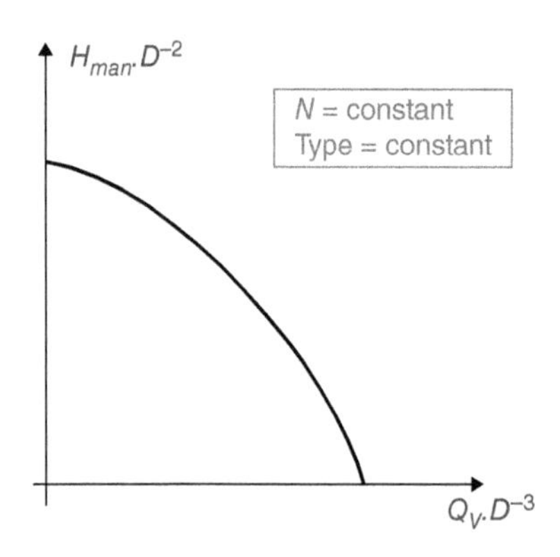

Figure 3.108 One bundle.

This is represented in Figure 3.106.

Stay with the same homologous series of pumps but go a step further. Of the model of pump that was just studied, larger and smaller versions are made by working on D_2. If the pumps work at the same speed, their characteristics will look as they do in Figure 3.107.

In the same way all curves can be balled together into one curve by adjusting the values on the axis:

$$\frac{Q_V}{D^3}$$

See Figure 3.108. Now we get one curve because both values on the axis are only a function of the duty point.

The last step in the reasoning is to consider all pumps in the series together, whereby D_2 and N can vary, alone or together. Following the previous, can, if on the ordinate is placed:

$$\frac{H_{man}}{D^2 \cdot N^2}$$

and on the abscissa:

$$\frac{Q_V}{D^3 \cdot N}$$

Indeed, they will only be functions of the duty point (Figure 3.109).

The efficiency curves can also be put on those diagrams. The dots are the BEP (see Figures 3.110 and 3.111).

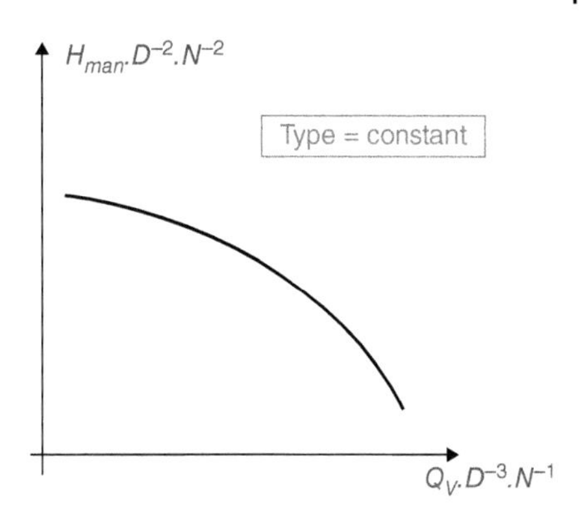

Figure 3.109 Global characteristic.

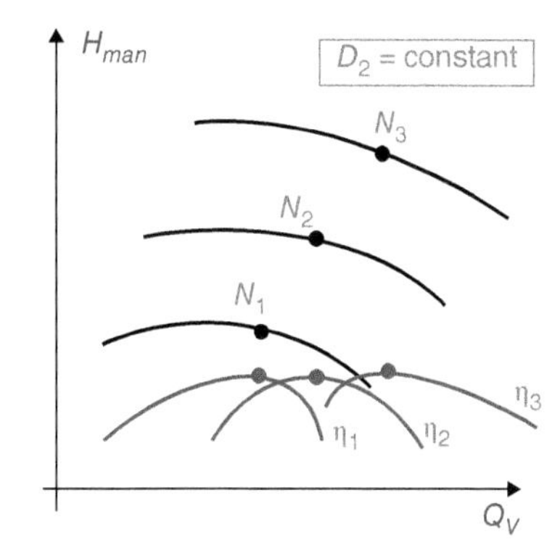

Figure 3.110 Efficiency curves.

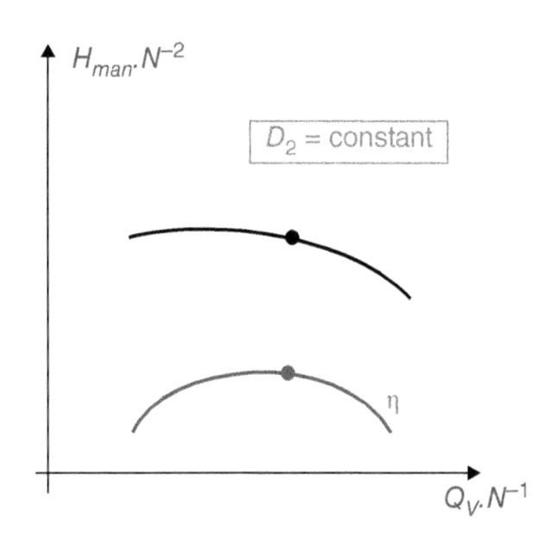

Figure 3.111 Characteristics bundled together.

This graphic can be extended to the $\Pi_1 - \Pi_2$ graphic because it can be accepted that the efficiency does not depend on the dimension of the pump, determined by the diameter. This is confirmed by practice.

The general graphic for a series of homologues pumps is presented in Figure 3.112.

3.3.4 Optimal Homologous Series

In the past those characteristics (as in Figure 3.108) of homologous series of pumps were drawn. It turned out that for one type of pump (a certain centrifugal pump) various

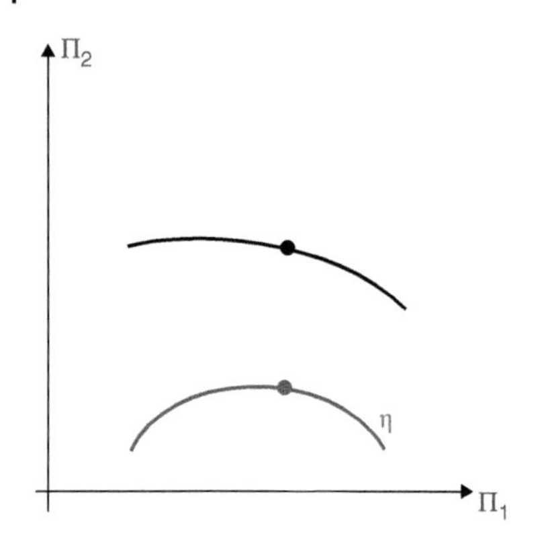

Figure 3.112 General graphic homologous series of pumps.

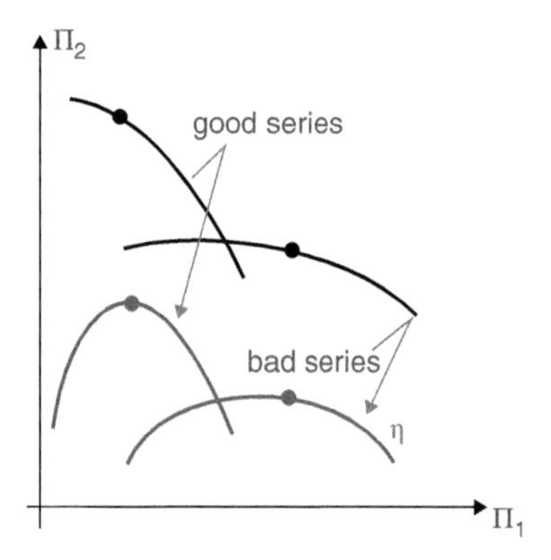

Figure 3.113 Good and bad series of pumps.

characteristics were found on the market. But these series didn't fit at their BEP. In other words, for a series of homologous pumps only one Π_1–Π_2 graphic was the best. Such a series can be represented by two constant value of Π_1 and Π_2 be represented if one takes the values at their BEP (Figures 3.113 and 3.114).

3.3.5 Rateau Numbers with Axial Pumps

Figure 3.114 represents the characteristics for radial, screw (francis-vane), diagonal (mixed flow), and axial pumps.

3.3.6 The Specific Speed

As discussed, if a manufacturer wants to design a whole series of pumps of various dimensions of a determined type, he is confronted with the problem of the dimension D_2.

Therefore, we look for a number that is characteristic for a series of pumps. In this number the variable "diameter" is no more present. This can be done as follows: consider the duty point with coordinates Π_1 and Π_2 (that are constants) and eliminate D_2:

$$\Pi_1 = Q_V \cdot N^{-1} \cdot D_2^{-3}$$

$$\Pi_2 = H_{man} \cdot N^{-2} \cdot D_2^{-2}$$

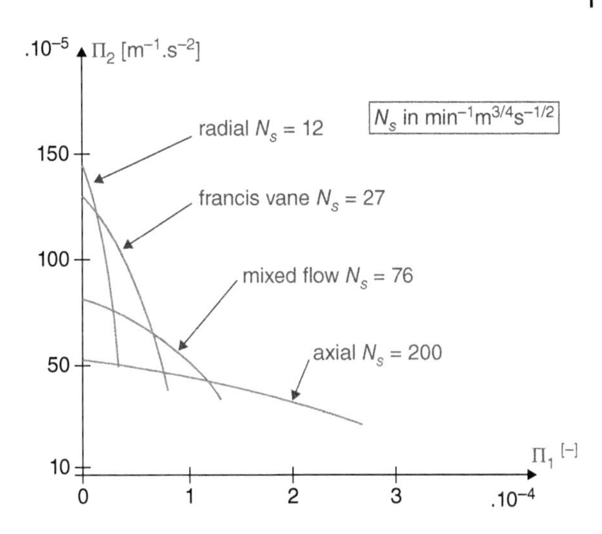

Figure 3.114 Characteristic per pump.

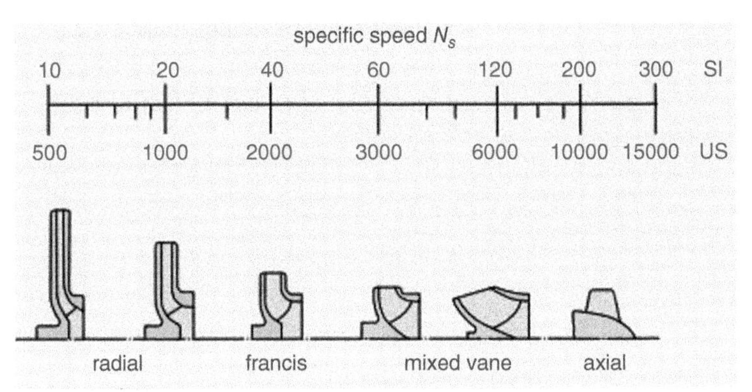

Figure 3.115 Specific speed (SI unit with N in rpm). Source: SKF.

$$(\Pi_1^{1/2}) = (Q_V^{1/2}) \cdot (N^{-1})^{1/2} \cdot (D_2^{-3})^{/2}$$

$$(\Pi_2)^{-3/4} = (H_{man}^{-3/4}) \cdot (N^{-2^{-3/4}}) \cdot (D_2^{-2^{-3/4}})$$

$$\Pi_1^{1/2} \cdot \Pi_2^{-\frac{3}{4}} = H_{man}^{-3/4} \cdot N \cdot Q_V^{1/2}$$

The last equation is a constant for a series of homologous pumps and is called the *specific speed of the pump* (Figure 3.115).

$$N_s = N \cdot Q_V^{1/2} \cdot H_{man}^{-3/4} \tag{3.21}$$

One of the first problems is that this is not a speed at all, in SI and MKSA the units are (where *m* stands for meter and *s* for second):

$[N_s] = \mathrm{m}^{3/4} \cdot \mathrm{s}^{-3/2}$

In the metric system the speed N is mostly expressed in rotations per minute, so that the units are:

$[N_q] = \mathrm{min}^{-1} \cdot \mathrm{m}^{3/4} \cdot \mathrm{s}^{-1/2}$

There are some other definitions:

- Sometimes the angle velocity ω is used instead of the speed N.
- Sometimes US units are used.
- The specific speed can made dimensionless by using in the expression for Π_2 $(g \cdot H_{man})$ instead of H_{man} (where g stands for the gravitation acceleration) and by expressing rotational speed in rad/1:

$$N_\omega = \omega \cdot Q_V^{1/2} \cdot g^{\frac{-3}{4}} \cdot H_{man}^{-3/4}$$

$$[N_\omega] = [\,\omega]] \cdot [Q_V^{1/2}] \cdot [H_{man}]^{\frac{3}{4}} = rad \cdot s^{-1} \cdot m^{\frac{3}{2}} \cdot s^{\frac{-1}{2}} \cdot m^{\frac{-3}{4}} \cdot (s^{-2})^{\frac{-3}{4}} \cdot m^{\frac{-3}{4}}$$

$$[N_\omega] = s^{-1} \cdot s \cdot s^{\frac{3}{2}} \cdot m^{\frac{3}{2}} \cdot m^{\frac{-3}{4}} \cdot m^{\frac{-3}{4}} = \text{(nihil)}$$

When one deals with multistage pumps, the specific speed per stage is calculated.

With US units, gallons per minute is used, for manometric head the feet, and for the speed the rpm. The conversion from US units to SI (with rpm) is made by dividing by 51.68.

When the efficiency of pumps is set out against the specific speed (Figure 3.116) it seems that the various types of pumps can be distinguished.

Sometimes, as mentioned above, one uses the angle velocity ω (in rad/s) instead of the speed N and $g \cdot H_{man}$ instead of H_{man}. In this way, one becomes a dimensionless definition of the specific speed:

$$N_\omega = \omega \cdot Q_V^{\frac{1}{2}} \cdot (g \cdot H_{man})^{-\frac{3}{4}} \tag{3.22}$$

This definition allows for British and SI units to use their own units and to come to the same value of the specific speed, at least if one uses a coherent units system.

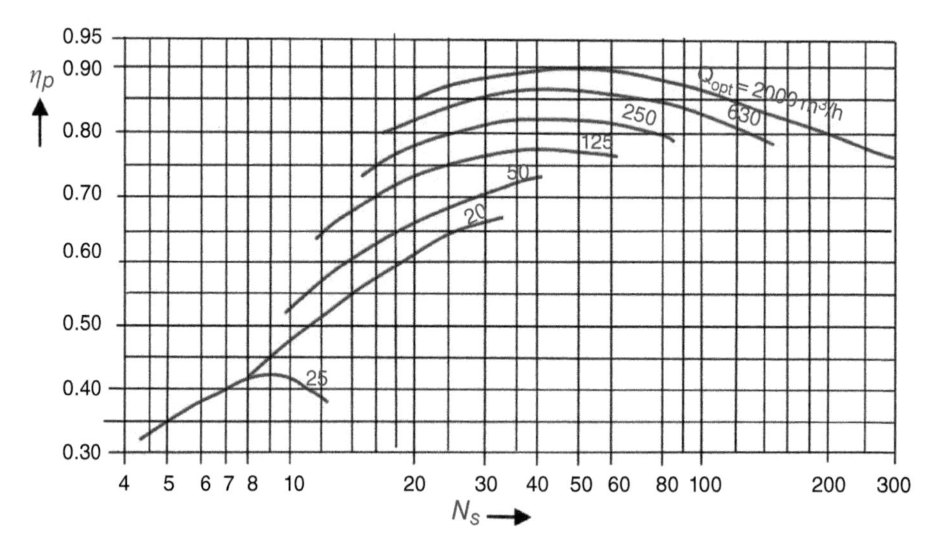

Figure 3.116 Pump efficiency and specific speed.

The numeric ratio between N_ω and N_s is:

$$\frac{N_s}{N_\omega} = \frac{N \cdot Q_V^{\frac{1}{2}} \cdot H_{man}^{-\frac{3}{4}}}{\omega \cdot Q_V^{\frac{1}{2}} \cdot (g \cdot H_{man})^{-\frac{3}{4}}}$$

$$\frac{N_s}{N_\omega} = \frac{N \cdot 60 \cdot (g)^{\frac{3}{4}}}{2 \cdot \pi \cdot N}$$

$$\frac{N_s}{N_\omega} = \frac{30 \cdot (9,81)^{\frac{3}{4}}}{\pi}$$

$$\frac{N_s}{N_\omega} = 52.9$$

Be careful, this is not dimensionless. It has the dimensions of N_s because N_ω is dimensionless.

1 [US gallon] = $3.785 \cdot 10^{-3}$ [m^3]

1 [foot] = 0.3 [m]

1 [hour] = 60 [minutes]

The relationship between the specific speed in US and SI units is:

$1\, N_s(SI) = 1\,[\text{min}^{-1}] \cdot 1\,[\text{m}^3/\text{s}] \cdot 1\,[\text{m}] = \frac{60}{3.785 \cdot 10^{-3}} \cdot \frac{1}{0.3} N_s(US)$

$$1\, N_s(SI) = 50\, N_s(US)$$

The specific speed of a series of homologous pumps was determined at BEP. This number characterizes a determined series of turbopumps in their BEP. The specific speed is determined at optimal volumetric flow and optimal manometric head.

Take an example. Consider a turbopump rotating at an rpm of 1450 and at its BEP has a volumetric flow of 2000 [l/min] at a manometric head of 10 [m].

Calculate the specific speed N_s of the pump:

$N_s = 1450 \cdot \left(\frac{2}{60}\right)^{\frac{1}{2}} \cdot (10)^{\frac{-3}{4}} = 47$

Suppose now that with a frequency regulator the speed is changed to 11 000 [rpm]. According to the affinity laws, the flow and the head will be:

$$\frac{Q_{v,1}}{Q_{v,2}} = \frac{N_1}{N_2}$$

$$\frac{H_{man,1}}{H_{man,2}} = \left(\frac{N_1}{N_2}\right)^2$$

Where *1* is the state at speed 1450 rpm and state *2* at 1100 [rpm].

where from:

$$Q_{v,2} = \frac{N_2}{N_1} \cdot Q_{v,1} = \frac{1100}{1450} \cdot \frac{2}{60} = 0.0253\,[\text{m}^3/\text{s}]$$

$$H_{man,2} = \left(\frac{N_2}{N_1}\right)^2 \cdot H_{man,2} = \left(\frac{1100}{1450}\right)^2 \cdot 10 = 5.76\,[\text{m}]$$

For the specific speed:

$$N_{s,2} = N_2 \cdot Q_{V,2}^{\frac{1}{2}} \cdot H_{man,2}^{-\frac{3}{4}} = 1100 \cdot (0.0253)^{\frac{1}{2}} \cdot (5.76)^{\frac{-3}{4}} = 47$$

Figure 3.117 Attacked rotor centrifugal pump by cavitation. Source: Utwente.

Eventually the same value becomes as before. Why does the change of speed of the pump have no influence on the specific speed? This is because use is made of the affinity laws. But these laws say only what will be the characteristic of the pump at another speed. That does not mean that this has a relation with the new duty point, because this one is dependent on the system curve!

This specific speed characterizes the pump in its BEP, not in other duty points. In practice the new duty point will probably work with a worse efficiency. The specific speed is a measure of the BEP and its efficiency.

3.3.7 Cavitation

Cavitation is caused when the static pressure of a liquid decreases below a critical value that is close to the vapor pressure. In areas of low static pressure, this is on the suction side; gas bubbles will form: they are dragged with the pumped liquid to areas of high pressure. There they implode. This comes down to a rapid (ms) disappearance of the bubbles and local pressure peaks that can attain 20 000 [bar]. The fixed walls of the pump housing and the vanes are eroded by pressure peaks of 20 [kHz] that eat the surfaces.

Cavitation is a continuous process. At low damage the surfaces exhibit a porous to spongey structure. The flow and the efficiency will decrease. At serious damage the impeller can be completely destroyed after a few days.

Cavitation can be remarked by:

- Decrease and fluctuation of pressure and flow.
- *Vibrations and noise*: from a light rustle to a heavy sputter. The noise sounds like hammer beats (metal on metal) and can be clearly distinguished from air bubbles in the pumped liquid.

Figures 3.117 and 3.118 show for radial and axial turbopumps the places that can be attacked. Cavitation is not limited to these pumps. It can occur with other pumps too.

To prevent the liquid from evaporating it must be ensured that the pressure losses in the suction pipe on one side and the suction pressure on the other side are not too high. Pumping hot liquids will generally be done by letting the liquid flow to the pump.

Practically, the value of a technical concept, the NPSH (net positive suction head) determines the allowed suction head for a determined pump (Figure 3.118).

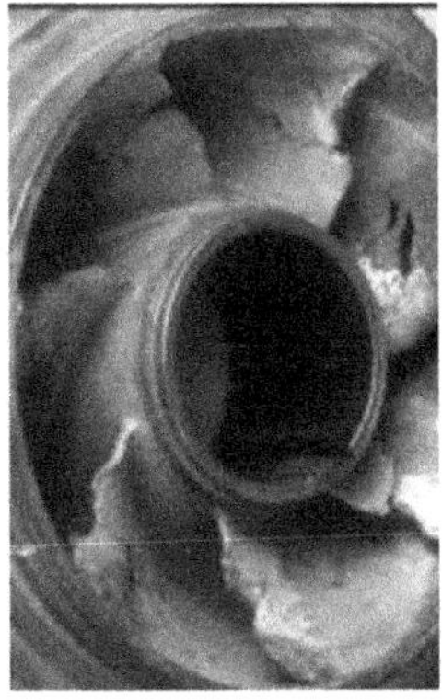

Figure 3.118 Axial pump damaged by cavitation. Source: Wikipedia and Utwente.

The static pressure differential up- and downstream of clearance gaps – especially the impeller clearance gap between the pump casing and the vane tips of open axial and mixed flow impellers (not fitted with outer shroud) and the clearances at the suction eye of closed radial and mixed flow impellers – may, in combination with the sharp edges at the clearance gap, result in extremely high local flow velocities. As a consequence, a correspondingly low static pressure develops in the clearance gap which may decrease as far as the vapor pressure of the fluid, even if the $NPSH_a$ (see next section) is sufficient.

3.3.8 NPSH

Consider a centrifugal pump. At the suction flange is the static pressure p_1' and on the discharge flange the static pressure p_2'. The progress of the static pressure in the pump is shown in Figure 3.119. On the figure the static vapor pressure p_v of the liquid, at the current temperature of the liquid, is set out. It is possible that this vapor pressure line intersects the pressure in the pump. In the case of intersection there is cavitation.

The static pressure p_1' can be calculated by the user, but the pressure just before the impeller cannot be calculated. Why? When the liquid enters the impeller via the suction

Figure 3.119 Mounting (after pompendokter.nl).

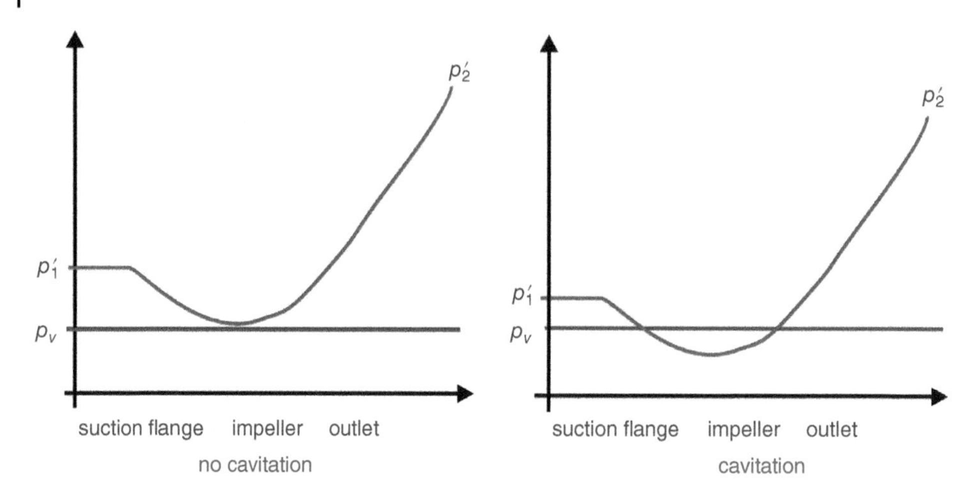

Figure 3.120 Suction pressure.

flange it will on its way to the impeller undergo some pressure losses. How much precisely the static pressure p_1' at the suction flange should be in order to prevent cavitation must be determined by the manufacturer.

A first concept is given by the NPSH. The user calculates *the total* pressure at the inlet of the suction flange and converts this pressure into meters, just by dividing by $\rho \cdot g$ (Figure 3.120).

The user calculates the available $NPSH_a$, where the index *a* stand for *available*. The user will thus deliver the liquid with a certain total pressure, expressed in meters, and that is the pressure that he makes available at the suction flange of the pump.

Good. But that doesn't mean yet that no cavitation will occur. That's why the manufacturer must give the required pressure so that no cavitation will occur. And that is given by the concept $NPSH_r$, the required pressure, where the index *r* stands for *required*. The manufacturer must perform some complicated measurements in its pump, whereby the inlet pressure loss is measured. This is the pressure lost from the suction flange to the lowest pressure at the inlet of the impeller.

Practically it comes down to the fact that the available $NPSH_a$ must always be greater than the required $NPSH_r$, in order to avoid cavitation. Mostly a safety margin is taken:

$$NPSH_a > NPSH_r + 0.5\,[\text{mlc}]\ (\text{meter liquid column})$$

Consider the situation in Figure 3.121.

The *static* pressure p_1' at the suction flange is given by:

$$p_1' = p_a - \rho \cdot g \cdot H_s - p_{fs} - \frac{1}{2} \cdot \rho \cdot c_1'^2$$

The *total pressure* at the suction flange:

$$p_1' + \frac{1}{2} \cdot \rho \cdot c_1'^2$$

If the vapor pressure is subtracted from the total pressure, the resultant pressure can be used to overcome the losses at the entrance of the pump, thus between the suction flange and the lowest pressure at the entrance of the impeller.

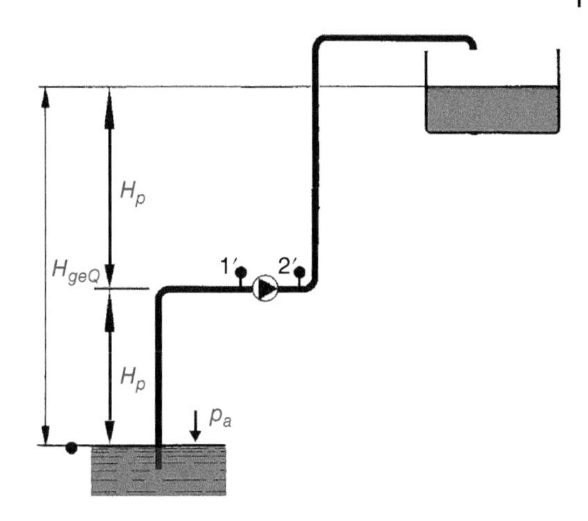

Figure 3.121 Pump installation.

This is the net available pressure:
$p'_1 + \frac{1}{2} \cdot \rho \cdot c'^2_1 - p_v$
And if this is converted to the unit meter we get the $NPSH_a$, expressed in height, and independent of the kind of liquid, because the variable ρ does not occur in the expression:

$$NPSH_a = \frac{p'_1}{\rho \cdot g} + \frac{1}{2} \cdot \frac{c'^2_1}{g} - \frac{p_v}{\rho \cdot g}$$

The manufacturer indicates by its value of the $NPSH_r$ how much pressure loss there will be at entrance in the pump. It is, of course, the intention that the user has a greater value of its $NPSH_a$

Also, with piston pumps and rotating displacement pumps cavitation problems occur. With piston pumps the acceleration losses must be taken into account. In general, it is so that the user consult sits manufacturer, or uses software provided by the manufacturer.

3.3.9 NPSH Characteristics

In Figure 3.122 pump characteristic with efficiency curves, power curves for various diameters. The *NPSH* is only one line.

In Figure 3.123 the characteristic of a centrifugal pump is given, at various speeds. Here we see that the *NPSH* is dependent on the motor speed.

3.3.10 Counteracting Cavitation

The appearance of cavitation can be heard by the typical noise of an irregular flow. What can one do?

- The vapor pressure p_v decrease by lowering the temperature.
- The pressure at the suction flange can be increased:
  - Increasing the level of the suction reservoir or putting the pump lower
  - Increasing the pressure in the suction reservoir
- Decreasing the friction in the suction pipe line by increasing the diameter of the suction line, by decreasing the flow, cleaning the suction filter...

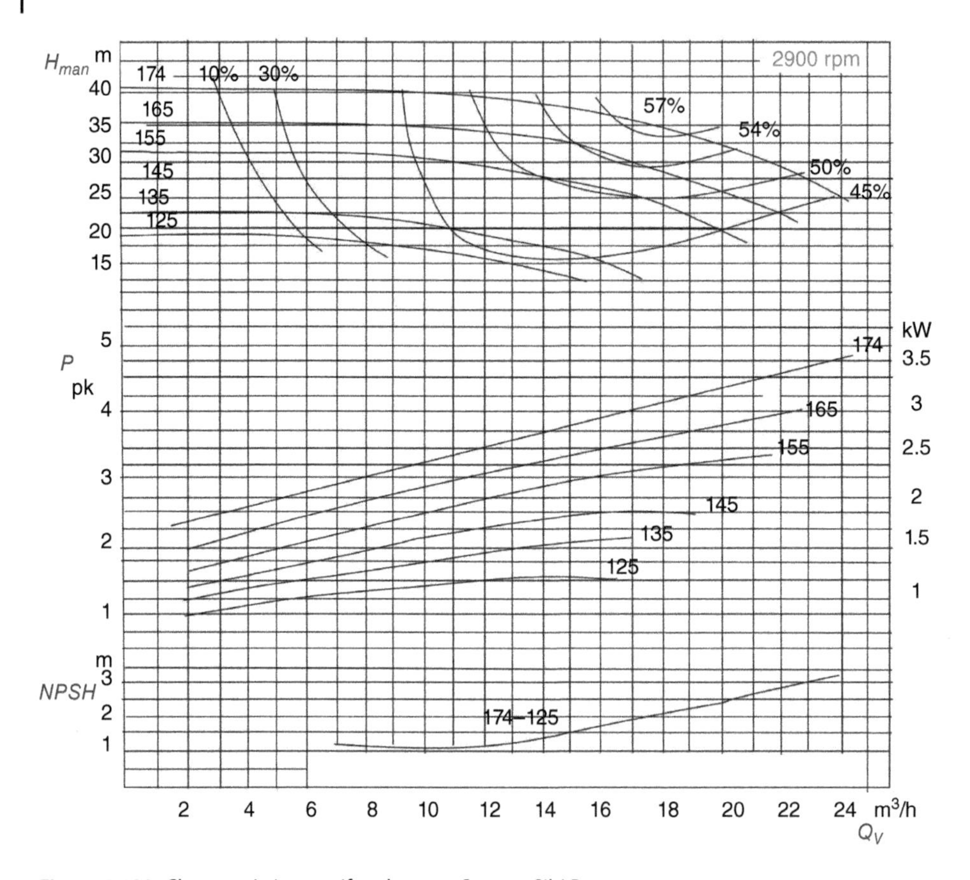

Figure 3.122 Characteristic centrifugal pump. Source: Sihi Pumps.

- Never place a throttle valve in the suction line for regulation of the flow, always in the discharge line (be careful: this is only valid for non-displacement pumps).
- Changing the flow so that the $NPSH_r$ becomes smaller.
- Changing the Cr pump: choose for a pump with a lower $NPSH_a$, choose for a lower speed and thus a bigger pump.
- In some cases, cavitation is not excluded. In that case one has to choose sustainable materials like CR, CrNi, or CrNiMo cast iron. Plastic is a less sensible choice for cavitation than metal.

3.3.11 Inducers

Impellers with a specific speed lower than 38 are equipped with an inducer (Figures 3.124–3.127). This is a spiral form impeller on the suction side of the impeller. The *NPSHr* characteristic is lowered and thus the cavitation sensibility is decreased. This works in an area between 20 and 90% of the BEP. Be careful: outside this area the $NPSH_r$ characteristic becomes worse than without an inducer.

An inducer is an axial impeller with vanes that form a helicoidal impeller. Such an inducer functions as a booster pump for the main impeller. Mostly the inducer has two to four blades. The inducer gives the sucked liquid enough pressure so that the main

Kennlinien zu Abwasser-Kreiselpumpe
Characteristic curves for effluent pump
Courbe pour pompe à large passage

3920 DN 200

Zweikanalrad
Two-channel impeller
Roue à deux canaux

Kennlinie Curve / Courbe	Drehzahl (1/min) R.p.m. / Tr/min	Freier Durchgang Max. size solids / passage roue
57 388	960–1760	45 x 65 mm

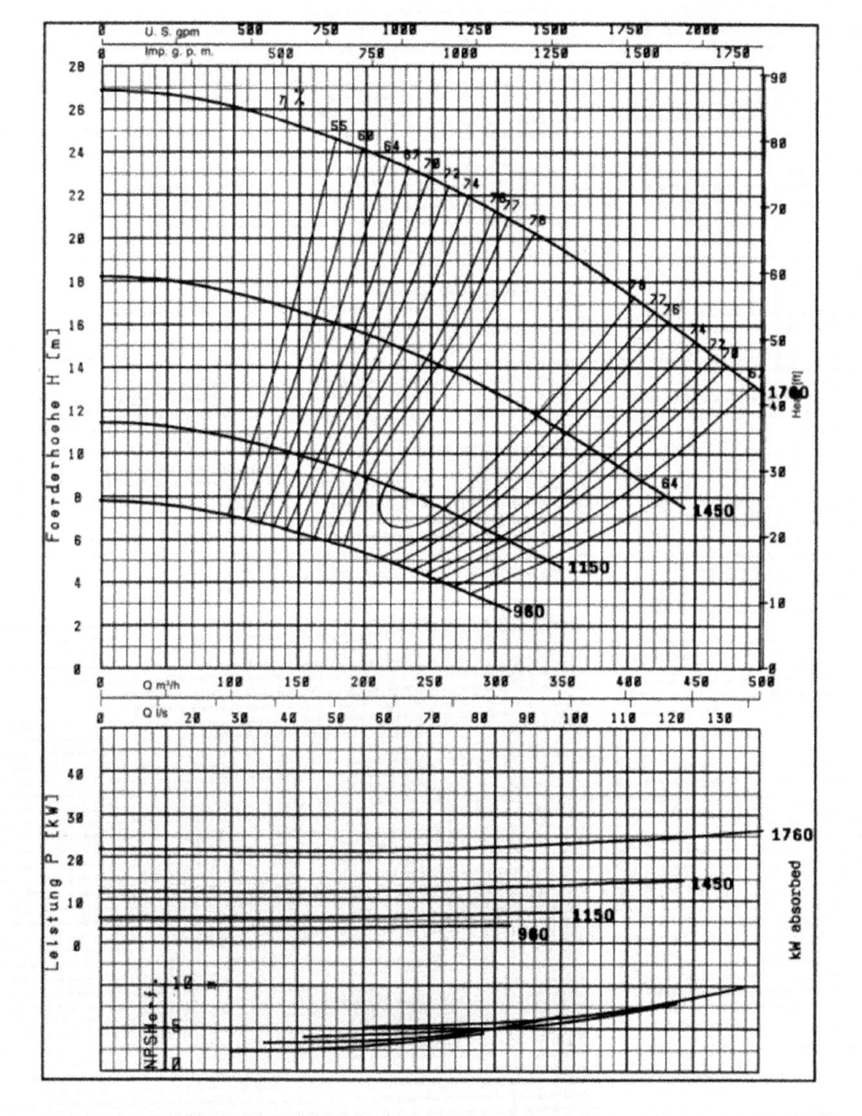

Gültig für p=1 kg/l und Kinematische Zähigkeit bis max. 20 mm²/s. Förderwert und Wirkungsgradgarantie nach DIN 1944, III.
Valid for p=1 kg/l and kinematic viscosity up to max. 20 mm²/s. Delivery capacity and efficiency guarantee according to DIN 1944, III.
Valable pour p=1 kg/l et viscosité cinématique jusqu'à maxi. 20 mm²/s. Garantie du et rendenement suivant DIN 1944, III.

Figure 3.123 Characteristic centrifugal pump at various speeds. Source: Andritz.

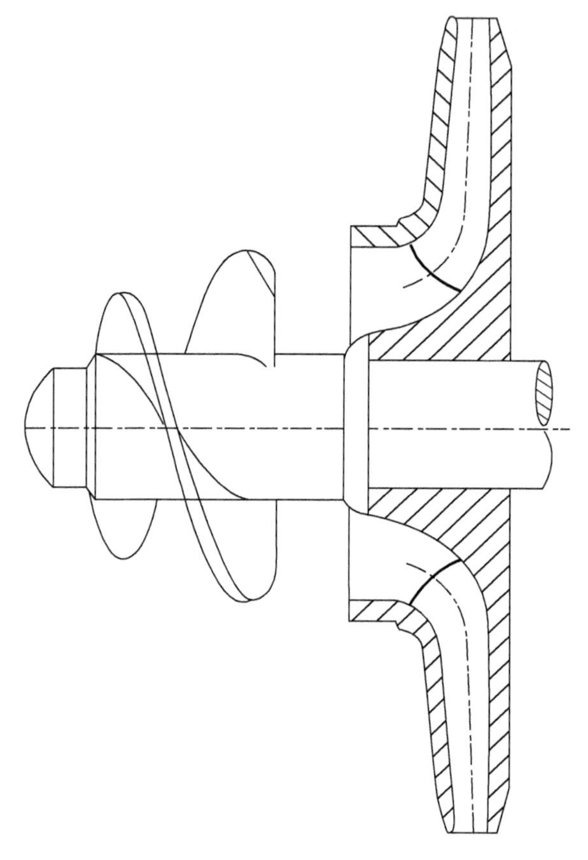

Figure 3.124 Inducer before impeller.

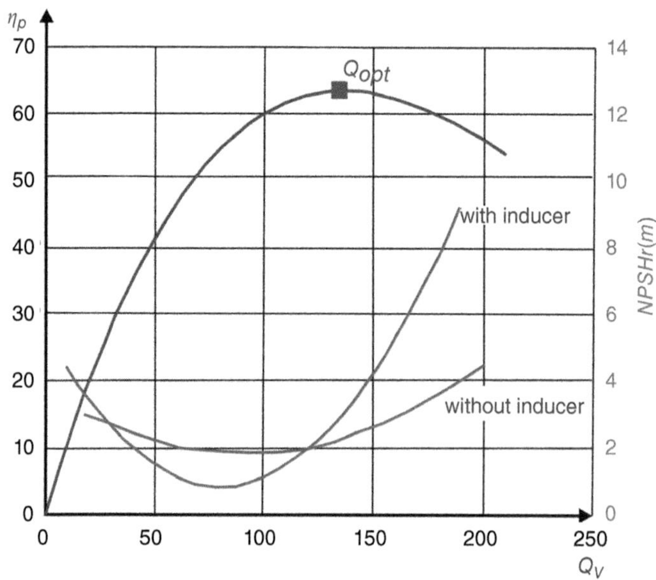

Figure 3.125 Characteristic with and without inducer.

Figure 3.126 Inducer. Source: Lawrence Pumps.

Figure 3.127 Inducers.

impeller is satisfied with its delivered pressure. Although the inducer has a lower $NPSH_r$ than the main impeller, it can, and mostly it is so, cavitate under normal conditions.

The interesting thing is that the axial impeller consumes little power and makes practically no noise, or vibrations, and the mechanical problem that comes with that. Meanwhile, the main impeller receives a sufficient high pressure so that it can work without cavitation.

Figure 3.128 Double sided inlet. Source: Rateau.

3.3.12 Double Sided Entry

In Figure 3.128 a double-sided centrifugal pump is shown. In a pump the liquid exercises a force, because of its pressure at the discharge side, on the rotor. This axial force is transmitted to the bearing. An axial bearing is necessary. In a doubled-sided pump the liquid exercises an axial force in both directions on the rotor. No axial bearing is then necessary.

3.3.13 Characteristics of Pumps

This section looks for the characteristic curves in an $H_{man} = \varphi(Q_V)$ diagram for constant specific speed.

Because:

$$N_s = N \cdot Q_V^{1/2} \cdot H_{man}^{-3/4}$$

$$H_{man} = (N_s)^{\frac{-4}{3}} \cdot (N)^{\frac{4}{3}} \cdot \left((Q_V)^{\frac{1}{2}}\right)^{\frac{4}{3}} = (N_s)^{\frac{-4}{3}} \cdot (N)^{\frac{4}{3}} \cdot (Q_V)^{\frac{2}{3}}$$

As:

$$N \div Q_V$$

And:

$N_s = \text{constant}$

$$H_{man} \div (Q_V)^{\frac{4}{3}+\frac{2}{3}} \div Q_V^2$$

At constant specific speed the characteristic is a parabola through the origin (Figure 3.129). For different values of the specific speed there are different parabolas. These parabolas were already drawn on the pump characteristic in the chapter on radial turbopumps, but were not explained (in Section 3.1.15).

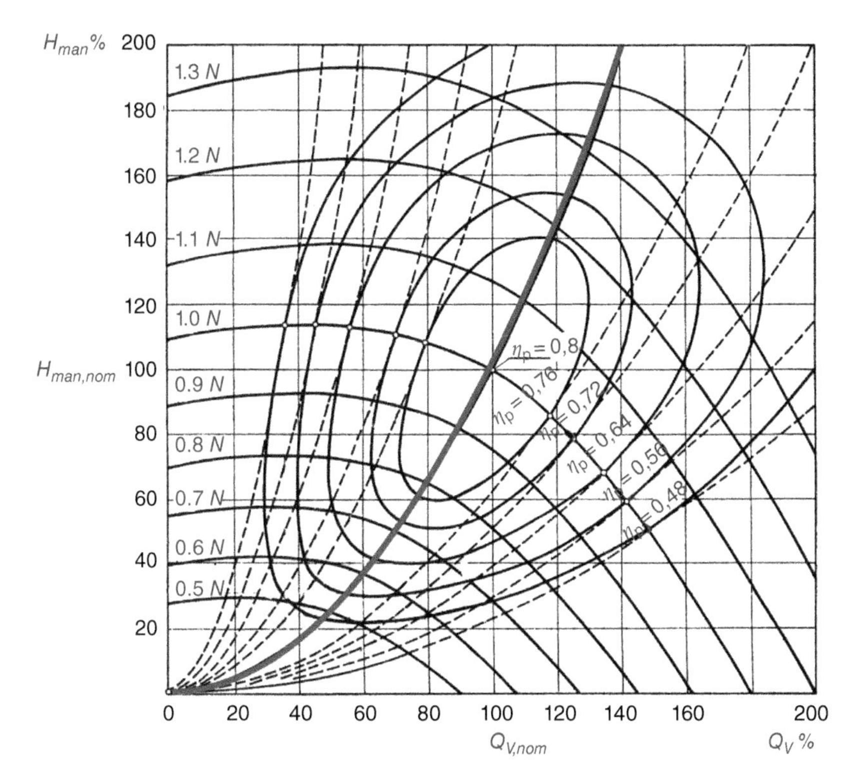

Figure 3.129 Dashed lines are parabola for changing speed N and constant specific speed.

3.3.14 Suction Specific Speed

Suction specific speed – N_{ss} – can be useful when evaluating the operating conditions on the suction side of pumps. Suction specific speed is used to determine what pump geometry – radial, mixed flow, or axial – to select for a stable and reliable operation with maximum efficiency without cavitation. N_{ss} can also be used to estimate safe operating ranges.

The suction specific speed can be calculated as:

$$N_{ss} = N \cdot (Q_V)^{\frac{1}{2}} \cdot (NPSH_r)^{\frac{-3}{4}}$$

As a rule of thumb the specific suction speed should be *below* 180 to avoid cavitation and unstable and unreliable operation. Empirical studies indicate that safe operating ranges from the BEPs are narrower at higher suction specific speeds.

For a double suction pump the flow is divided by two (Figure 3.130).

$$N_s = \frac{\sqrt{\frac{Q_V}{2}}}{H_{man}^{\frac{3}{4}}} \cdot N \, N_s = \frac{\sqrt{Q_V}}{H_{man}^{\frac{3}{4}}} \cdot N$$

At first sight a large eye diameter seems beneficial. The flow velocity is smaller, resulting in a higher localized static pressure. This gives a safety margin against cavitation (Figure 3.130). This is why a larger suction pipe at the pump inlet seems a good idea. This can be seen at the value of the $NPSH_r$ (Figure 3.131).

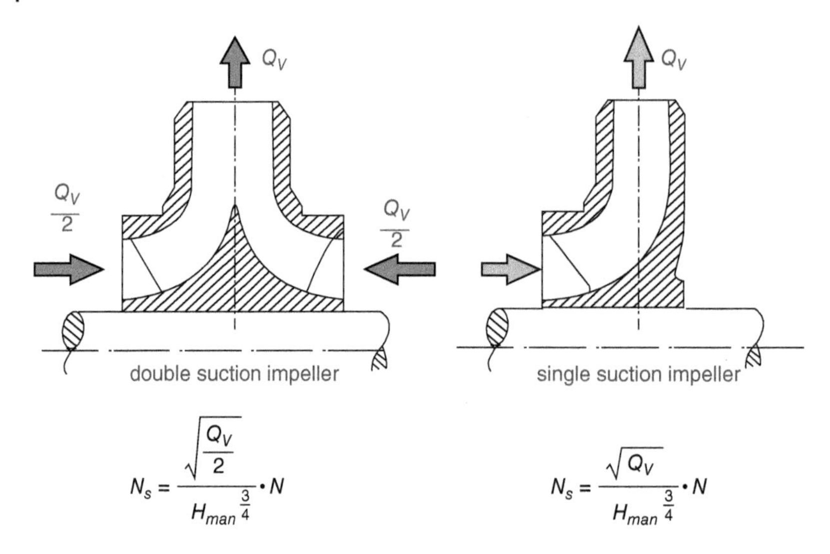

Figure 3.130 Double vs single suction pump. Source: Engineering Toolbox.

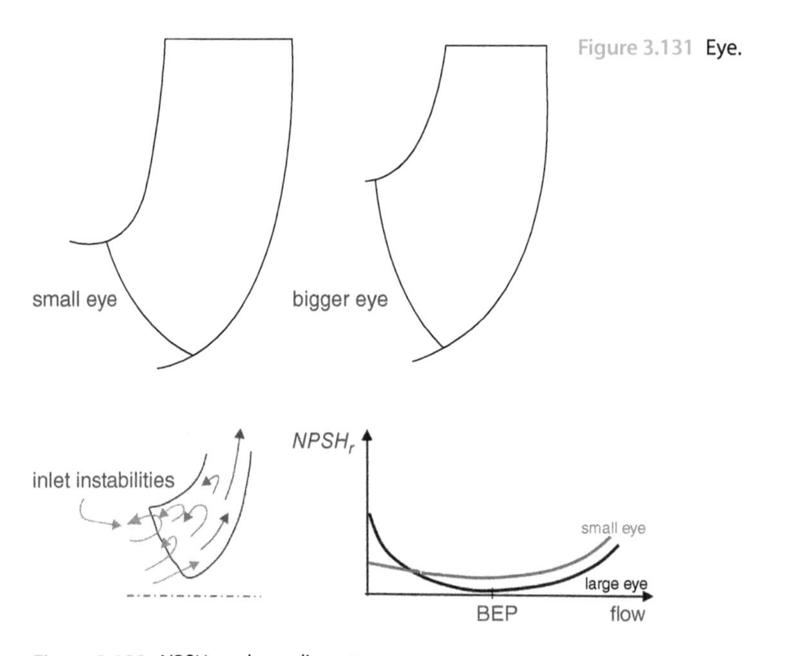

Figure 3.131 Eye.

Figure 3.132 $NPSH_r$ and eye diameter.

Unfortunately, the flow of liquid at the impeller region is not so simple and uniform as that. At the BEP there is no problem, but at smaller flows the fluid doesn't follow the path indicated by the vanes. The result is a vortex flow (see Figure 3.132 left), a recirculation area appears at the eye with cavitation and vibration as a result. A high $NPSH_r$ is only better as long as the pump does not operate significantly far below its BEP (Figure 3.132 right).

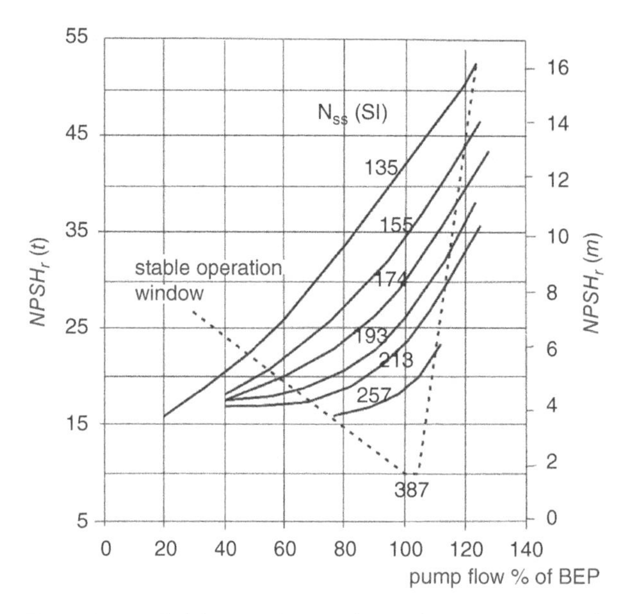

Figure 3.133 Stable operation window without cavitation, in function of N_{ss}.

The effect of the impeller eye diameter on potential suction recirculation can be evaluated by the N_{ss}. Research shows that impellers with an N_{ss} greater than 180 (SI) for water and 220 (SI) for hydrocarbons reach the dangerous zone (Figure 3.133).

3.3.15 Series Connection

For a determined volumetric flow and speed the feed pressure of an impeller is limited.

The manometric feed pressure can be increased by increasing the impeller diameter. When at the same time the flow would remain constant, the impeller should be made smaller. A very small impeller eye in combination with a big diameter has a low efficiency. Thus, there is a physical limit.

The feed head of a centrifugal pump can also be increased by placing impellers in series (see Section 3.1.18).

Placing different pumps in series is used by, for instance, the fire brigade, when a long hose must be built. First a pump is placed, then a hose, then a pump, then a hose, and so on. If all pumps were placed directly one after the other, the pressure in the hose would be too high for the strength of the hose.

When separate pumps are placed in series the pump with the best suction properties is placed first (cavitation).

The combined pump curve is found graphically by adding the manometric heads of the in series connected pumps for every flow (this is every vertical line). The intersection point of the combined pump curve and the system curve is the duty point (Figure 3.134).

Among other things such a series connection is used when a bigger pump cannot work at the available NPSH of the system. A smaller pump, a booster pump, is then placed before the bigger pump to increase the suction pressure of the bigger pump.

The only important thing with series connection is that both pumps must be able to deliver the necessary flow.

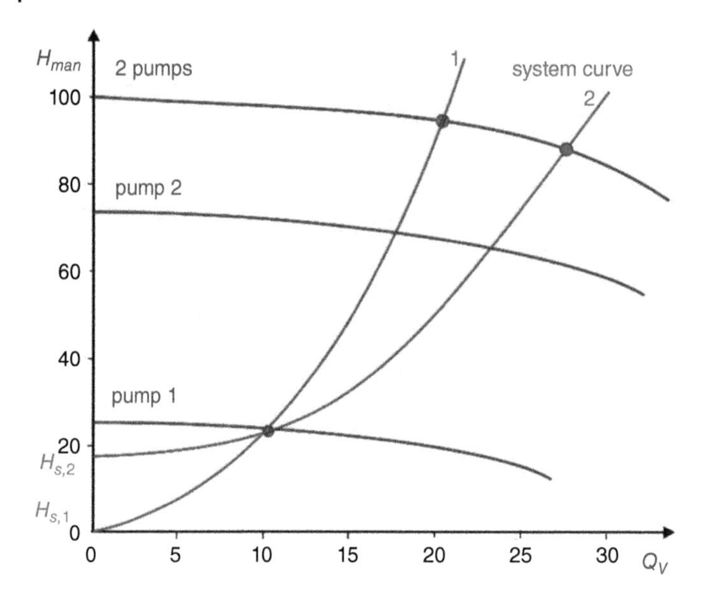

Figure 3.134 Series connection.

3.3.16 Parallel Connection

3.3.16.1 Simple Case

Parallel connection can have two purposes:

- Increase reliability, one working and one in reserve.
- Deliver a bigger flow.

Parallel pumps in connection must be used with a check valve (Figure 3.135).

Consider first the a simple case, whereby both pumps have a decreasing pump curve.

As long as the weakest pump, pump 1, hasn't reached the strength of the strongest pump, pump 2, only pump 2 delivers a flow (see magenta line). From the moment the weaker pump reaches its maximum pressure, both pumps will work together and deliver the same manometric head.

By adding the flows Q_{V1} and Q_{V2} on the horizontal line of both pumps (Figure 3.136) one finds a point of the common characteristic pump curve.

By repeating for other manometric head, more points can be found; they can be connected to the common pump curve.

Consider now the system curve (Figure 3.137). It intersects the common pump curve in an duty point P. The horizontal line through P gives the flows Q_{V1} and Q_{V2} separately. Only from this point may the efficiency, the power, and the NPSH be read.

The total flow $Q_{V1} + Q_{V2}$ is always smaller than the sum of the flows in the case that every pump would work separately:

$$Q_{V1} + Q_{V2} < Q_{VA} + Q_{VB}$$

3.3.16.2 Case with Increasing Pump Curve

Start from a volumetric flow equal to zero (Figure 3.138). At low flows pump 2 delivers a higher head than pump 1 can reach. As a result only pump 2 will work and pump 1

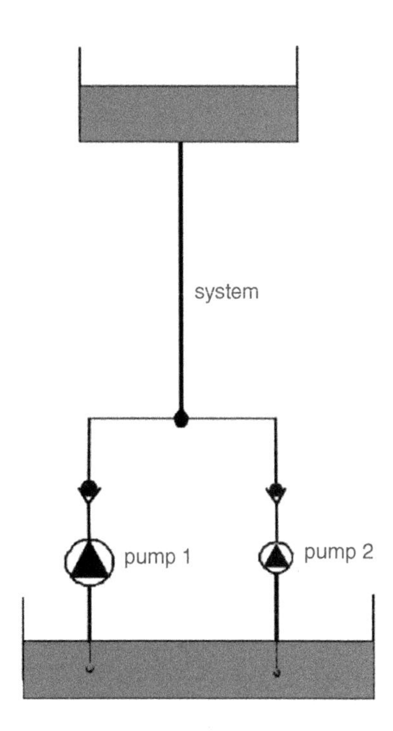

Figure 3.135 Parallel connection.

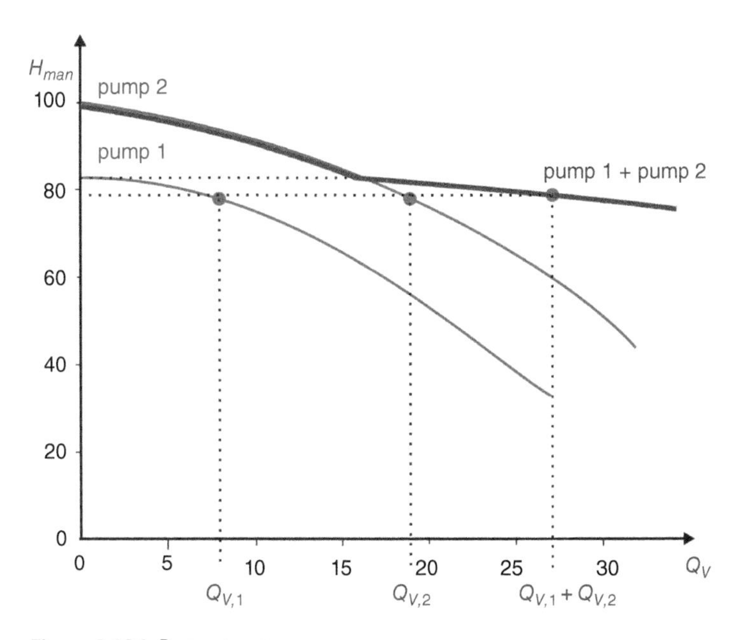

Figure 3.136 Pump curve.

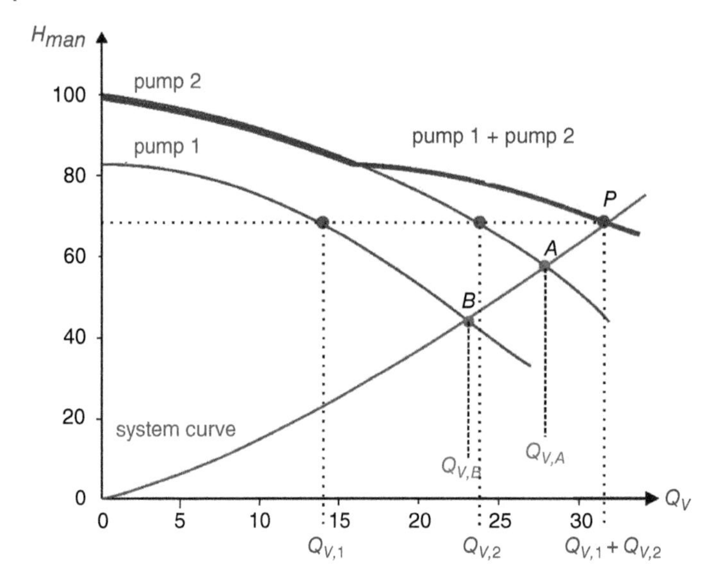

Figure 3.137 Duty point.

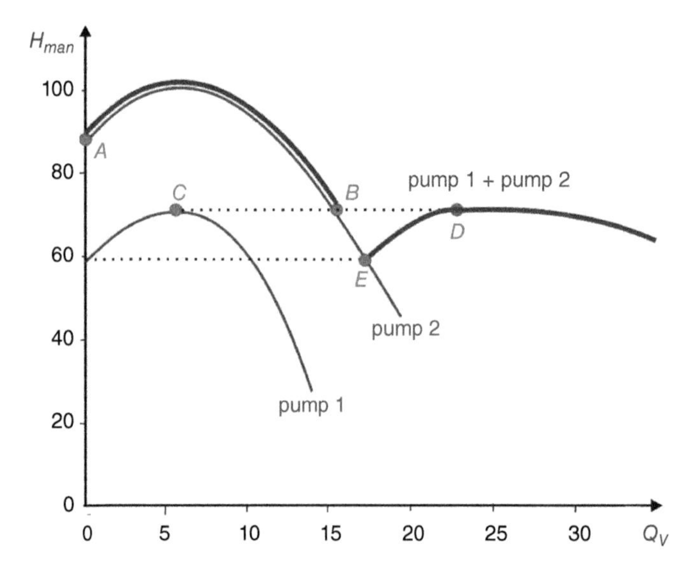

Figure 3.138 Parallel connection.

is jobless. The common pump curve is given by the curve of pump 2 and this continues as long as a point is reached where pump 1 has a pressure equal to that of pump 2. This is represented in the graphic by the magenta curve AB. Once the pump curve reaches point B, pump 1 has a working point C.

Pump 1 will deliver a flow according to point C and this is expressed in the common pump curve by the jump of point B to point D.

When the required head decreases further, pump 1 can deliver a flow according to the left part of its curve. This is represented in the graphic by the curve DE. But these heads are also reachable in the right part of the curve.

Now, from point D on, the common pump curve must be extended to the right by taking the locus of the partial flows at a certain head.

If one looks at the common pump curve, it is clear that the curve DE is slowing down. And the BE curve is not even defined. It is then advised to stay out of that area because of its unstable character. At regular changes of the pump curve it can happen that pump 1 sees its work point balance from one place in the decreasing part to a point in the increasing part of its curve, a phenomenon that is accompanied with shock behavior of its flow. This leads to pressure peaks and a surge.

3.3.17 Influence Viscosity

Consider Newtonian liquids.

The characteristics of turbopumps (H_{man}, η_p, and P_t in function of Q_V) change only significantly from a kinematic viscosity of $\nu > 20 \cdot 10^{-6}$ [m²/s] and from this value on the characteristics must be recalculated with empirical means. The two most known methods are the one from the Hydraulic Institute (HI) (Figure 3.139) and the one from the pump manufacturer KSB (Figure 3.140). Both methods use diagrams to calculate conversion factors, which are admittedly used in the same way. They differ with KSB apart from the values H_{man}, η_p, and Q_V also takes into account the specific speed.

The HI-diagram was only measured at $N_s = 15$–20 and leads in this limited area to the same results as that of KSB. This last one works in an area 6.5–45 and with viscosities ν from 4000·to 10^{-6} [m²/s].

The characteristic values H_{man}, η_p, and Q_V of a single stage centrifugal pump known for water (index w) can be recalculated for viscous liquids (index ν) as follows:

$$Q_{V,\nu} = f_Q \cdot Q_{V,w}$$

$$H_{man,\nu} = f_H \cdot H_{man,w}$$

$$\eta_{p,\nu} = f_\eta \cdot \eta_{p;w}$$

The factors f are named k in HI diagrams (Figure 3.140). The values of the flow and manometric head are values in the BEP.

The conversion of Figure 3.139 is valid for the range:

$$0.8 \cdot Q_{V,opt} < Q_V < 1,2 \cdot Q_{V,opt}$$

where $Q_{V,\,opt}$ is the value in the BEP.

At $Q_V = 0.8 \cdot Q_{V,\,opt}$ must be posed that $H_{man,\,\nu} = 1.03 \cdot f_H \cdot H_{man,\,w}$.

At flow zero: $H_{man,\,\nu} = H_{man,\,w}$ and also $\eta_{p,\,\nu} = \eta_{p,\,w} = 0$.

This way four points of the characteristic curve are determined.

The pump power for the viscous liquid becomes (Figure 3.141):

$$P_\nu = \frac{\rho_\nu \cdot g \cdot H_{man,\nu} \cdot Q_{V,\nu}}{\eta_{p,\nu}}$$

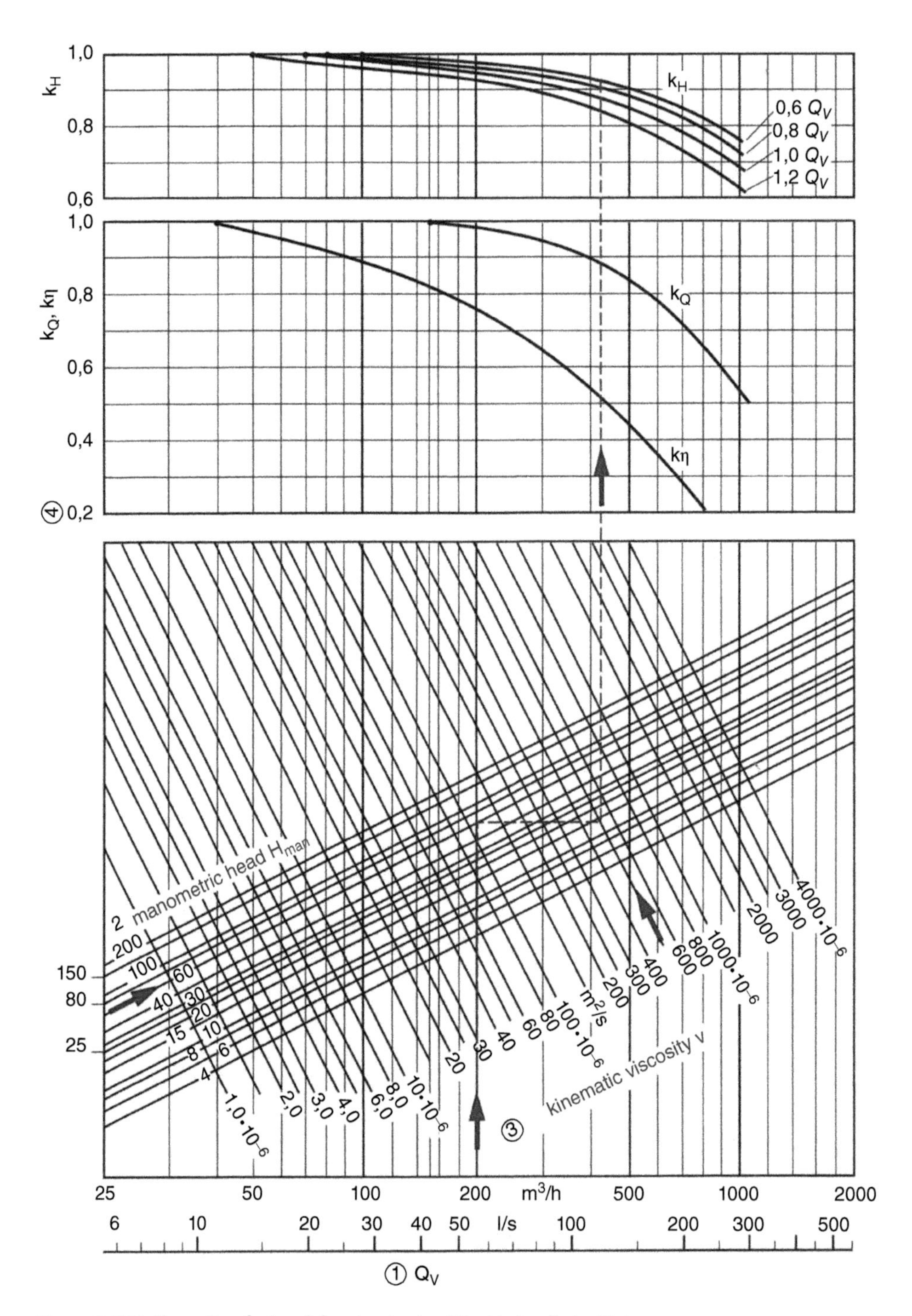

Figure 3.139 Correction factors k for standards of the Hydraulic Institute.

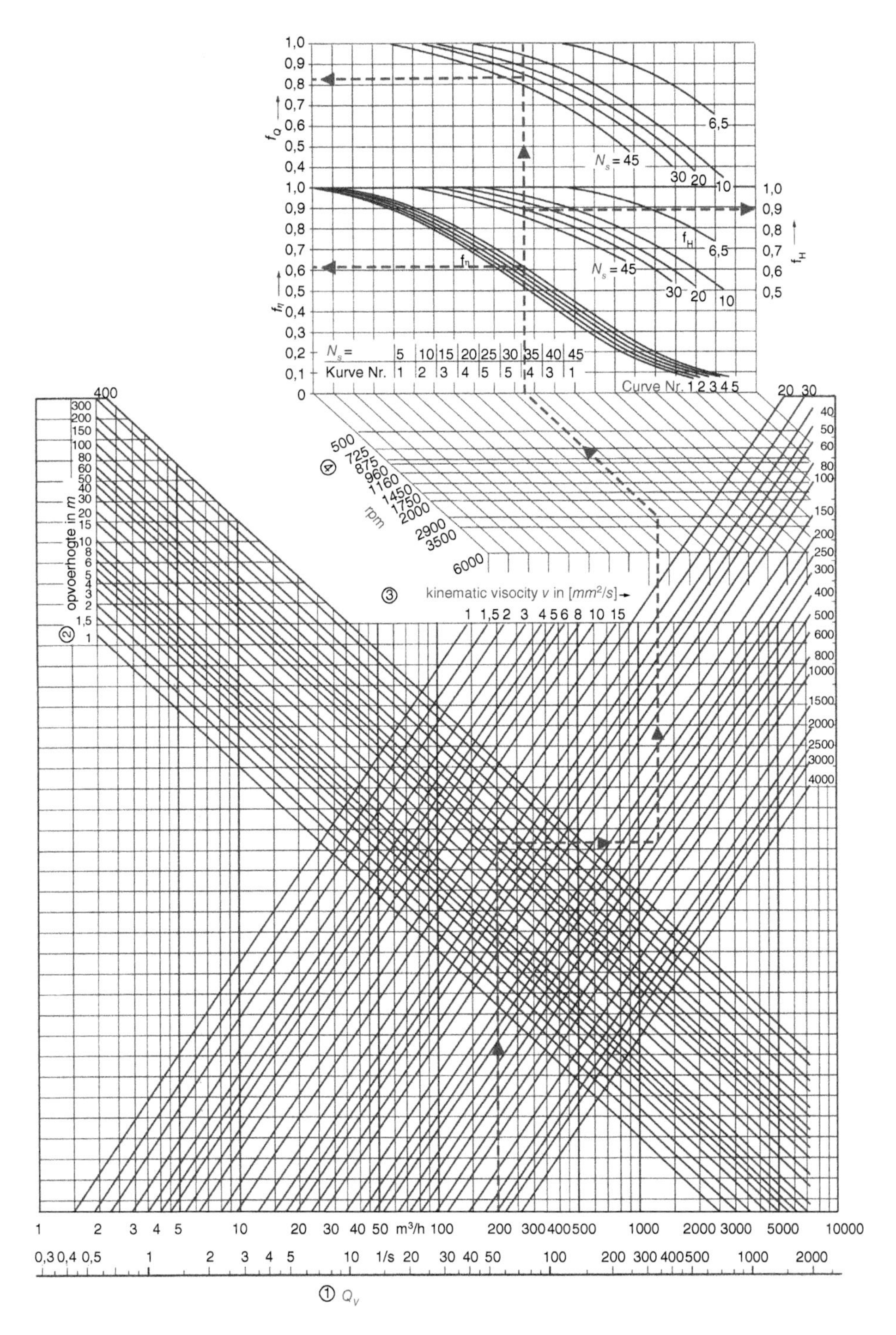

Figure 3.140 Correction factors *f* according to KSB.

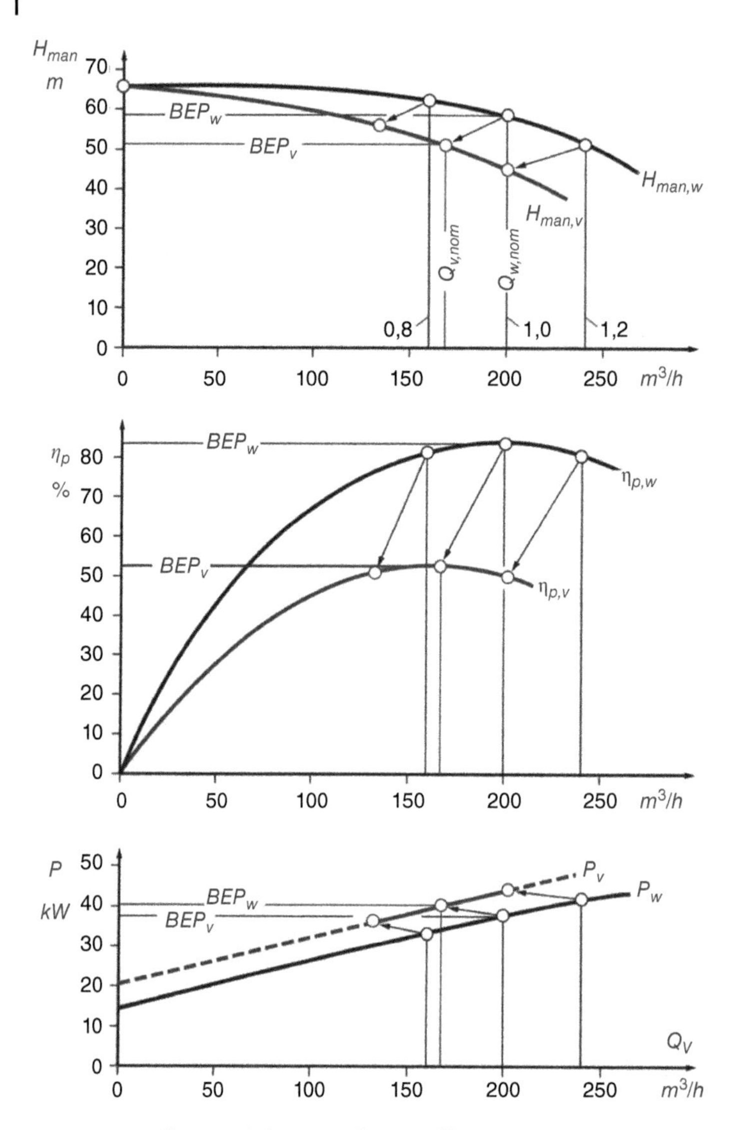

Figure 3.141 Characteristic curves. Source: KSB.

3.3.18 Special Turbopumps

3.3.18.1 Submersible Pumps

Drilled wells, water, and oil are exploited by special pumps. Mostly these are multistage turbopumps with small diameters that are erected vertically. In Figures 3.142–3.146 such a pump is represented. The pump rests on the bottom of the well. The pump rests on the bottom of the well. It is ready to work, because the impeller and pump housing are dropped in the liquid. These pumps have a very long transmission shaft that is mostly

Figure 3.142 Submersible pump.

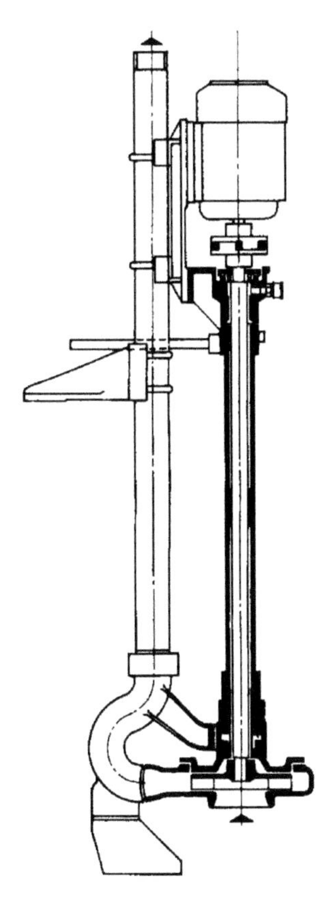

Figure 3.143 Multi cells submersible pump. Source: Sulzer Photo Gallery.

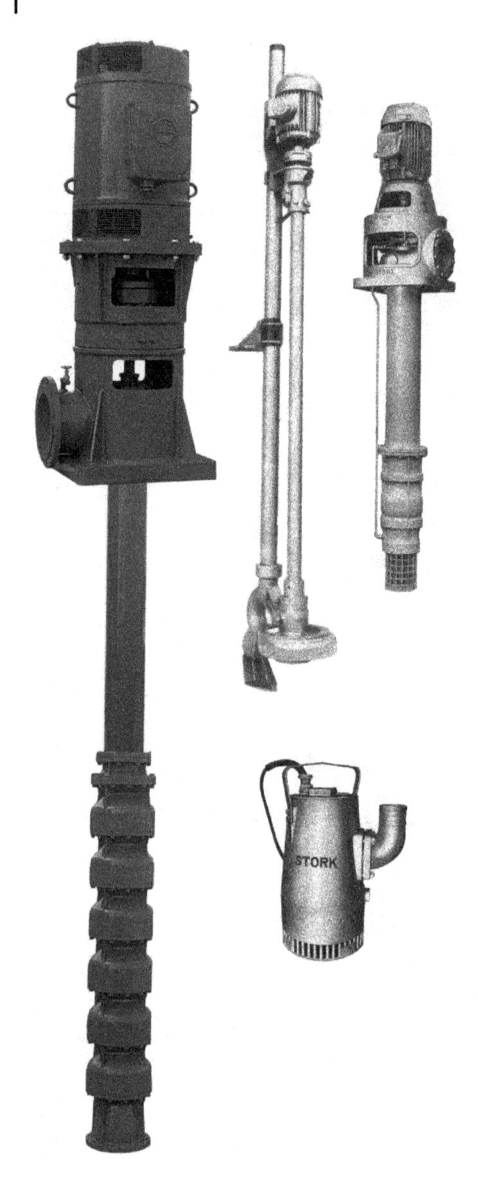

Figure 3.144 Submersible pumps.

composed of pieces of 2–3 [m]. The electromotor is set up on the surface. Typically, such a transmission shaft can attain 100–200 [m].

3.3.18.2 Electropumps

Here the pump is directly coupled with the motor. The combination pump motor is let down into the well (Figure 3.147).

The motor is placed under the pump, and the stator windings are isolated and impermeable to water (or oil). They are run through by the liquid that takes care of the cooling.

Presentation: 3000 [rpm] and upwards of 2000 [m] head.

Figure 3.145 Multi cell submersible pump. Source: Sulzer.

3.3.19 Contaminated Liquids

In order to make turbopumps suitable for contaminated liquids use is made of one-, two-, and three-channel pumps. In this way, large passages are offered to the liquid. Some typical implementations are shown in Figures 3.148–3.150.

Sometimes one channel pumps are offered with precisely the aim of transporting solid substances (Figures 3.149 and 3.150). The solid parts will be mixed with water. Mostly,

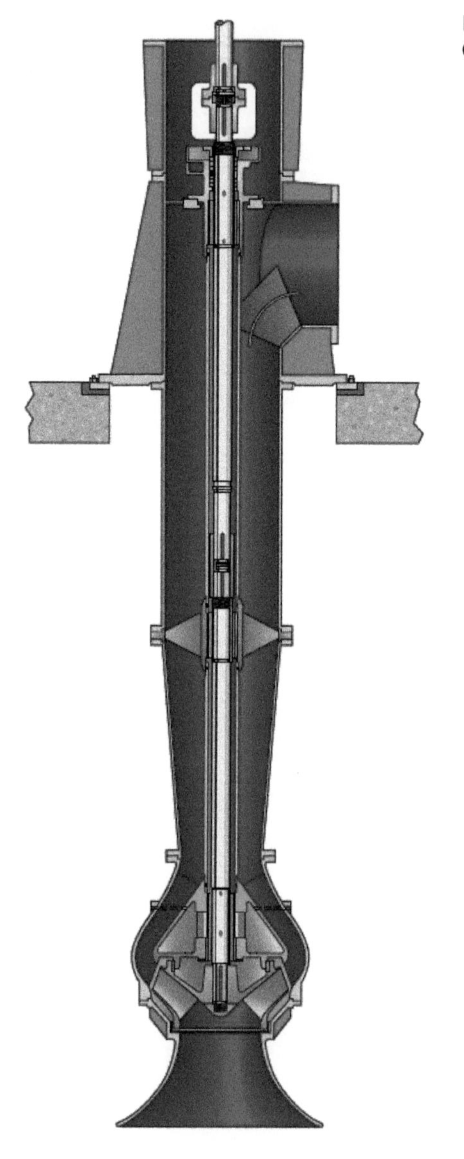

Figure 3.146 Half-axial submersible pump. Source: Gould Pumps.

the aim is to transport alimentation and, of course, to damage food as little as possible. A yield of 98% is normal (Figures 3.151 and 3.152).

3.3.20 Cutter Pumps

Sometimes the impellers are constructed is such a way that they work as cutters, and shatter the coarse parts in the transported liquid (Figures 3.153 and 3.154).

Figure 3.147 Electropumps.

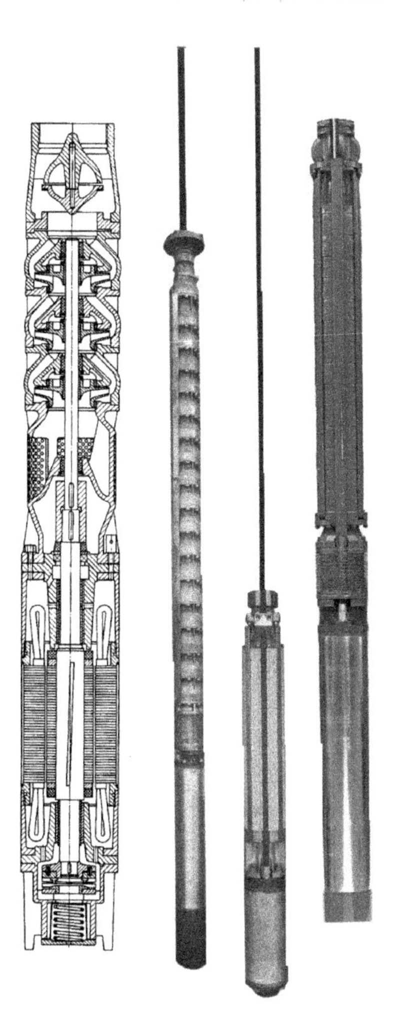

3.3.21 Mounting

Now we look at the various possible mountings of turbopumps, except for submersible pumps and electropumps.

The classic way is the horizontal mounting. The support of the pump can be under the bearing bracket (Figures 3.155 and 3.156.) for heavy pumps, or under the pump housing. This is called being *foot mounted*.

For pumps for hot liquids the previous support is not adequate because the temperature expansion would wrap the pump on the foundation and the shaft of the pump and motor would not be in line anymore. That's why one prefers a support of the pump along the horizontal shaft. Such a mounting is called being *centerline mounted* (Figure 3.157).

Another construction method is the one where pump and motor are placed on one continuous shaft so that a compact total construction is obtained. The motor is not supported, only the pump. This mounting is called *close coupled* (Figure 3.158). Figure 3.159 presents a vertical execution of a closed coupled mounting.

Figure 3.148 Turbopumps for contaminated liquids. Source: KSB.

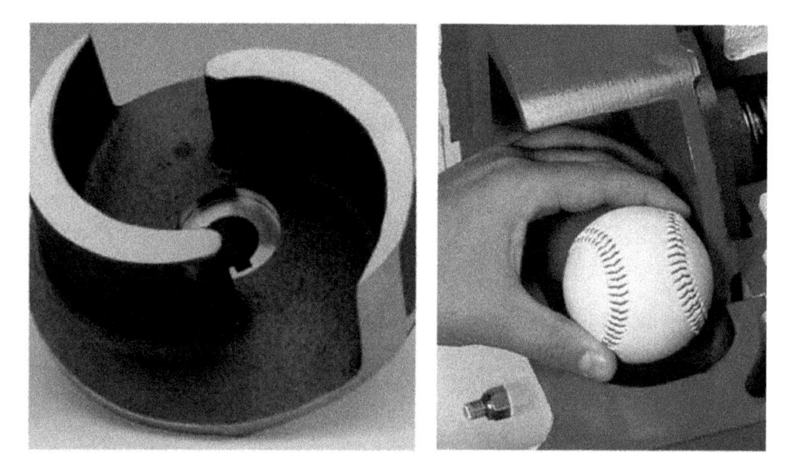

Figure 3.149 Two-channel pump. Source: Gould Pumps.

Figure 3.150 Three-channel impeller.

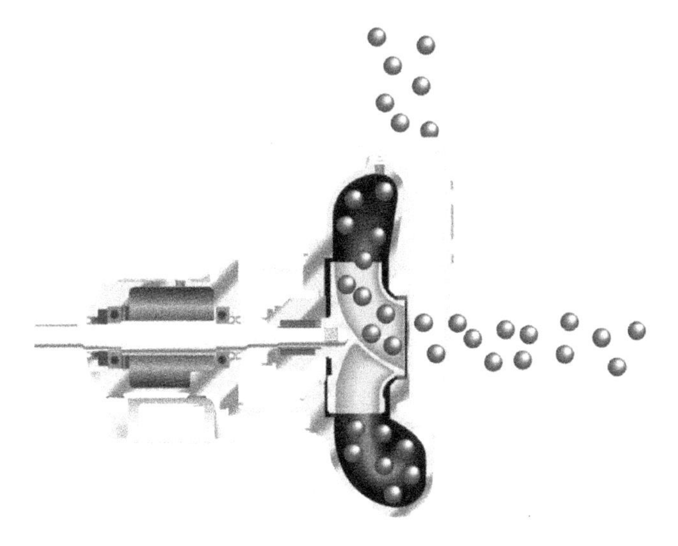

Figure 3.151 One-channel pump. Source: Cornell.

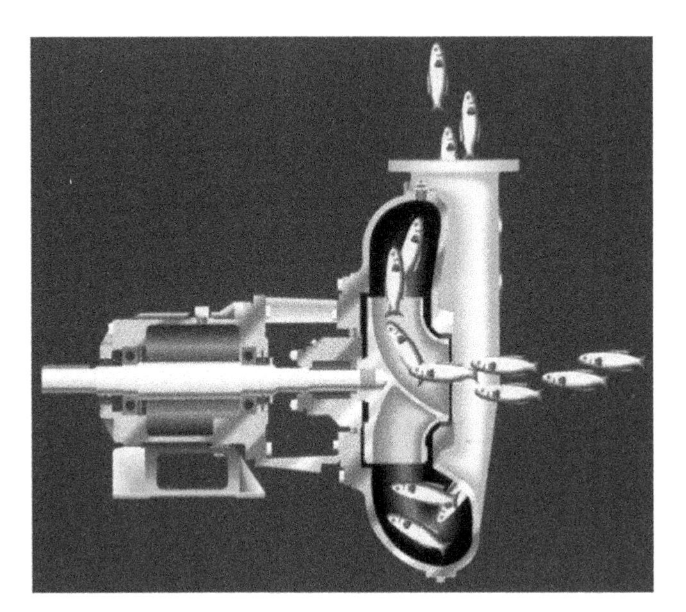

Figure 3.152 One-channel pump. Source: Cornell.

When suction and discharge line are in each other's extension this is called *in line* pumps.

- *Overhung impeller type*: the impeller is mounted on the end of a shaft which is "overhung" from its bearing supports.
- Constructions with *spacer* coupling (*back pull out*) (Figures 3.160 and 3.161): the construction is made of a motor, a "spacer", and the pump. The spacer can simply be removed so that afterwards the bearing and shaft with the impeller can be removed.

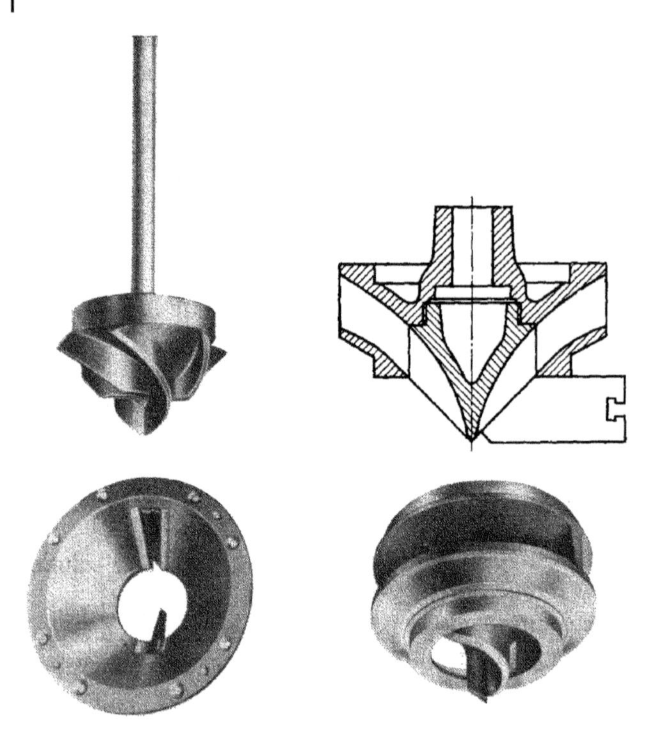

Figure 3.153 Cutter impellers.

Figure 3.154 Cutter pumps. Source: ABS.

The motor and pump housing stay on the foundation, and the pipe lines do not have to be disconnected (Figures 3.160 and 3.161).

As far as the mounting and demounting of centrifugal pumps is concerned, for:

- *radially split pumps* see Figure 3.162.
- *axially split* see Figures 3.163.

Figure 3.155 Support under the bearing. Source: Gould Pumps.

Figure 3.156 Support under the bearing. Source: Gould Pumps.

Figure 3.157 Centerline mounted. Source: Ensival Morret.

Figure 3.158 Close coupled. Source: Gould Pumps.

Figure 3.159 Vertical mounting close coupled, in line. Source: Gould Pumps.

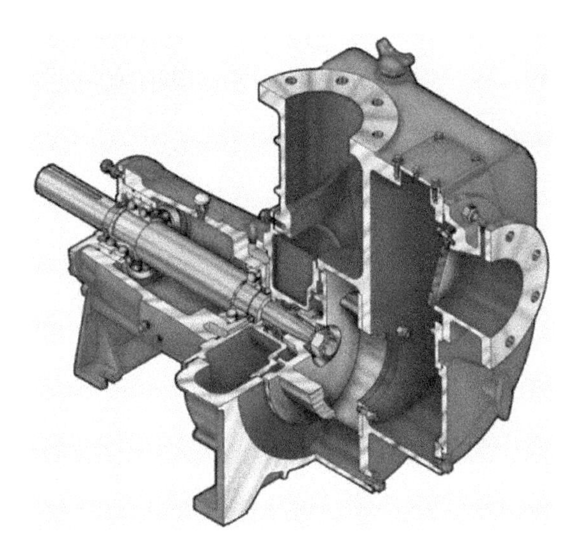

Figure 3.160 Back pull out construction.

Figure 3.161 Sight on the pump of Figure 3.160. Source: Gould Pumps.

Figure 3.162 Radially split centrifugal pump. Source: Gould Pumps.

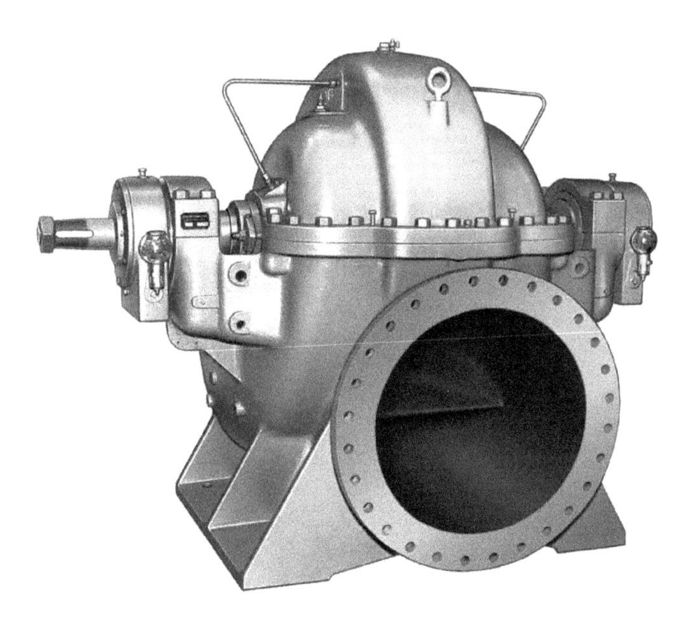

Figure 3.163 Axially split pumps. Source: Gould Pumps.

4

Flow-Driven Pumps

CHAPTER MENU

4.1 General

Jet pumps are devices to transport, compress, or mix gasses, liquids, or solids. When a gas or liquid is used to power the pump, we speak of *flow-driven pumps*. They convert static pressure energy into dynamic pressure. They are pumps without moving parts.

The basic principle of jet pumps is that a liquid or gas jet flows out of a jet pipe at a high velocity and low pressure and drags the liquid, gas, or solid out of its surroundings and accelerates. The result is a mixture of flow-driven fluids and transported substances, from which the velocity is reduced in a second jet pipe and the pressure again increased.

Figure 4.1 shows the three main parts:

1. the jet pipe
2. the diffusor
3. the head.

The channel of the diffusor consists of the inflow direction narrowing part, the run-up cone, a cylindric part, the neck, and a widening part, the spout cone.

Normative for the action of jet pumps are the pressures on in- and outlet and the mass flow (see Figure 4.2).

Pumps and Compressors, First Edition. Marc Borremans.

Companion website: www.wiley.com/go/borremans/pumps

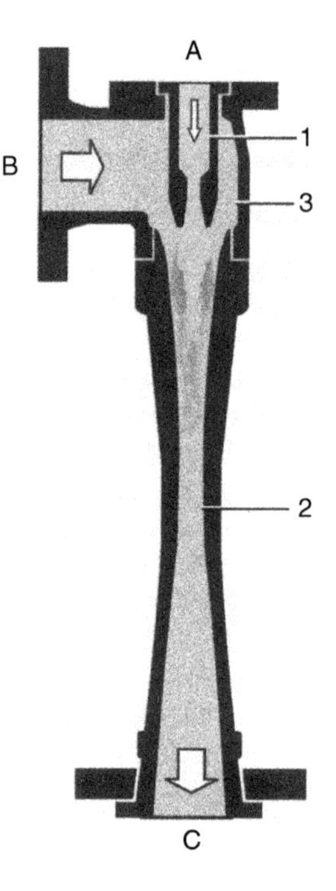

Figure 4.1 Jet pump. Source: GEA Wiegand.

According to DIN 24 290 jet pumps are classified into the following categories:

Gas jet pumps	**Steam jet pumps**	**Liquid jet pumps**
Gas jet fan	Steam jet fan	Liquid jet fan
Gas jet compressor	Steam jet compressor	Liquid jet compressor
Gas jet vacuum pump	Steam jet vacuum pump	Liquid jet liquid vacuum pump
Gas jet liquid pump	*Steam jet liquid pump*	*Liquid jet liquid pump*
Gas jet solid pump	*Steam jet solid pump*	*Liquid jet solid pump*

For the part pumps in this book only the italic versions are of importance. The other jet pumps will be discussed in the part on compressors.

4.2 Liquid Jet Liquid Pump

By all jet pumps (Figure 4.2) the flow-driving fluid with flow Q_{M1} has the highest pressure p_1, the suction flow Q_{M0} the lowest pressure p_0, and the mix Q_{M2} has a pressure p_2 lying in between the two other pressures: $p_1 > p_2 > p_0$.

Some implementations are given in Figure 4.3.

Figure 4.4 shows its application, thus:

- At a demineralization of water, the salts in the water will be removed. First in a cation exchanger the positive ions as Na^+, Ca^{2+}, and Mg^{2+} are removed. Then in a deaerator the CO_2 is removed. At last the negative ions like Cl^- and SO_4^{2-} are removed.

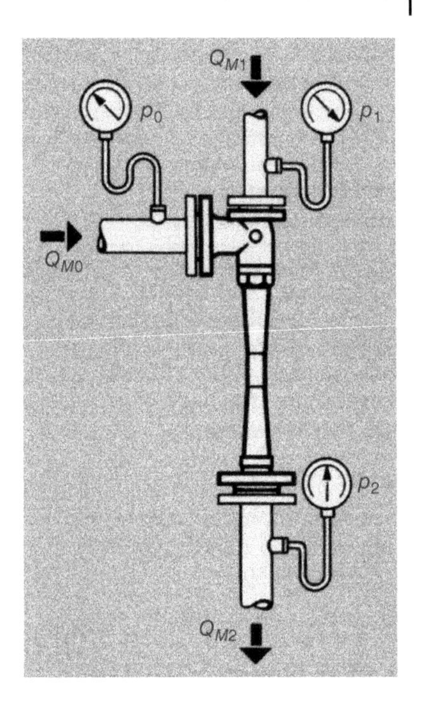

Figure 4.2 Values. Source: GEA.

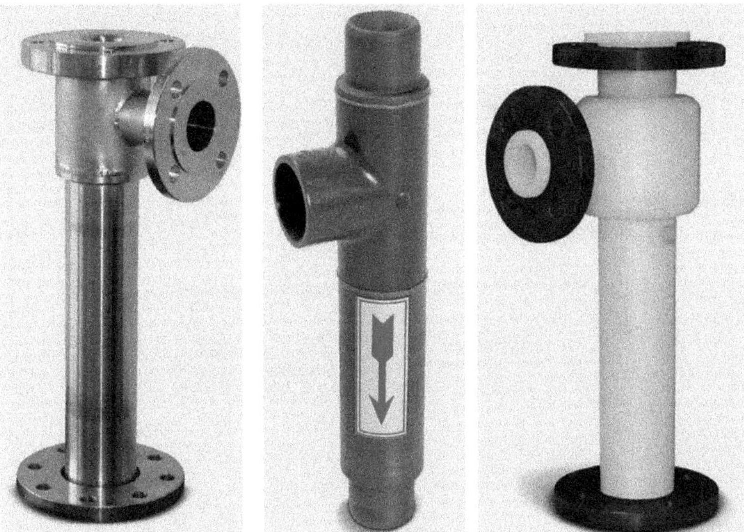

Figure 4.3 Liquid jet pump (left: stainless steel, middle: PP, right: PTFE for aggressive media). Source: GEA.

- Such a cation and anion exchanger consist of a set of resin grains that catch the wanted ions and replace it respectively with H^+ and OH^-. But after some time, the exchanger is saturated with caught ions and it should be regenerated. This happens by rinsing with HCl and NaOH. Hereby the liquid jet pumps suck respectively the acid or base. The driving liquid is the water itself.

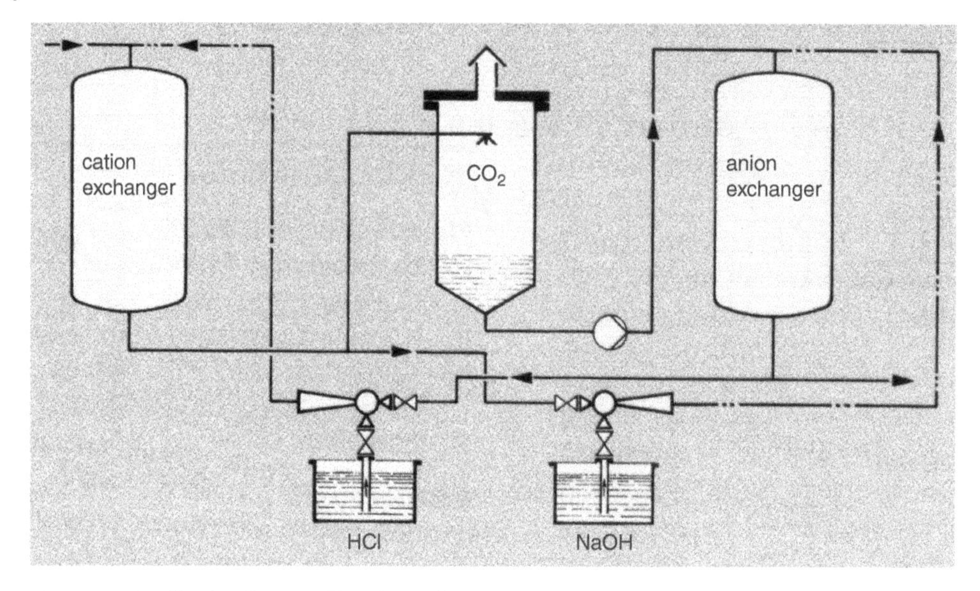

Figure 4.4 Application. Source: GEA Wiegand.

4.3 Liquid Jet Solid Pump

This pump concerns the transport of granular substances. The material that should be transported is supplied with a hopper to the jet pump. The driving fluid, mostly water, is supplied at high velocity to the mix room of the pump and drags the material that should be transported in this room. Dependent on the material a quantity of water has to be supplied.

Liquid jet pumps for solid substances can be designed as mobile units (Figure 4.5) or stationary installations with a hopper and rinsing water connection.

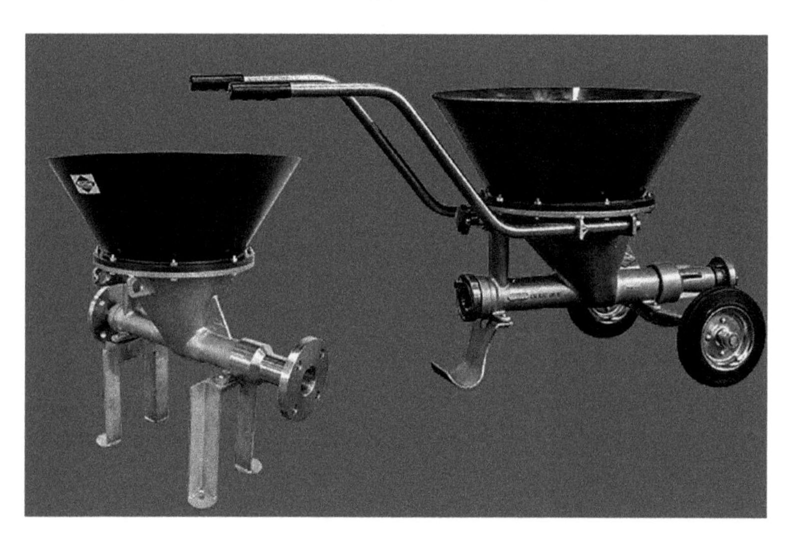

Figure 4.5 Mobile unit.

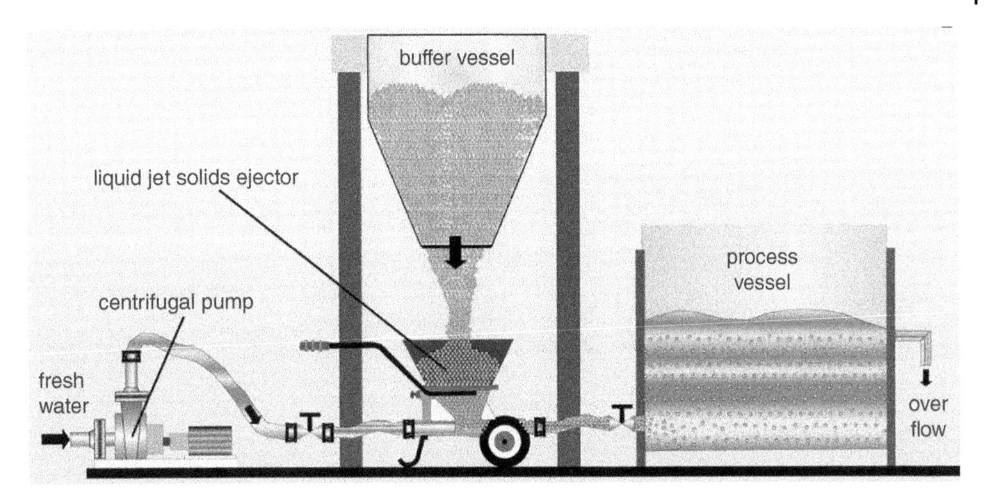

Figure 4.6 Application.

Liquid jet pumps for solid substances are applied for the transportation of sand, gravel, salts, and or such to fill reactor vessels.

Figure 4.6 shows its application.

4.4 Liquid Jet Mixers

In Figure 4.7 a typical example of a neutralization basin is shown.

A neutralization basin is a big volume where the pH is corrected in order to attain the desired pH.

Typically it concerns an acid or base originating from an ion exchanger that must be brought to pH = 7 before being evacuated.

But, in fact, the case must be seen wider. In general, one speaks of a homogenization basis. There waste water can be homogenized before evacuation or various batches of a product can be homogenized with each other (for instance parties diesel and kerosene).

So all processes whereby a fluid has to be brought homogeneously or held can be realized with mixers. The only condition is that the viscosity < 500 [mm^2/s, or cSt], and that there not too big density differences between the fluids to mix.

4.5 Steam Jet Liquid Pump

Here steam is used as the driving fluid. In the jet pump the steam sucks liquid, mixes with it, and condensates. The liquid to transport may not be too hot. This is because the driving steam must be able to condensate so that its volume decreases and the whole mix can easily flow through the pipe.

4.6 The Feedback Pump

Such a water jet pump (Figure 4.8) can be used to pump water out of a deep well. First, a quantity of liquid (the same as the one to pump) is pressed underpressure in a jet

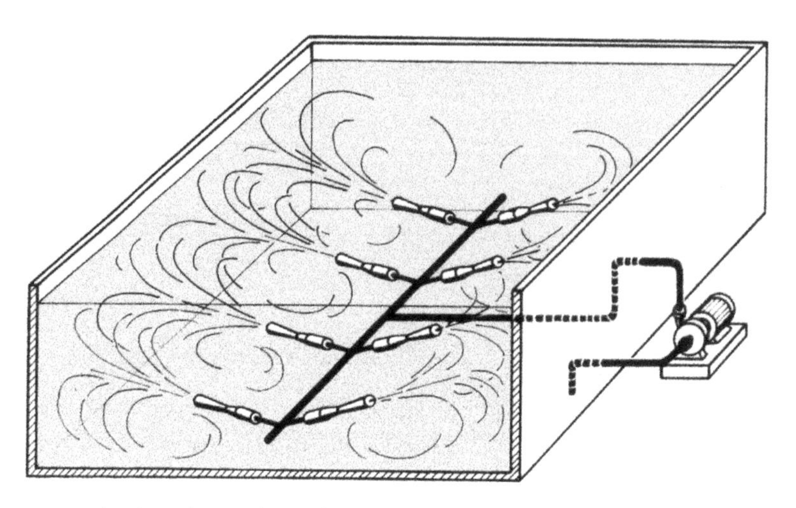

Figure 4.7 Neutralization basin. Source: GEA Jet Pumps GmbH, Ettlingen.

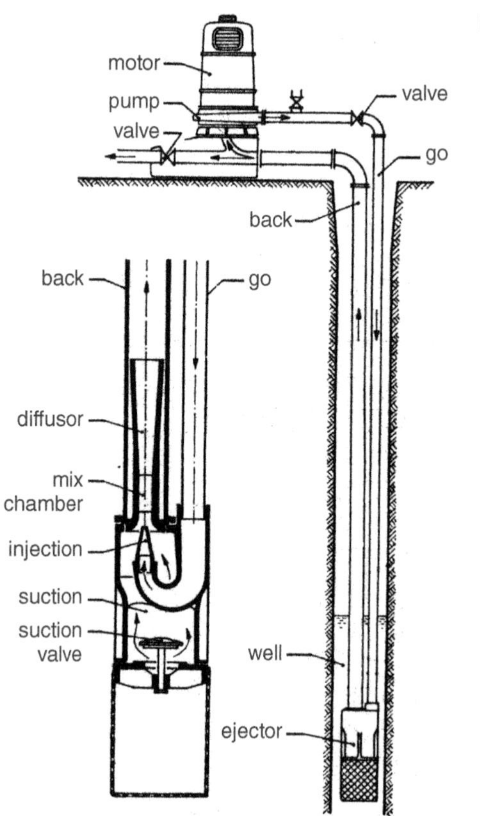

Figure 4.8 Feedback pump.

pipe. This causes the water jet pump to work and this water drives the well water to the surface. A part of the water is used again as the driving fluid.

By placing more such water jet pumps under each other it is possible to attain very high suction heads (Figure 4.8).

4.7 Air Pressure Pump

This pump type has no moving parts and is insensible for contaminations. Drawback: a compressor is needed (Figure 4.9).

The equipment consists primarily of a U-tube with legs of unequal length (Figure 4.9). At the bottom of the long leg compressed air can be fed. The short leg is completely submersed in water.

In the left drawing no air is supplied. The liquid in both legs is on the same level. In the right drawing air is supplied in the right leg. Little bubbles dispense in the liquid mass and this causes the specific mass in the right leg to decrease. The static pressure decreases. This causes a thrust between the left and right leg (Archimedes). The liquid in the right leg rises. This way a pumping action is created.

An application is this one where deep bore wells are made and the disconnected groundwater mix is pumped to the surface.

There are implementations to 600 [m]. For this reason, this pump is called the *mammoth pump*.

The need to use a compressor has evidently a bad influence on the pump efficiency (Figure 4.10).

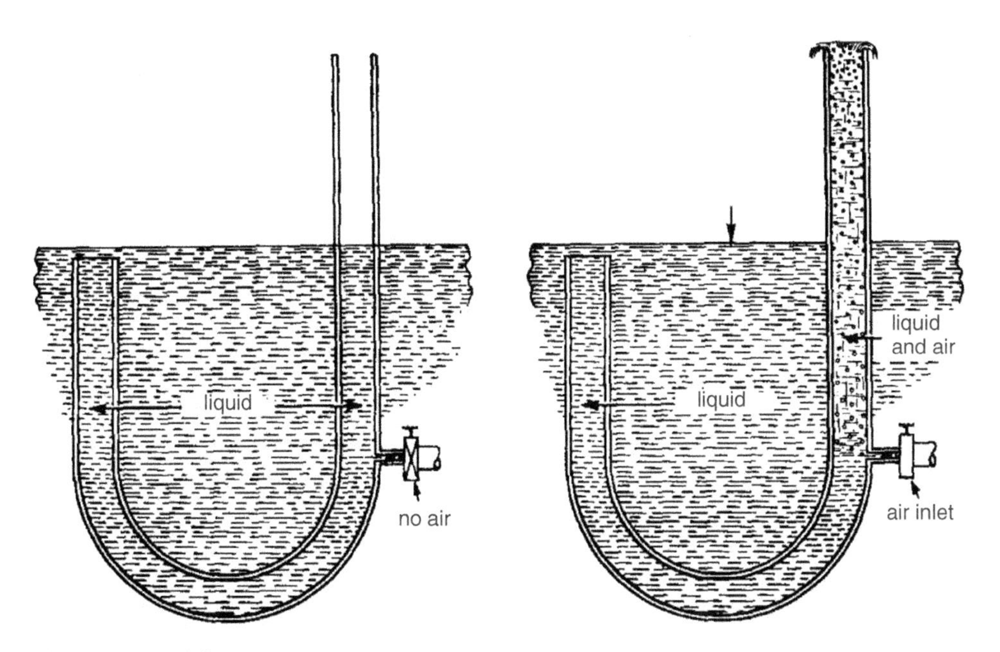

Figure 4.9 Principle.

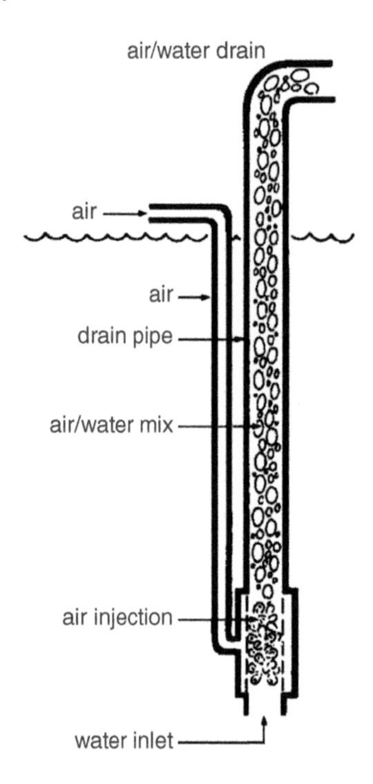

Figure 4.10 Air pressure pump.

5
Sealing

CHAPTER MENU

5.1 Labyrinth Sealing

Almost everywhere pumps with rotating shaft are used a sealing is needed. The shaft seal forms a barrier between what is in the pump and the atmosphere. A seal is necessary to keep the lubrication inside and the contamination outside a cavity where the oil is kept. A typical seal application comprises a stationary element, a rotating element, and a controlled cavity.

A pump where the shaft goes through the seal is not completely sealed. It is a challenge to minimize the leak. In its simplest form, a seal comprises a rotating element and a stationary element (Figure 5.1).

To enhance the effectiveness of this seal, one or more concentric grooves can be machined in the housing bore at the shaft end (Figure 5.2). The grease emerging through the gap fills the grooves and helps to prevent the entry of contaminants. For labyrinth seals, the best rule of thumb is that clearance should be a little more than the bearing clearance. The closer the seal is to the shaft, the better it will seal. Of course, if it touches the shaft both may be damaged, typically resulting in rapid increases in temperature and vibration levels (especially axially on the end that is rubbing). Rather than determining labyrinth seal clearance from the shaft diameter, it's better to work from the sleeve-bearing clearance. A good guideline that several manufacturers use is 0.05–0.10 [mm].

Pumps and Compressors, First Edition. Marc Borremans.

Companion website: www.wiley.com/go/borremans/pumps

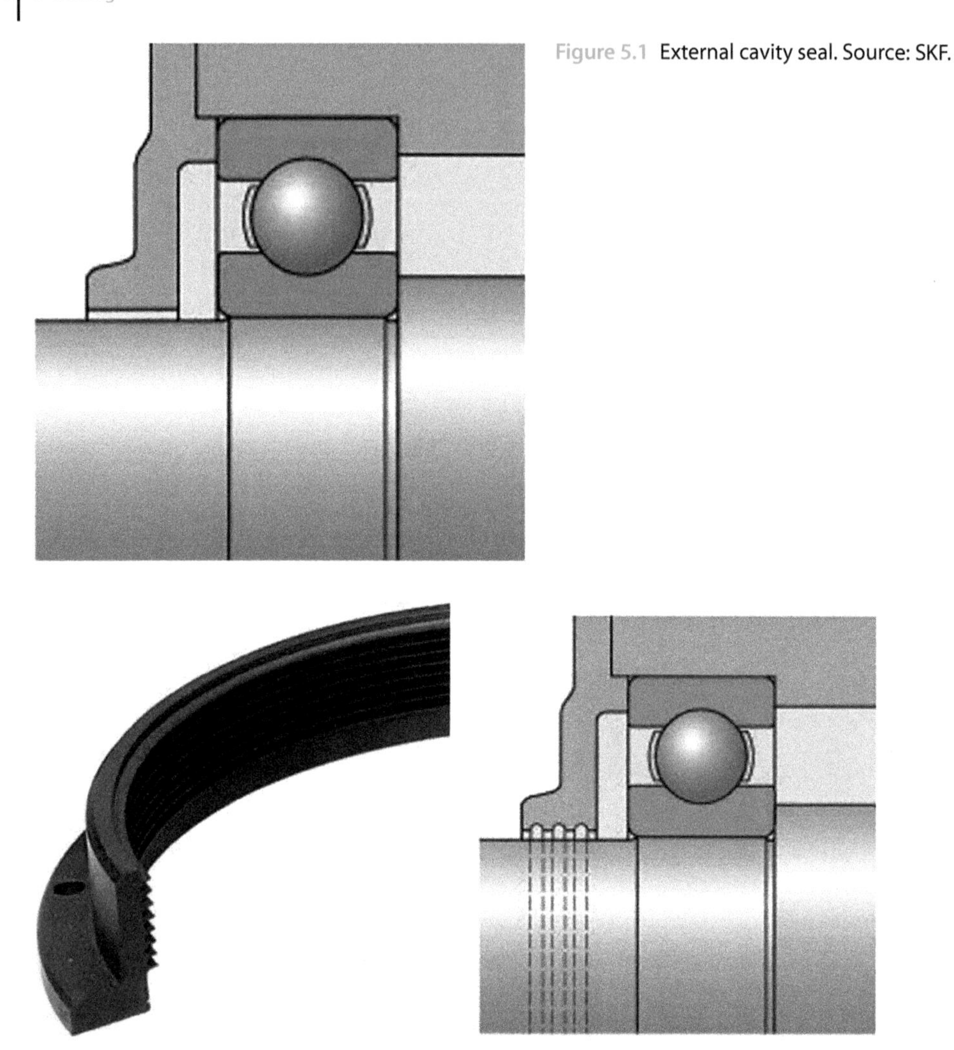

Figure 5.1 External cavity seal. Source: SKF.

Figure 5.2 External gap type seal with concentric grooves. Source: SKF.

With oil lubrication and horizontal shafts, helical grooves – right-hand or left-hand depending on the direction of shaft rotation – can be machined into the shaft or housing bore (Figure 5.3). These grooves are designed to return emerging oil to the bearing position. Therefore, it is essential that the shaft rotates in one direction only.

Single or multistage labyrinth seals, typically used with grease lubrication, are considerably more effective than simple gap-type seals, but are also more expensive. Their effectiveness can be further improved by periodically applying a water-insoluble grease via a duct to the labyrinth passages. The passages of the labyrinth seal can be arranged axially (Figure 5.4) or radially (Figure 5.5), depending on the housing type (split or non-split), mounting procedures, available space, etc. The width of the axial passages of the labyrinth (Figure 5.4) remains unchanged when axial displacement of the shaft occurs in operation and can therefore be very narrow. If angular misalignment of the shaft relative to the housing can occur, labyrinths with inclined passages can be used (Figure 5.6).

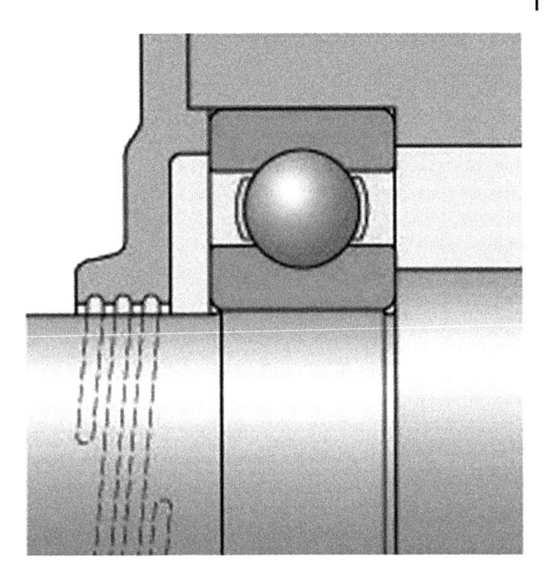

Figure 5.3 External gap seal with helical grooves. Source: SKF.

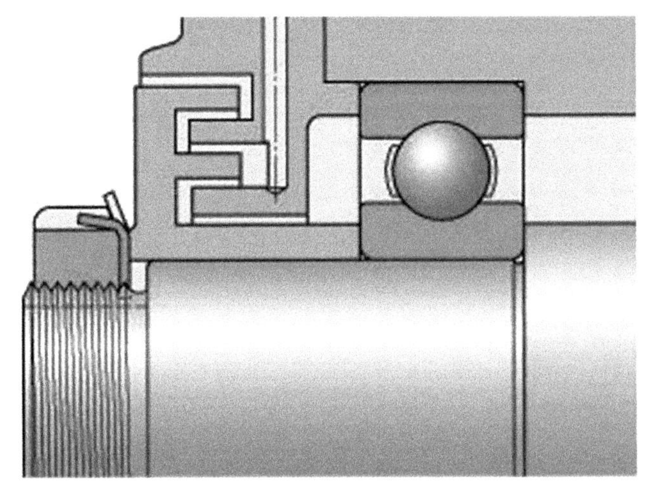

Figure 5.4 External labyrinth seal, passages arranged axially. Source: SKF.

Effective and inexpensive labyrinth seals can be made using commercially available products, such as sealing washers (Figure 5.7). Sealing effectiveness increases with the number of washer sets and can be further improved by incorporating flocked washers.

Rotating discs (Figure 5.8) are often fitted to the shaft to improve the sealing action of shields. Flingers, grooves, or discs are used for the same purpose with oil lubrication.

The labyrinth seal is constrained in its ability to function with a wide variety of fluids, owing to its limited ability to centrifugally push the lubricants back into the controlled fluid cavity. In addition, it can only handle a limited volume of lubricating fluid in the controlled fluid cavity, and it requires high rotational speeds to seal adequately. The seal is not able to keep out external environmental contaminants entering the controlled fluid cavity when the seal is not rotating, and some fluid seepage is expected from the seal. In general, the standard labyrinth type seals have limited abilities.

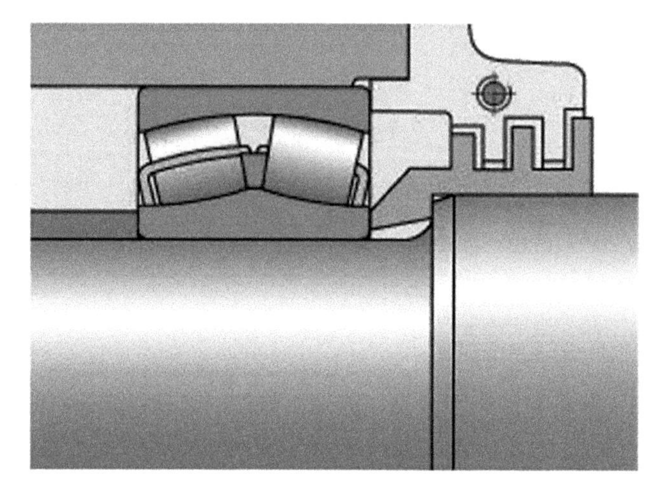

Figure 5.5 External labyrinth seat; passages arranged radially.

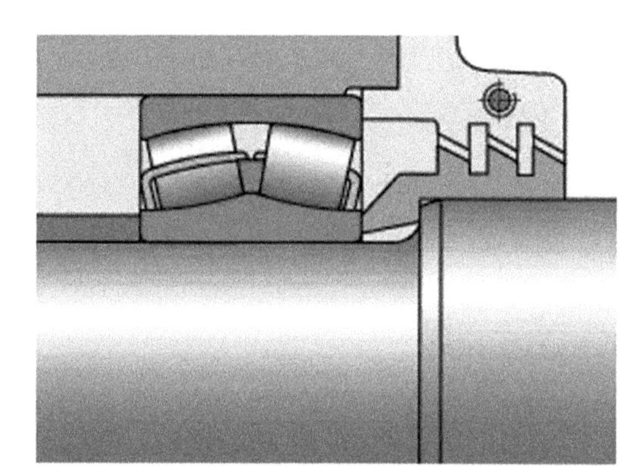

Figure 5.6 Inclined passages. Sources: SKF.

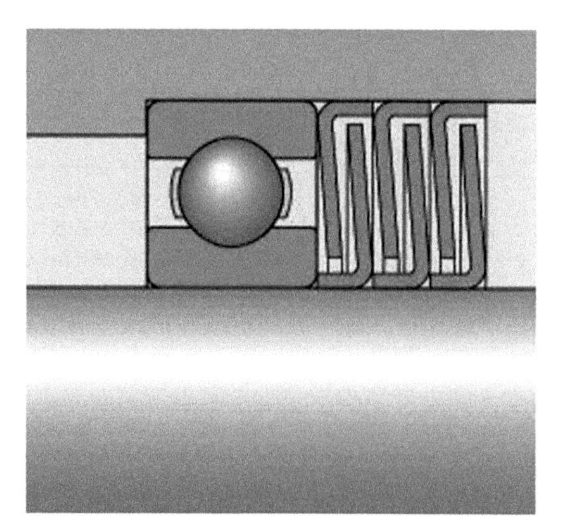

Figure 5.7 Labyrinth seal consisting of multiple sealing washers.

Figure 5.8 Rotating disc acting as a shield.
Sources: SKF.

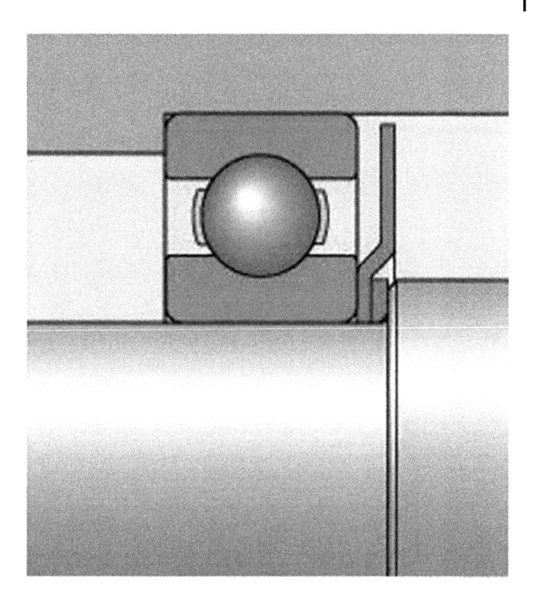

The *hybrid labyrinth seal* has been modified by adding a contact seal component designed to centrifugally lift off the rotating shaft when the seal is operating at normal speeds. This modification solves some of the limitations of the standard labyrinth seal, and makes it function more like a contact lip seal. In addition, a lip seal component can be incorporated such that it will only make contact at lower speeds and will centrifugally lift from the contact surface at higher speeds. While the hybrid labyrinth seal does solve the problem of contaminants entering the controlled fluid cavity when the seal is not rotating, it adds challenges in mounting the seal and controlling axial movement, and reduces the efficiency of the seal by adding a contact element.

5.2 Lip Seals

Initially, labyrinth seals were developed as a sealing method that would operate in extremely high shaft speed environments, or in applications where the durability of a traditional contact seal was limited. The contact lip seal has the stationary sealing element, or lip, in contact with the rotational element. This constant contact creates wear on both the lip and the rotating element, and adds significant parasitic drag to the system. In addition, operating speed is limited in these applications, owing to overheating and wear on the contact lip of the seal, as a result of rubbing between the rotating and nonrotating surfaces. Ultimately, the seal lip will melt or centrifugally lift from the contact surface as operating speeds increase.

The primary purpose of a lip seal is simple: exclude contaminants while retaining lubricants. By nature, lip seals function by maintaining friction (Figure 5.9).

The principle of a common oil lip seal is given in Figure 5.10.

This type has a carbon steel insert and a rubber exterior. The rubber gives a good sealing capability, even when the housing is not fully in tolerance. The sealing lip with spring provides interference on the shaft for effective sealing. The outside diameter, with inner metal reinforcement case, allows press-fitting in the housing, with sufficient interference

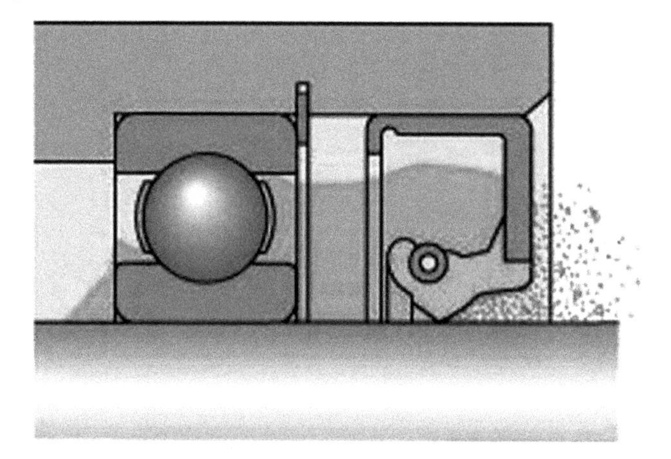

Figure 5.9 Purpose of lip seal. Source: KSB.

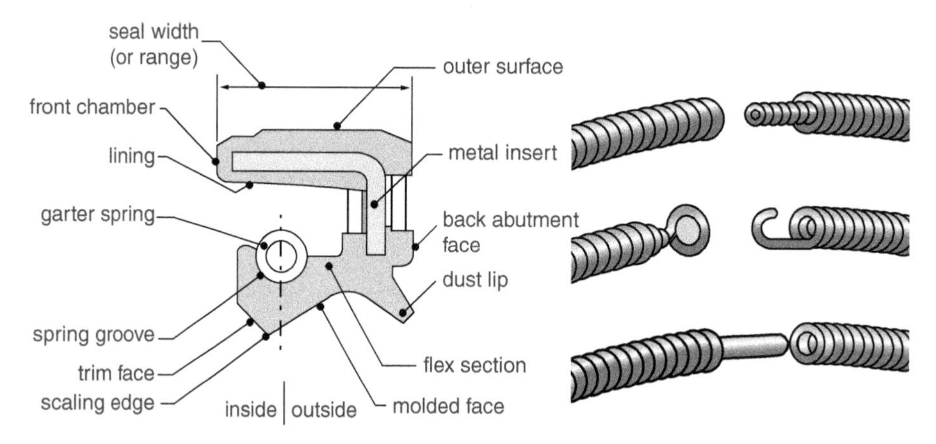

Figure 5.10 Principle of lip seal (Source: Eriks) and garter springs (Source: SKF).

on the rubber to provide static sealing. The sealing element is produced from a high performance nitrile rubber. This in combination with a high quality galvanized steel garter spring gives the oil seal an optimum life. In order to prevent leakages due to a hydrodynamic pumping effect it is necessary that the sealing lip contact area on the sleeve or shaft has no tracks of wear. The reinforcing case of carbon steel is standard but stainless steel or brass can also be used. This type is fully covered with rubber on the inside of the reinforcing case. It is fitted with a stainless steel garter spring. The seal lip is pressed against the shaft by the garter spring.

Depending on the seal material and medium to be retained and/or excluded, commonly used materials for radial shaft seals can be used at temperatures between −55 °C (−65 F) and +200 °C (375 F) (Figure 5.11).

The seal counter face, that part of the shaft where the seal lip makes contact, is of vital importance to sealing effectiveness. The surface hardness of the counter face should be at least 45 [HRC] at a depth of at least 0.3 [mm]. The surface texture should be in

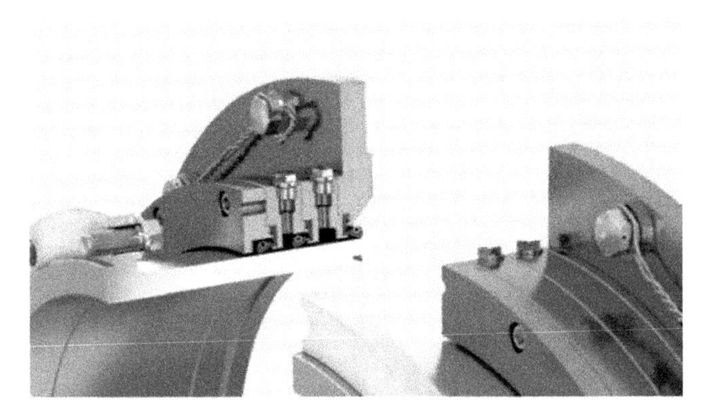

Figure 5.11 Lip seals: stern tube aft seal. Source: Lagersmit.

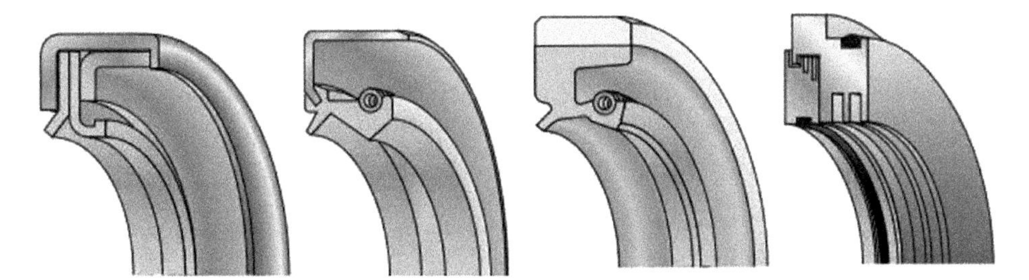

Figure 5.12 Lip seals: yellow: rubber, gray: metal reinforcement, 4th figure: with labyrinth seal. Source: Parker.

accordance with ISO 4288 and within the guidelines of Ra = 0.2–0.5 [μm]. In applications where speeds are low, lubrication is good, and contamination levels are minimal, a lower hardness can be acceptable (Figure 5.12).

If the primary purpose of the radial shaft seal is lubricant retention, the seal should be mounted with the lip facing inward (Figure 5.13). If the primary purpose is to

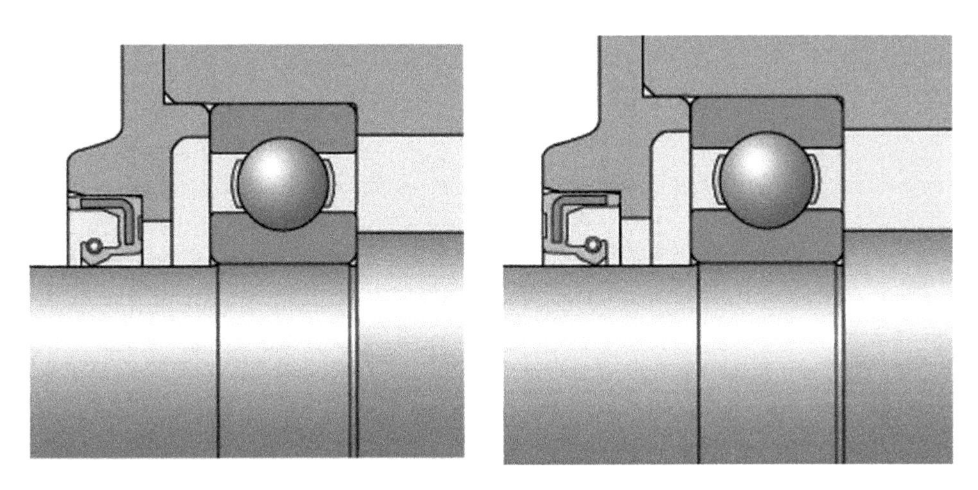

Figure 5.13 Lip facing inward or outward. Source: SKF.

Figure 5.14 Single lip and double lip.

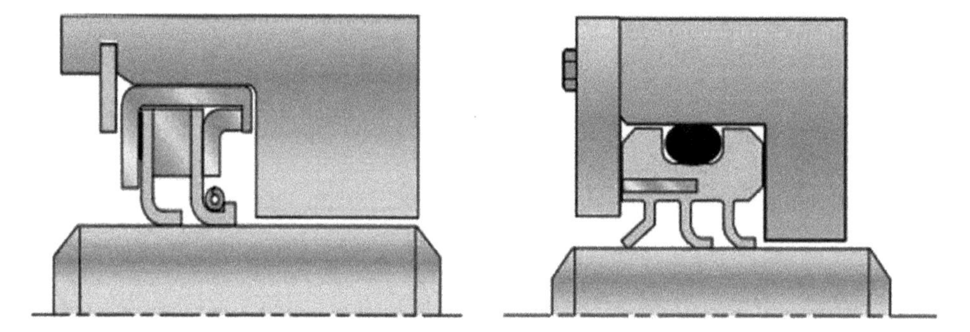

Figure 5.15 Double seal, one reinforced, and triple seal. Source: Parker.

exclude contaminants, the lip should face outward, away from the bearing (Figures 5.14 and 5.15).

5.3 V-Ring Seals

The sealing principle of a "V-seal" is an all rubber one-piece seal which is mounted on a shaft and seals axially against a counter face. Normally the seal rotates with the shaft. The construction of the V-seal is divided into three parts: (a) the body; the body of the V-seal which clamps itself on the shaft; (b) the conical self-adjusting lip; (c) the hinge.

Through its elastic fit the V-ring clamps itself in position on the shaft and rotates with it. Because of this, the static sealing between the shaft and the V-ring is ensured. The flexible sealing lip contacts the counter face – settles tolerances and deviations – thus producing a dynamic seal. The V-seal serves two purposes: a dynamic sealing element as well as a flinger. Because of the peripheral speed, dirt and liquids are removed and will not attach to the sealing lip. As Figure 5.16 shows, the sealing lip prevents dirt and liquids from entering. Moreover, V-rings show some flexibility concerning alignment of shafts (Figure 5.17).

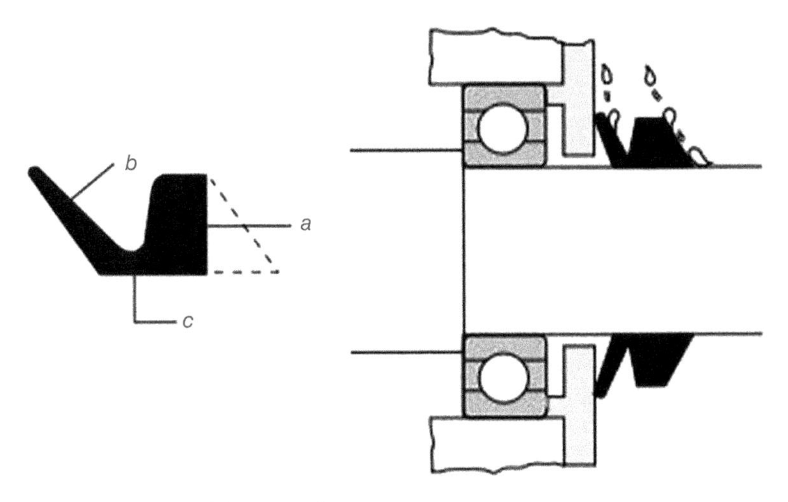

Figure 5.16 Lip seal. Source: Eriks.

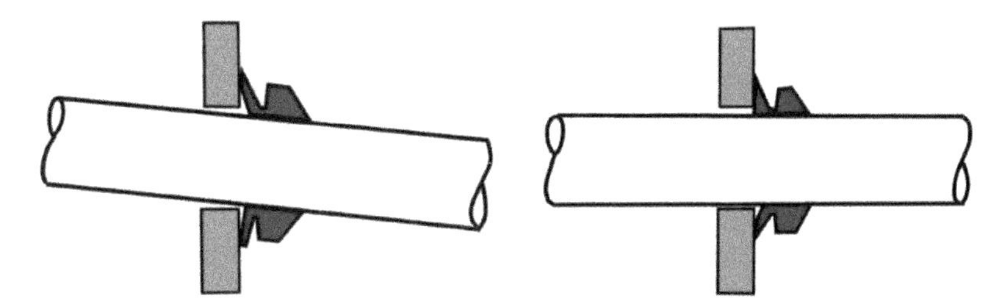

Figure 5.17 Flexibility of V-ring seals.

V-rings are suitable for both grease and oil-lubricated applications. For sealing grease lubricated bearing arrangements and protecting against contaminants, the V-ring should be arranged outside the housing cover or housing wall. Dust, water spray, and other contaminants can be excluded in this position (Figure 5.18a). The V-ring can also act as a grease valve, where used grease or excess new grease can escape between the counter face and the sealing lip (Figure 5.18b). The installation of two opposing V-rings

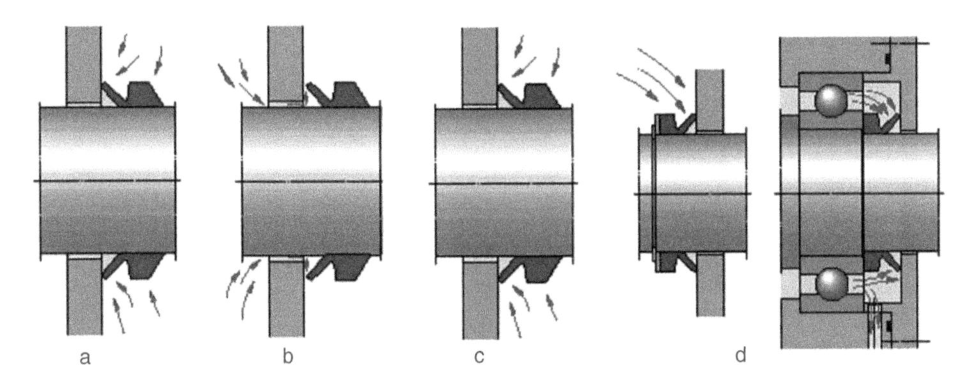

Figure 5.18 V-ring seals. Source: KSB.

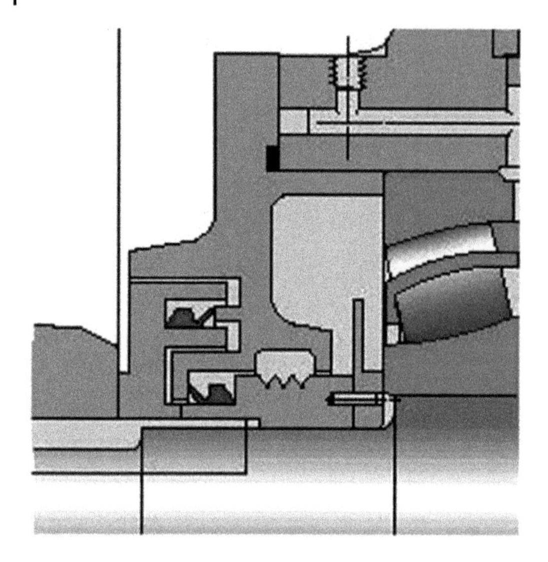

Figure 5.19 Combination two V-rings and a labyrinth seal.

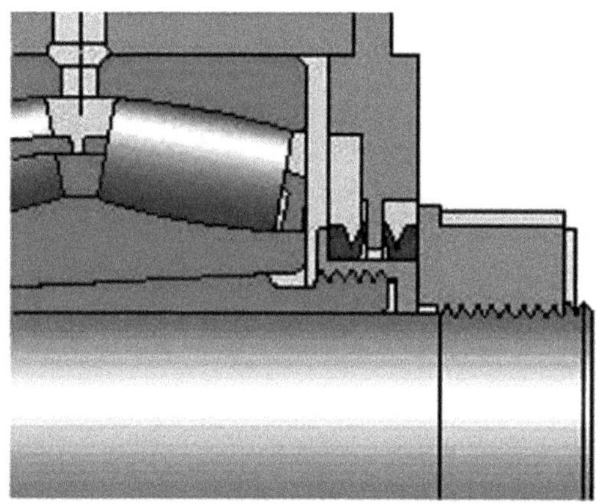

Figure 5.20 Labyrinth and two opposing V-seals. Sources: SKF.

can be used in applications where lubricant retention and contaminant exclusion are of equal importance (Figure 5.18c). If V-rings are used to retain oil, they should always be located axially on the shaft on the lubricant side (Figures 5.18d–5.20).

5.4 Gland Packing

The gland packing's application limit is primarily determined by the extent to which heat developed due to friction can be dissipated. For heavy-duty gland packings, leakage water is actually pre-cooled by means of an internally cooled shaft protecting sleeve and a cooling jacket.

The packing materials generally employed are braided cords made from asbestos-free yarn such as ramie, aramid, PTFE (e.g. Teflon™), graphite fibers, or cotton, which are processed on special machines to form endless square braids.

The packings can be adjusted and are suitable for higher pressures and circumferential speeds than lip seals (Figure 5.21).

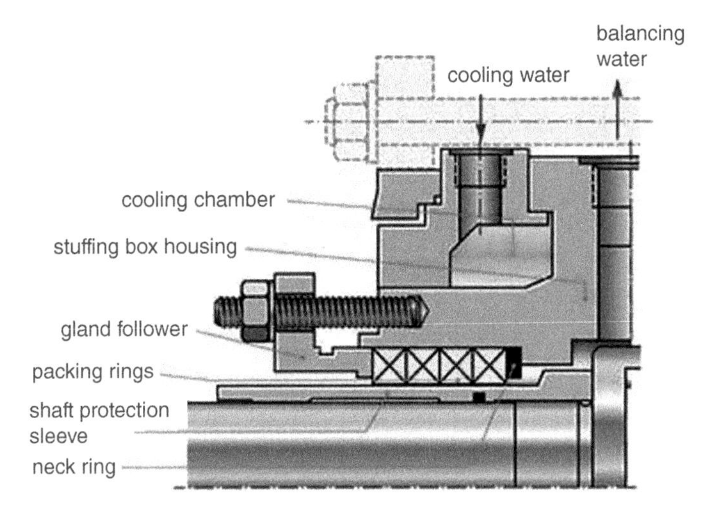

Figure 5.21 Four-shaft seal: water-cooled gland packing. Source: KSB.

Different packing variants are used depending on whether the pump is run (in suction head or suction lift operation) or whether it handles clean or contaminated fluid.

In the case of positive pressure, the gland packing is equipped with three to five packing rings. These packing rings are pressed together axially via the *gland follower*. As a result they expand radially which means the pressure on the shaft protecting sleeve is increased. This has an influence on the *clearance gap width* and the *leakage rate* at this location.

Various excellent gland materials are available (Figures 5.22). They are sold as strings; it suffices to cut them to the desired length. With the exception of graphite glands, all glands are twisted from isolating threads. In other words, these threads do not conduct the friction heat. For a gland packing to function, but primarily survive, it is necessary that a leak is allowed. The leak must be big enough for the friction heat to be evacuated. Often this is a personal matter; one mechanic may try to keep the leakage as small as a few drops per minute; another may only be pleased if there is a continuous fluid flow.

In any case, a few basic rules can be stipulated. It is clear that by choosing a gland packing this should be one that is resistant to the used process fluid (Figure 5.23).

Figure 5.22 Gland materials. Source: Eriks.

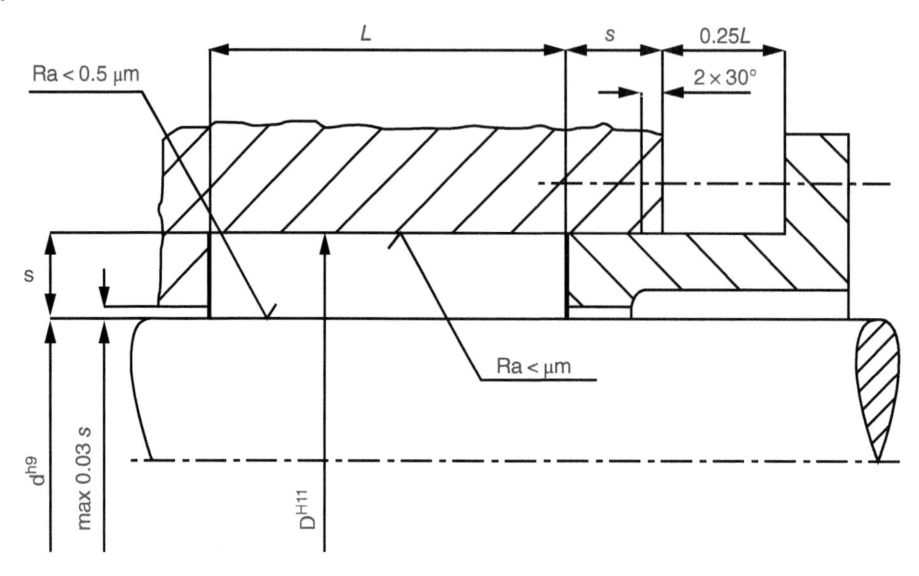

Figure 5.23 Constructive dimensions. Source: Eriks.

One has to consult the catalogs of reputable dealers. Chemical durability, the highest allowable pressure and the maximum allowable circumferential velocity are parameters to consider. Constitutively one takes into account the dimensions, tolerances, and roughness, as shown in Figure 5.24, where *L*: length gland, *s*: section gland, *D*: gland diameter, and d: shaft or shaft bus dimension.

Apart from these requirements the choice of the material of the shaft or shaft bus upon which the gland has to be fixed is also very important. The chosen material must be not only resistant to the process liquid but also hard enough and have the right surface hardness and roughness for a limited wear. A hardness of more than 55 [HRC] and a roughness of Ra < 0.5 [μm] is recommended.

Above that the material of the shaft or shaft bushing may not be isolating. It must be able to evacuate the friction heat. Hafts or shaft bushing with sprayed ceramic layers (high surface hardness, but bad heat conduction) are not recommended. Carbides, chrome carbide, or wolfram carbide are preferred. Of course, there are a few applications where ceramic layers can be used: very low circumferential velocities, very low pressures, and reciprocating movement (plunger pumps).

The mounting must be done with caution. The section of the gland is given by:

$$\frac{(D-d)}{2}$$

The length of every gland ring can be found by using the formulae:

$$L = 1.03 \text{ to } 1.07\,(d+s)\cdot 3.14$$

The factor 1.03–1.07 is dependent on the linear compressibility of the twisted gland. Very loose glands can be pressed easily, strong twined glands much less so.

The number of gland rings that can be mounted is more dependent on the pressure than on the length of the gland chamber. In most cases four to five rings will suffice

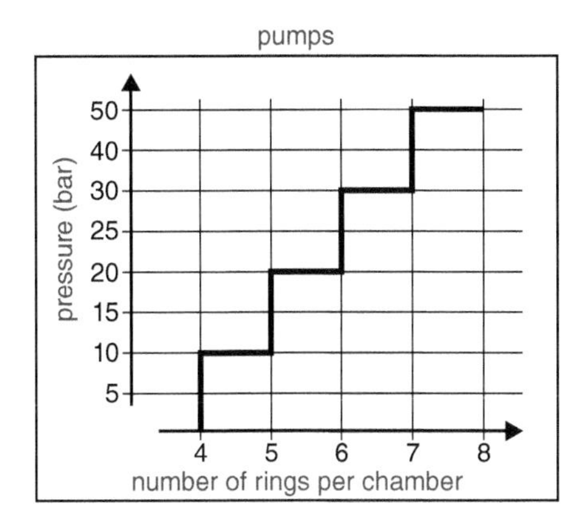

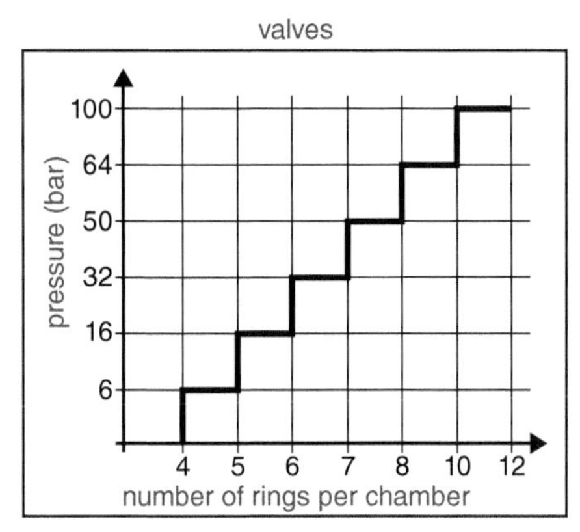

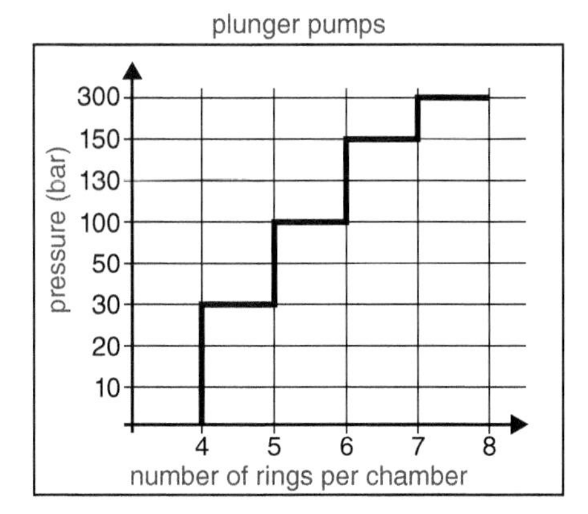

Figure 5.24 Number of rings. Source: Eriks.

(Figure 5.24). Every cut of a gland ring is 45–60 rotated to one another for no leakage channel to arise.

As the leakage with gland packings is relatively high compared with mechanical seals, the former are mostly employed for environmentally friendly fluids only.

The gland packing can be operated without cooling for fluid temperatures up to 120 °C.

When used with hot water up to 180 °C, the gland packing must be fitted with a cooling jacket. For higher temperatures cooling is ensured via a combination of an internally cooled shaft protecting sleeve and a cooling jacket.

To enlighten the pressure on the glands and to balance the axial thrust, back vanes are used (Figure 5.25). These are narrow vanes in a radial arrangement on the rear side of a centrifugal pump's impeller. The sealing pressure is higher than the suction pressure and lower than the discharge pressure. Although the exact values differ from one pump to another, the following formulae have proven sufficiently accurate for determining the sealing pressure of a single stage centrifugal pump:

For a pump speed of $N = 1500$ [rpm]:

$$p_A \cong p_s + 2/3 \cdot H_{man} \cdot \rho$$

where: p_A: sealing pressure and p_s: suction pressure.

Figure 5.25 Counter vanes. Source: Gwyneth.

For a pumps speed of $N = 3000$ [rpm]:

$$p_A \cong p_s + 1/3 \cdot H_{man} \cdot \rho$$

5.5 Lantern Rings

If the process fluid contains abrasive particles these will rather wear the bushing, which will lead to lubrification. In that case one uses a lantern ring, placed in the middle of the gland rings (Figure 5.26 and 5.27). Lantern rings are made of bronze or brass. When corrosive fluids are used PTFE is used.

The gland packing in Figure 5.29 has four gland rings. The packing material is compressed with a packing press, kept firm by bolts. It is important that these bolts are not over-tightened. The packing has to be able to leak in order to cool the packing and the shaft and lubricate it.

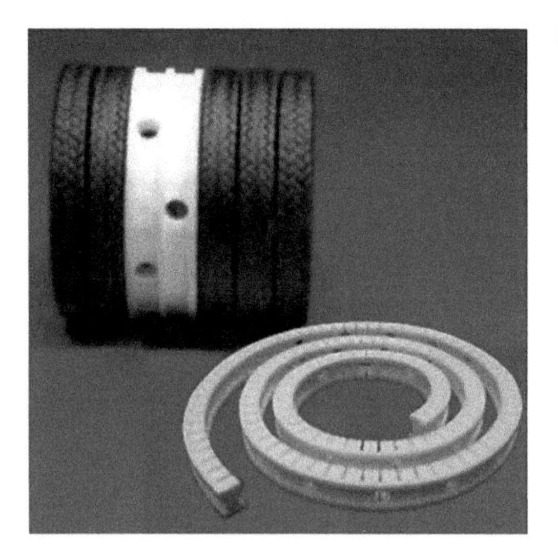

Figure 5.26 Lantern rings.

Figure 5.27 Lantern ring and gland rings.

Clean liquid in over pressure is injected above the lantern ring. This liquid flows back to the pump and keeps abrasive particles away from the packing rings. The leak to the atmosphere will then just be pure liquid.

Above that the lantern ring has a cooling function.

As the pump discharge pressure is higher than the atmospheric pressure, air cannot penetrate the pump.

The barrier fluid pressure should normally be approx. 10% or at least 2 [bar] higher than the highest pressure to be sealed.

A barrier fluid connection is also required under suction head conditions (positive pressure), if the fluid is contaminated. If this were not the case, the contaminants would be forced through the packing with the leaking fluid. The contaminants would settle at the contact face of the gland packing and rapidly destroy the shaft protecting sleeve due to their abrasive effect.

In this case, an external barrier fluid supply is the only suitable option. The lantern ring would then be fitted as the innermost ring (Figure 5.28).

As the barrier fluid pressure is higher than the pump pressure, a certain amount of the barrier fluid mixes with the fluid handled inside the pump, so that compatibility between the barrier fluid and the fluid handled should be ensured.

When the gland packing is serviced, both the packing cord and shaft protecting sleeve must be assessed for wear.

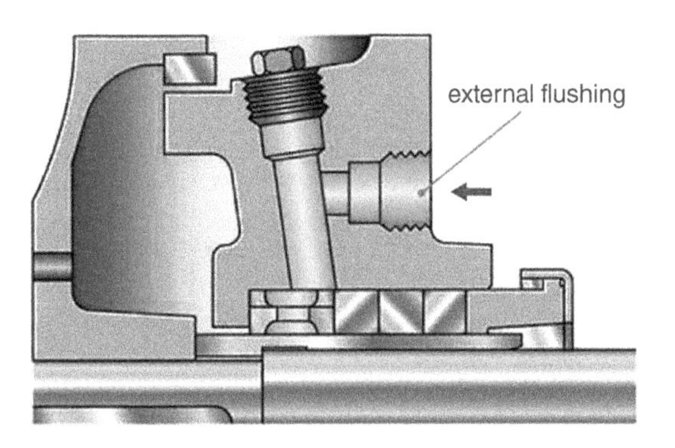

Figure 5.28 External flushing of glands via lantern ring. Source: KSB.

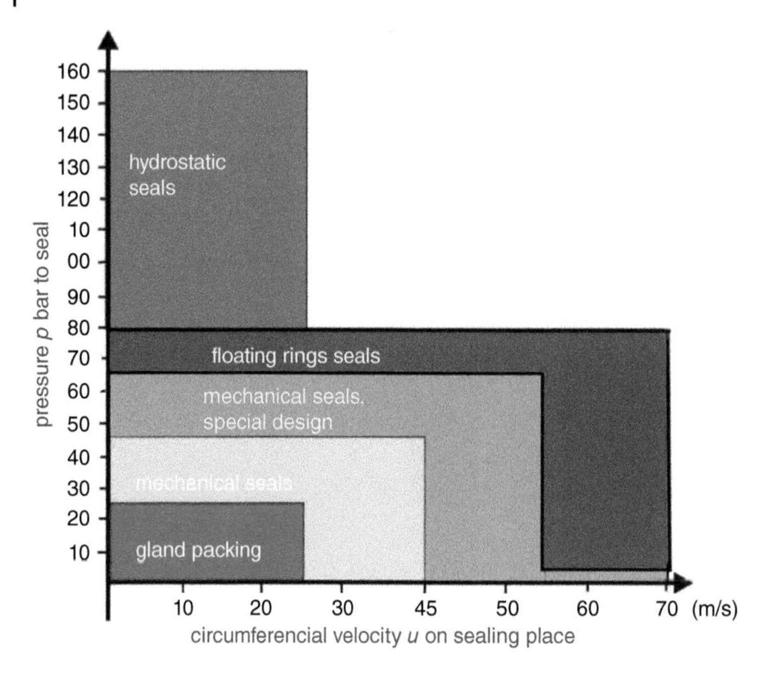

Figure 5.29 Application fields for type of seals. Source: KSB.

This method is particularly important when the fluid's purity cannot be guaranteed at all times.

Shaft seals (Figure 5.29) are by their nature susceptible to leakage, and with some types leakage is actually essential to ensure proper sealing functioning. The suggestion that a seal shaft provides "zero leakage" is therefore misleading. However, depending on the seal type chosen, the amount of leakage can vary considerably. A volute casing pump with at the sealing area of 20 [m/s] and a pressure to be sealed of 15 [bar] which uses a *gland packing* for sealing has a leakage rate of about 5–8 [l/h], while the leakage rate of *a mechanical seal* used under the same conditions is only approx. 6 [cm^3/h] (0.006 [l/h]).

Figure 5.29 shows different types of sealings and their application field.

5.6 Mechanical Seals

5.6.1 Fundamentals

A mechanical seal is a precision device. Applications are centrifugal pumps, compressors, turbines, mixers, marine propellers shafts, and aircraft engines. Although mass-produced mechanical seals as those used in automotive water pumps and household apparatus, cost in the range of $1–2, most mechanical seals range in price from a few hundred to several hundred thousand dollars (for a nuclear reactor coolant pump seal system).

Mechanical seals have the purpose of preventing leakage of a fluid (liquid or gaseous) through the clearance between a shaft and the fluid container. That doesn't mean that a mechanical seal does not leak. It does – it should – because there must be lubrification, but leakage is so minimal that it can't be detected (vaporization).

The main components of a mechanical seal are the seal rings on which a mechanical force is acting, generated by springs or bellows, and an hydraulic force, generated by the process of fluid pressure (Figure 5.30–5.32).

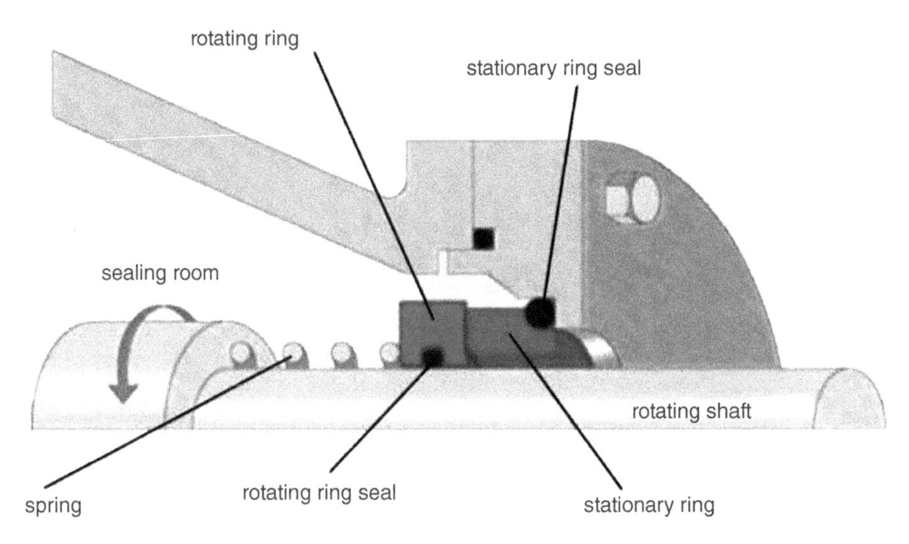

Figure 5.30 Mechanical seal: mating faces. Source: Fluiten.

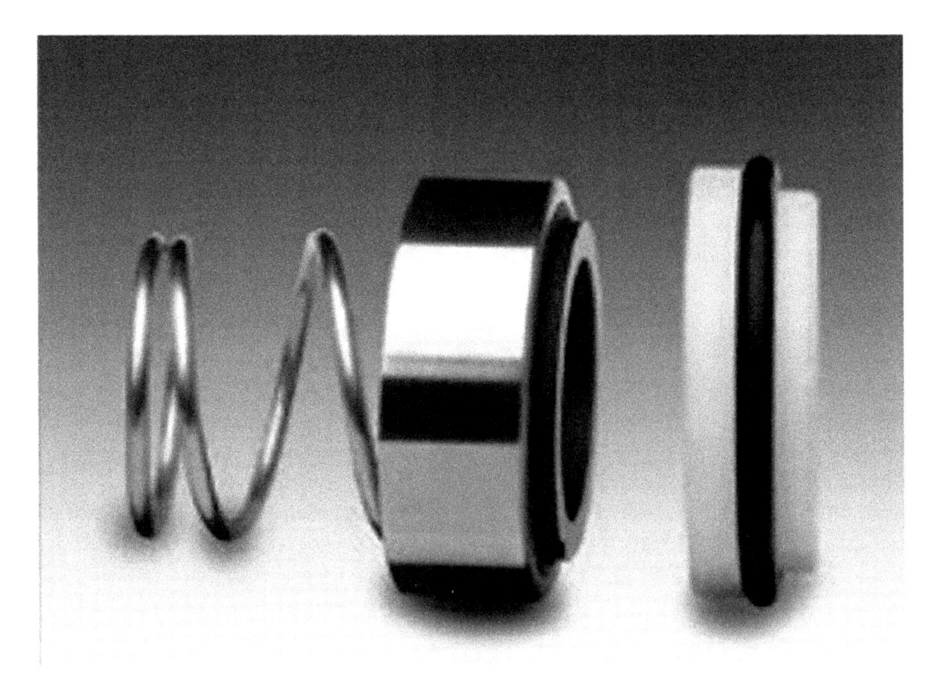

Figure 5.31 Mechanical seal. Source: Flowserve.

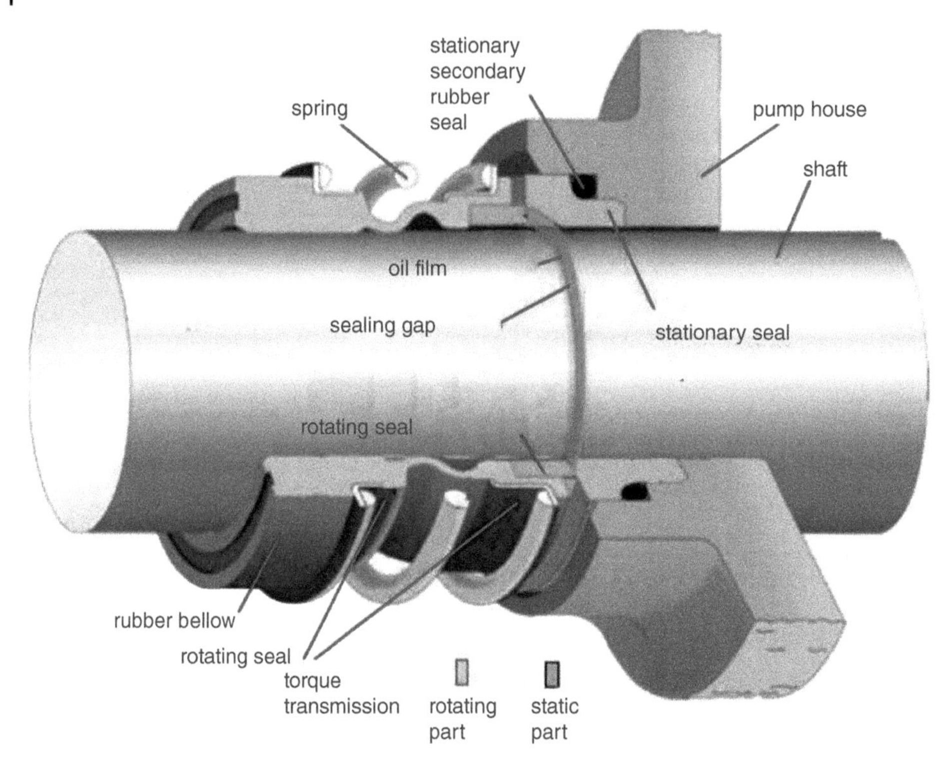

Figure 5.32 Global view on mechanical seal. Source: Grundfos.

The two annular seal faces have flat mating surfaces. One of those faces, the *rotating ring*, is mounted on the rotating shaft, while the other, the *stationary ring*, is mounted on the housing. One of the two is free to float in the axial direction, because it is loaded by a spring or bellows, and forces the seal face toward the fixed face. The interface between the two faces is lubricated by the process fluid, or, with double mechanical seals, by a proper auxiliary fluid (Figure 5.31).

Secondary seals are required to perform static sealing between rotary rings and shafts and also between stationary rings and the casing of the machinery.

Elastomeric O-rings are usually used as secondary seals but alternative systems can be used, as is described in the following sections.

During the pump standstill the compression force of the spring keeps the two seals together and there is no leak.

When the shaft starts to rotate, the pressure pushes the liquid against the spring and opens the sealing face. This generates a lubricating film.

A mechanical seal has the ability to tolerate some degree of eccentricity, or misalignment, while maintaining a relatively low leakage rate. Even if the two faces are not parallel and concentric, they will still track each other because of the two faces are flexibly mounted and float. The thickness of the lubricating film in the interface is typically in the order of microns (μm).

The pressure in the film decreases from the pressure of the process fluid to the atmospheric pressure (Figure 5.33).

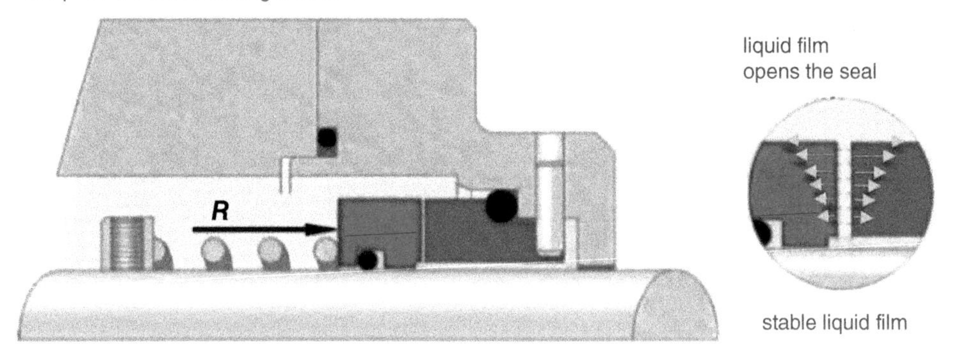

Figure 5.33 Mechanical seal. Source: Fluiten.

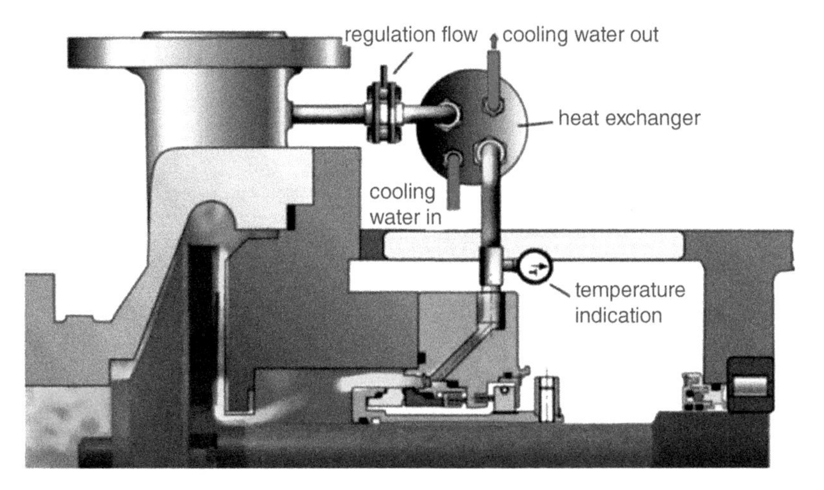

Figure 5.34 Cooling of the seal faces. Source: AESSAL.

5.6.2 Unbalanced Seals

In mechanical seals, in addition to the closing force generated by the springs or bellow, a hydrostatic force generated by the fluid pressure acts on the seal ring. The spring force is small, so we shall ignore it for this analysis.

In this section ways to remove the friction heat generated in the sealing are considered. One can also look for means to decrease the heat (Figure 5.34). Factors that influence this are: roughness of the two sliding surfaces, the pressure on these surfaces, the kind of liquid, temperature of the liquid, speed of the pump, etc.

The term *hydraulic balance* refers to the equilibrium between the opposing hydraulic locking forces on the sliding surfaces that pinch the liquid film in order to obtain a stable hydraulic film with low pressures on the sliding faces. This will decrease the friction heat and prolong the lifecycle of the seal.

Consider the situation in Figure 5.35.

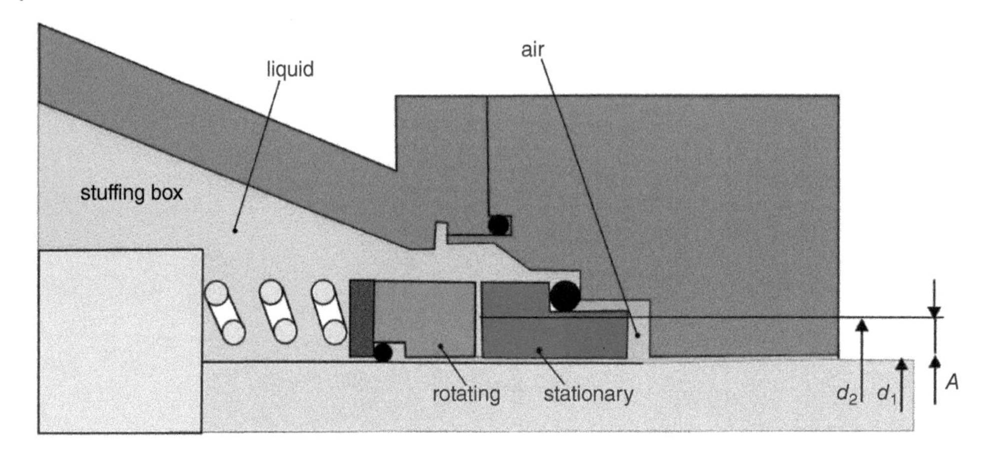

Figure 5.35 Unbalanced seal. Source: Fluiten.

As previously discussed the fluid pressure also penetrates between the seal faces, producing a lubrication film and generating an opening force. The process fluid also penetrates around the spring and acts as a hydraulic force on the rotating ring, better: on the *hydraulic surface* (see Figure 5.36).

If R is the total force of the rotating ring and S the total force of the stationary ring, A the facing surface, A_h the hydraulic surface, b the liquid film area, s the pressure in the liquid film between the two faces, and p the pressure in the process liquid, take for instance:

$$p = 10 \text{ [bar] and } A = \pi \cdot \frac{(d_2^2 - d_1^2)}{4}$$

Then: $R = p \cdot A_h$ Suppose in a first case that:

$$A_h = A$$

$$S = s \cdot A = s \cdot A_h$$

Because: $R = S$ (equilibrium):

$$s \cdot A_h = p \cdot A_h$$

From where:

$$s = p = 10 \text{ [bar]}$$

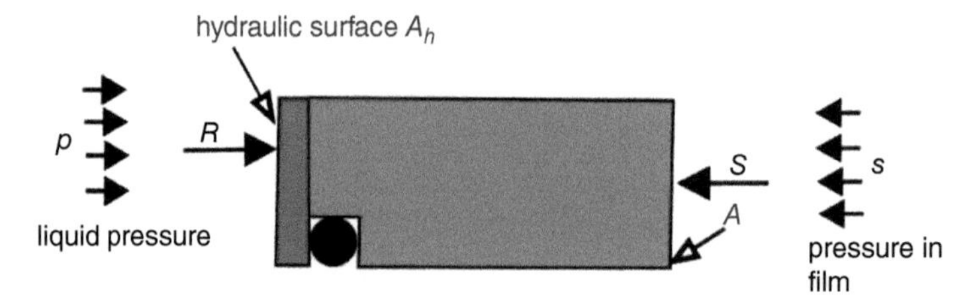

Figure 5.36 Equilibrium of forces.

This is a hydraulic unbalanced seal. There is an equilibrium between the hydraulic closing pressure p and the pressure s on the mating faces.

Suppose now that $A = 10$ [cm^2].

The closing force R amounts to:

$$\mathrm{R} = p \cdot A_h = p \cdot A = 10 \cdot 10^5 \cdot 2 \cdot 10^{-4} = 200\ [\mathrm{N}]$$

In fact, in an unbalanced seal the hydraulic surface is too high compared with A.

Another case is the one is which $A_h > A$. Take, for instance, $A_h = 1.25{\cdot}A$.

Then:

$$R = p \cdot A_h.$$

$$\mathrm{S} = \mathrm{s} \cdot A = \mathrm{s} \cdot \frac{A_h}{1.25}$$

Because: $R = S$ (equilibrium):

$$s \cdot \frac{A_h}{1.25} = p \cdot A_h$$

From where:

$$s = 1.25 \cdot p = 12.5\ [\mathrm{bar}]$$

Now the closing force R amounts to:

$$\mathrm{R} = p \cdot A_h = p \cdot 1.25 \cdot A = 10 \cdot 10^5 \cdot 1.25 \cdot 2 \cdot 10^{-4} = 250\ [\mathrm{N}]$$

If the stuffing box pressure is very high, typically over 15 [bar], then the closing force may be too great to allow the boundary layer liquid that lubricates the faces to be sufficient and the faces will wear prematurely. That's why unbalanced seals are only adequate for low pressure applications.

Also, when the sliding velocities of the mating faces are low, the friction heat in the gap cannot be transferred enough to the shaft housing and to the atmosphere, because the convection heat transfer is low. That's why unbalanced seals are only used with sliding velocities up to 15 [m/s].

Generally, unbalanced seals have good performance when subjected to vibrations, misalignments, or cavitation; they are cheaper and their application does not require shaft or sleeve notching (see the next section).

5.6.3 Balanced Seals

Here the shaft has a notch (Figure 5.37).

Suppose again the *hydraulic surface* A_h equal to the mating surface A (Figure 5.38):

$$A_h = A$$

And:

$$p = 10\ [\mathrm{bar}]$$

Then:

$$R = p \cdot A_h - p \cdot 0.25 \cdot A = p \cdot A - 0.25 \cdot p \cdot A = 0.75 \cdot p \cdot A$$

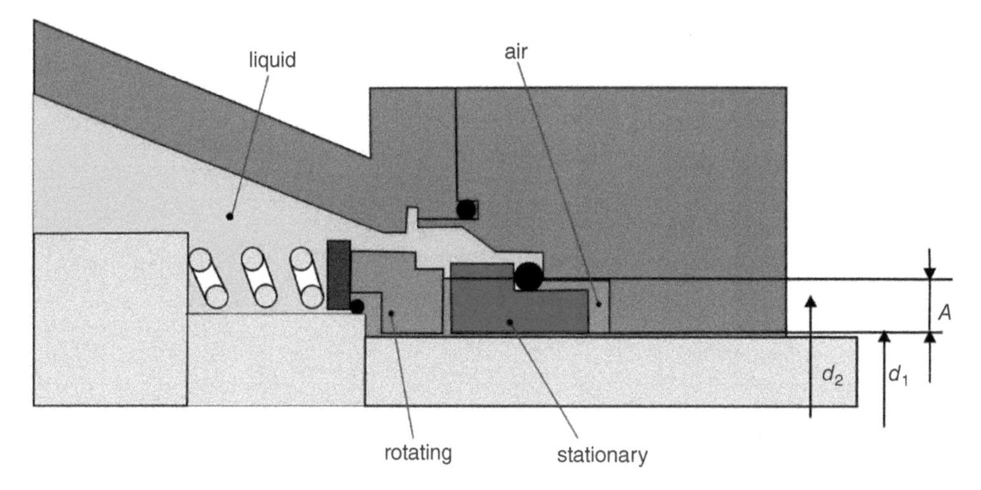

Figure 5.37 Unbalanced seal. Source: Fluiten.

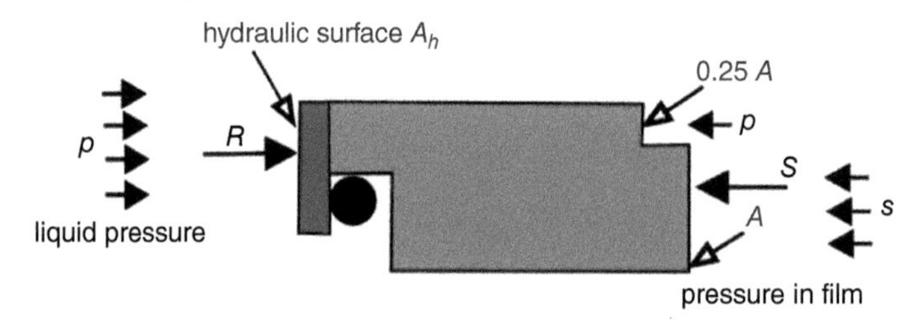

Figure 5.38 Equilibrium of forces. Source: Fluiten.

And:

$$S = s \cdot A$$

Because: $R = S$ (equilibrium):

$$s \cdot A = 0.75 \cdot p \cdot A$$

From where:

$$s = 0.75 \cdot p = 7.5\ [\text{bar}]$$

The closing force R now amounts to:

$$\text{R} = p \cdot A_h = p \cdot A - 0.25 \cdot p \cdot A = 0.75 \cdot p \cdot A = 0.75 \cdot 10 \cdot 10^5 \cdot 2 \cdot 10^{-4}$$
$$= 150\ [\text{N}]$$

If the hydraulic surface is chosen smaller, for instance:

$$A_h = 0.75\,A$$

Then:

$$R = p \cdot A_h - p \cdot 0.25 \cdot A = p \cdot 0.75\,A - 0.25 \cdot p \cdot A = 0.5 \cdot p \cdot A$$

Because: $R = S$ (equilibrium):

$$s \cdot A = 0.5 \cdot p \cdot A$$

From where:

$$s = 0.5 \cdot p = 5\,[\text{bar}]$$

The closing force R now amounts to:

$$\text{R} = 0.5 \cdot p \cdot A = 0.5 \cdot 10 \cdot 10^5 \cdot 2 \cdot 10^{-4} = 100\,[\text{N}]$$

This gives a stable hydraulic film with lower pressure on the mating faces. The dissipated heat will decrease because the pressure in the seal gap is quiet low and so the life cycle of the seal is high.

Balanced seals are used for high pressure applications and high sliding velocities of the mating faces. Because the seal gap is higher they have a higher leaking rate.

Also in cases where a high value of vapor pressure has to be considered, a balanced mechanical seal is the right choice (Figure 5.39).

5.6.4 The Configurations

5.6.4.1 Single Internal Seal

This is the most popular and efficient configuration for the most applications.

It is called internal because of its being completely submerged in the product. It is an unbalanced sealing designed for pressure acting outside the seal; therefore, usually, if it is installed as an external seal, the fluid pressure will cause translation of the stationary ring and excessive separation of the seal faces (Figure 5.40).

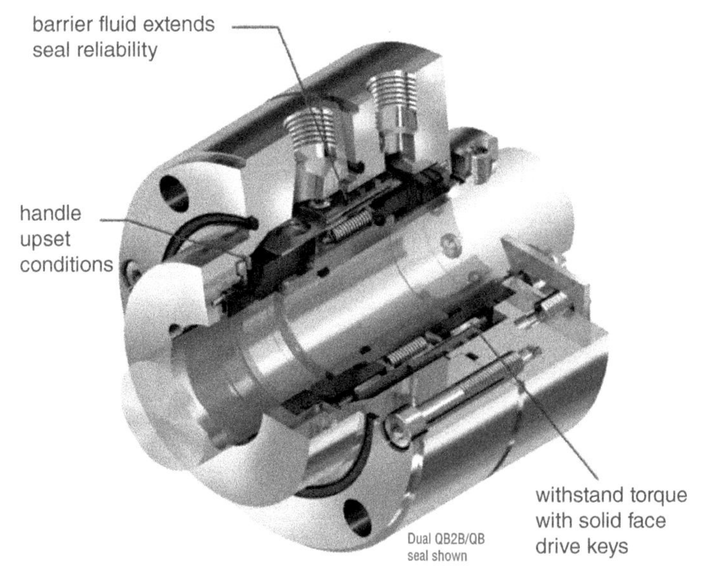

Figure 5.39 Balanced mechanical seal. Source: Fluiten.

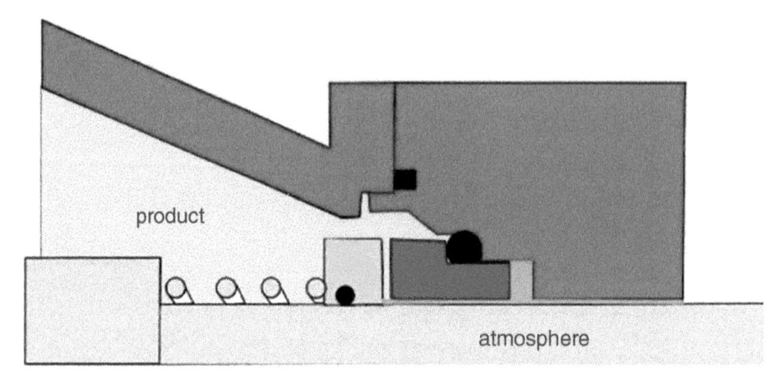

Figure 5.40 Single internal seal. Source: Fluiten.

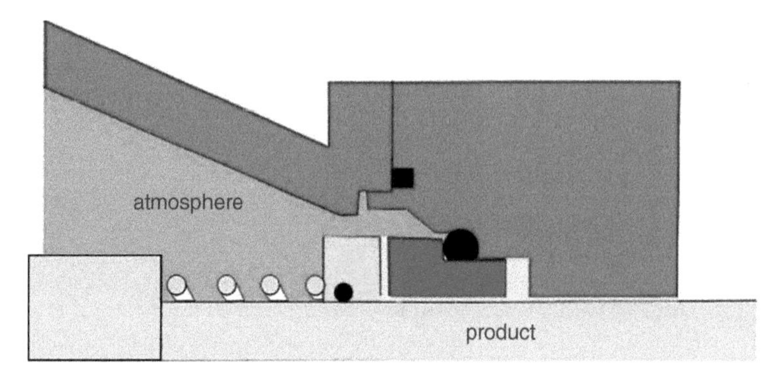

Figure 5.41 Single external seal. Source: Fluiten.

5.6.4.2 Single External Seal

In this execution the sealed product is inside the seal and the outside part of the rotary ring is exposed to the atmosphere (Figure 5.41).

It is employed with aggressive fluids which can chemically attack materials commonly used for internal seals or when the use of special materials is considered too expensive.

In this type of seal often there are no metallic parts in contact with the product or, if there are any, special materials such as Hastelloy or titanium are used.

The rotary ring and the stationary ring (in contact with process fluid) can be made of graphite, ceramic, or silicon carbide.

Gaskets can be made of fluoroelastomer or PTFE (Figure 5.42).

5.6.4.3 Back-to-back Double Seal

This configuration is recommended with critical products (i.e. gaseous, abrasive, toxic, or lethal) and generally where no emissions in the atmosphere are permitted. The back-to-back layout, so called because the two seals are placed literally back-to-back, allows a barrier made of a pressurized auxiliary fluid that is not harmful to the environment to be created.

The lubrication of the seal faces is carried out by the auxiliary fluid, which should be compatible with the process fluid (Figure 5.43).

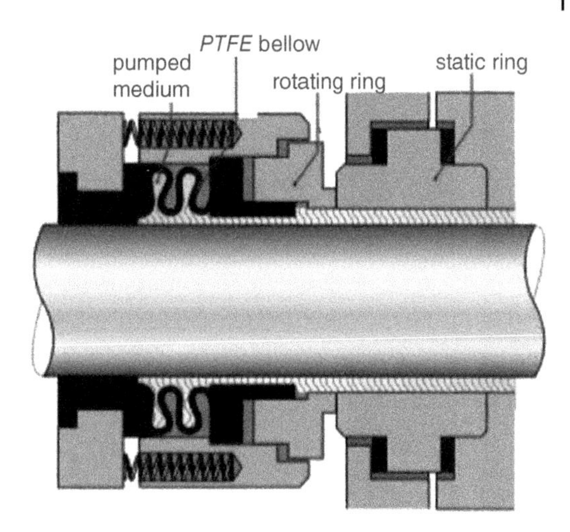

Figure 5.42 PTFE bellows instead of spring external seal. Source: Grundfos.

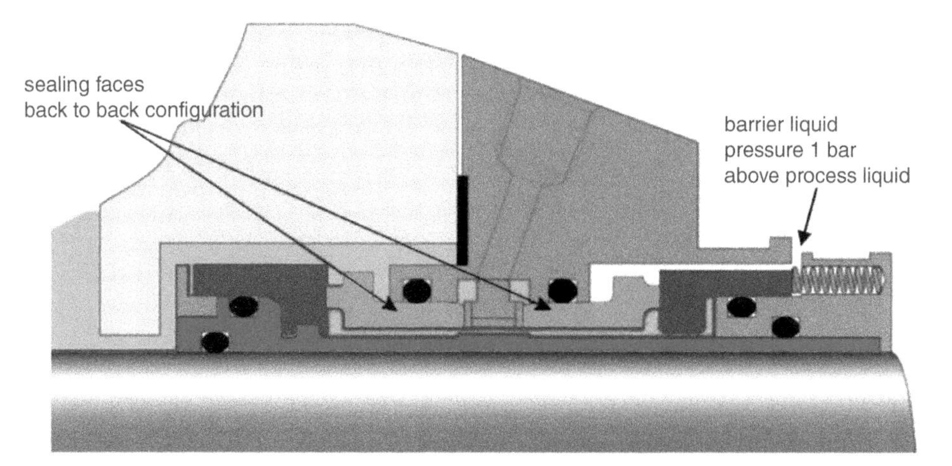

Figure 5.43 Back to back. Source: Fluiten.

In a back-to-back configuration an internal pressurization having a value greater than the process fluid (at least 1 [bar] or 10% more) is required in order to avoid opening the seal (as explained in Section 5.6.4.1) and to provide an efficient barrier against leakage of process fluid into the atmosphere (Figure 5.44).

5.6.4.4 Tandem Double Seal (Face-to-back Seal)

In this configuration the two seals are assembled with the same orientation. The right seal in the drawing seal is lubricated by the process liquid and a recirculation pipe line is provided. A second seal, identical as the first, is placed before (or behind) the first one ("in tandem"). The auxiliary (*buffer*) fluid often is at a lower pressure than the process fluid but also pressurized systems can be implemented with suitable seal rings (see dual seals, Figure 5.45 and 5.46).

In an unpressurised configuration there is the advantage of avoiding relatively costly pressurization systems obtaining a performance equivalent to the back-to-back layout's,

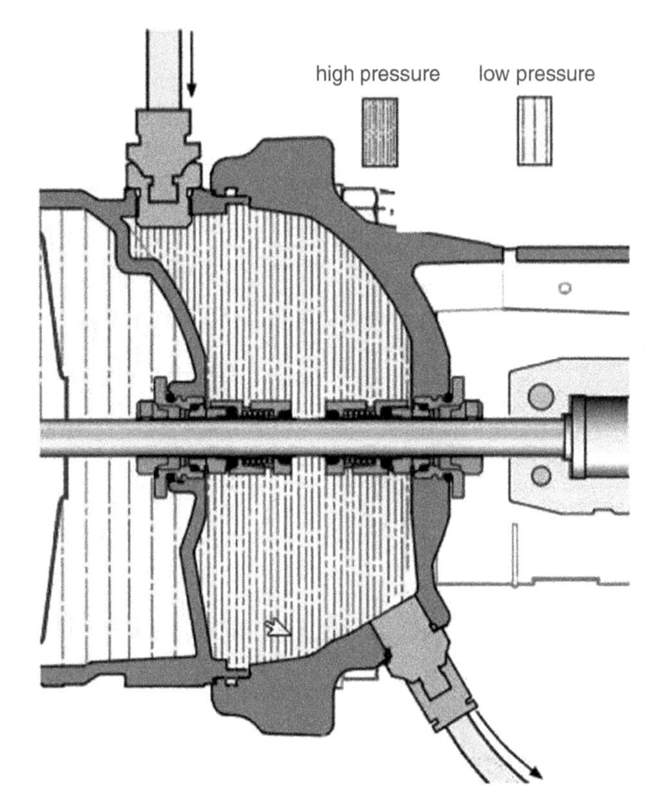

Figure 5.44 Back to back arrangement. Source: Grundfos.

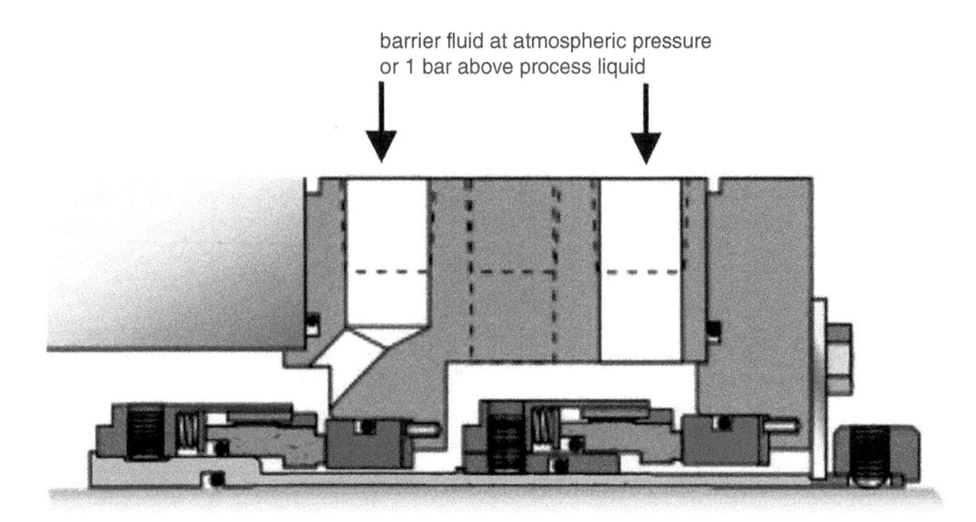

Figure 5.45 Tandem double seal configuration.

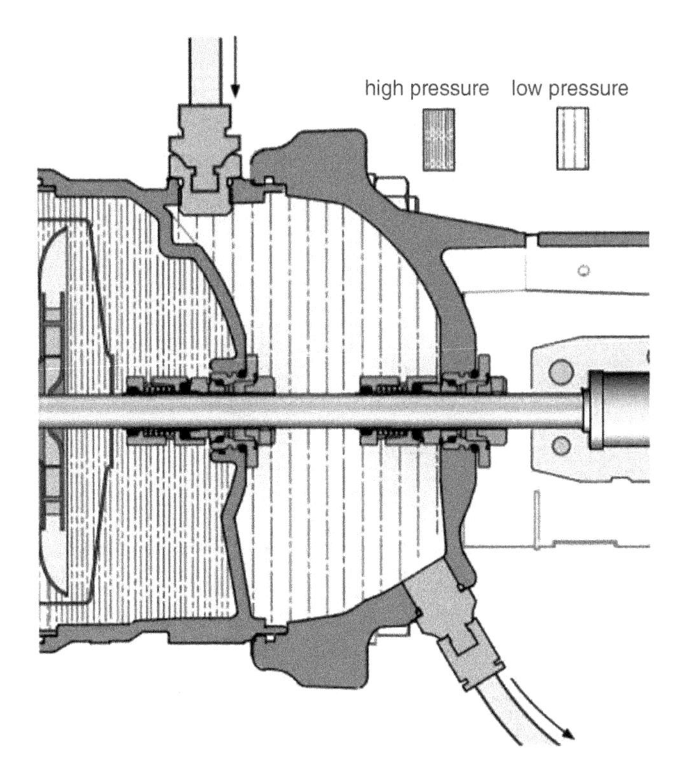

Figure 5.46 Tandem seal. Source: Grundfos.

which consists of no leakage of the process fluid into the atmosphere and good lubrication and cooling of the seal rings.

This configuration, however, is not suitable for toxic, abrasive, or highly viscous process fluids, prone to create sticking of seal rings; in these cases the back-to-back configuration should be used. Or there must be a safe place, a vapor recuperation system, or a flare provided.

Tandem double seals are usually employed in petrochemical and refinery plants, where service with high vapor pressure and low specific weight on centrifugal pumps is required.

5.6.4.5 Dual Seal

This is a new configuration foreseen by API 682 standard, where the two seals are assembled in a tandem layout (Figure 5.47).

A special design of the seal rings gives the possibility to operate both in an unpressurized system and in a pressurized system (as with the back-to-back configuration), obtaining the advantages of the two previous configurations.

Only a cartridge assembly is allowed by API 682 in this configuration.

5.6.4.6 Face-to-face Seal

This last double seal configuration is composed of a unique central stationary ring and two opposite rotary rings.

It can work in the same way as a dual seal (pressurized and unpressurized system).

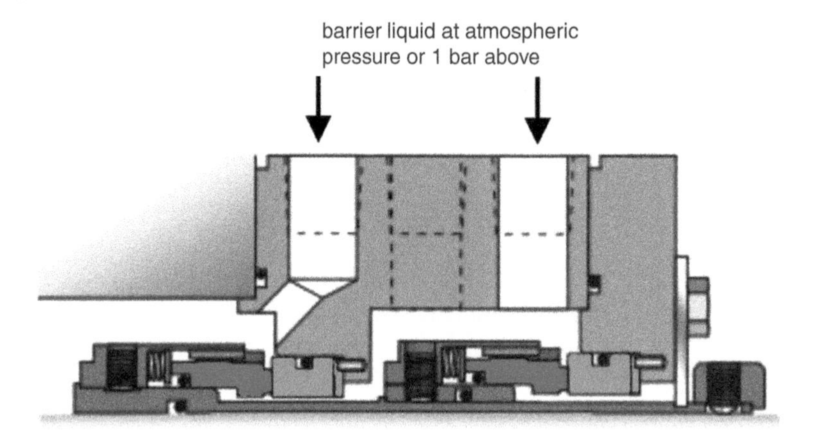

Figure 5.47 Dual seal configuration.

Less used than some of the previous configurations, it has some interesting features like reduced overall length and the spring not in contact with the process fluid (Figure 5.48).

5.6.5 Calculation of Liquid Flow

Let b be the gap between the seals (Figure 5.35). The surface A where the fluid runs through in the seal gap is:

$$A = \pi \cdot d_2 \cdot b$$

The volumetric flow of the leak flow is then:

$$Q_V = c \cdot \pi \cdot d_2 \cdot b$$

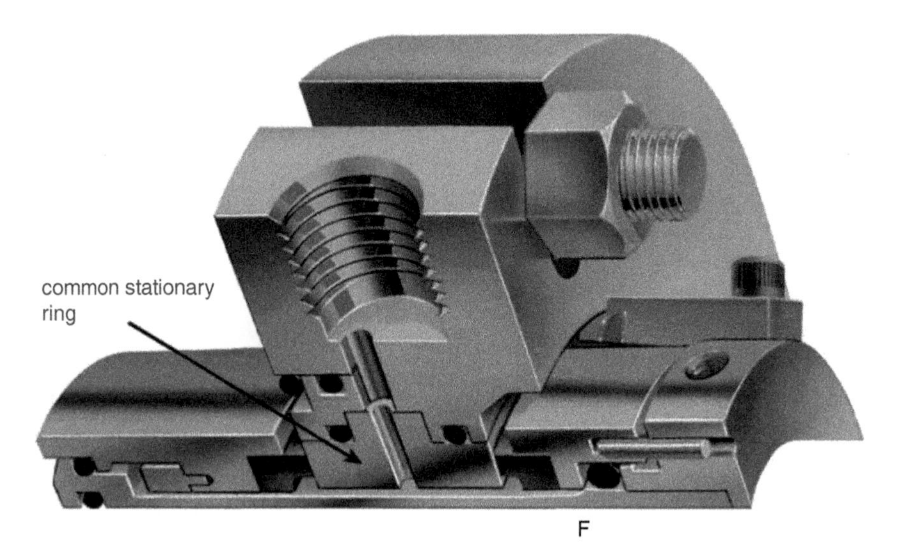

Figure 5.48 Face-to-face seal.

The hydraulic diameter of the seal gap:

$$D_H = \frac{4 \cdot A_W}{P_W} = \frac{4 \cdot \pi \cdot d_2 \cdot b}{2 \cdot \pi \cdot d_2} = 2 \cdot b$$

The overpressure on the seal gap is:

$$\Delta p = \frac{64}{Re} \cdot \frac{d_2 - d_1}{2 \cdot b} \cdot \rho \cdot \frac{c^2}{2} = \frac{64 \cdot \upsilon}{c \cdot 2 \cdot b} \cdot \frac{d_2 - d_1}{2 \cdot b} \cdot \rho \cdot \frac{c^2}{2} = \frac{8 \cdot \eta}{b} \cdot \frac{d_2 - d_1}{b} \cdot c^2$$

$$\Delta p = \frac{8 \cdot \eta \cdot d_2 - d_1 \cdot c}{b^2}$$

From where:

$$c = \frac{b^2 \cdot \Delta p}{8 \cdot \eta \cdot d_2 - d_1}$$

Filling in the values for c in that for Q_V :

$$Q_V = \pi \cdot d_2 \cdot b \cdot \frac{b^2 \cdot \Delta p}{8 \cdot \eta \cdot d_2 - d_1}$$

Or:

$$Q_V = \pi \cdot d_2 \cdot \frac{b^3 \cdot \Delta p}{8 \cdot \eta \cdot d_2 - d_1}$$

Let:

- $b = 0.2 \cdot 10^{-6}$ [m]
- $d_2 = 22$ [mm] $= 0.022$ [m]
- $d_1 = 17$ [mm] $= 0.017$ [m]
- $\Delta p = 10$ [bar] $= 10^6$ [Pa]
- $\eta = 0.001$ [Pa $\cdot$ s] for water on 20C

Then:

$$Q_V = 1.4 \cdot 10^{-12} \left[\frac{m^3}{s}\right] = 1.4 \cdot 10^{-12} \cdot 10^6 \cdot 3600 \left[\frac{ml}{h}\right] = 0.005 \left[\frac{ml}{h}\right]$$

A volute casing pump with a circumferential speed at the sealing area of 20 [m/s] and a pressure to be sealed of 15 [bar] which uses a gland packing for sealing has a leakage rate of about 5–10 [l/h].

5.7 Hydrodynamic Seal

5.7.1 Hydrodynamic Seal with Back Vanes

Hydrodynamic seals are used when there must be absolutely no leakage. This is especially the case in the chemical industry. Gland packings and mechanical seals are mostly susceptible to wear and need therefore regular maintenance and eventual replacement. Hydrodynamic seals of the shaft have no contact between shaft and seal during the operation of the pump and therefore there can be no wear at all.

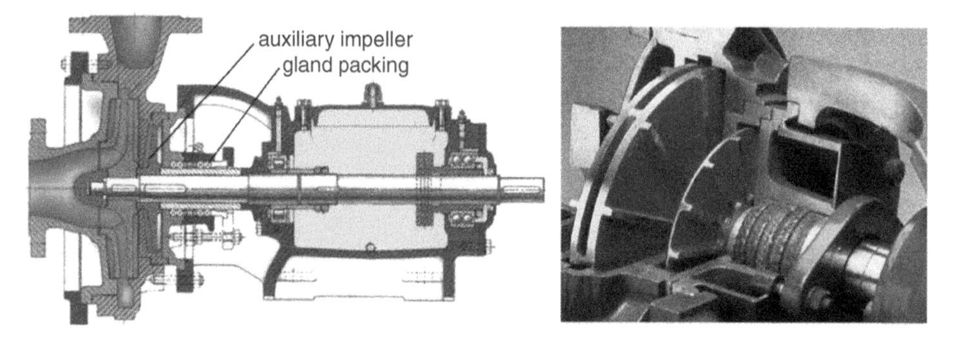

Figure 5.49 Hydrodynamic seal. Source: Glynwed/Rheinhütte Pumpen.

Figure 5.49 is a sectional view of a centrifugal pump with hydrodynamic sealing. The pump consists of an impeller with vanes on the rear side, one or more back vanes, and a classic gland packing (Figure 5.50).

During the operation of the pump, back vanes rotate and the leak fluid in the impellers form a liquid ring so that liquid cannot leak into the atmosphere. Additional advantage is that eventual solid parts are kept away from the shaft seal.

The hydrodynamic seal, however, only works when the pump rotates. When stationary the gland packing takes over the role of the seal.

During start-up and stopping of the pump the hydrodynamic seal will not do its work perfectly and a small leak will occur because of rotational speed reasons. Therefore, a drain must be provided.

5.7.2 Journal Bearing

A journal bearing is a device that supports a shaft (Figure 5.50 and 5.51). The shaft must be hard, for instance stainless steel. Round the shaft fits the journal, very tightly in the housing, that consists of soft material of antimony, tin, or copper. Because of its looks it is often called white metal (tin, antimony, and copper but sometimes including lead, called *babbit*; see Figure 5.52). Lead alloys have been virtually eliminated in modern applications due to lower strength and environmental considerations. The babbit can

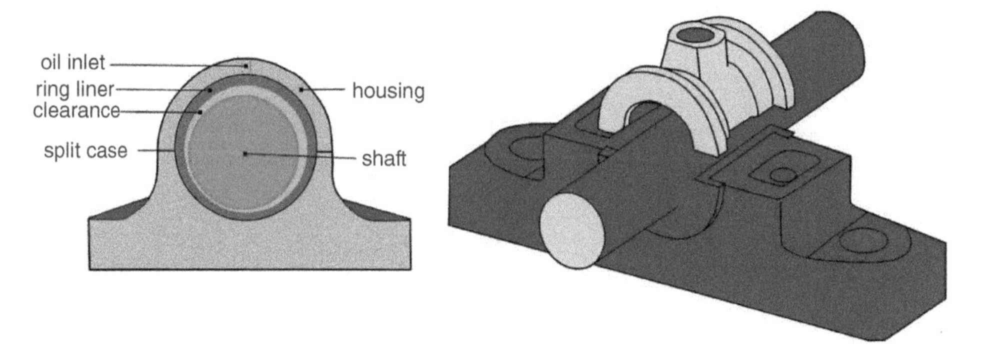

Figure 5.50 Journal bearing.

Figure 5.51 Journal.

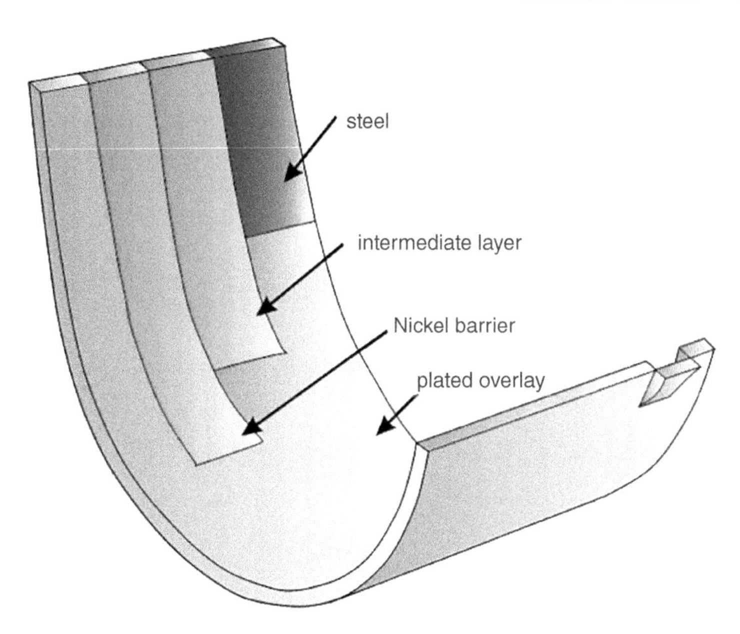

Figure 5.52 Babbit.

wear out and can be replaced. Because it is soft, the babbit can also absorb abrasive parts. In the housing, made of cast iron, is an opening to let the lubrication oil in (Figure 5.53.)

5.7.3 Hydrodynamic Effect Converging Gap

In the case of two very narrow parallel faces (a few micrometers, from which one has a velocity u_0, the fluid will adhere on both the faces. There is a viscous effect and the

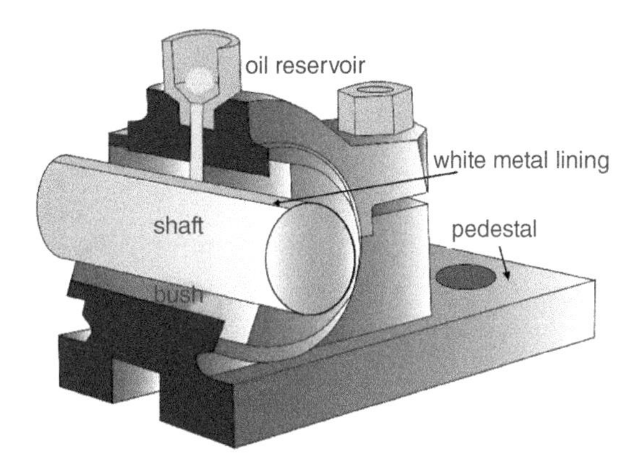

Figure 5.53 Oil supply.

velocity distribution is like in Figure 5.54. It is supposed that there is *no pressure drop*. This flow is called a Couette flow. The velocity distribution is linear in this case.

If one of the faces is tilted *slightly* the fluid will be forced in the smaller area and be pumped into a smaller cross-section. *It can be proven* that in that case the pressure will first increase and then decrease. This pumping action is delivered by the torque on the shaft, it must come from somewhere. This creates a pressure distribution like in Figure 5.55. The moving face is tilted upwards. For this pumping action an external force is needed. Mostly the wedge is not diverging but converging. The pressure distribution is a *hydrodynamic* pressure, because it is the result of the velocity of the journal bearing.

The pressure distribution will lift the journal and counteract the weight W of the shaft. Suppose still that there is no pressure loss caused by friction. The inclination of the bearing is very small compared with the length in x direction. We still can assume that it are parallel faces. Eliminate a very small infinitesimal cube of dimensions dx, dy, dz. The tensions working on the faces are represented in Figure 5.56 according to the law of Newton in x-direction:

$$(p - p - dp) \cdot dy \cdot dz - \tau \cdot dx \cdot dz + (\tau + d\tau) \cdot dx \cdot dz = \rho \cdot dx \cdot dy \cdot dz \cdot \frac{d^2u}{dx^2}$$

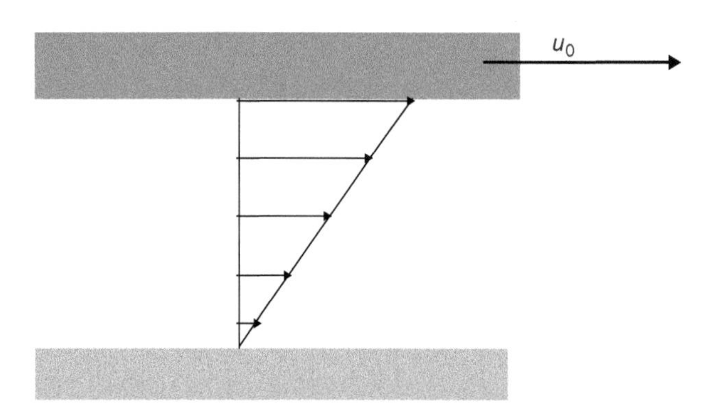

Figure 5.54 Two parallel faces.

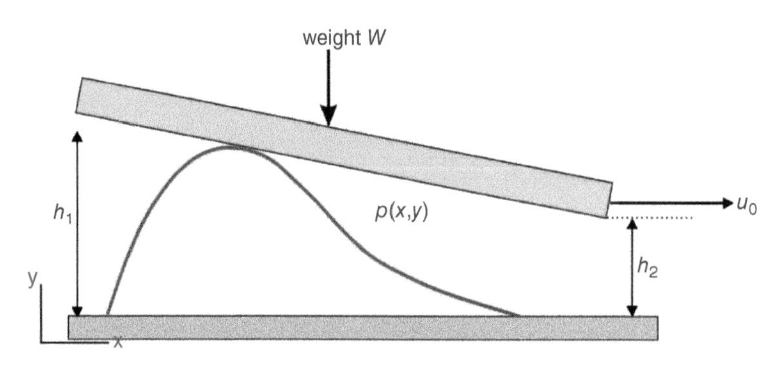

Figure 5.55 Tilted bearing.

Figure 5.56 Infinitesimal volume of liquid.

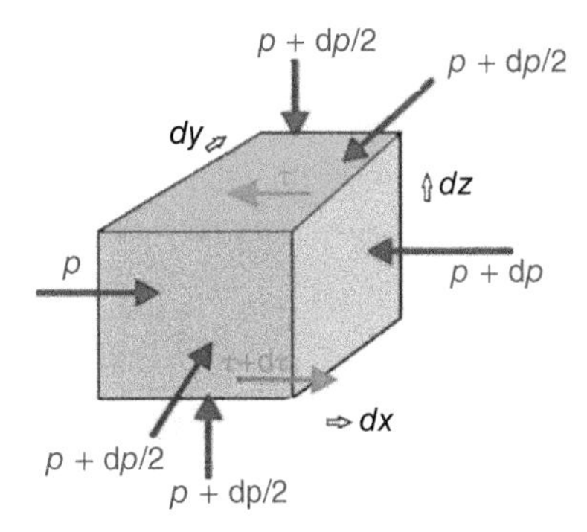

With

$$d\tau = \eta \cdot d\left(\frac{du}{dy}\right)$$

$$\frac{d^2u}{dx^2} = 0$$

because u is not dependent on x, for the case of two parallel plates.

Or:

$$\frac{dp}{dx} = \eta \cdot \frac{d^2u}{dy^2}$$

$$-dp \cdot dy + \eta \cdot \frac{d}{dy}\left(\frac{du}{dy}\right) \cdot dx = 0$$

where u is the velocity of the fluid, function of x and y. The term $\frac{dp}{dx}$ is independent of y.

Integrating the above formulae:

$$\int \frac{dp}{dx} \cdot dy = \int \eta \cdot \frac{d^2u}{dy^2} \cdot dy$$

Or:

$$\frac{dp}{dx} \cdot y = \eta \cdot \frac{du}{dy} + C_1$$

C_1 is an integration constant.

Integrating again:

$$\int \frac{dp}{dx} \cdot y \cdot dy = \eta \cdot \int \frac{du}{dy} \cdot dy + \int C_1 \cdot dy$$

leads to:

$$\frac{dp}{dx} \cdot \frac{y^2}{2} = \eta \cdot u + C_1 \cdot y + C_2$$

Calculating the integrating constant from the conditions:

$$u = 0 \quad y = 0$$

$$u = u_0 \quad y = h$$

leads to:

$$C_2 = 0$$

$$C_1 = \frac{1}{h} \cdot \frac{dp}{dx} \cdot \frac{h^2}{2} - \frac{\eta \cdot u_0}{h}$$

The velocity u can then be written as:

$$u = \frac{u_0}{h} \cdot y - \frac{h^2}{2 \cdot \eta} \cdot \frac{dp}{dx} \frac{y}{h} \cdot \left(1 - \frac{y}{h}\right)$$

The term $\left(1 - \frac{y}{h}\right)$ is always greater than or equal to 1. $\frac{dp}{dx}$ can be positive of negative. The first term in the expression for u is the case for parallel planes, the second term is a correction for the inclined plane and its pressure mountain.

If only the first term is considered, the velocity profile looks like in Figure 5.57. The mean velocity u_m is the same at the beginning and at the end of the wedge, because of its linear character of the distribution:

$$u_m = \frac{u_0}{2}$$

But the flow at entrance:

$$Q_{V1} \div h_1 \cdot u_m$$

And at outlet:

$$Q_{V2} \div h_2 \cdot u_m$$

Thus the law of conservation of flow is not respected. That's because the second term of the expression for u is not been taken into account.

This second term is the Poiseuille flow:

$$P = -\frac{h^2}{2 \cdot \eta} \cdot \frac{dp}{dx} \frac{y}{h} \cdot \left(1 - \frac{y}{h}\right)$$

If $P < 0$ then $\frac{dp}{dx} > 0$, the pressure is increasing. This is at the entrance of the gap. For $P > -1$ the velocity u can become negative, there is a backflow. The motion of the

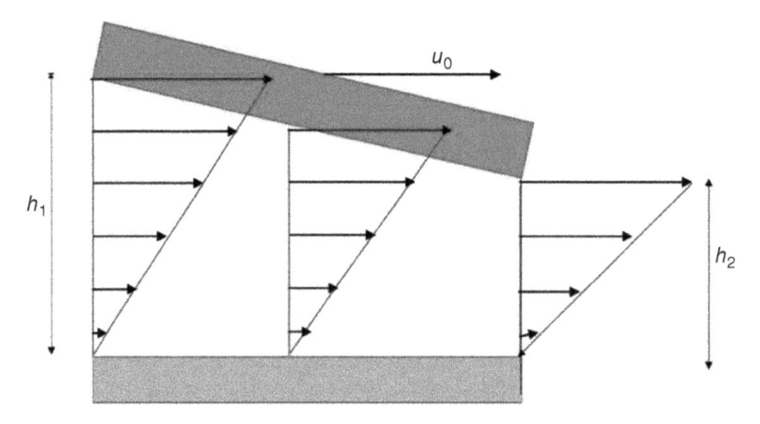

Figure 5.57 Converging gap.

journal bearing is not strong enough in the low area of the entrance area to overcome the pressure gradient.

For $P > 0$ it is the inverse.

Let's take for instance P = 2 at outlet and P = −2 at entrance of the gap. The term $\frac{-P \cdot \eta}{h^2}$ is set out against $\frac{y}{h}$ in the diagram of Figure 5.58 and accordingly the velocity diagram for the second term u looks like in Figure 5.59.

Finally, the velocity profile for the converging gap will be the sum of the Couette and the Poiseuille flow (Figure 5.60).

This way the conservation of volumetric flow will be respected.

5.7.4 Journal Bearing Lift Force

Figure 5.61 represents a journal bearing, first at rest. At startup there is metal-to-metal contact between the journal and the housing. The bearing rolls up to the left. When the speed is high enough, the oil is circulated and begins to make a converging gap between the journal and the housing. The journal bearing is lifted and rolls to the right, where it comes to rest. In the converging gap a pressure is created by the bearing that is responsible for the lift (Figure 5.62).

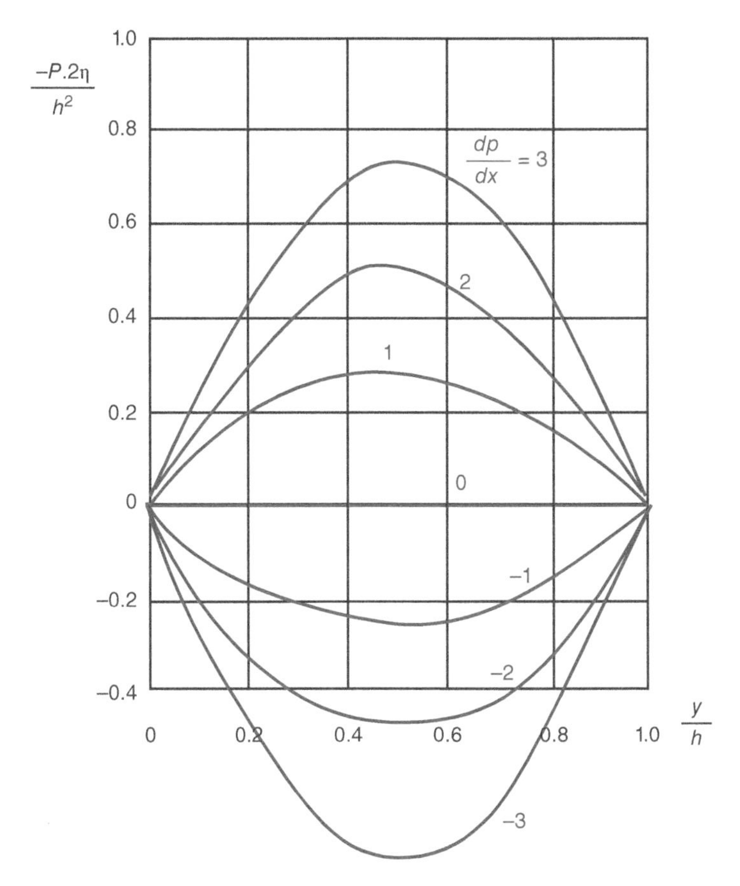

Figure 5.58 Second term of u for P = 2.

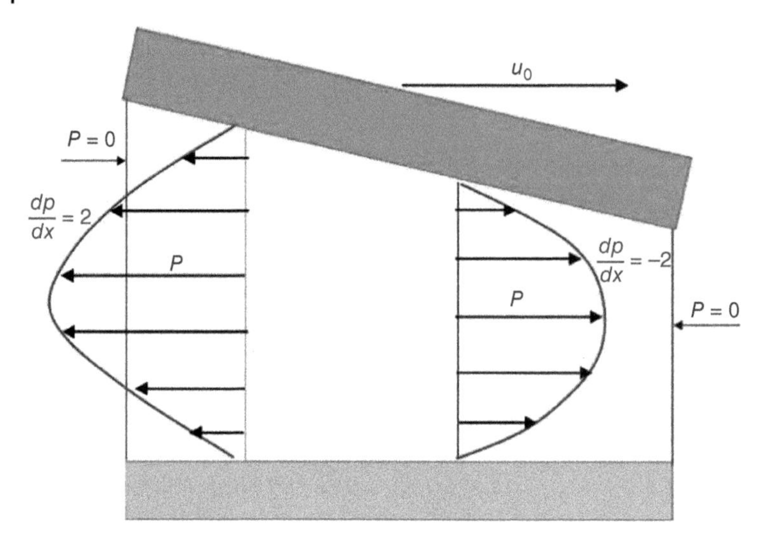

Figure 5.59 Velocity distribution (second term).

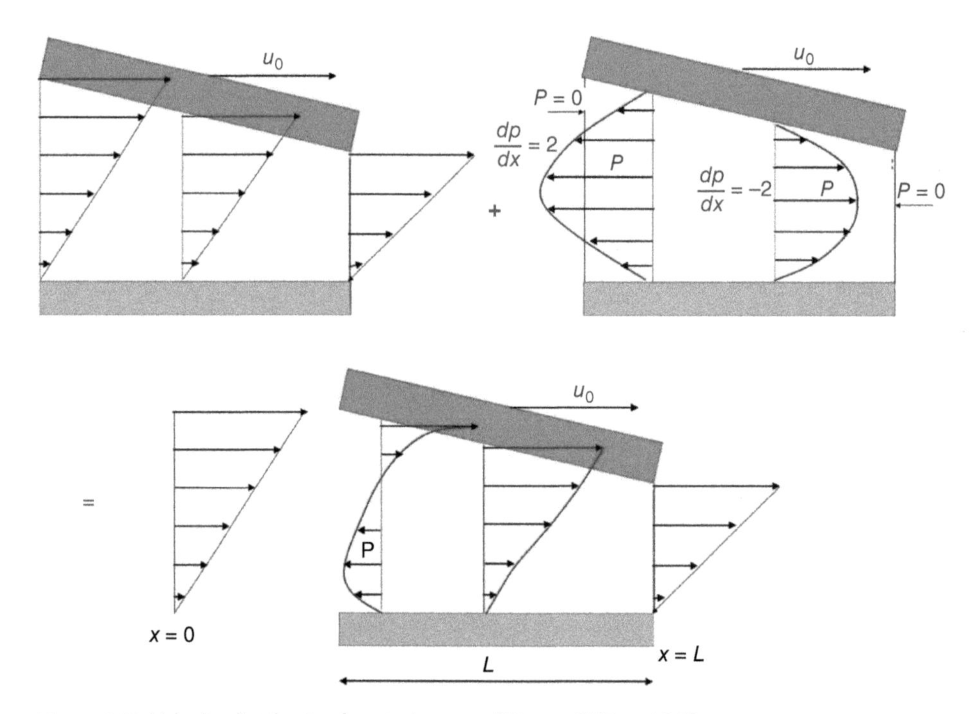

Figure 5.60 Velocity distribution for u is the sum of Figures 5.57 and 5.59.

5.7.5 Hydrodynamic Mechanical Seals

Some mechanical seals, especially for high pressure applications, have a converging gap machined into the faces during manufacturing. This process is known as *pre-coning*, and works to give further assurance that a converging gap will be maintained during

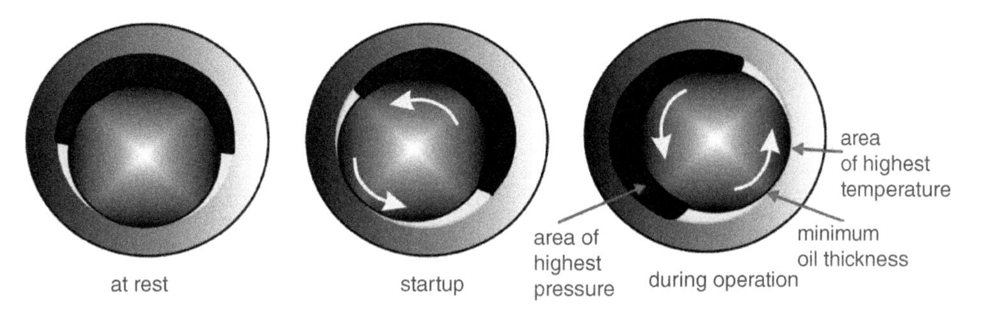

Figure 5.61 Journal bearing: lift force.

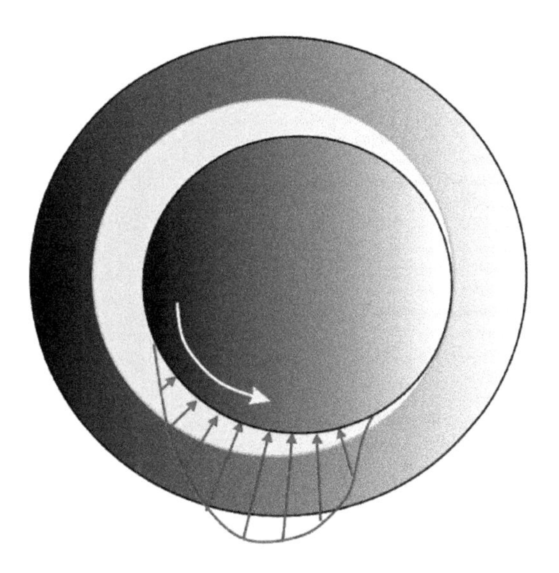

Figure 5.62 Pressure distribution in journal bearing.

operating conditions. The seal faces are still subject to thermal and mechanical deformation, so the actual coning during operation will vary from the specified pre-coning given at unpressurized conditions.

Hydrodynamic seals evolve a pressure distribution in the face gap through the relative motion between the seal faces. This pressure, generated by motion, is caused by the circumferential variation in the film thickness, in contrast to the axisymmetric film thickness of classical hydrostatic mechanical seals. This converging gap generates elevated pressure when the seal faces slide relative to one another, as in a slider bearing. Typically, hydrodynamic seals operate with full-film lubrication, and the film thickness tends to be proportional to the speed of rotation. Some hydrodynamic seals contain hydropaths, which are grooves or slots machined into the seal face that act as a pump and create an air film that the opposing sealing surface will ride on. These hydropaths create an asymmetric seal face that deforms differently under mechanical and thermal loads.

Also in a diverging gap a pressure mountain will be created (Figure 5.63). Think about a waterski.

On this principle the hydrodynamic tracks in the seal rings are based (Figure 5.64).

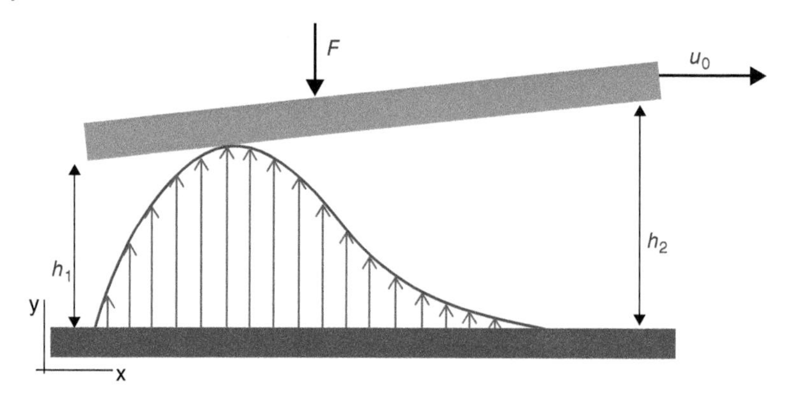

Figure 5.63 Diverging gap.

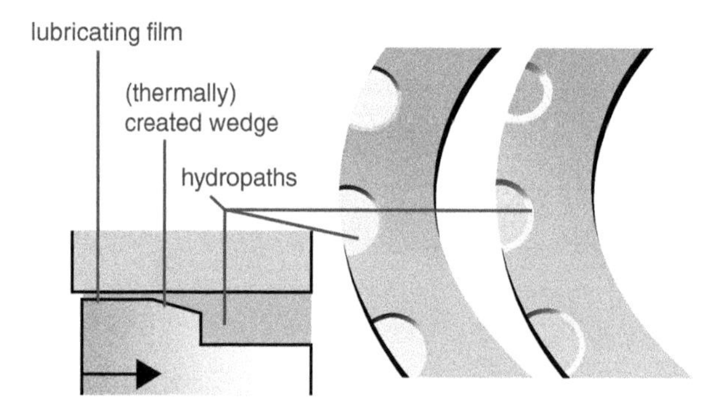

Figure 5.64 Hydrodynamic tracks (hydropaths) in seal rings. Source: Grundfos.

5.8 Floating Ring Seals

Carbon floating seals (chamber seal) are reliable alternatives, when mechanical seals reach their limits. If the application has a circumferential speed above 40 [m/s], normal for high pressure pumps, turbine and compressor applications, carbon (or another hard material) floating seals are the alternatives to use. A special design allows speeds of up to 250 [m/s], while the sealing effect is not decreased. If applications show a high volume of radial misalignment the split design of the seal rings allows a movement of up to 5 [mm]. This radial movement is compensated for by the segmented design of the seal rings and the circulating garter spring,. If a radial movement occurs, the garter spring puts the ring back in position. High pressure applications are sealed with an armored carbon ring (nonsplitted) which has a high mechanical strength. Owing to the metal bandage the sealing gap between the shaft and the seal is constant over different temperature areas. Applications up to 800 °C are sealed with carbon floating seals.

As depicted in Figure 5.65, the seal consists of a ring, which is free to move radially against the rotor while the downstream, axial face is always in contact with the stator. The barrier between the high (upstream) pressure and the low (downstream) pressure consists of a primary seal and a secondary seal. The primary seal is represented by the annular clearance between the inner surface of the floating ring and the rotor; the

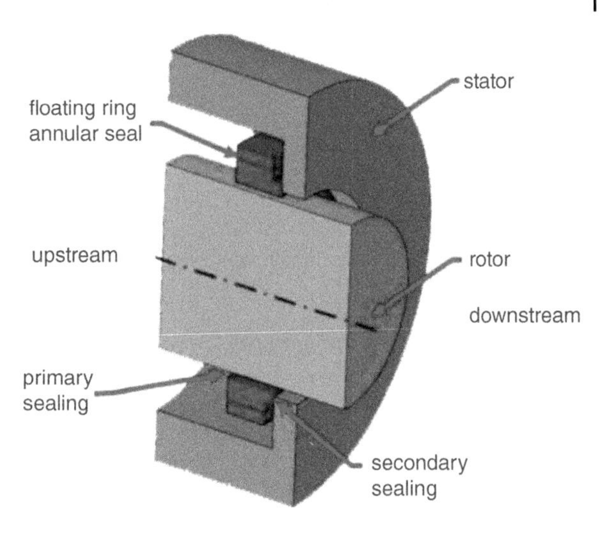

Figure 5.65 Simple image of a floating ring seal.

very small clearance, of the order of 20 [μm], ensures a low leakage (0.01 [ml/h] for a seal diameter of 200 [mm]) between the upstream and downstream zones. However, the hydrodynamic forces engendered by the axial flow in the annular clearance (Figure 5.66) will make the ring "float" above the rotor, i.e. the seal will follow the static and dynamic displacements (vibrations) of the rotor. The secondary seal is the contact surface between the downstream face and the stator. The pressure difference between the upstream and the downstream chamber will press the seal against the stator and engender friction forces that oppose the radial displacements of the floating ring. Such a seal contains mostly several floating rings. The seal is segmented (Figures 5.67–5.68).

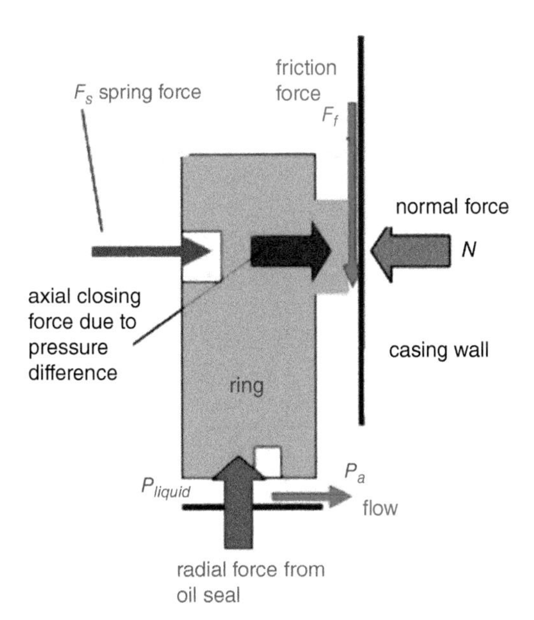

Figure 5.66 Forces on a floating ring.

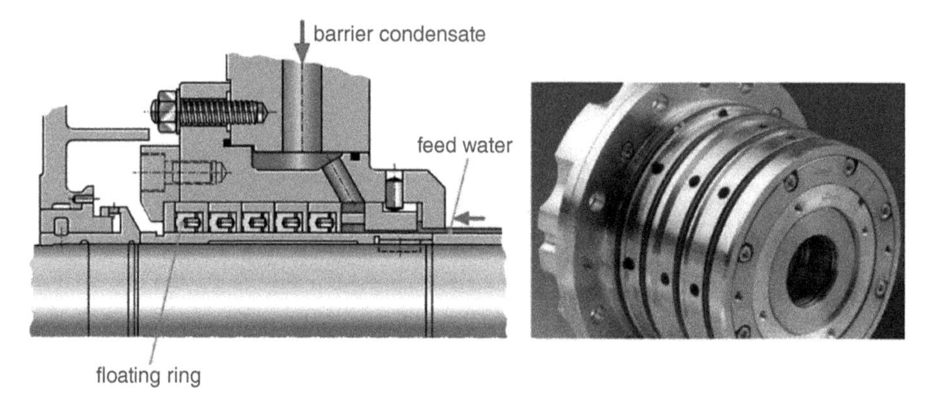

Figure 5.67 Segmented floating ring seals in housing. Source: KSB.

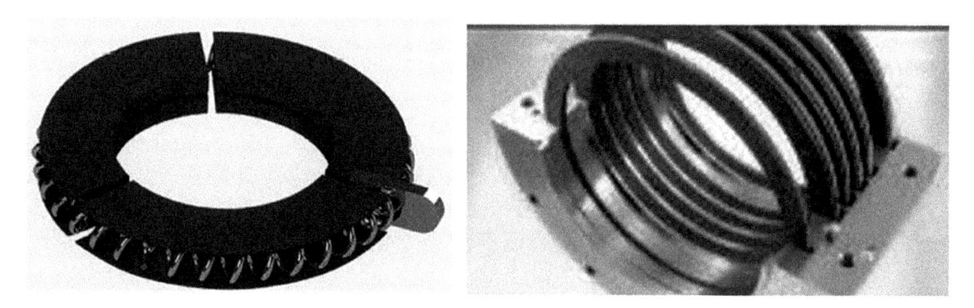

Figure 5.68 Garter spring, and segmented floating rings. Source: Flowserve.

5.9 Hermetic Pumps

5.9.1 Magnetic Coupling

A magnet pump is often called a *magnetic coupled pump* or *magnetic driven pump*. The pump part is separated from the drive by means of a magnet coupling. This magnet coupling consists of an inner rotor (magnet rotor) and an outer rotor; in between there is a separation jacket (Figures 5.69–5.72). More traditional pumps have a mechanical seal that seals the shaft. A magnet pump has a separation jacket that isolates the pump/motor completely from the atmosphere. This is a seal less pump. The big advantage of this is that it is safer to use and kinder to the environment and offers substantial maintenance-cost savings.

The coupled magnets are attached to two concentric rings on either side of the containment shell on the pump housing (Figure 5.69). The outer ring is attached to the motor's drive shaft; the inner ring, to the driven shaft of the pump. Each ring contains about the same number of identical, matched, and opposing magnets, arranged with alternating poles around each ring. The magnets are often made of rare earth metals such as samarium or neodymium alloyed with other metals. The most common combinations are samarium–cobalt and neodymium–iron–boron. These complex alloys have

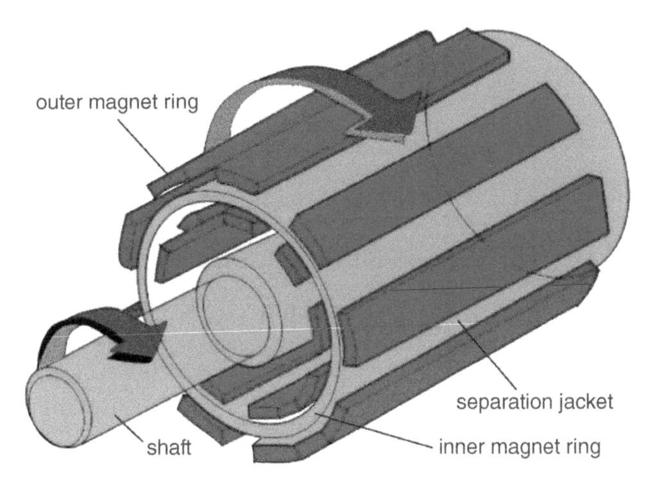

Figure 5.69 Rotor.

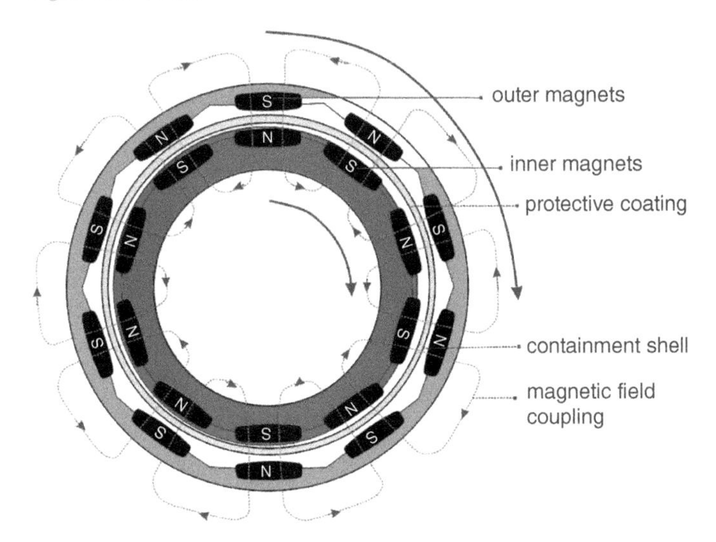

Figure 5.70 Cross-section of magnet rings and containment shell.

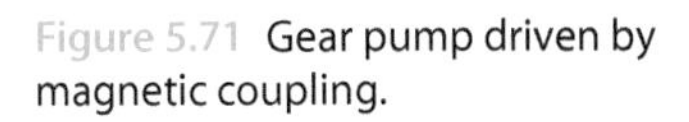

Figure 5.71 Gear pump driven by magnetic coupling.

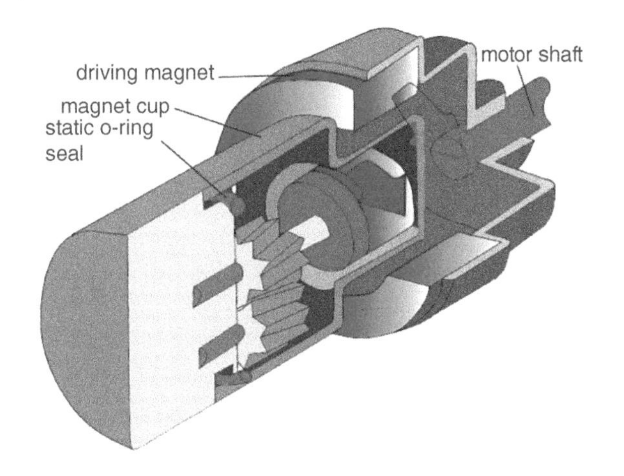

Figure 5.72 Hermetic vane pump. Source: Corken.

two main advantages over traditional magnets: lower mass required to maintain a specific torque – hence smaller and less complex pumps, and greater temperature stability – and the magnetic torque reduces with increases in temperatures but less so with rare earth alloy magnets than with traditional iron magnets.

Rotary lobe, circumferential piston, and timed two-screw (two-spindle) pumps are not available with magnetic drives. They have external timing gears to prevent contact between the two rotating elements. Both of the shafts would have to have separate magnetic couplings, with timing gears external to the magnetic couplings, which would be prohibitively expensive.

Progressive cavity pumps (Mono-pumps) are typically not offered with magnetic drives because of the difficulties in controlling the eccentric shaft motion, which usually requires external bearings to compensate for both radial and axial thrust.

Based on design of magnetic drive pumps, there is an extremely low risk of fluid and vapor emissions from being leaked. As a result, people working on or near the pump are not exposed to dangers from hazardous, corrosive, flammable, and/or explosive fluids, and other toxic chemicals. In addition, expensive liquids are not wasted. Magnetic drive pumps are also reliable. When choosing a reliable brand, you can be confident the pump will operate as specified. A pump of this type requires very little maintenance. The primary reason is that the design is actually quite simple. In fact, when used for normal operations it is common to see magnetic drive pumps go 10 years or longer before needing any kind of repair. Even then, repairs are often inexpensive. Another huge benefit associated with magnetic drive pumps is, because of an easy coupling, there is no need for an alignment of the pump or motor.

For applications that involve solids, even a small percentage, magnetic drive pumps are not suitable because they only work with clean liquids free of solids.

Magnet-driven pumps require higher power absorption than conventional pumps.

From a cost perspective, magnetic drive pumps tend to be more expensive; however, because maintenance is very low a quick payback can also be attained.

Figures 5.70 and 5.71 represent a gear pump and a vane pump with magnetic sealing.

5.9.2 Canned Motor Pump

A canned motor pump has a rotor, of the driving motor that is "canned" in a stainless steel bus. Round the bus is the stator of the motor. The axis is hollow. The stator is dry, round the rotor of the motor flows the process liquid, and it is also subjected to liquid pressure. The cooling liquid can come from the process liquid or from an external source (Figures 5.72–5.76).

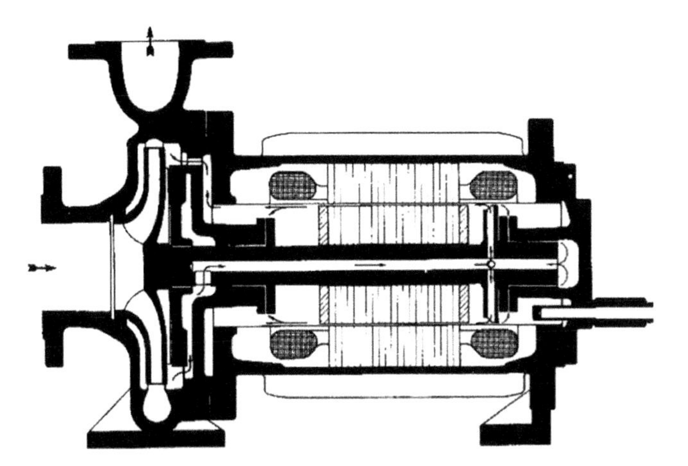

Figure 5.73 Principle of a canned motor pump.

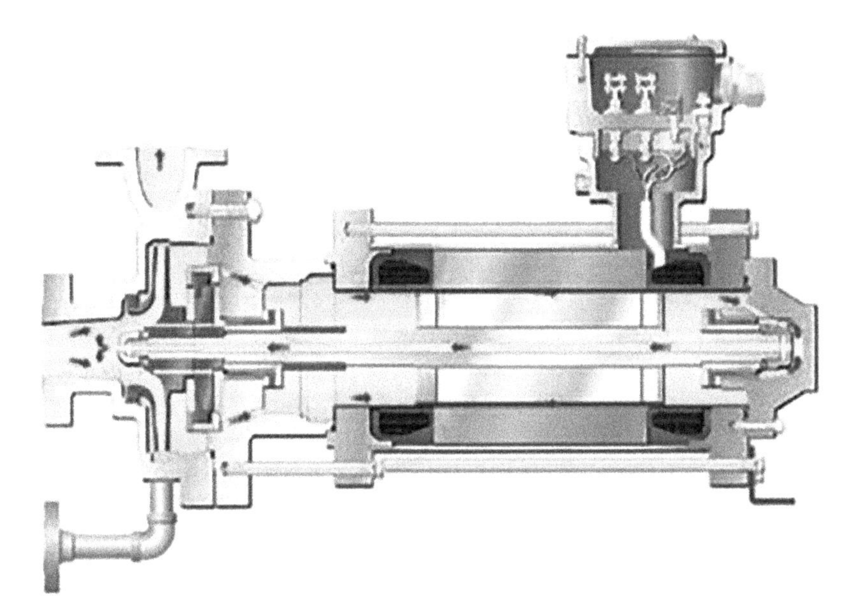

Figure 5.74 Canned motor pump type CAM with internal cooling. Source: Lederle-Hermetic-Pumpen GmbH.

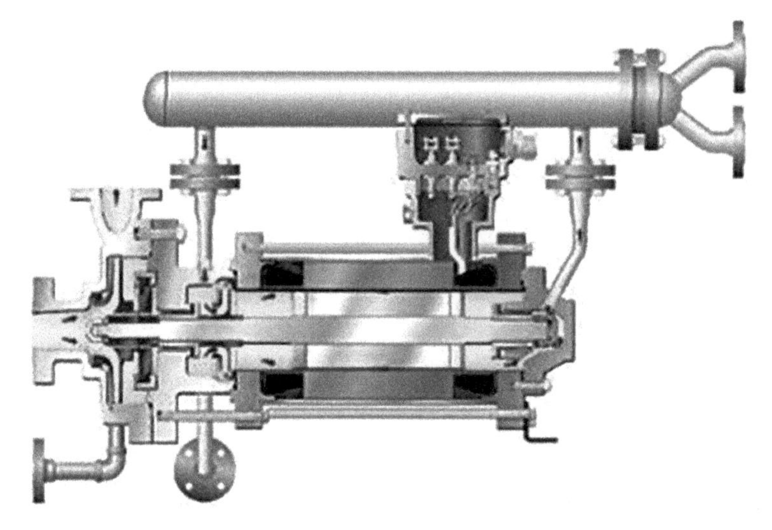

Figure 5.75 Canned motor pump type CAM with external cooling. Source: Lederle-Hermetic-Pumpen GmbH.

Figure 5.76 Multistage canned motor pump type CAM. Source: Hermetic-Pumpen GmbH.

Part II

Compressors

6

General

CHAPTER MENU

6.1 Terminology

A compressor (Figure 6.1) is a device that compresses a gas to a pressure higher than the atmospheric pressure. This rises the temperature of the gas, its specific mass ρ (absolute density), its volumetric flow Q_V decreases, while its mass flow Q_M stays constant (equality of mass per units of time through the compressor: what comes in must come out).

As the mass flow is given by:

$$Q_M = \rho \cdot Q_V = \rho \cdot c \cdot A = constant$$

ρ increases while section A decreases (Figure 6.1), so c has to increase.

If the pressure of the gas in a barrel is lowered to a pressure lower that the atmospheric, one speaks about a *vacuum pump*. Dependent on the delivered pressure one speaks about *fans* if the pressure is 1.25 or lower and high-pressure fans (blowers) at compression ratios of 1.25–2. Above that one speaks of *compressors*. This division is purely arbitrary; some people have other norms.

6.2 Normal Volume

For gasses on various pressure and temperature it is difficult to compare the quantities. The concept of volumetric flow has no sense standing alone as long as is not specified at which pressure and temperature this flow is measured. That's why a normal situation

Pumps and Compressors, First Edition. Marc Borremans.

Companion website: www.wiley.com/go/borremans/pumps

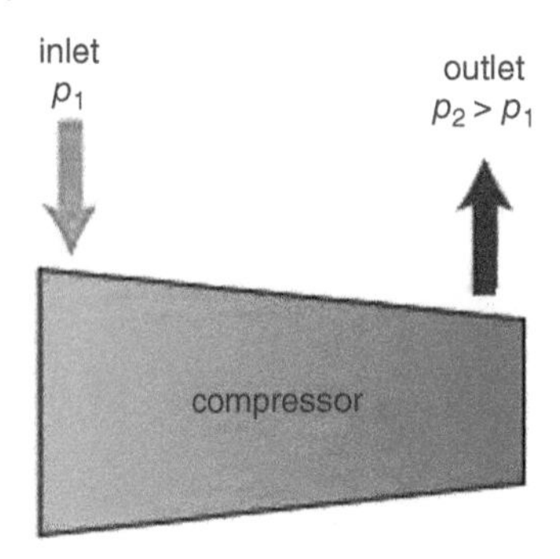

Figure 6.1 Compressor.

is defined at 0 °C and 1.013 [bar] (1013 [hPa]). A cubic meter gas in the *normal* state is called a *normal cubic meter* and is noted as:

$$1\ [\mathrm{Nm^3}] \text{ or } 1\ [\mathrm{m_n^3}] \text{ or } 1\ [\mathrm{m^3(n)}]$$

6.3 Ideal Gasses

An ideal gas is one that obeys the following law:

$$p \cdot v = \frac{R}{M} \cdot T$$

where:

p: the static pressure in Pascal [Pa]
v: the specific volume in $\left[\frac{\mathrm{m^3}}{\mathrm{s}}\right]$
R: universal gas constant 8315 $\left[\frac{\mathrm{J}}{\mathrm{kg \cdot kmole}}\right]$
M: atomic mass [kg]
T: temperature [K]

An ideal gas is characterized by its isentropic exponent γ:

$$\gamma = \frac{c_p}{c_v}$$

With:

- c_p: specific heat at constant pressure $\left[\frac{J}{kg \cdot K}\right]$, where J stands for Joule.
- c_v: specific heat at constant volume $\left[\frac{J}{kg \cdot K}\right]$

A very interesting formulae is that of Mayer:

$$c_p = c_v + \frac{R}{M}$$

It is easy to prove that:

$$c_p = \frac{\gamma}{\gamma - 1} \cdot \frac{R}{M}$$

And:

$$c_v = \frac{1}{\gamma - 1} \cdot \frac{R}{M}$$

For air: $\gamma = 1.4$ at atmospheric conditions, but this is dependent on the temperature. For noble gasses: γ is not dependent on the temperature.

6.4 Work and Power

6.4.1 Compression Work

The external work W_{12} [Joule] can be in different forms. We only consider mechanical work here. In the case of Figure 6.2 a piston cylinder filled with gas is considered. A *system* is the substance being studied. In a *closed system* no mass can enter or leave the system border. How much work delivers the surrounding to the gas when its volume changes? This work is called *compression work* and noted W_c.The unit is the Joule.

The pressure on the piston surface A is p.

If the gas expands, the piston moves to the right. Work on the piston is called *compression work*. If the piston moves an infinitesimal amount dx, the compression work on the piston, which is received by the environment, is:

$$dW_c = -F \cdot dx = -p \cdot A \cdot dx$$

where F is the force that the gas exercises on the piston. The minus sign is in accordance with the convention of the thermodynamics, to the system given work is positive.

The increase of the volume of the gas is dV so that: $dx = dV/A$ and:

$$dW_c = -p \cdot dV$$

If we consider just 1 [kg] of the gas, then this becomes, in specific values:

$$dw_c = \frac{dW_c}{m} = -p \cdot \frac{dV}{m} = -p \cdot d\left(\frac{V}{m}\right) = -p \cdot dv$$

Consider a restricted state change $1 \to 2$ like in Figure 6.2 then the *specific external work* is:

$$w_{c12} = \int_1^2 dw_c = \int_1^2 -p \cdot dv$$

This work can be interpreted as the surface under the curve $1 \to 2$ (Figure 6.3):

Figure 6.2 Compression work.

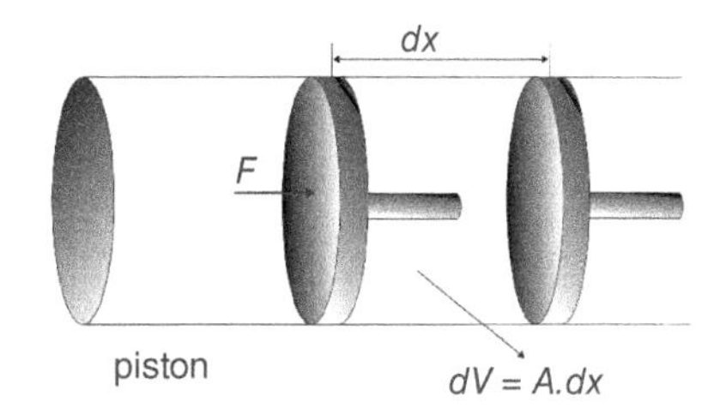

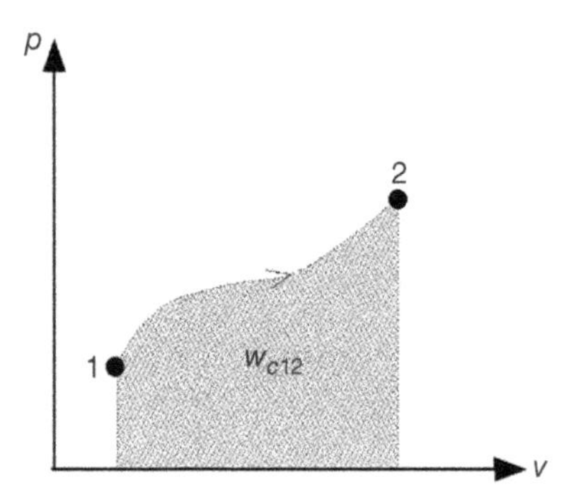

Figure 6.3 Graphical interpretation of the specific compression work.

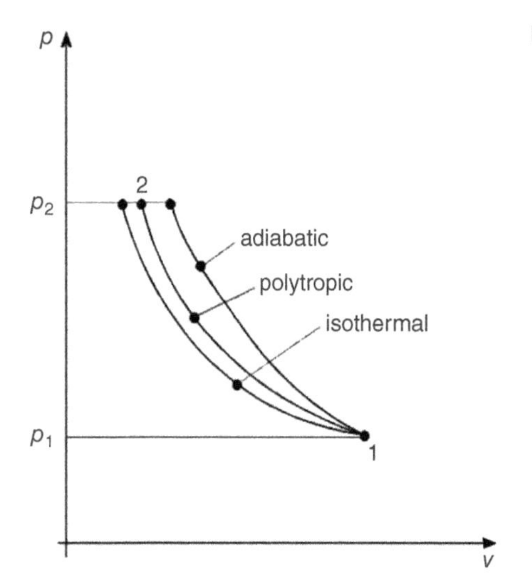

Figure 6.4 Compression.

6.4.2 Technical Work

In Figure 6.4 the compression of a gas is represented in a thermodynamic *diagram*: on the abscissa the specific volume of the gas (in m^3/kg) and on the ordinate the pressure of the gas (in Pa of bar…) are set out. The curve $1 \rightarrow 2$ is the state change of a gas when this is compressed from pressure p_1 to p_2. Remember that an adiabat (state change without exchange of heat) is steeper than an isotherm (state change with constant temperature). A polytropic compression is one where there is heat loss to the environment and the temperature increases.

The area under the curve till the p axis is the compression work. This means, as we will see later, that an isothermal compression is less energy consuming! Cooling the compressor will be beneficial.

A random state change $1 \rightarrow 2$ of the gas is called a *polytropic*. It is assumed that:

$$p \cdot v^n = \text{Constant}$$

Or:

$$p_1 \cdot v_1^n = p \cdot v^n = p_2 \cdot v_2^n$$

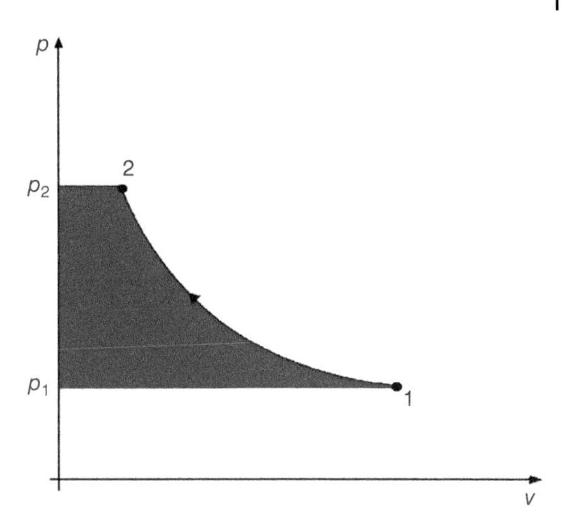

Figure 6.5 Specific technical work (colored).

The polytropic exponent n:

$$1 \leq n \leq \gamma$$

where $\gamma = \frac{c_p}{c_v}$ is the *is entropic exponent.*

c_p and c_v are respectively the specific heats at constant pressure and at constant volume, expressed in the unit [J/kg·K]. The isentropic and polytropic exponents are both dimensionless.

When $n = 1$ the compression is *isothermal* and for $n = \gamma$ *adiabatic.*

Two other laws form together with the above mentioned, the laws of Poisson:

$$\frac{T_2}{T_1} = \left(\frac{p_2}{p_1}\right)^{\frac{n-1}{n}}$$

$$T_2 \cdot {v_2}^{n-1} = T_1 \cdot {v_1}^{n-1}$$

According to thermodynamic theory, the *specific technical work* w_{t12} is the work delivered to the shaft of the compressor to realize the state change $1 \rightarrow 2$:

$$w_{t12} = \int_1^2 v \cdot dp + w_{f,12} + \frac{1}{2} \cdot (c_2^2 - c_1^2) + g \cdot (z_2 - z_1)$$

The changes of kinematic and potential energy are neglected. They are always too small (in this book).

$w_{f,12}$ is the specific friction work in the compressor. This term is always positive.

The *specific technical work* – this is *the work on the shaft of an open* system – can be interpreted as the surface under the curve $1 \rightarrow 2$, with reference to the ordinate (Figure 6.5).

The technical work is received by the open system when the flow of mass enters in the open system and to deliver work to let the flow back out, together with the compression work. The first term is called the *displacement work.*

In thermodynamics various sorts of work are studied: *mechanical work, compression work (volume change work), pressure change work, displacement work, technical work.*

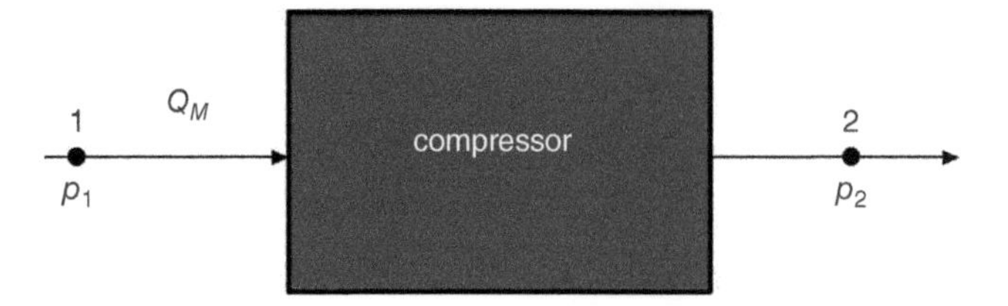

Figure 6.6 Flow.

For the study of compressors, it is, as for other devices, finally the technical work on the shaft that is of importance. That's why in the literature the prefix *technical* is often omitted, and so the letter t in in w_{t12}. The same counts for the indexes 1 and 2.

6.4.3 Technical Power

The technical power P_{t12} on the shaft of the compressor for the state change $1 \rightarrow 2$ is proportional to the compressed mass flow Q_M and is given by (Figure 6.6):

$$P_{t12} = Q_M \cdot w_{t12} \text{ [}Watt\text{]}$$

With:

$$Q_M = \rho_1 \cdot Q_{v1} = \rho_2 \cdot Q_{v2} \text{ [kg/s]}$$

$$Q_M = \frac{Q_{v1}}{v_1} = \frac{Q_{v2}}{v_2} \tag{6.1}$$

Where:

About the symbols and names of the concept of technical power the same remarks are valid as was done for the work.

6.5 Nozzles

Consider a tube like in Figure 6.7. For this moment the pressure p_2 is supposed to be higher than the *counterpressure* p_1. The first part of the tube is converging, then there is a so-called *throat*, and finally there is the diverging part. Suppose further that the gas entering the tube has a velocity lower than the speed of sound in that gas; this is a subsonic speed. For air at 20 °C the speed of sound is 343 [m/s]. In the converging nozzle the velocity of the gas will increase, the specific volume v will increase slowly, and the static pressure p will decrease. This is like in a liquid, according to the law of Bernoulli. In the throat the gas will reach the speed of sound. Once that happens the likeness with the law of Bernoulli is no longer valid. The velocity of the gas will further decrease. It will be higher than the speed of sound. This is a supersonic flow. The specific volume will increase very fast and the static pressure will further decrease (Figure 6.8). This is not the same as Bernoulli would predict. It has a contrary behavior. So, beware of a nozzle and a supersonic flow!

If, on the other hand, the gas enters the converging nozzle with a supersonic velocity, then its static pressure will increase, its velocity will decrease until the speed of sound is

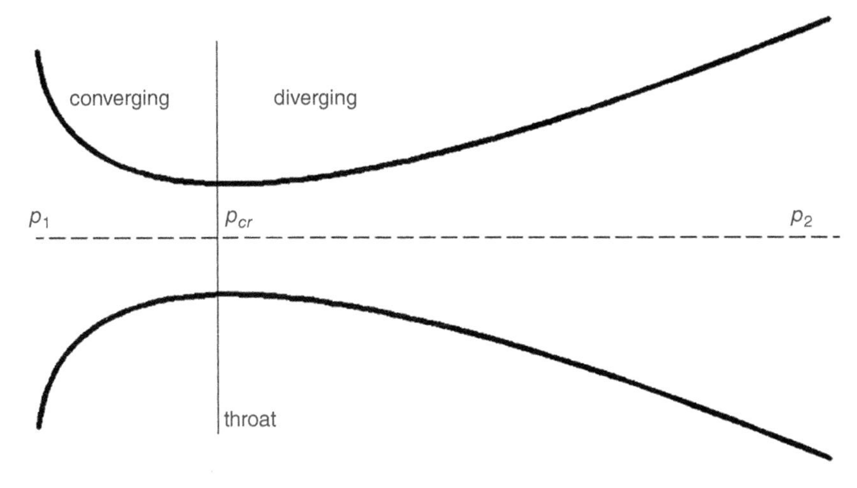

Figure 6.7 Converging–diverging nozzle.

Figure 6.8 Velocity, static pressure, and specific volume.

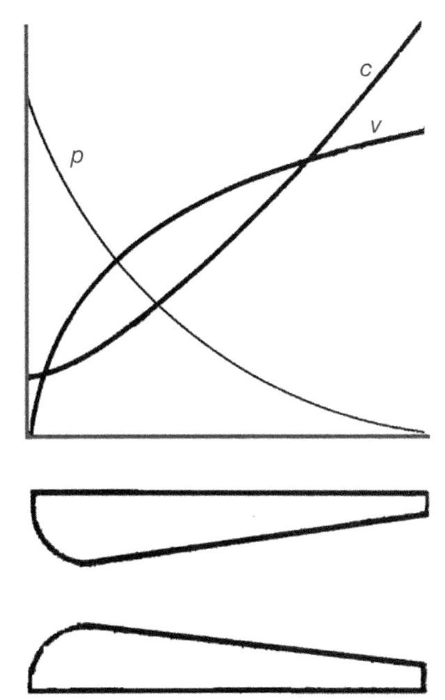

reached in the throat, and the specific volume will decrease. In the diffusing nozzle the specific volume will increase further, the static pressure will decrease, and the velocity of the gas will increase.

Why doesn't this happen with a liquid? Well, the speed of sound is a phenomenon that relies on the elasticity of the medium. The pressure wave from a sound wave is faster in an elastic medium. A liquid is much lesser elastic than a gas. For water the speed of sound is 1510 [m/s] at room temperature. This order of magnitude of the velocity of a liquid is never attained in practice. Remember the order of magnitude of velocities of liquids in practice, given earlier in Chapter 1: 1.5 [m/s] in pipe lines till 30 [m/s] in centrifugal pumps.

6.6 Flow

It will be clear that, according to Equation 6.1, speaking about the volumetric flow of a compressor is senseless if the conditions are not given. What can be done? The flow in normal cubic meters can be given. The volumetric flows in normal cubic meters are equal at the inlet and outlet of the compressor (not taken into account the leak losses).

In the compressed air technique, the concept *free air, FAD,* or *free air delivery* is often handled.

- Atmospheric pressure at 1 [atm] (1013 [bar])
- Atmospheric temperature at 20 °C
- Relative humidity of the air 0% (*dry air*)
- The motor speeds at 100% of its nominal value
- See also norm ISO 1217 annex C, or Cagi-Pneurop PN 2 CPT 2.

The technical power on the shaft of the compressor can be calculated theoretically (see later).

This calculation is based on:

- A frictionless flow of the gas in the machine.
- The hypothesis that the aspired mass flow is equal to the discharged flow.

In reality there is power loss caused by a viscous friction of the gas in the compressor on one hand and a certain amount of leak flow that flows back from the press to the suction side. This means that more power must be delivered than theoretically predicted.

There are various possibilities to define the efficiency of a compressor. In this book we use a qualitative definition. The efficiency of a compressor is the ratio of the power taken by the gas in the compressor and the power that must be delivered on the shaft of the compressor. What power is lost in the compressor? Well, the friction power inside the compressor and the power wasted for leak losses in the compressor. Just like in a pump.

Example 6.1
A compressor compresses 300 [m^3/h] air from 15 °C and 1–6 [bar]. The compression may be supposed to be adiabatic. Kinetic energies are negligible. Air behaves in this conditions as an ideal gas. Its molecular mass is 29 $\frac{kg}{kmol}$, and $\gamma = 1.4$. Neglect friction in the compressor.

Calculate:

- The mass flow Q_M.
- The volumetric flow at the inlet and the outlet.
- The specific displacement and compression work.
- The technical power that the electrical motor has to exercise on the shaft of the compressor.

Solution

- The volumetric flow at the inlet:

$$Q_{V,in} = \frac{300}{3600} = 8.33 \cdot 10^{-2} \left[\frac{m^3}{s}\right]$$

The specific mass at entrance is:

$$\rho_{in} = \frac{p_{in} \cdot M}{R \cdot T_{in}} = \frac{10^5}{8315} \cdot \frac{29}{288} = 1.21 \left[\frac{kg}{m^3}\right]$$

Remember that the temperature in the ideal gas law is in Kelvin.
From where the mass flow:

$$Q_M = \rho_{in} \cdot Q_{v;in} = 1.21 \cdot 8.33 \cdot 10^{-2} = 0.1 \left[\frac{kg}{s}\right]$$

- The volumetric flow at the outlet:
  Calculate first the final temperature:

$$T_{out} = T_{in} \cdot \left[\frac{p_{in}}{p_{out}}\right]^{\frac{\gamma-1}{\gamma}} = 288 \cdot \left[\frac{6}{1}\right]^{\frac{14-1}{1.4}} = 480\,[K]$$

From where:

$$\rho_{out} = \frac{p_{out} \cdot M}{R \cdot T_{out}} = \frac{6 \cdot 10^5 \cdot 29}{8315 \cdot 480} = 4.36 \left[\frac{kg}{m^3}\right]$$

From where:

$$Q_{V,out} = \frac{Q_M}{\rho_{out}} = \frac{\cdot 10^{-2}}{4.36} = 2.3 \cdot 10^{-2} \left[\frac{m^3}{s}\right]$$

- The technical work done for the change of state in → out:
  Let's calculate it first for a polytropic compression met factor n.
  For any state between in and out:

$$p \cdot v^n = p_{in} \cdot v_{in}^n$$

From where:

$$v^n = \frac{p_{in} \cdot v_{in}^n}{p}$$

Or:

$$v = p_{in}^{\frac{1}{n}} \cdot v_{in} \cdot p^{-\frac{1}{n}}$$

And:

$$w_{t,\text{in}\to\text{out}} = \int_{in}^{out} v \cdot dp = \int_{in}^{out} p_{in}^{\frac{1}{n}} \cdot v_{in} \cdot p^{\frac{1}{n}} \cdot dp = p_{in}^{\frac{1}{n}} \cdot v_{in} \cdot \int_{in}^{out} p^{-\frac{1}{n}} \cdot dp$$

$$w_{t,\text{in}\to\text{out}} = p_{in}^{\frac{1}{n}} \cdot v_{in} \cdot \left[\frac{p^{\left(-\frac{1}{n}+1\right)}}{\left(-\frac{1}{n}+1\right)}\right]_{in}^{out} = p_{in}^{\frac{1}{n}} \cdot v_{in} \cdot \frac{n}{n-1} \cdot \left[p^{\frac{n-1}{n}}\right]_{in}^{out}$$

$$= p_{in}^{\frac{1}{n}} \cdot v_{in} \cdot \frac{n}{n-1} \cdot \left(p_{out}^{\frac{n-1}{n}} - p_{in}^{\frac{n-1}{n}}\right)$$

$$w_{t,\text{in}\to\text{out}} = p_{in}^{\frac{1}{n}} \cdot p_{in}^{\frac{n-1}{n}} \cdot v_{in} \cdot \frac{n}{n-1} \cdot \left(\frac{p_{out}^{\frac{n-1}{n}}}{p_{in}^{\frac{n-1}{n}}} - 1\right)$$

Finally:

$$w_{t,\text{in}\to\text{out}} = p_{in} \cdot v_{in} \cdot \frac{n}{n-1} \cdot \left(\frac{p_{out}^{\frac{n-1}{n}}}{p_{in}^{\frac{n-1}{n}}} - 1 \right)$$

For an adiabatic compression: $n = \gamma$:

$$w_{t,\text{in}\to\text{out}} = \frac{\gamma}{\gamma-1} \cdot p_{in} \cdot v_{in} \cdot \left(\frac{p_{out}^{\frac{\gamma-1}{\gamma}}}{p_{in}^{\frac{\gamma-1}{\gamma}}} - 1 \right)$$

For an isothermal compression the denominator becomes zero. Somewhere in the calculation we divided by something that might become zero, we supposed that $(n-1)$ could not be zero, and that is intolerable. This must be recalculated.

$$w_{t,\text{in}\to\text{out}} = \cdots = p_{in}^{\frac{1}{n}} \cdot v_{in} \cdot \int_{in}^{out} p^{-\frac{1}{n}} \cdot dp$$

$$= p_{in}^{1} \cdot v_{in} \cdot \int_{in}^{out} \frac{dp}{p} = p_{in} \cdot v_{in} \cdot [\ln p_{out} - \ln p_{in}]$$

For an isothermal compression:

$$w_{t,\text{in}\to\text{out}} = p_{in} \cdot v_{in} \cdot \ln \frac{p_{out}}{p_{in}}$$

Now, for the adiabatic, numerical:

$$w_{t,\text{in}\to\text{out}} = \frac{\gamma}{\gamma-1} \cdot p_{in} \cdot v_{in} \cdot \left(\frac{p_{out}^{\frac{\gamma-1}{\gamma}}}{p_{in}^{\frac{\gamma-1}{\gamma}}} - 1 \right)$$

$$= \frac{1.4}{1.4-1} \cdot 10^5 \cdot \frac{1}{1.21} \cdot \left(\left(\frac{6}{1}\right)^{\frac{0.4}{1.4}} - 1 \right) = 193\ [\text{kJ/kg}]$$

The compression work: $w_{c,\text{in}\to\text{out}} = \int_{in}^{out} p \cdot dv = \int_{in}^{out} \frac{p_{in} \cdot v_{in}^{\gamma}}{v^{\gamma}} \cdot dv = p_{in} \cdot v_{in}^{n} \cdot \int_{in}^{out} v^{-\gamma} \cdot dv$

$$w_{c,\text{in}\to\text{out}} = p_{in} \cdot v_{in}^{n} \cdot [v^{-\gamma+1}]_{in}^{out}$$

$$= p_{in} \cdot v_{in}^{\gamma} \cdot \frac{-1}{\gamma-1} (v_{in}^{-\gamma+1} - v_{out}^{-\gamma+1}) = \frac{-1}{\gamma-1} \cdot p_{in} \cdot v_{in}^{\gamma} \cdot v_{in}^{-\gamma+1} \cdot \left(1 - \frac{v_{out}^{-\gamma+1}}{v_{in}^{-\gamma+1}} \right)$$

$$= \frac{1}{\gamma-1} \cdot p_{in} \cdot v_{in} \cdot \left(\frac{v_{in}^{\gamma-1}}{v_{out}^{\gamma-1}} - 1 \right)$$

Numerical: $w_{c,\text{in}\to\text{out}} = \frac{1}{0.4} \cdot 10^5 \cdot \frac{1}{1.21} \cdot \left(\frac{4.36^{1.4-1}}{1.21^{1.4-1}} - 1 \right) = 138\ [\text{kJkg}]$

Displacement work: $w_{disp,\ \text{in}\to\text{out}} = w_{t,\ \text{in}\to\text{out}} - w_{c,\ \text{in}\to\text{out}} = 303 - 138 = 165\ \text{kJ/kg}$

Technical Power: $P_{t,\ \text{in}\to\text{out}} = Q_M \cdot w_{t,\ \text{in}\to\text{out}} = 0.1 \cdot 193 = 19.3\ [\text{kW}]$

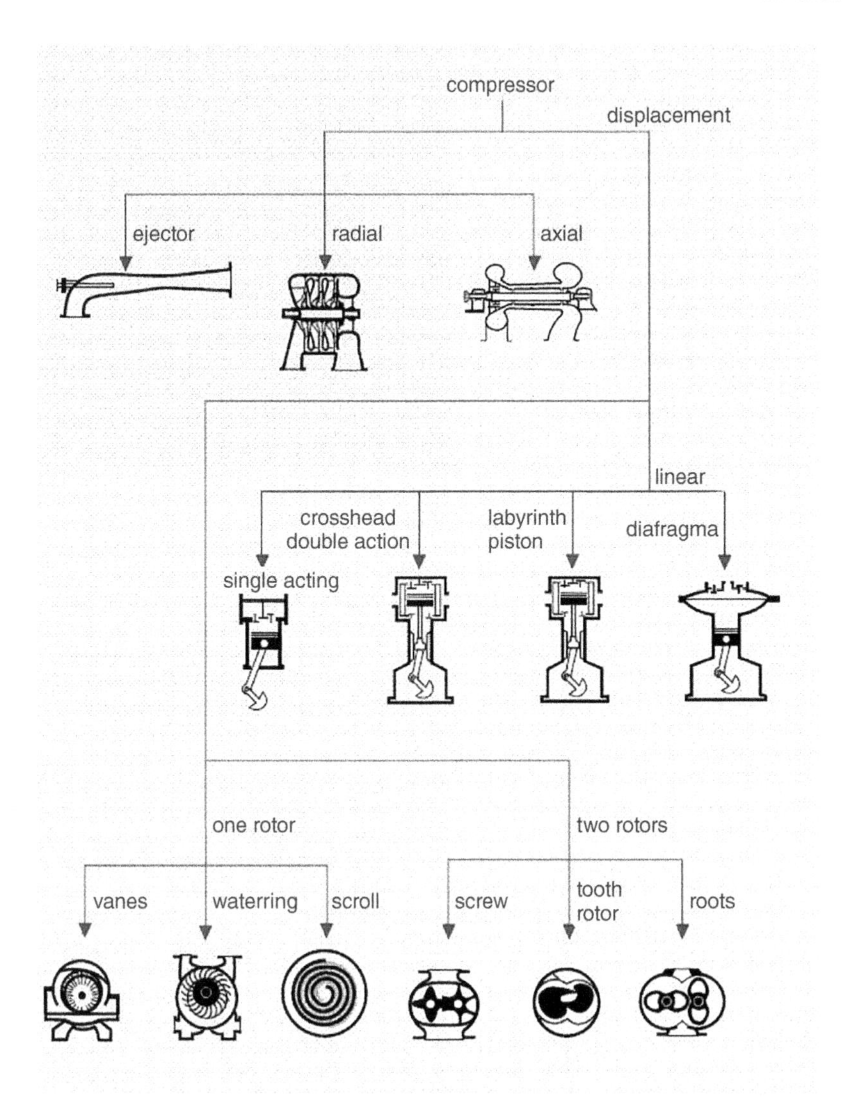

Figure 6.9 Types of compressors. Source: Atlas Copco.

6.7 Choice and Selection

The different types of compressors are represented in Figure 6.9.

The most important parameters when making a choice of a compressor are the necessary flow and pressure.

Another criterion is that the compressor works oil-free or oil-injected. This depends on the kind of application and the budget.

Till 20 [kW] use is made of a belt drive or direct coupling, above 20 [kW] a gearbox or an elastic coupling is used.

Scroll compressor	1.5–15 [kW] Max 10 [bar]	Continuous and intermittent use Compressed air (oil-free) Refrigerators
Tooth compressor	18.5–55 [kW] Max 8.5 [bar]	Continuous use Compressed air (oil-free)
Screw compressor	5–2000 [kW] Max 10 [bar] Also small pressure designs	Continuous use Compressed air (oil-free and oil injected)
Small piston compressor	2–15 [kW] Max 30 [bar]	Intermittent use Compressed air (oil-free and oil injected)
Industrial piston compressor	30–300 [kW] 7–500 [bar]	Continuous use Not for air Process applications Gas compression
Turbo's	From 250 [kW] Good efficiency from 500 [kW] (three stages) 1.5–200 [bar]	Best is continuous charged Process applications
Roots blower	Max 1 [bar] 5–300 [kW]	Continuous use Oil-free
Fans	Centrifugal: big flows Axial: Very high flows Small feed heads	Coolers Household devices

6.8 Psychrometrics

When several compressors are placed in series there is no problem when compressing gasses as NH_3, but with moist air problems can arise with the water content of the air. The science that treats moist air is called *psychrometrics*. A part of it, just enough for what is needed in the following chapters, will be treated here.

6.8.1 Partial Pressure

Suppose a vessel with volume V contains different ideal gasses. The temperature of the gas in the vessel is T. The state of the substance can be described not only by its total pressure p and the temperature T but also by values that represent the composition. According to Dalton's laws:

- In a mix of (ideal) gasses every partial gas behaves like it was alone in the volume.
- The partial gasses is spread over the volume and exercises a pressure, the *partial pressure*, that it would exercise if it were alone in that space.

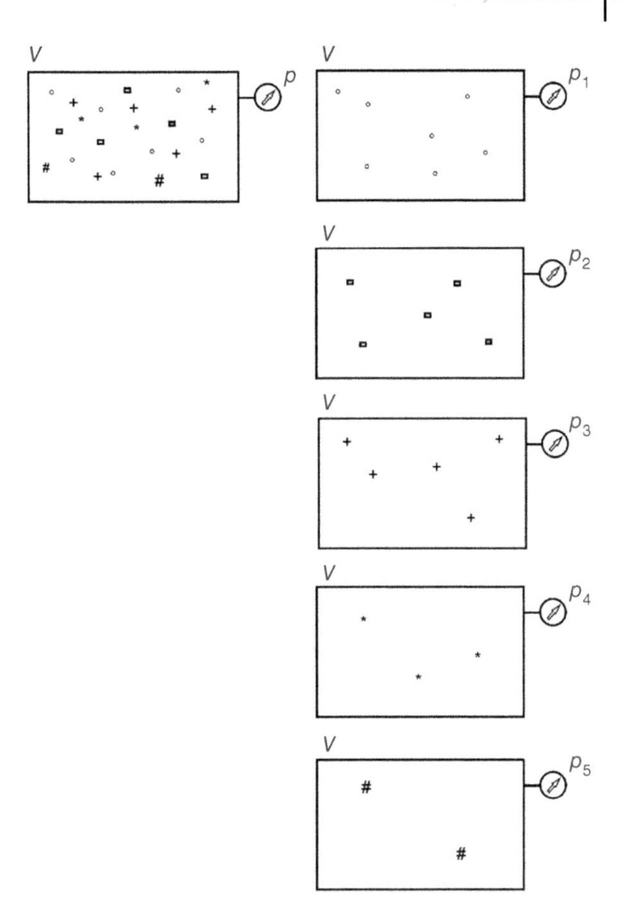

Figure 6.10 Partial pressure.

So, if the masses of the partial gasses are represented by m_i, where i is a number indicating the partial gas, and the same notation is used for the molar masses M_i (Figure 6.10):

$$p_1 \cdot V = m_1 \cdot \frac{R}{M_1} \cdot T \tag{6.2}$$

$$p_2 \cdot V = m_2 \cdot \frac{R}{M_2} \cdot T$$

And so on.

The total mass of the mix of partial gasses:

$$m = \sum_{i=1}^{n} m_i$$

where the index i starts at 1 and ends at $i = n$. n is the number of partial gasses.

The total pressure p of the mix is equal to the sum of partial pressure of the composing gasses:

$$p = \sum_{i=1}^{n} p_i$$

6.8.2 Equivalent Molar Mass

Summation of Equation (6.2) leads to:

$$V \cdot \sum_{i=1}^{n} p_i = R \cdot \sum_{i=1}^{n} \frac{m_i}{M_i} \cdot T$$

Or:

$$p \cdot V = R \cdot \sum_{i=1}^{n} \frac{m_i}{M_i} \cdot T$$

And the general law for a mixture of ideal gasses is:

$$p \cdot V = m \cdot \frac{R}{M} \cdot T$$

where:

$$m = \sum_{i=1}^{n} m_i$$

Identifying the last three expressions:

$$M = \frac{m}{\sum_{i=1}^{n} \frac{m_i}{M_i}}$$

The case now for dry air: at sea level the constitution of dry air is:

Sort	m_i in %
N_2	75.47
O_2	23.19
Ar	1.29
CO_2	0.05
H_2	0.00

where $m_i\% = \frac{m_i}{m} \cdot 100$.

The molar masses of the partial gasses can be looked up in the table of Mendeleev. Finally, for dry air the result is:

$$M(\text{dry air}) = 29\ [\text{kmol/kg}]$$

Dry air can now be treated as an ideal gas with an equivalent molar mass of 29.

6.8.3 Moist Air

Moist air is a mix of two ideal gasses: dry air ($M_{DA} = 29$) and water in superheated state ($M_W = 18$).

Mixing the two ideal gassers leads to a new ideal gas. The total pressure of that gas can be written as:

$$p = p_{DA} + p_W$$

where W stands for water vapor and DA for dry air.

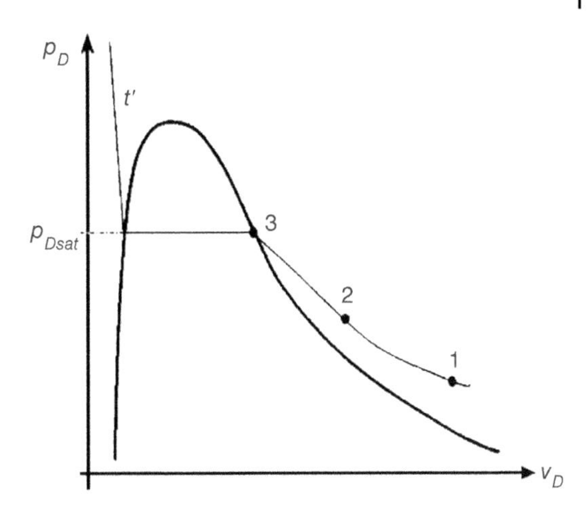

Figure 6.11 *pv* diagram of water.

p_A and p_W are the partial pressures of dry air and superheated steam. If T is the temperature of the moist air and V the volume where the moist air is locked up in, then:

$$p_W \cdot V = m_W \cdot \frac{R}{M_W} \cdot T$$

$$p_{DA} \cdot V = m_{DA} \cdot \frac{R}{M_{DA}} \cdot T$$

One very important conclusion: *the more water vapor in the air, the higher its partial pressure.*

6.8.4 Water Content

Water content is defined as:

$$x = \frac{m_W}{m_{DA}}$$

This variable is expressed in kg/kg dry air, but because the magnitude of x is in the order of gram, x will mostly be expressed in gram/kg dry air.

6.8.5 Saturated and Unsaturated Air (with Water)

In order to explain these concepts the reasoning is followed in a *pv* diagram of water (Figure 6.11).

Mostly the water vapor in the moist air is in a superheated state. Suppose such a state represented by state 1 (Figure 6.10). Add now in *an isothermal* way water vapor, at constant temperature t'. Then the evolution is 1→ 2. Continue this procedure. There will come a point, represented by point 3, when the water vapor condensates. It seems that the moist air, at temperature t', cannot contain more water vapor than corresponds with the partial pressure of point 3. One talks now of *saturated air*. One means here that the air is saturated with water vapor. In fact, this is a poor choice of words, because in thermodynamics a saturated state of air means a state whereby the air is in an equilibrium of liquid and gas. When the air is not saturated (in the psychrometric sense) one speaks of unsaturated air.

The value of the partial pressure corresponding with state 3, at temperature t', is the *saturation pressure,* noted as:

$$p_3 = p_{W,sat}(t')$$

This value can be read in the steam tables.

6.8.6 Relation Between x and p_W

The gas laws state:

$$p_W \cdot V = m_W \cdot \frac{R}{M_W} \cdot T$$

$$p_{DA} \cdot V = m_{DA} \cdot \frac{R}{M_{DA}} \cdot T$$

$$p = p_{DA} + p_W$$

Eliminating p_A out of these expressions, with:

$$x = \frac{m_W}{m_{DA}}$$

And:

$$\frac{M_W}{M_{DA}} = 0.622$$

Gives:

$$x = 0.622 \cdot \frac{p_W}{p - p_W}$$

Special case, saturation:

$$x_{sat} = 0.622 \cdot \frac{p_{W,sat}}{p - p_{W,sat}}$$

Conclusion: If x and p stay constant (this means, at a constant atmospheric pressure), and no water is added, then the partial pressures will stay constant too.

7

Piston Compressors

CHAPTER MENU

7.1 Indicator Diagram

In Figure 7.1 a horizontal piston compressor is shown, together with its indicator diagram. Vertically the pressure in the cylinder is set out; horizontally, the volume in the cylinder, that is taken by the gas that has to be compressed.

The task is to compress a gas from inlet pressure p_1 to an outlet pressure p_2 (mostly p_2 is the pressure of a press reservoir).

It is supposed that these pressures are constant.

Like with piston pumps a crank connecting rod mechanism is used. The valves are *freely floating*. This means that they open by a pressure difference over the valve (not by a cam like with fuel motors). Springs press the valves on their seat and help to close them.

In *a* the piston is in the "dead" point on the left side. Both valves, suction and discharge, are closed. The "dead" volume V_0 is filled with gas on end compression pressure p_2.

The piston moves to the right so that in the cylinder the rest gas of the precious cycle is expanded. The pressure in the cylinder decreases, see a → b. When the pressure in the cylinder in state *b* becomes lower than the suction pressure p_1 the suction valve opens. The gas is sucked isobar. The state of the gas does not change, only the mass in the cylinder increases.

The piston ends in the dead point on the right side. At state *c* the movement of the piston reverses. The suction valve is then closed by the action of the bas because. the piston exercises and overpressure on the gas. The compression is given by the curve c → d.

Pumps and Compressors, First Edition. Marc Borremans.

Companion website: www.wiley.com/go/borremans/pumps

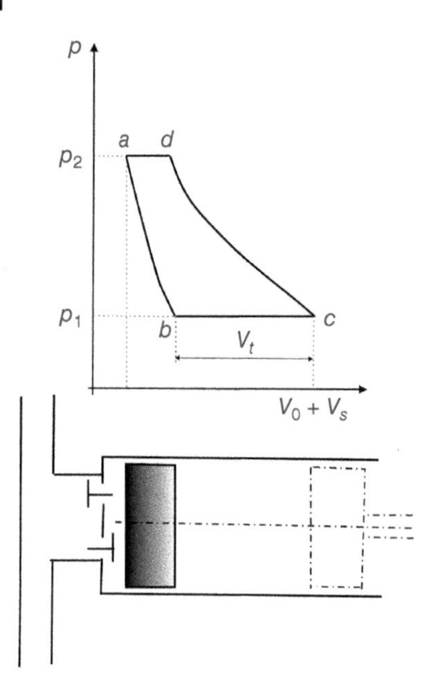

Figure 7.1 Indicator diagram.

In *d* the pressure in the cylinder has become greater than the pressure p_2 of the press reservoir. The discharge valve opens and the compressed gas streams out of the cylinder. On the left, when the piston is in extreme position at state *a*, there is still the rest gas that cannot be pushed away.

The cycle is repeated.

7.2 Parts

7.2.1 Cylinders

In Figure 7.2 a section of a two-cylinder piston compressor is represented. The compressor starts up unloaded to avoid peak currents. The same thing happens with the stopping cycle: the relief valve is gradually discharged in order to cause no abrupt stop.

In Figures 7.3 and 7.4 the parts of a heavy piston compressor with crosshead mechanism are shown.

The compressor can, as a piston pump, be double acting (Figure 7.5).

7.2.2 Sealing

Piston compressors are divided into oil-lubricated and oil-free compressors. In the case of oil-free compressors, the piston is sealed by Teflon piston rings (Figures 7.6–7.10).

Oil free compressors are also designed with *labyrinth* sealing between piston and cylinder wall (Figure 7.11). The piston does not touch the cylinder wall, but the surface of the piston, and eventually the cylinder wall, has consecutive notches. The gas that wants to leak outside has to travel the difficult passage through that labyrinth and undergoes a big resistance. The leak flow is very small. With a double acting compressor,

Figure 7.2 Section two-cylinder piston compressor V. Source: Bitzer.

Figure 7.3 Components of compressor with cross head.

the gas leaks to the underside during the press stroke. This loss does not mean any loss of gas, but a loss of energy.

7.2.3 Valves

The valves together with the sealing are the two critical links with a piston compressor. In Figure 7.12 a double-action piston is shown with four disc valves.

Figure 7.4 Crankshaft hyper-compressor. Source: Burckhardt.

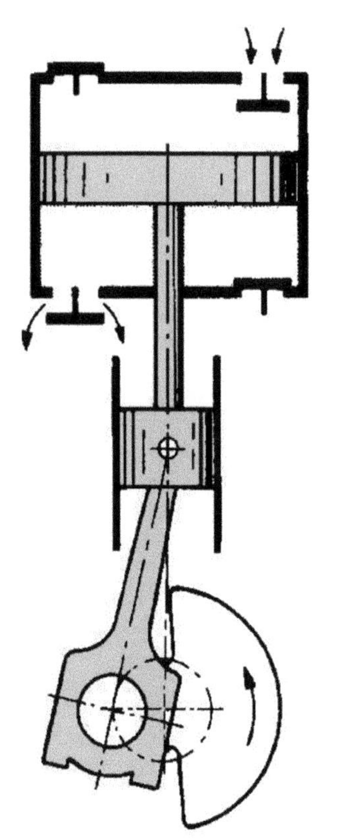

Figure 7.5 Principle of a double acting compressor.

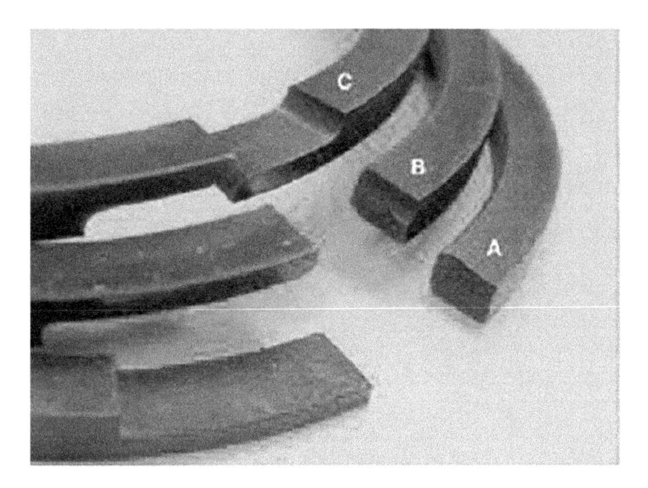

Figure 7.6 Teflon piston rings. Source: Burckhardt.

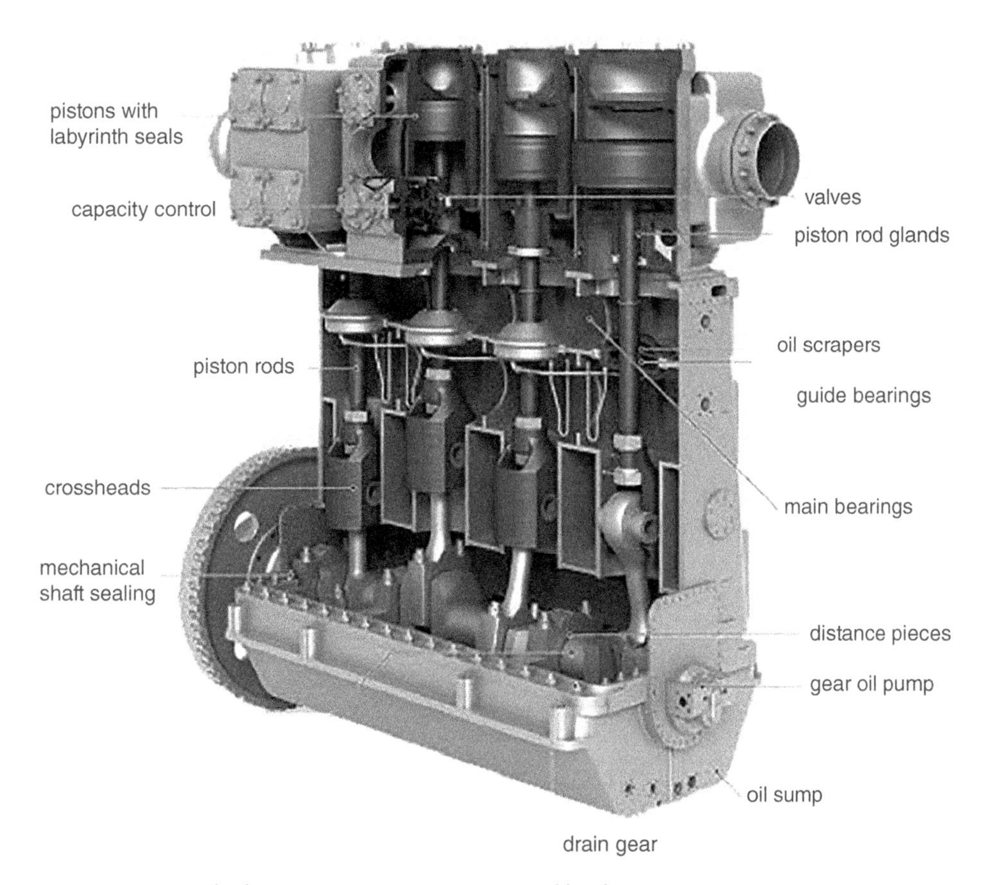

Figure 7.7 Four-cylinder piston compressor. Source: Burkhardt.

Figure 7.8 Vertical cylinder, labyrinth piston.

In Figures 7.13–7.16 valves are shown for clean gasses. The valve discs are operated by springs that keep the discs against their seat.

7.3 Volumetric Efficiency

With the sucking of fresh gas, a volume V_t, the *allowable volume*, is sucked.

The *volumetric efficiency* is defined as:

$$\lambda = \frac{V_t}{V_s}$$

where V_s is *the stroke volume.*

The factor of the dead volume, where V_0 represents the *dead volume* (clearance volume):

$$\varepsilon = \frac{V_o}{V_s}$$

In practice this is 4–8%. The highest values for ε are found with small compressors.

The numerical values of ε determines the maximum pressure ratio that can be attained. Let's look now at the functional relationship between the pressure ratio p_2/p_1, λ and ε.

Figure 7.9 Vertical cylinder, labyrinth piston.

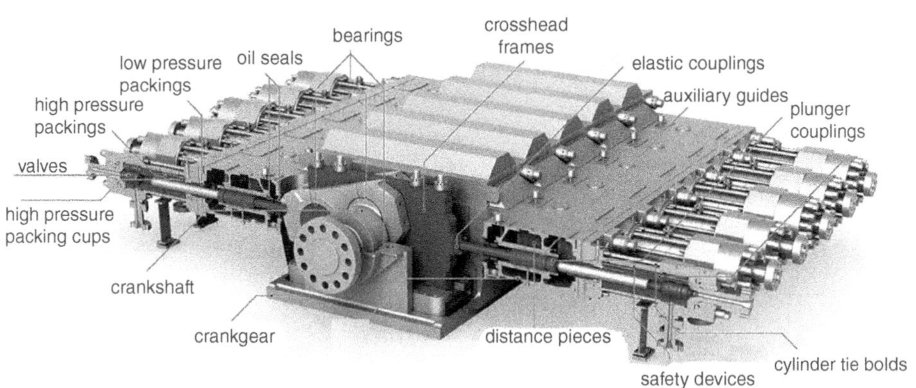

Figure 7.10 Horizontal two-cylinder piston compressor. Source: Burkhardt.

In the indicator diagram the lines a → b and c → d respectively represent an expansion of the dead gas and a compression of the allowable gas. When these state changes elapse in a polytropic way the following thermodynamic equations can be applied:

$$p_a \cdot v_a^n = p_b \cdot v_b^n \tag{7.1}$$

$$p_c \cdot v_c^n = p_d \cdot v_d^n$$

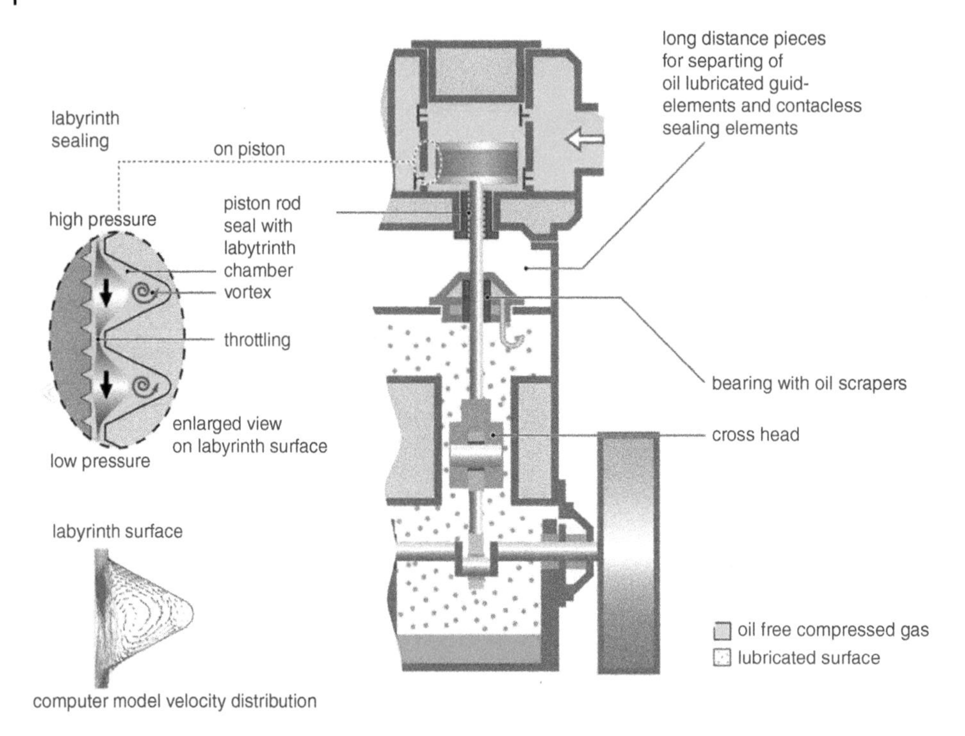

Figure 7.11 Piston compressor with labyrinth pistons. Source: Burkhardt.

Figure 7.12 Valves double-acting piston. Source: Hoerlinger.

Figure 7.13 Valve. Source: Hoerlinger.

Figure 7.14 Plate valve. Source: Burckhardt.

where n is the polytropic factor (here equal assumed for expansion and compression); v is the specific volume of the gas, and the index points to the state.

It is Equation (7.1) that is of interest:

$$\frac{p_a}{p_b} = \left(\frac{v_b}{v_a}\right)^n = \left(\frac{V_b}{V_a}\right)^n = \left(\frac{V_b}{V_0}\right)^n = \left(\frac{V_0 + V_s - V_t}{V_0}\right)^n$$
$$= \left(1 + \frac{V_s}{V_0} - \frac{V_t}{V_0}\right)^n = (1 + (1 - \lambda)/\varepsilon)^n \tag{7.2}$$

Equation (7.2) shows that:

$$\lambda = 1 - \varepsilon \cdot \left(\left(\frac{p_2}{p_1}\right)^{\frac{1}{n}} - 1\right) \tag{7.3}$$

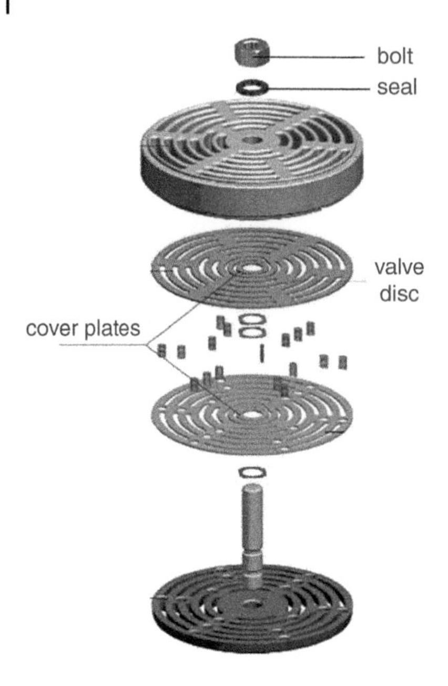

Figure 7.15 Plate valve. Source: Burckhardt.

Figure 7.16 Valve of a hyper-compressor. Source: Burckhardt.

Equation (7.3) states that the volumetric efficiency λ:

- Decreases with increasing pressure ration.
- Decreases with increasing value of ε.
- Increases as the compression happens more adiabatic.

In Table 7.1 Equation (7.3) is calculated for some values of the pressure ratio and for ε.
How should these values in Table 7.1 now be interpreted? As far as concerns the values of ε, we cannot change anything because this depends on the construction. That

Table 7.1 Values for λ.

$\frac{p_2}{p_1}$	$n = 1.3$		$n = 1.2$		$n = 1.1$	
	$\varepsilon = 4\%$	$\varepsilon = 6\%$	$\varepsilon = 4\%$	$\varepsilon = 6\%$	$\varepsilon = 4\%$	$\varepsilon = 6\%$
5	0.90	0.85	0.85	0.83	0.86	0.80
10	0.80	0.70	0.77	0.65	0.72	0.57
20	0.64	0.46	0.55	0.33	0.42	0.14

Table 7.2 Limit-pressure ratio.

	$n = 1.2$		$n = 1.1$	
	$\varepsilon = 4\%$	$\varepsilon = 6\%$	$\varepsilon = 4\%$	$\varepsilon = 6\%$
$\frac{p_2}{p_1}$	50	31	36	26

the volumetric efficiency increases as the process is more adiabatic is a disappointment, because later we will learn that an isothermal compression is better (lower power consumption). In the formula for the volumetric efficiency the pressure ratio appears. This ratio is given by the fact that the user knows perfectly well what the inlet and discharge pressures are. If a volumetric efficiency of 80% can be accepted as good, then, in the case of a single stage compression, we must limit ourselves to a compression ratio of 5.

In the case of a volumetric efficiency that is zero the compressor would suck no flow at all. The pressure ratio that could be reached in that case can be calculated from Equation (7.3), for $\lambda = 0$:

$$\left(\frac{p_2}{p_1}\right)_{\lambda=0} = \left(\frac{1+\varepsilon}{\varepsilon}\right)^n$$

$\left(\frac{p_2}{p_1}\right)_{\lambda=0}$ is the *limit-pressure ratio.*

Table 7.2 gives some values.

The limit pressure ratio is of great importance in the vacuum technique because the reachable vacuum pressure is limited in this way. When a piston compressor is used to make a space vacuum, the reachable vacuum is limited (Figure 7.17).

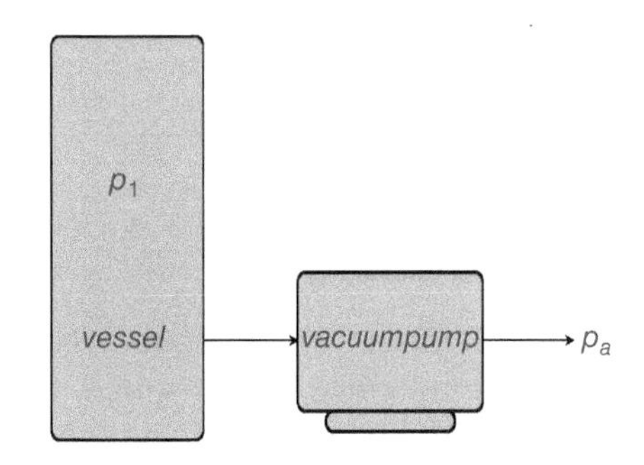

Figure 7.17 Piston compressor as vacuum pump.

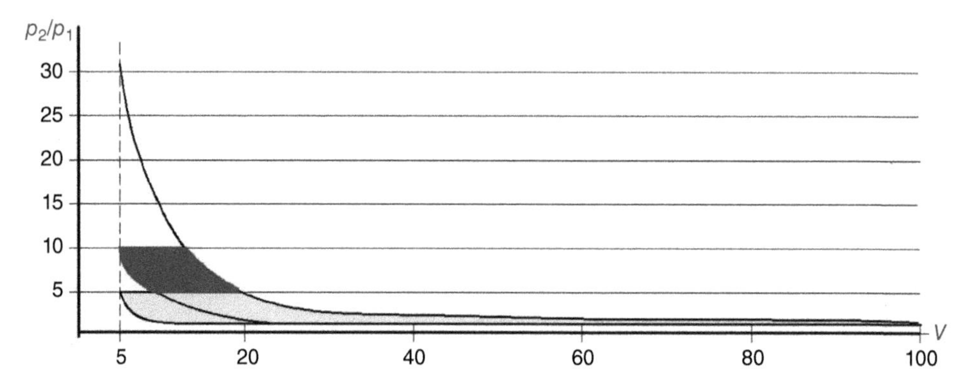

Figure 7.18 Influence allowable volume on the pressure ratio.

Suppose a piston compressor with properties $\varepsilon = 6\%$ and $n = 1.2$ then the limit pressure ratio is, according to Table 7.2, the value 31.

In other words, the deepest pressure is:

$$p_1 = \frac{p_a}{31} \cong \frac{10^5}{31} \cong 3 \cdot 10^3 \text{ [Pa]} \cong 30 \text{ [mbar]}$$

But in that case there is no flow. The normal reachable vacuum pressure will thus be somewhat higher (Figure 7.18).

As an illustration Figure 7.18 shows a compressor with a changing pressure ratio.

At a pressure ration of 5 one gets the yellow colored indicator diagram. At a pressure ration of 10 the allowable volume becomes smaller; see the blue diagram. At pressure ratio 31 the indicator diagram is reduced to one line that runs till 31 [bar] and then lowers again to 1 [bar], and that repeats itself. The expansion curve of the dead gas is now equal to the compression curve of it. The piston moves without any sense. No gas is delivered.

7.4 Membrane Compressor

As with pumps there are also membrane compressors. The membrane is driven electrically, pneumatically, hydraulically, or mechanically (Figure 7.19).

7.5 Work and Power

7.5.1 Technical Work

In this section the *technical work* is examined.

For simplicity's sake, we assume that at the rear of the piston there is a vacuum. In practice must be worked with effective pressures, but for the expression of the technical work this leads to the same result.

- Study first the expansion of the dead gas in the dead volume (Figure 7.20). It concerns the state change a → b. Because it is assumed that behind the piston there is a vacuum, the gas will deliver work (in thermodynamically sense) to the piston. It concerns here

Figure 7.19 Hydrogen diaphragm compressor 1000 [bar_g].

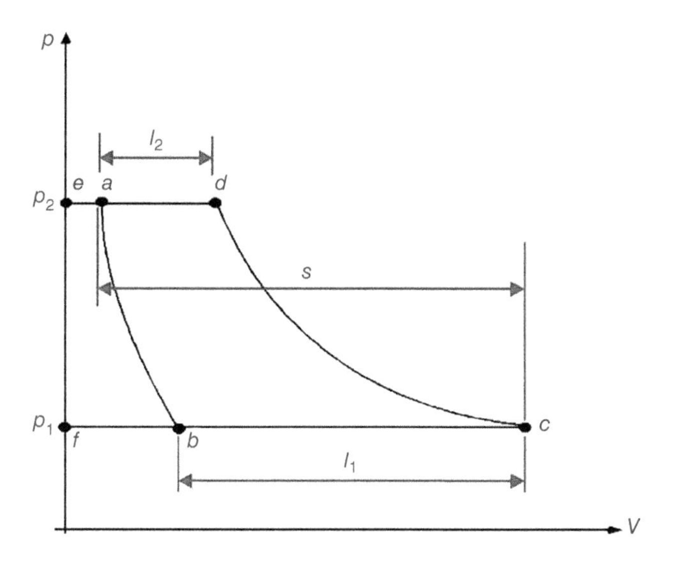

Figure 7.20 Technical work.

a closed volume, so the work done by the gas is a *compression work* (of volume change work:

$$-\int_a^b p \cdot dV \tag{7.4}$$

The algebraic value of this equation is negative because $dV > 0$. This is OK, the gas delivers work.

The work can be interpreted as the surface under the curve a → b, with respect to the abscissa.

- Then the suction valve opens and we get the state change b → c. During the suction change, the gas flows into the cylinder under pressure p_1. At the end of this state change the volume V_s of the cylinder is completely filled. Assume that the piston has run a distance l.
  Because the gas exercises on the piston a pressure p_1 and on the rear of the piston there is no pressure (by hypothesis) the gas exercises a work on the piston, with *work = force x distance traveled*:

$$-p_1 \cdot l_1 \cdot A = -p_1 \cdot V_t = -p_1 \cdot (V_c - V_b) \tag{7.5}$$

  where:
  - A: the piston surface.
  - l_1: the distance traveled by the piston from b to c.

  The minus sign is in accordance with the thermodynamic agreement that work delivered by the gas at the surroundings is considered negative.
- Then the gas in the cylinder is compressed from state c to state d. Both valves are closed and the surrounding exercises on the closed system of the gas a *compression work* (see Section 6.4.1 in Chapter 6):

$$-\int_c^d p \cdot dV \tag{7.6}$$

- At last the compressed gas is expelled. It is now on pressure p_2 and has a volume V_d. The piston presses with pressure p_2 on the gas. The distance to cover is $(s - l_2)$, where s is the stroke length. The surrounding delivers a work on the gas:

$$+p_2 \cdot A \cdot (s - l_2) = +p_2 \cdot (V_d - V_a) \tag{7.7}$$

The *technical work* W_{t12}, this is the work on the shaft of the compressor, delivered by the surrounding, needed to suck the gas, to compress it, and to press it way, is given by the summation of Equations (7.5)–(7.7)

$$W_{t12} = -\int_c^d p \cdot dV + p_2 \cdot (V_d - V_a) - p_1 \cdot (V_c - V_b) - \int_a^b p \cdot dV$$

$$= -\int_c^d p \cdot dV + p_d \cdot V_d - p_c \cdot V_c + p_b \cdot V_b - p_a \cdot V_a - \int_a^b p \cdot dV$$

$$W_{t12} = -\int_c^d p \cdot dV + |p \cdot V|_c^d + |p \cdot V|_a^b - \int_a^b p \cdot dV$$

Before working this out, let's split this in:

$$|p \cdot V|_c^d - \int_c^d p \cdot dV$$

$$|p \cdot V|_a^b - \int_a^b p \cdot dV$$

The first term is the *displacement work*; this is necessary the let the gas in and out of the device. The second term is the *compression work*; this is needed to compress the close volume gas (or expand).

According to the rule of the *partial integration* known from the calculation techniques of the integral calculation:

$$W_{t12} = \int_c^d V \cdot dp + \int_a^b V \cdot dp \tag{7.8}$$

The two together form the *technical work*; this is the work done by the surrounding, on a *translating* or *rotating shaft*.

The first term in Equation (7.8) is (thermodynamically) positive; the second one, negative. The first term is represented by the surface $O_{cd} = e \rightarrow d \rightarrow c \rightarrow f$; the second term, by the surface $O_{ab} = e \rightarrow a \rightarrow b \rightarrow f$. The difference between both surfaces is equal to the surface of the loop $a \rightarrow d \rightarrow c \rightarrow d$. For one cycle of the piston compressor one comes to:

$$W_{t12} = \oint V \cdot dp \tag{7.9}$$

Equation (7.9) represents the technical work W_{t12} needed to compress a quantity gas equal to the allowable volume V_t with the piston compressor from pressure p_1 to pressure p_2. This can be interpreted as the surface under the loop in the indicator diagram. The indexes 1 and 2 point onto the state change $1 \rightarrow 2$ and are often omitted.

7.5.2 Isothermal Compression

First the technical work W_{t12} is calculated for the case that the expansion $a \rightarrow b$ and compression $c \rightarrow d$ is isotherm.

$$W_{t12} = \oint V \cdot dp = O_{cd} + O_{ab}$$

Calculate first $O_{cd} = \int_c^d V \cdot dp$.

Because of the isotherm character:

$$p_1 \cdot V_c = p_2 \cdot V_d = p \cdot V$$

$$V = \frac{p_1 \cdot V_c}{p}$$

$$O_{cd} = \int_c^d \frac{p_1 \cdot V_c}{p} \cdot dp = p_1 \cdot V_c \int_c^d \frac{dp}{p} = p_1 \cdot V_c \cdot [ln\, p]_{p_1}^{p_2}$$

$$O_{cd} = p_1 \cdot V_c \cdot ln\frac{p_2}{p_1}$$

Analogous, for the state change $a \rightarrow b$:

$$p_2 \cdot V_a = p_1 \cdot V_b = p \cdot V$$

$$V = \frac{p_1 \cdot V_b}{p}$$

$$O_{ab} = \int_a^b V \cdot dp = \int_a^b \frac{p_1 \cdot V_b}{p} \cdot dp = p_1 \cdot V_b \int_{p_2}^{p_1} \frac{dp}{p} = p_1 \cdot V_b \cdot [ln\, p]_{p_2}^{p_1}$$

$$O_{ab} = p_1 \cdot V_b \cdot ln\frac{p_1}{p_2} = -p_1 \cdot V_b \cdot ln\frac{p_2}{p_1}$$

So:

$$W_{t12} = \oint V \cdot dp = O_{cd} + O_{ab} = p_1 \cdot V_c \cdot ln\frac{p_2}{p_1} - p_1 \cdot V_b \cdot ln\frac{p_2}{p_1}$$

$$= p_1 \cdot (V_c - V_b) \cdot ln\frac{p_2}{p_1}$$

$$W_{t12} = p_1 \cdot V_t \cdot ln\frac{p_2}{p_1} \tag{7.10}$$

Per cycle, this is one rotation of the crank, a quantity of volume equal to the allowable volume V_t is processed. This volume is determined by the pressure p_1.

If the number of revolutions of the crank axis per minute is represented by N then the volumetric flow Q_{V1} sucked by the compressor is given by:

$$Q_{V1} = V_t \cdot \frac{N}{60} \; [\text{m}^3/\text{cycle} \cdot \text{number of cycles/second} = \text{m}^3/\text{second}]$$

The delivered power P_{t12} to the shaft of the compressor is calculated as follows. Per cycle, this is per rotation of the crankshaft, a work W_{t12} must be delivered. The number of cycles per minute is N.

$$P_{t12} = W_{t12} \cdot \frac{N}{60} \; [\text{W}]$$

$$P_{t12} = p_1 \cdot V_t \cdot ln\frac{p_2}{p_1} \cdot \frac{N}{60}$$

$$P_{t12} = p_1 \cdot Q_{V1} \cdot ln\frac{p_2}{p_1} \tag{7.11}$$

7.5.3 Polytropic Compression

In the case of a polytropic compression:

$$O_{cd} = \int_c^d V \cdot dp$$

$$p_1 \cdot V_c^n = p_2 \cdot V_d^n = p^n \cdot V^n$$

$$V = \left(\frac{p_1}{p}\right)^{\frac{1}{n}} \cdot V_c$$

After calculation:

$$O_{cd} = \frac{n}{n-1} \cdot p_1 \cdot V_c \cdot \left(\left(\frac{p_2}{p_1}\right)^{\frac{n-1}{n}} - 1\right)$$

Analogous for O_{ab}:

$$O_{ab} = -\frac{n}{n-1} \cdot p_1 \cdot V_b \cdot \left(\left(\frac{p_2}{p_1}\right)^{\frac{n-1}{n}} - 1\right)$$

where:

$$W_{t12} = O_{cd} + O_{ab} = \frac{n}{n-1} \cdot p_1 \cdot (V_c - V_b) \cdot \left(\left(\frac{p_2}{p_1}\right)^{\frac{n-1}{n}} - 1\right)$$

$$W_{t12} = \frac{n}{n-1} \cdot p_1 \cdot V_t \cdot \left(\left(\frac{p_2}{p_1} \right)^{\frac{n-1}{n}} - 1 \right)$$

$$P_{t12} = \frac{n}{n-1} \cdot p_1 \cdot Q_{v1} \cdot \left(\left(\frac{p_2}{p_1} \right)^{\frac{n-1}{n}} - 1 \right)$$

7.5.4 Conclusions

The expression for the work per cycle for a polytropic compression:

$$W_{t12} = \frac{n}{n-1} \cdot p_1 \cdot V_t \cdot \left(\left(\frac{p_2}{p_1} \right)^{\frac{n}{n-1}} - 1 \right)$$

If m^* is the transported mass per cycle and making use of the ideal gas law, it can also be written as:

$$W_{t12} = \frac{n}{n-1} \cdot m^* . \frac{R}{M} \cdot T_1 \cdot \left(\left(\frac{p_2}{p_1} \right)^{\frac{n}{n-1}} - 1 \right)$$

A first conclusion is: *aspire the gas as cold as possible.*
Now we convert the found formulas to those one finds in a course on thermodynamics.
In the case of an *isothermal* compression one found:

$$W_{t12} = p_1 \cdot V_t \cdot ln \frac{p_2}{p_1}$$

This relation can easily be converted to specific technical work w_{t12} in the case of an isothermal compression. It suffices to consider not the allowable volume V_t but the volume of 1 [kg] gas; this is the specific volume v_1:

$$W_{t12} = p_1 \cdot v_t \cdot ln \frac{p_2}{p_1} [\text{J/kg}] \tag{7.12}$$

One analogous way for polytropic compression is:

$$W_{t12} = \frac{n}{n-1} \cdot p_1 \cdot V_t \cdot \left(\left(\frac{p_2}{p_1} \right)^{\frac{n-1}{n}} - 1 \right)$$

$$w_{t12} = \frac{n}{n-1} \cdot p_1 \cdot v_1 \cdot \left(\left(\frac{p_2}{p_1} \right)^{\frac{n-1}{n}} - 1 \right) \tag{7.13}$$

In the thermodynamic *pv* diagram Equations (7.12) and (7.13) are found by the integrated surface under the curve $1 \rightarrow 2$ with respect to the ordinate. From this we see that the most advantage compression is the isothermal compression; see Figure 7.21.

A numerical example will make this clear. Consider the compression of an air volumetric flow of 0.1 [m^3/s] with a piston compressor from 1 [bar(a)] to 7 [bar(a)].

Isothermal:

$$P_{t12} = p_1 \cdot Q_{V1} \cdot ln \tfrac{p_2}{p_1} = 10^5 \cdot 0.1 \cdot ln 7 = 19 \cdot 10^3 \text{ [W]} = 19 \text{ [kW]}$$

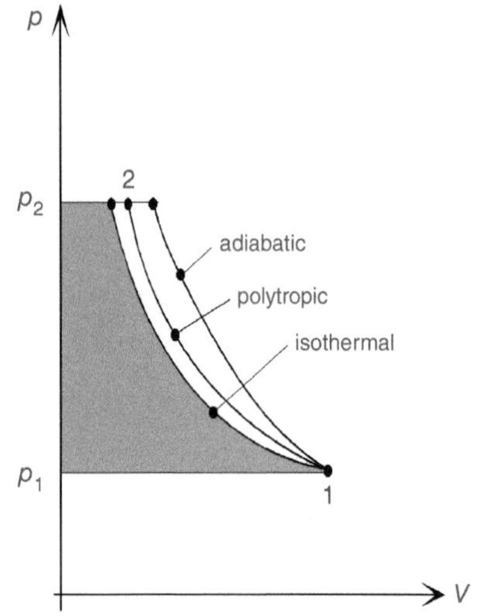

Figure 7.21 Various types of compression.

Polytropic with factor $n = 1.3$:

$$P_{t12} = \frac{n}{n-1} \cdot p_1 \cdot Q_{v1} \cdot \left(\left(\frac{p_2}{p_1} \right)^{\frac{n-1}{n}} - 1 \right)$$

$$= \frac{1.3}{1.3-1} \cdot 10^5 \cdot 0.1 \cdot \left((7)^{\frac{1.3-1}{1.3}} - 1 \right) \text{ [W]} = 24 \text{ [kW]}$$

Adiabatic, with for air: $n = \gamma = 1.4$:

$$P_{t12} = \frac{n}{n-1} \cdot p_1 \cdot Q_{v1} \cdot \left(\left(\frac{p_2}{p_1} \right)^{\frac{n-1}{n}} - 1 \right) = \frac{1.4}{1.4-1} \cdot 10^5 \cdot 0.1 \cdot \left((7)^{\frac{1.4-1}{1.3}} - 41 \right) \text{ [W]} = 26 \text{ [kW]}$$

At an ideal isothermal compression, the temperature during the compression is constant. Mostly this is the surrounding temperature. To reach an isothermal compression care must be taken to get rid of all compression heat to the surrounding area.

This can be reached by:

- *Cooling ribs* on the compressor; this is the case with small compressors (Figure 7.22).
- *Cooling ribs in combination with a fan*; this is the use with middle big compressors (Figure 7.23).
- Perform the compressor *double-walled* and make use of water cooling; the investment is responsible for big compressors.

7.5.5 Efficiency of a Piston Compressor

As said before, an isothermal compression is energetically the best. Take this as a reference and define the isothermal efficiency of a compressor as:

$$\eta_{iso} = \frac{w_{t,iso}}{w_{t,pol}}$$

where $w_{t,iso}$ is the specific technical energy for an isothermal compression and $w_{t,pol}$ the specific technical work for a polytropic compression.

Figure 7.22 Cooling ribs. Source: Compare – Gardner Denver.

Figure 7.23 Piston compressor with pressure vessel, fan, and cooling ribs. Source: Compare – Gardner Denver.

This can also be written as:

$$\eta_{iso} = \frac{w_{t,iso}}{w_{t,pol}} = \frac{w_{t,iso}}{w_{t,pol}} \cdot \frac{Q_M}{Q_M} = \frac{P_{t,iso}}{P_{t,pol}}$$

where $P_{t,\,iso}$ is the technical power for an isothermal compression and $P_{t,pol}$ the specific technical power for a polytropic compression.

In the example given the results were:

Isothermal compression:

$$P_{t12} = 19\ [\text{kW}]$$

Polytropic compression with $n = 1.3$

$$P_{t12} = 24\ [\text{kW}]$$

Adiabatic, with for air: $n = \gamma = 1.4$:

$$P_{t12} = 26\ [\text{kW}]$$

The isothermal efficiency of the compressor in the case of a polytropic compression with exponent $n = 1.3$ is:

$$\eta_{iso} = \frac{19}{24} = 0.8$$

In the case of an adiabatic compression:

$$\eta_{iso} = \frac{19}{26} = 0.7$$

7.6 Two-stage Compressor

7.6.1 Motivation

In Section 7.1.3 it was demonstrated that the volumetric efficiency λ of the piston compressor becomes very small at high compression ratios. This leads automatically to big energy losses because an important portion of the cylinder volume is needlessly compressed and expanded.

- A second reason why in one stage we cannot endless compress is the end compression temperature. With oil-free compressors the temperature may only rise to 125 °C; with oil-lubricated compressors, to 75 °C (above that temperature the oil cracks and loses its lubricating properties).
- Assume that the compression takes place with polytropic exponent n, then, according to Poisson, with a state compression $1 \rightarrow 2$:

$$\frac{T_2}{T_1} = \left(\frac{p_2}{p_1}\right)^{\frac{n-1}{n}}$$

Take, for instance, $n = 1.2$ at a suction temperature of 10 °C. Then $T_1 = 283$ K.
At a pressure ration of 6: $T_2 = 381\ \text{K} = 108$ °C; this is tolerable with oil-free compressors. At a pressure ratio of 10 one gets: $T_2 = 415\ \text{K} = 142$ °C; this is not tolerable.

- The end compression temperature is strongly dependent on the gas that is compressed. Assume, for the sake of simplicity, that an isentropic compression at $T_1 = 27\,°C = 300\,K$. With a noble gas as He the isentropic exponent $\gamma = 1.67$. For a compression ratio of 4 one finds that $T_2 = 523\,K = 250\,°C$.
  With butane (C_4H_{10}) can be read on a *ph* diagram (Figure 7.24) what the end temperature will be. If one starts with $T_1 = 27\,°C = 300\,K$ at 1 [bar] can be found: $T_2 = 340\,K = 67\,°C$.
- A third reason for the use of multistage compressors is that if intercooling is used – see later – the isothermal compression is nearly reached, and this has a favorable effect on the compression energy.

7.6.2 Two Stages

7.6.2.1 General

Consider in Figure 7.25 a low-pressure compressor that sucks a gas at pressure p_1 and temperature T_1 and compresses this gas polytropically to a pressure p_2 and temperature T_2. After that the gas is led through a heat exchanger, *an inter-cooler*. Suppose that the cooler works ideally so that there are no friction losses: at the outlet the pressure is still p_2. Furthermore, it is assumed that the temperature at the outlet is brought back to T_1. To reach that it is recommended not to use fan cooling but water cooling. Water has a higher specific heat and a higher convection factor.

Then the gas is sucked by the high-pressure compressor and compressed to pressure p_3. The temperature now rises to T_3.

Assume for the calculation of the expansion that:

- The expansion of the dead volume is polytropic with a factor n.
- The polytropic factors for both low and high pressure compression are the same.

  The following symbols are used:

  - V_t': allowable volume low pressure compressor
  - V_t'': allowable volume high-pressure compressor
  - V_s': stroke volume low-pressure compressor
  - V_s'': stroke volume high-pressure compressor
  - ε': factor dead volume low-pressure compressor
  - ε'': factor dead volume high-pressure compressor
  - λ': volumetric efficiency low-pressure compressor
  - λ'': volumetric efficiency high pressure compressor.

  The total technical work W_{t12} for the low pressure compressor:

$$W_{t12} = \frac{n}{n-1} \cdot p_1 \cdot V_t' \cdot \left(\left(\frac{p_2}{p_1} \right)^{\frac{n-1}{n1}} - 1 \right) \tag{7.14}$$

The total technical work W_{t23} for the high-pressure compressor:

$$W_{t23} = \frac{n}{n-1} \cdot p_2 \cdot V_t'' \cdot \left(\left(\frac{p_3}{p_2} \right)^{\frac{n-1}{n}} - 1 \right) \tag{7.15}$$

The transported mass in V_t' is the same as that in V_t'' and we note it m^*:

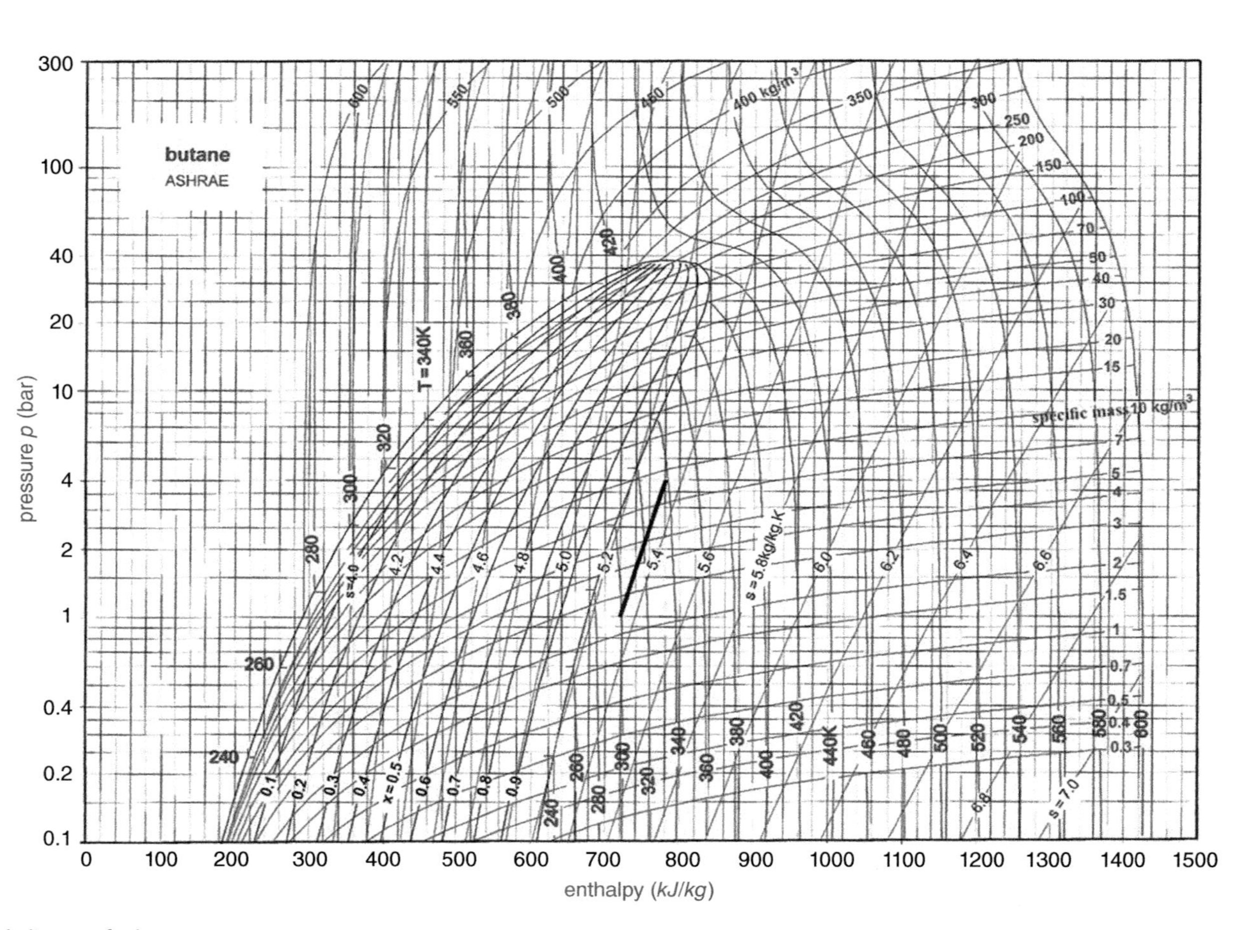

Figure 7.24 *ph* diagram for butane.

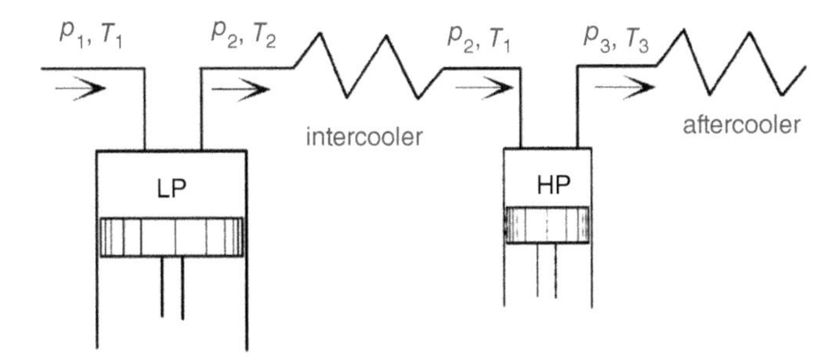

Figure 7.25 Two-stage compressor.

The LP compressor sucks a volume V_t'. The ideal gas law:

$$p_1 \cdot V_t' = \frac{m^*}{M} \cdot R \cdot T_1 \tag{7.16}$$

where M is the molar mass of the gas [kg/kmol] and R the universal gas constant [8315 J/kmol K).

The HP compressor sucks a volume V_t''. The ideal gas law applied to this volume:

$$p_2 \cdot V_t'' = \frac{m^*}{M} \cdot R \cdot T_1 \tag{7.17}$$

Taking into account Equations (7.16) and (7.17), on the one hand, and (7.14) and (7.15), on the other, the total technical work W_{t13} needed to compress the mass m^* from p_1 to p_3 is:

$$W_{t13} = W_{t12} + W_{t23}$$

$$W_{t13} = \frac{n}{n-1} \cdot p_1 \cdot V_t' \cdot \left(\left(\frac{p_2}{p_1} \right)^{\frac{n-1}{n}} + \left(\frac{p_3}{p_2} \right)^{\frac{n-1}{n}} - 2 \right) \tag{7.18}$$

7.6.2.2 Indicator Diagram

In Figure 7.26 the indicator diagram is represented for the case of a polytropic two-stage compression. If the compression was done polytropically in one stage the three colored surfaces would be used. With an intercooler shown in the pV diagram by the red and yellow surfaces alone, we see that there is a gain of the blue surface, in other words the isothermal compression is better approximated.

7.6.2.3 Intermediate Pressure

The compression task is given by the pressures p_1 and p_3. The value of the pressure p_2, the intermediate pressure, is determined by the requirement that the total to deliver technical work W_{t13} must be minimal:

$$\frac{\partial W_{t13}}{\partial p_2} = 0$$

This comes down to calculating the derivative:

$$\frac{\partial}{\partial p_2} \left(\left(\frac{p_2}{p_1} \right)^{\frac{n-1}{n}} + \left(\frac{p_3}{p_2} \right)^{\frac{n-1}{n}} \right) = 0$$

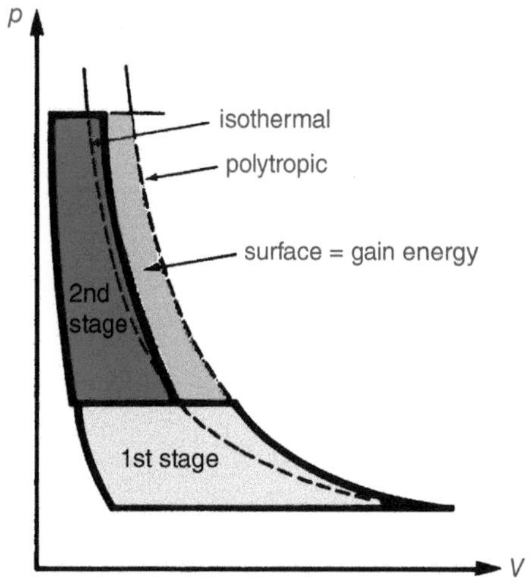

Figure 7.26 Influence two-stage compression on compression work.

Set $a = \frac{n-1}{n}$ and $x = p_2$ to become a simplified notation:

$$\frac{\partial}{\partial x}\left(\left(\frac{x}{p_1}\right)^a + \left(\frac{p_3}{x}\right)^a\right) = 0$$

$$+a \cdot x^{a-1} \cdot p_1^{-a} - a \cdot x^{-a-1} \cdot p_3^a = 0$$

Multiply both members with x^{-a+1}:

$$-x^{-2a} \cdot p_3^a + p_1^{-a} = 0$$

$$p_1^{-a} = x^{-2a} \cdot p_3^a$$

$$x^{2a} = (p_3 \cdot p_1)^a$$

$$x^2 = p_3 \cdot p_1$$

$$x = p_2 = \sqrt{p_3 \cdot p_1}$$

Or:

$$p_2^2 = p_2 \cdot p_2 = p_3 \cdot p_1$$

$$\frac{p_3}{p_2} = \frac{p_2}{p_1} \tag{7.19}$$

In other words, the ideal compression ratios of the first and second stages are equal. From Equation (7.19)

$$\frac{p_3}{p_1} = \left(\frac{p_2}{p_1}\right)^2$$

or:

$$\frac{p_2}{p_1} = \sqrt{\frac{p_3}{p_1}} \tag{7.20}$$

For three- and four-stage compressors the same conclusions apply.

Remark 1. The pressure ratio is calculated on the basis of the *absolute* pressure.
Remark 2. Pick up the difference between pumps and compressors.
For a series connection of k pumps will every pump i contribute with manometric feed pressure $p_{man,i}$ in equal amount to the total pressure rise p_{man}:

$$p_{man} = \sum_{i=1}^{k} p_{man,i}$$

For instance: 1→ 2→3 [bar]
For multistage compressors the pressure ratio $\frac{p_{i+1}}{p_i}$ of every compressor stage in principle is equal, so that the total pressure ratio is given by the multiplication of the partial pressure ratios:

$$\frac{p_k}{p_1} = \prod_{i=1}^{k} \frac{p_{i+1}}{p_i}$$

For instance: 1→3 →9 [bar]
Remark 3. The price of a two-stage compressor is, of course, higher than that of a single compressor, but after some time the gain in energy cost of the two steps will surpass the extra cost.

7.6.2.4 Work Per Stage

When the intermediate pressure is chosen according to Equation (7.19), Equation (7.16) for the technical work of the HP stage can be written as:

$$W_{t23} = \frac{n}{n-1} \cdot p_2 \cdot V_t'' \cdot \left(\left(\frac{p_3}{p_2} \right)^{\frac{n}{n-1}} - 1 \right)$$

$$W_{t23} = \frac{n}{n-1} \cdot p_1 \cdot V_t' \cdot \left(\left(\frac{p_2}{p_1} \right)^{\frac{n}{n-1}} - 1 \right) \tag{7.21}$$

Equation (7.21) is identical to Equation (7.14), in other words the technical works for the LP and HP stages are equal.

7.6.2.5 Compression Temperatures

Application of the law of Poisson for the polytropic changes of state; for the LP compressor:

$$\frac{T_2}{T_1} = \left(\frac{p_2}{p_1} \right)^{\frac{n-1}{n}} \tag{7.22}$$

For the HP compressor:

$$\frac{T_3}{T_1} = \left(\frac{p_3}{p_2} \right)^{\frac{n-1}{n}} \tag{7.23}$$

Identification of (7.22) and (7.23):

$$T_3 = T_2$$

This result is welcome. The ideal intermediate pressure is also the ideal intermediate pressure for one and the same end compression temperature after every stage. Indeed, the temperature in a compressor is in general limited to 125 °C.

7.6.2.6 Volumetric Efficiency

Consider a single-stage compressor with a compression ratio of 9, $\varepsilon = 4\%$ and a polytropic exponent of $n = 1.3$.

The volumetric efficiency λ is then:

$$\lambda = 1 - \varepsilon \cdot \left(\left(\frac{p_2}{p_1} \right)^{\frac{1}{n}} - 1 \right) = 1 - 0.04 \cdot \left(9^{\frac{1}{1.3}} - 1 \right) = 0.82$$

The same task can be done with a two-stage compressor with an intermediate pressure of 3 [bar], with the same values for ε and λ leads to equal values for the volumetric efficiencies λ' and λ'':

$$\lambda' = \lambda'' = 1 - \varepsilon \cdot \left(\left(\frac{p_2}{p_1} \right)^{\frac{1}{n}} - 1 \right) = 1 - 0.04 \cdot \left(3^{\frac{1}{1.3}} - 1 \right) = 0.94$$

7.6.2.7 Cylinder Dimensions

The dimensions of the cylinders are determined by their stroke volumes V_s' and V_s''.

Because:

$$\lambda' = \frac{V_t'}{V_s'}$$

$$\lambda'' = \frac{V_t''}{V_s''}$$

Suppose ε and n are equal for both compressors then: $\lambda' = \lambda''$

So:

$$\frac{V_t'}{V_s'} = \frac{V_t''}{V_s''}$$

$$\frac{V_t'}{V_t''} = \frac{V_s'}{V_s''} \tag{7.24}$$

Equation (7.16) = (7.17):

$$p_1 \cdot V_t' = p_2 \cdot V_t''$$

From where:

$$\frac{V_t'}{V_t''} = \frac{p_2}{p_1} \tag{7.25}$$

Equation (7.25) = (7.24) = (7.20):

$$\frac{V_s'}{V_s''} = \frac{p_2}{p_1} = \sqrt{\frac{p_3}{p_1}}$$

If both compressors are on the same crankshaft with same stroke length s, and their piston surfaces are A' and A'', then the ratio of the diameters D' and D'' can be determined, because:

$$\frac{V_s'}{V_s''} = \frac{A' \cdot s}{A'' \cdot s} = \frac{\pi \cdot \frac{(D')^2}{4}}{\pi \cdot \frac{(D'')^2}{4}} = \left(\frac{D'}{D''} \right)^2 = \sqrt{\frac{p_3}{p_1}}$$

See Figure 7.27.

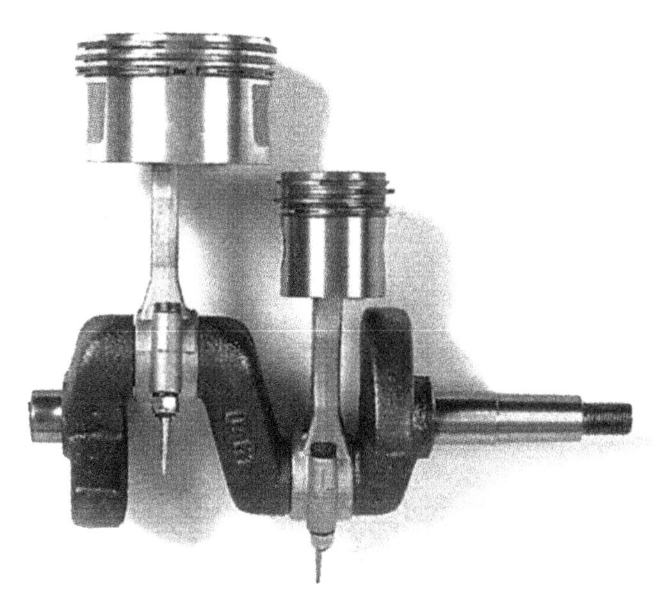

Figure 7.27 Two-stage compressor. Source: Shamal.

Example: Assume a two-stage compressor with suction pressure $p_1 = 1$ [bar] and end compression pressure $p_3 = 9$ [bar], then the construction of the pistons must be:

$$\frac{D'}{D''} = \sqrt[4]{9} = 1.7$$

7.6.2.8 Mounting

Another possibility is *tandem mounting* (Figure 7.28).

In Figure 7.27 a tandem mounting with two separate cylinders is shown.

7.7 Three or More Stages

In the same way as with the two-stage compressor an analysis can be made for a three- or four-stage compressor (Figures 7.29–7.32). The conclusion is the same: every stage should have the same compression ratio.

In practice:

- one stage : 7–15 [bar]
- two stages: ±100 [bar]
- three stages: ±1000 [bar]
- four stages: ±4000 [bar].

7.8 Problems with Water Condensation

Consider this time only a compression with moist air. The problem arises not with compression of a mono-molecular gas. The situation is a two-stage compression with a start

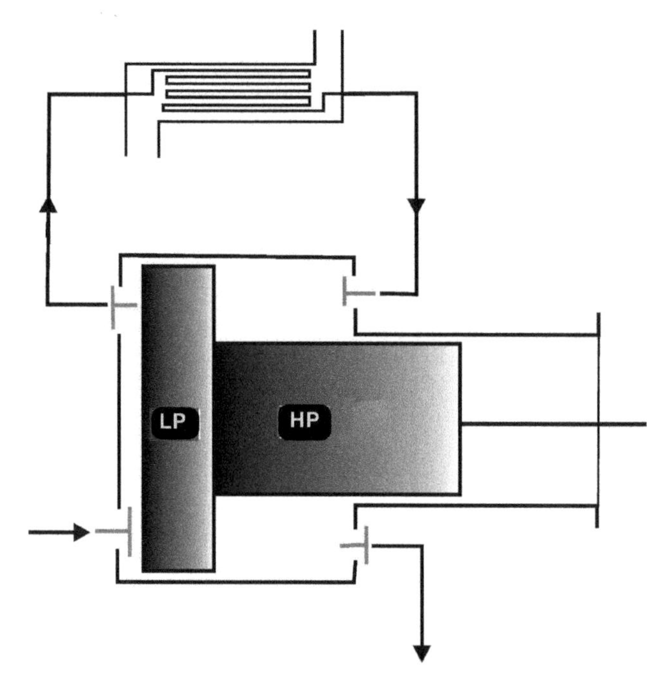

Figure 7.28 Two-stage compressor with intercooling in the middle.

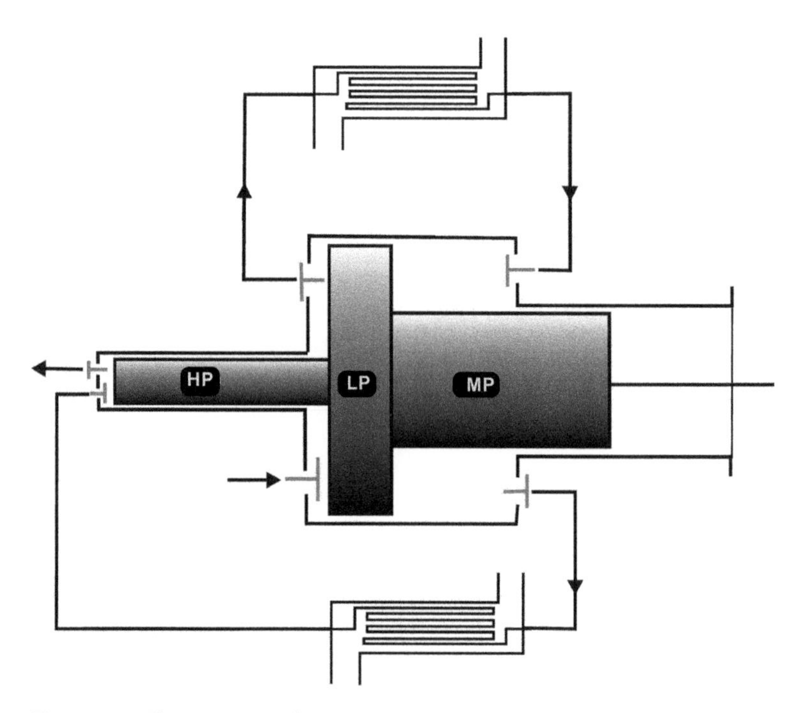

Figure 7.29 Three-stage tandem compressor.

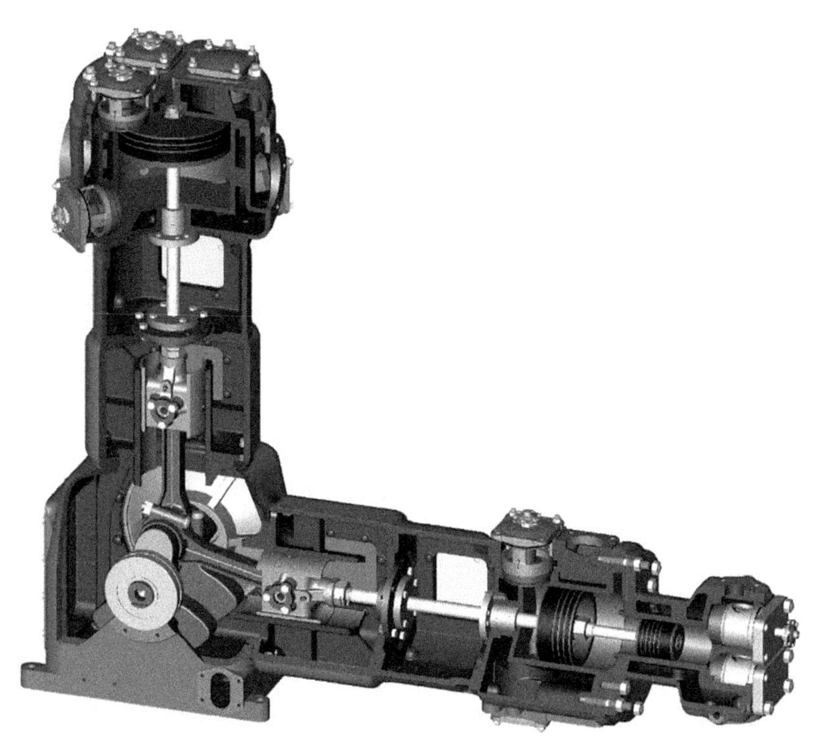

Figure 7.30 Three-stage compressor: vertical: double-acting, horizontal: two-stage tandem. Source: Ateliers François.

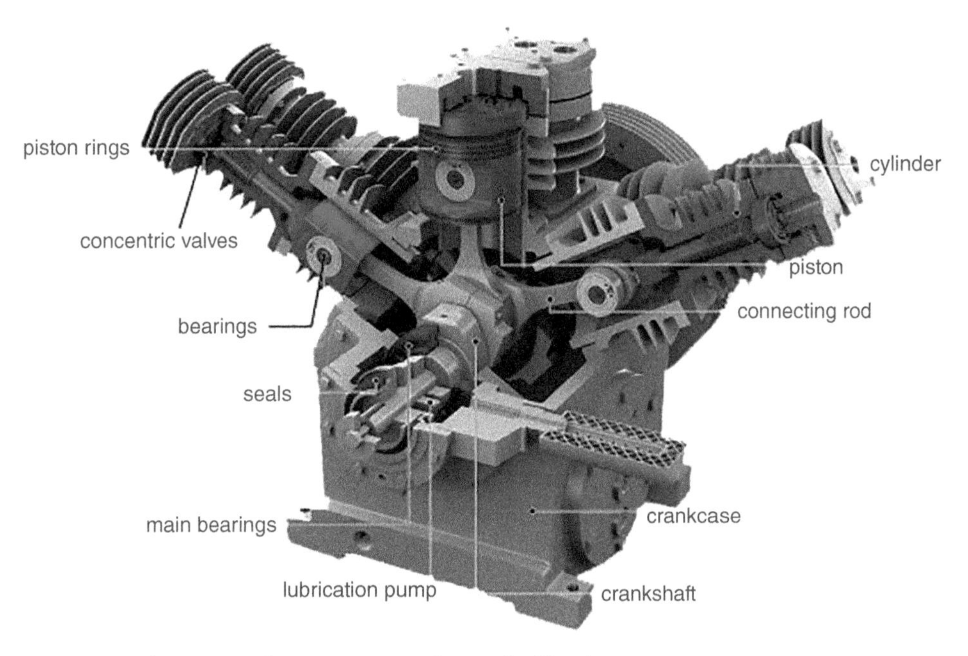

Figure 7.31 Three-stage piston compressor. Source: Burkhardt.

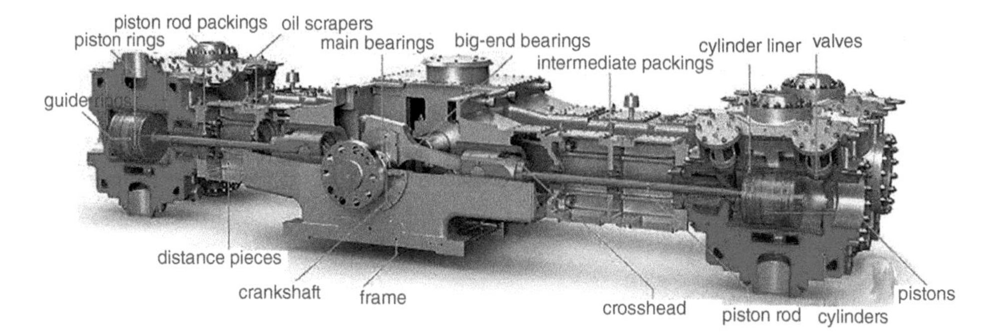

Figure 7.32 Hyper-compressor (six-stage), 3500 [bar]. Source: Burkhardt.

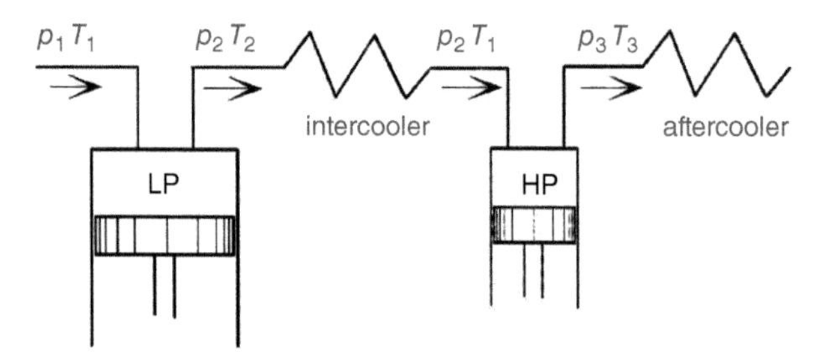

Figure 7.33 Two-stage compression with intercooler.

temperature $T_1 = 30°\text{C} = 303\,[\text{K}]$ and pressure $p_2 = 1[\text{bar}_\text{a}]$ to $p_2 = 5\ [\text{bar}_\text{a}]$. It's a polytropic compression with factor $n = 1.3$.

There's an intercooler, like in Figure 7.33.

Suppose for simplicity that the intake air is fully saturated. The partial pressure of the water in the air, at 30°C, from the saturated steam tables:

$$p_{W,sat}(30°\text{C}) = 0.0424\,[\text{bar}]$$

In that case the water content of the moist air is

$$x(30°C) = x_{sat} = 0.622 \cdot \frac{p_{W,sat}}{p - p_{W,sat}}$$

$$x(30°C) = x_{sat} = 0.622 \cdot \frac{0.0424}{1 - 0.0424} = 27.5\,[\text{g/kg}]$$

The temperature after the compression is:

$$T_2 = T_1 \cdot \left(\frac{p_2}{p_1}\right)^{\frac{n-1}{n}} = 303 \cdot \left(\frac{5}{1}\right)^{\frac{0.3}{1.3}} = 139\,K \cong 140°$$

The partial pressure of the water vapor after the compression is:

$$x(140°C) = 27.5\,\left[\frac{g}{kg}\right] = 0.622 \cdot \frac{p_W}{5 - p_W}$$

where:

$p_W = 0.21\,[\text{bar}]$

If the air is saturated, then this partial pressure is the saturation pressure, and this – see the saturated steam tables – means a temperature of:

$$p_w = p_{w,sat}(T_3) = 0.21 \ \ [\text{bar}] \Rightarrow \ T_3 \ \cong 60^\circ C$$

The air may not be cooled lower than 60 °C. What happens when this is not the case? Water condenses in the heat exchanger (intercooler) and this water enters the HP compressor. There it forms an emulsion with the lubricating oil (or it should be an oil-less compressor) and the lubrication is in danger.

7.9 Flow Regulation

7.9.1 Continuous Speed Regulation

For a single stage compressor:

$$Q_V = V_s \cdot \lambda \cdot N$$

We see that the volumetric flow is proportional to the speed of the drive.

If the speed of the compressor increased then the flow will increase too. Speed regulation is nowadays the common regulation. The reason is that the energy price increases all the time and the price of frequency inventers decreases.

7.9.2 Throttling Suction Line

In the suction line a regulator valve is placed. If the passage decreases the friction resistance in the suction pipe will increase and the suction pressure will decrease (Figure 7.34). This leads to a smaller allowable volume and lower pressure, so a smaller normal volume is sucked.

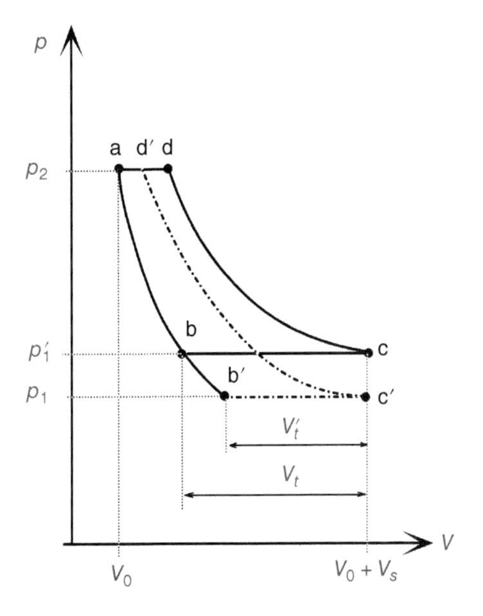

Figure 7.34 Throttling suction line.

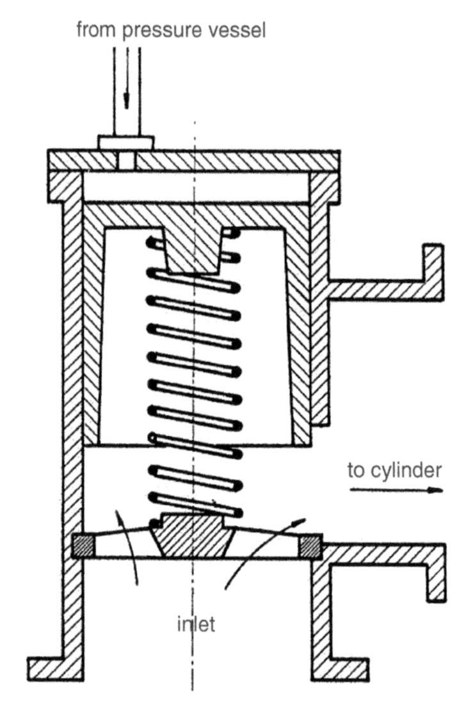

Figure 7.35 Regulation throttling suction line.

Express this in a formula. In the first state the suction pressure is p_1 and the allowable volume V_t. Make use of the ideal gas law, the sucked mass is m^*:

$$m^* = \frac{p_1 \cdot V_t \cdot M}{R \cdot T_1}$$

With throttling, the suction pressure is p_1' and the allowable volume V_t'; the sucked mass is m':

$$m'^* = \frac{p_1' \cdot V_t' \cdot M}{R \cdot T_1}$$

where:

$p_1' < p_1$ and $V_t' < V_t$

In many cases press gas is delivered to a press vessel until the pressure in the press vessel attains a certain value. This can be regulated automatically (Figure 7.35) or a pressiostate.

7.9.3 Keeping Suction Valve Open

The suction valve stays open during suction as well as press stroke. Examples of automatic regulation are shown in Figures 7.36 and 7.37.

7.9.4 Dead Volume

Here the flow is regulated by working on the size of the dead volume (Figure 7.38). According to:

$$\varepsilon = \frac{V_0}{V_s}$$

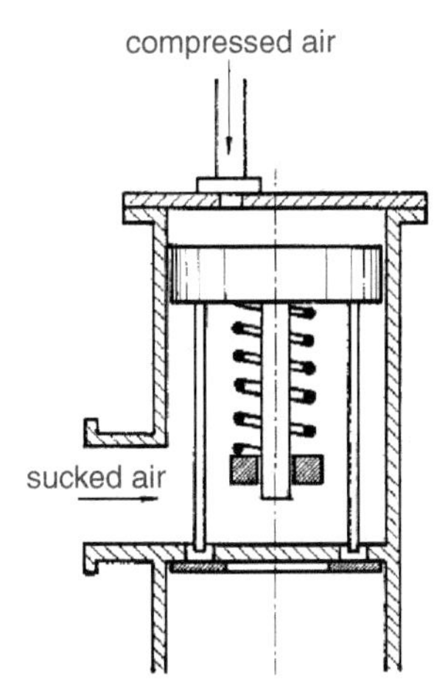

Figure 7.36 Regulation opening suction valve.

Figure 7.37 One-stage, multiple-cylinder piston compressor (refrigeration). Suction valve can be hydraulically lifted to start up unloaded but can also be used for regulation (see 3). 1: bearing, 2: lubrification system, 3: liftable valve, 4: suction and press valve, 5: piston, 6: welded carter. Source: Grasso.

$$\lambda = 1 - \varepsilon \cdot \left(\left(\frac{p_2}{p_1} \right)^{\frac{1}{n}} - 1 \right)$$

will λ decrease and so the flow of ε increases.

7.10 Star Triangle Connection

In Figure 7.38 a star connection of the three phase windings of the induction motor are represented. The line tension is 400 [V], the phase voltage is 230 [V]. Only 400 [V] is applied over the windings of the motor.

With a star connection, phase voltage is low as $1/\sqrt{3}$ times of the line voltage, whereas with a delta connection, phase voltage is equal to the line voltage (see Figure 7.39).

Another way to connect the line tension to the windings is a triangle connection (Figure 7.40).

The relation between the line voltage U_l and the phase voltage U_f for a triangle connection is given by (Figure 7.41):

$$|\overline{U_l}| = |\overline{U_f}| \cdot \sqrt{3}$$

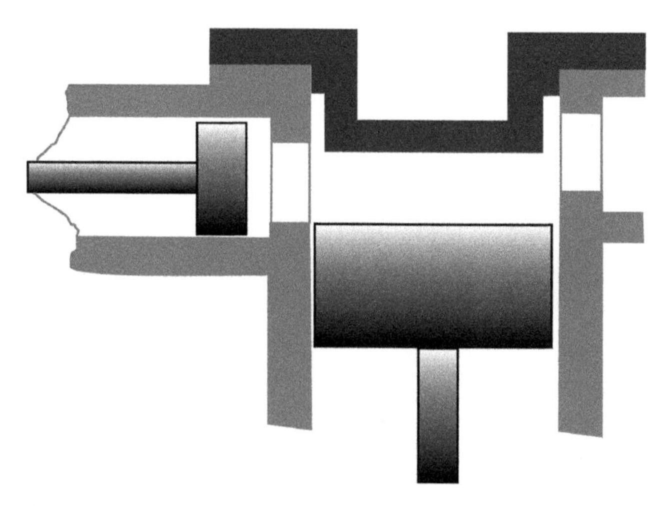

Figure 7.38 Increasing size dead volume.

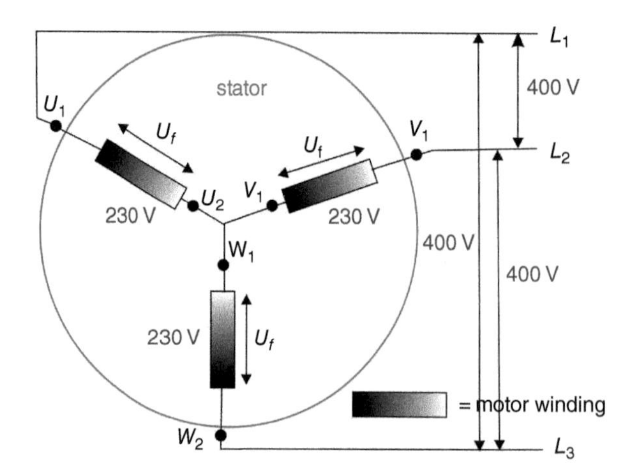

Figure 7.39 Star connection.

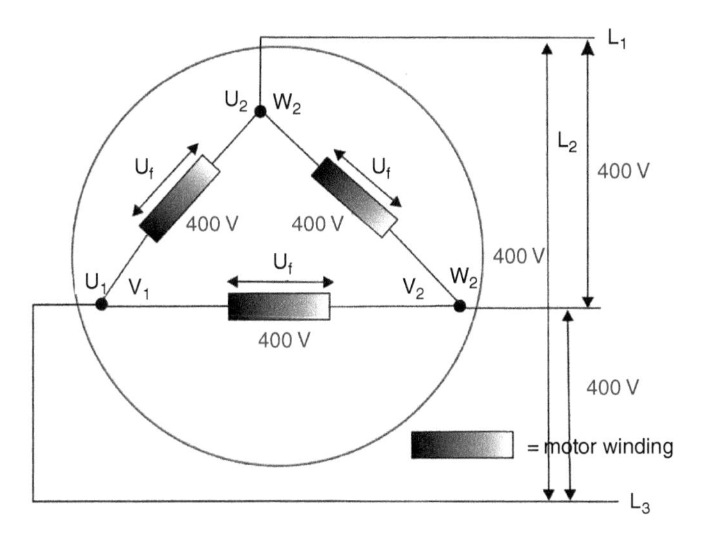

Figure 7.40 Triangle connection.

Figure 7.41 Phase and line voltage are vectors.

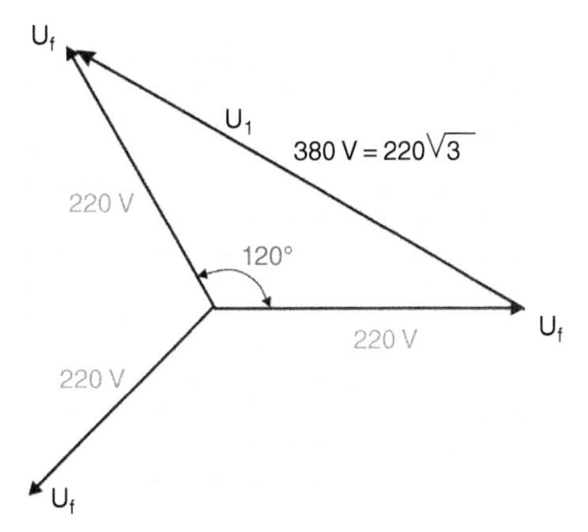

Consider now Figure 7.41 L_1, L_2, and L_2 are the three phases of the electrical network, for instance 400 [V] each. When the switches are closed (start) the windings of the motor stator are in a star connection and have a voltage of 400 [V]. When the switches are released the voltages over the windings are in a triangle connection and have a voltage of 230 [V] (Figure 7.42).

In Figure 7.43 torque versus speed for an induction motor is represented. When the motor starts the speed is zero. The torque delivered by the motor is given by M_{start}. This torque is necessary to overcome the nominal couple of the load M_{nom} and to overcome the inertia of the rotating mass of the rotor of the motor. The operating point of the motor shifts to the right on the curve until finally meeting the nominal couple.

To start the compressor the motor is first started with the compressor unloaded. The star connection is applied (Figure 7.44). The motor just has to overcome the inertia of its rotor. When the speed is about 90% of the maximum speed the triangle connection is connected to the motor and the compressor loaded. The operating point shifts very

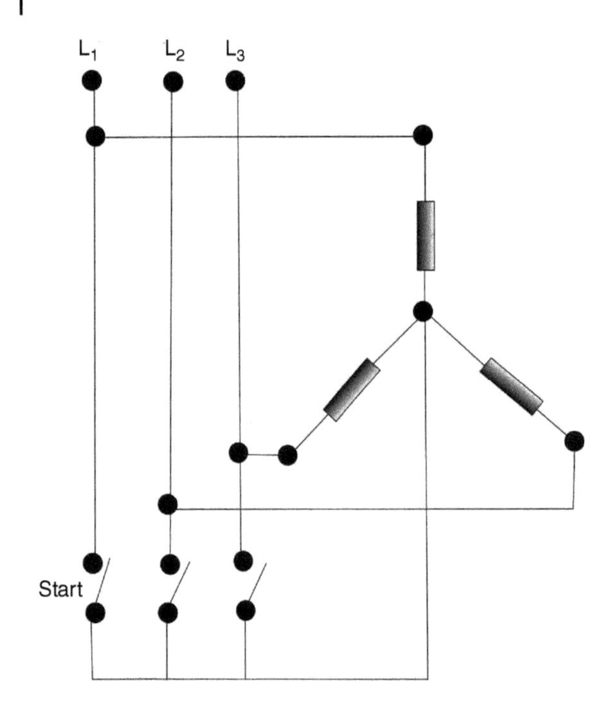

Figure 7.42 Triangle–star transformation.

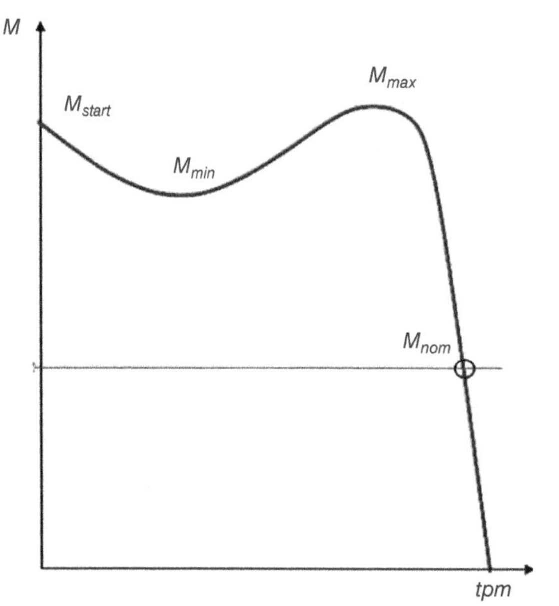

Figure 7.43 Torque versus speed for an induction motor.

quickly to a couple above the nominal couple. The torque is necessary to overcome the inertia mass of the compressor and its nominal couple. The speed increases and the operating point shifts to the right until finally it meets the nominal couple and an equilibrium is reached.

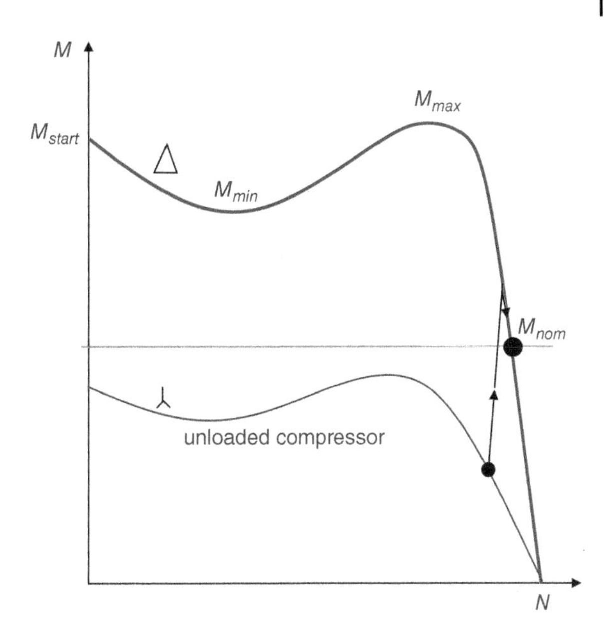

Figure 7.44 Regulation of the torque for an induction motor.

7.10.1 Speed Regulation with VFD

A frequency convertor is a precise electronic device that controls the speed of AC induction motors without affecting the energy consumption, torque, magnetic flux, etc. The speed of the motor is regulated by changing the voltage without changing energy consumption, etc. This way the speed can be lowered or increased.

How does this work?

The torque of an induction motor is given by:

$$T = C_1 \cdot \Phi \cdot I$$

where C_1 is a construction constant of the motor, I is the current through the stator windings, and Φ is the magnetic flux in the motor. This flux must remain constant, because when it increases it comes in the saturation region of the magnetic field, and that must be avoided. If the voltage over the windings is changed, so will the current and the torque. On the other hand, the power of the motor is given by:

$$P = T \cdot \omega$$

where ω is the pulsation, which is proportional to the speed N (in rpm) of the motor, according to:

$$\omega = 2 \cdot \pi \cdot f = \frac{2 \cdot \pi \cdot N}{60}$$

where f is the frequency of the voltage.

The voltage over the windings of the motor is given by:

$$U = k \cdot N \cdot \Phi - R_a \cdot I$$

where k is a construction factor of the motor, N is the speed, and R_a is resistance of the stator windings of the motor. The term $R_a \cdot I$ counts for about 5% so that:

$$U \cong k \cdot N \cdot \Phi \cong f \cdot \Phi$$

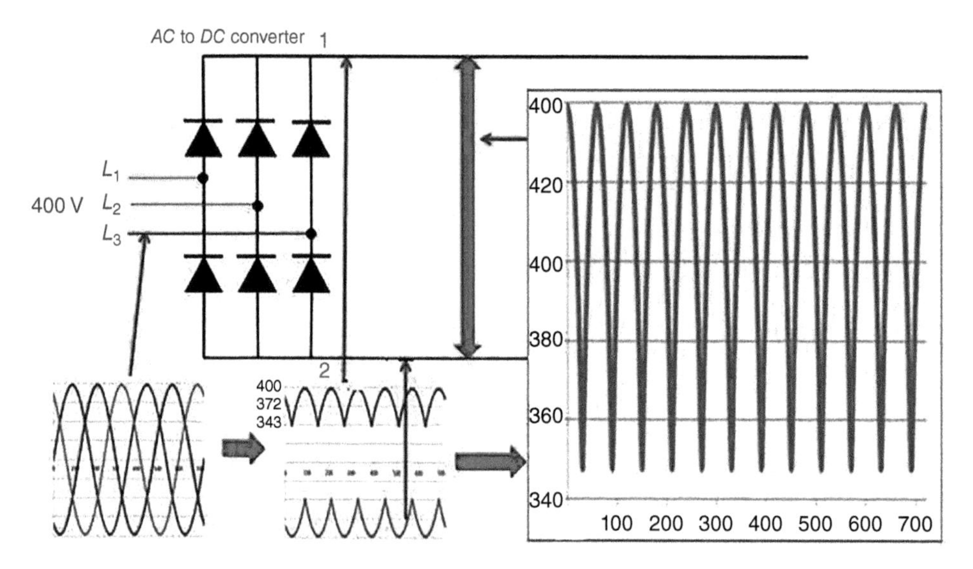

Figure 7.45 First part of the VFD.

As Φ must remain constant during a regulation:

$$\frac{U}{f} = constant$$

This is the condition for a *variable frequency drive* (VFD). The VFD must keep the voltage U in track with the frequency f so that their ratio remains constant. A logical electronic circuit is a digital one (one with "logic" devices). An alternative is an analog electronic circuit.

It's time to look at the first part of a VFD (Figure 7.45).

As shown in Figure 7.45, an alternating current (AC) line supply voltage of, for instance, 400 [V] is brought into the input section. From here, the AC voltage passes into a converter section that uses a diode bridge converter. The diodes are electronic devices that let the current pass in one direction only: the direction indicated by the arrow in the symbol of the element. If the side of the horizontal line in the symbol is positive the diode will let the current pass, and will be positive; on the other side, the voltage is negative. In the other case, no current will pass the diode.

The voltage of the line are shown in cyan, brown, and green in Figure 7.44. The result of the bridge is given in the figure in the middle of the figure. It shows what parts of the AC voltage is used. Those two waves have to be counted together to have the voltage on the line 1. The result is shown in the figure on the right.

Let's go now a step further. A large capacitor is added to the diode bridge. A capacitor is an element that doesn't like voltage changes over it. It tries to keep the voltage constant and will make a DC voltage of the AC voltage.

The result is shown in Figure 7.46.

The third and final part of the VFD is represented in Figure 7.47.

The DC voltage passes into the inverter section usually furnished with switches. In reality they are electronic devices that let the current pass when a pulse is set to the gate of it. These pulses are generated by an electronic logic circuit. In reality they are

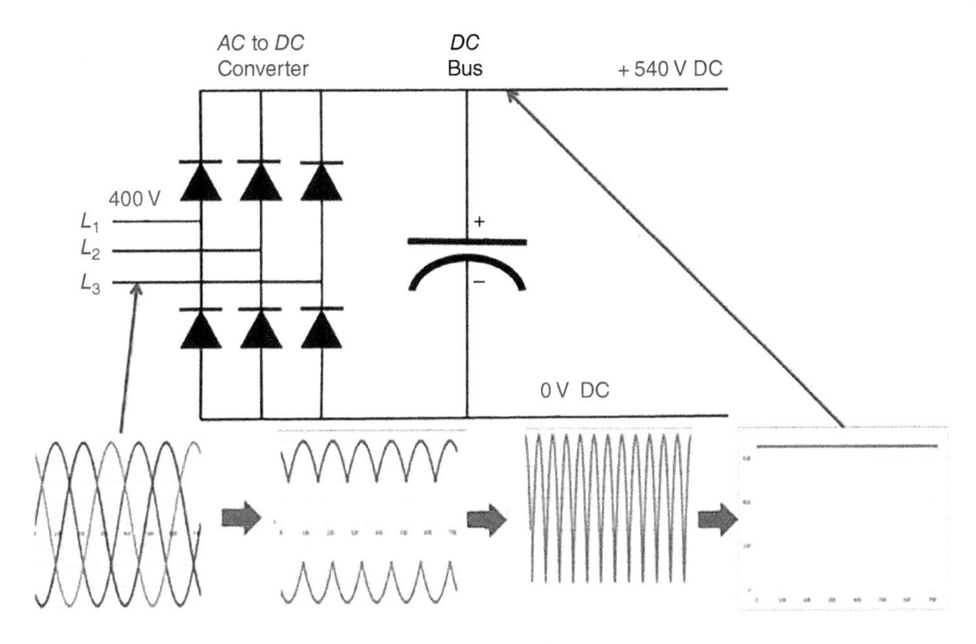

Figure 7.46 Adding a lager capacitor makes a DC voltage form the AC voltage.

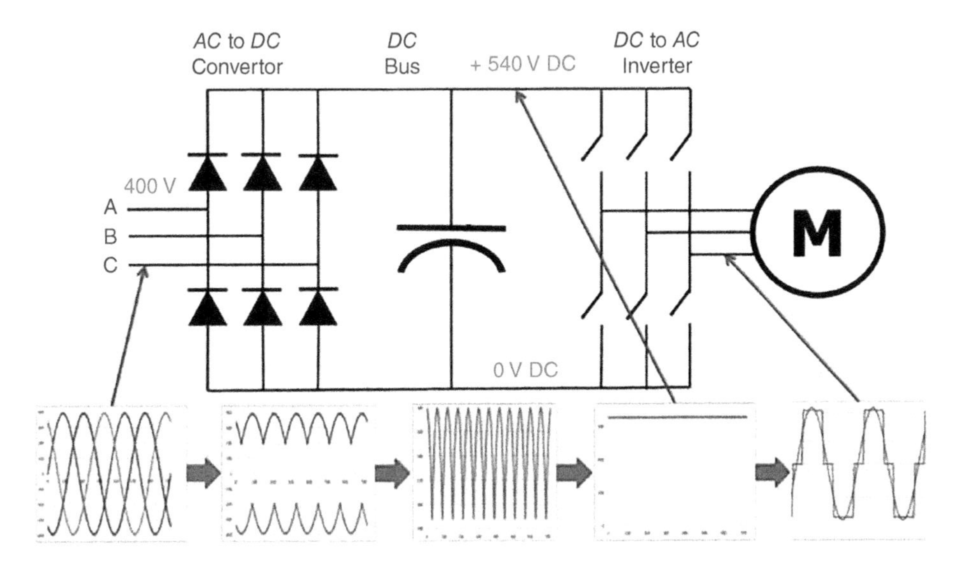

Figure 7.47 Third part of VFD.

electronic devices called insulated gate bipolar transistors (IGBTs), which regulate both voltage and frequency to the motor to produce a near sine wave like output.

Switching speeds of the IGBTs in a pulse width modulation (PWM) drive can range from 2 to 15 [kHz]. By having more pulses in every half cycle, the motor voltage is reduced. Also, the current wave shape to the motor is smoothed out as current spikes are removed. Figure 7.48 shows the voltage and current waveform outputs from a PWM

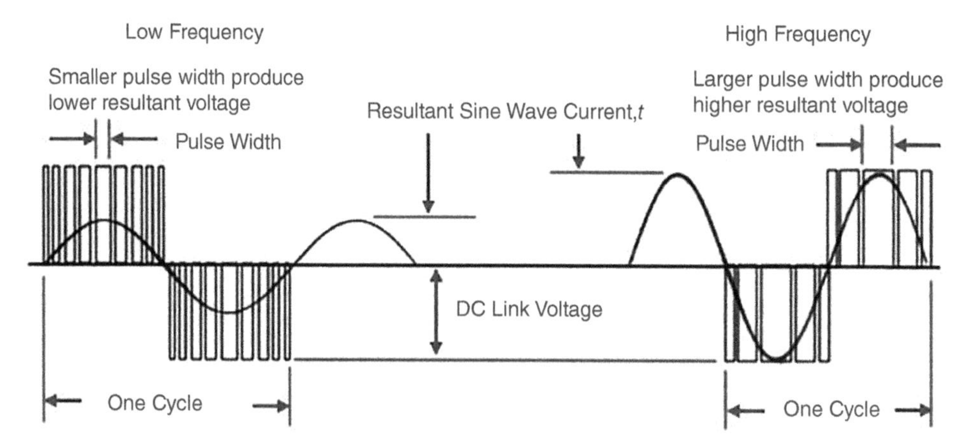

Figure 7.48 Pulse width modulation (PWM).

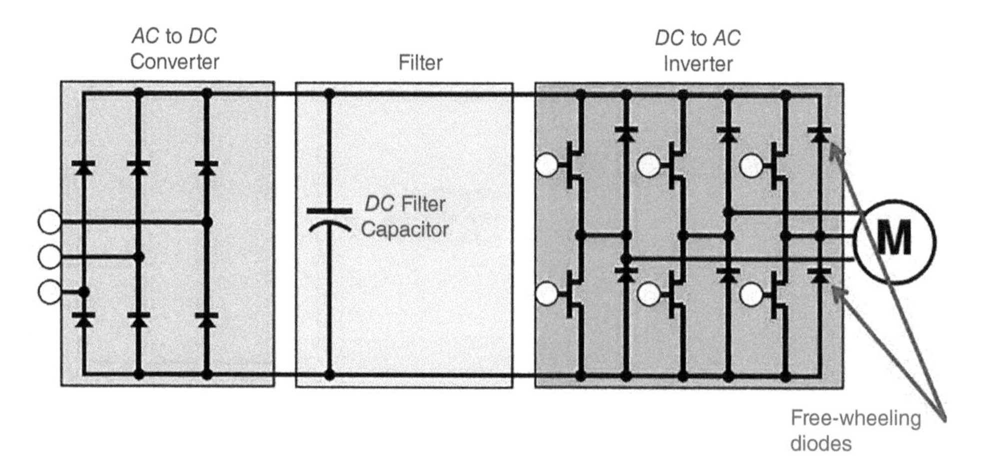

Figure 7.49 A VFD.

drive. The AC voltages are shown. As can be seen, the voltage is altered at the same time as the frequency or the speed of motor, in such a way that the flux in the motor remains constant. The rhythm at which the pulses are generated determines the frequency of the wave.

The three parts together are shown in Figure 7.49. The freewheeling diodes protect the IGBTs, obstructing current in the wrong direction.

Why don't all compressor regulations use a VFD? Because it's an expensive device.

7.11 Refrigeration Piston Compressor

Refrigerant compressors are divided into various types. The difference is on the basis of the sealing.

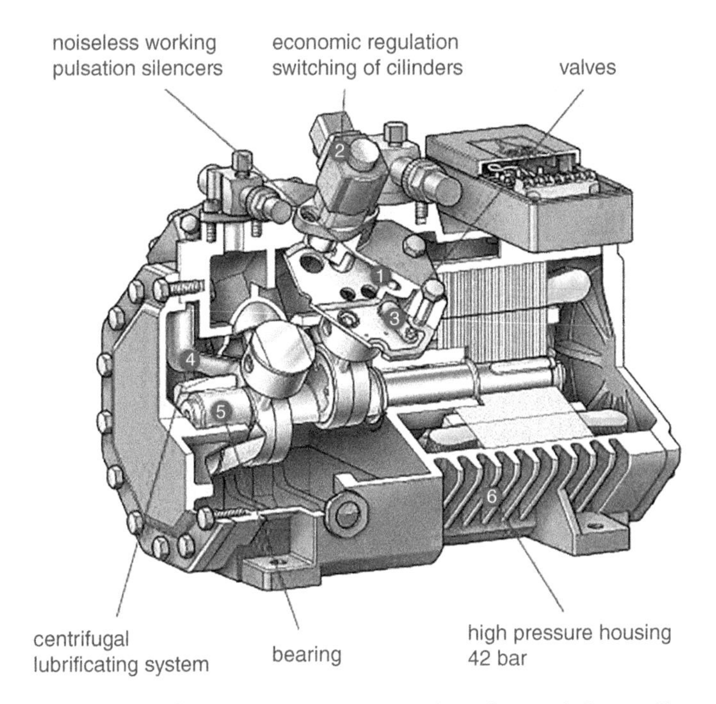

Figure 7.50 Refrigerant piston compressor (semi-hermitic). Source: Bitzer.

- *Open type compressor*: the shaft of the compressor runs through a seal to connect to the external drive of the motor.
- *Hermetic compressor:* the motor and the compressor are mounted in one pressure vessel; the motor shaft and the compressor shaft form one whole; the stuffing box gasket lacks; the motor is in contact with the refrigerant.
- *Semi-hermetic*: the housing is flanged so that everything is accessible for eventual reparation (Figure 7.50).
- *Welded hermetic compressor*: motor and compressor are mounted in a steel frame that is welded.

Hermetic compressors were developed to eliminate leaks through the sealing, a problem that is especially important with halogen refrigerants. For ammonia only an open design is used because of the threat of corrosion associated with ammonia and copper.

Piston compressors for refrigerant purposes are built in various sizes and types from 2-cylinder units to 12-cylinder machines with capacities of 250 [W] to many hundreds of kilowatts.

8

Other Displacement Compressors

CHAPTER MENU

8.1 Roots Compressor

8.1.1 Operation

The roots compressor (Figures 8.1 and 8.2) is composed of two rotors in the form of a figure of eight that rotate in opposite directions in a housing in the form of a flattened cylinder. While rotating, at first a gas is aspired out of the suction line, then locked up in the space between the rotor and housing and then transported to the discharge side, where it is displaced. During the transporting in the compressor the gas is not compressed, there is no *internal compression* like this, as is the case with a piston compressor and many other displacement compressors. At the moment the rotor head rotates along the edge of the discharge port compressed gas flows back from the discharge pipe. Then the transported gas is displaced against the discharge pressure to the discharge side.

The operation principle of the roots compressor is essentially different from that of the piston compressor because there is no internal compression and there are no valves.

8.1.2 Technical Work

We study the specific technical work w_{t12} needed to compress 1 [kg] of gas from pressure p_1 to p_2.

In Figure 8.3 a polytropic compression of a gas in a *thermodynamic pv* diagram is represented for a piston compressor by the state change a → b.

Pumps and Compressors, First Edition. Marc Borremans.

Companion website: www.wiley.com/go/borremans/pumps

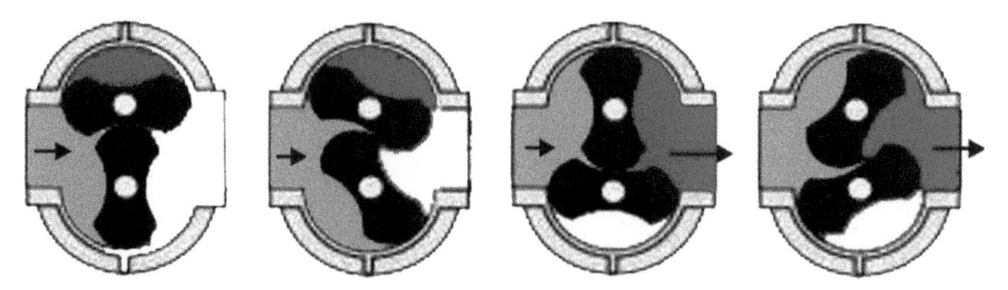

Figure 8.1 Operation principle roots compressor. Source: Howden.

Figure 8.2 Roots compressor (the biggest – 98.000 [m³/h]). Source: Aerzener.

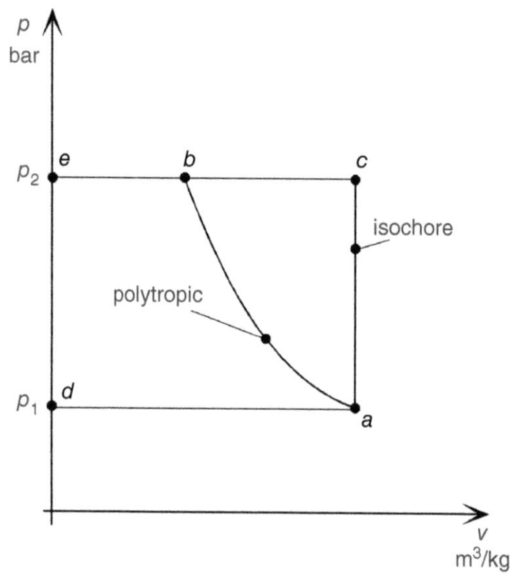

Figure 8.3 Technical work.

In a root compressor the beginning state of the gas, given by the specific volume (specific volume $v_1 = v_a$ and pressure p_1), is retained until the discharge port is reached. At that moment the gas in the discharge pipe flows back to the compressor and this leads to the pressure of the gas suddenly rising to the pressure p_2 in the discharge pipe. The state change can be represented by the isochore change a → c.

The specific technical work w_{t12} is given by:

$$w_{t,12} = \int_1^2 v \cdot dp$$

In Figure 8.2 it appears that:

$$\text{surface d} \rightarrow \text{a} \rightarrow \text{b} \rightarrow \text{e}\; w_{tab} = \int_a^b v \cdot dp =$$

$$\text{surface d} \rightarrow \text{a} \rightarrow \text{c} \rightarrow \text{e}\; w_{tac} = \int_a^c v \cdot dp$$

Because surface d → a → b → e < surface d → a → c → e the use of a root compressor is energetically not interesting. How bad it is can be demonstrated with an example (Figure 8.3).

Compare an adiabatic compression of air with a piston compressor to that with a roots compressor ($\gamma = 1.4$):

$$w_{tab} = \frac{n}{n-1} \cdot p_1 \cdot v_1 \cdot \left(\left(\frac{p_2}{p_1} \right)^{\frac{n-1}{n}} - 1 \right)$$

In the case of a roots compressor:

$$w_{tac} = \int_a^c v \cdot dp = v_1 \cdot (p_2 - p_1)$$

The ratio:

$$\frac{w_{tab}}{w_{tac}} = \frac{\frac{n}{n-1} \cdot p_1 \cdot v_1 \cdot \left(\left(\frac{p_2}{p_1} \right)^{\frac{n-1}{n}} - 1 \right)}{v_1 \cdot (p_2 - p_1)}$$

$$\frac{p_2}{p_1} = 1.5 : \frac{w_{tab}}{w_{tac}} = 0.87$$

$$\frac{p_2}{p_1} = 2 : \frac{w_{tab}}{w_{tac}} = 0.77$$

The extra energy for the roots compressor, represented by the surface a → b → c, increases with higher compression ratios. That's why roots compressors are used for small pressure increases, some tenths of a bar. The roots compressor is a transport compressor, and is sometimes called a *roots blower*.

This doesn't mean that there are no roots compressors; a company has a version that delivers 9 [bar].

8.1.3 Properties

The rotors fit accurately together, but make no contact with each other or the housing. A synchronous gear transmission is used to reach that (Figure 8.4).

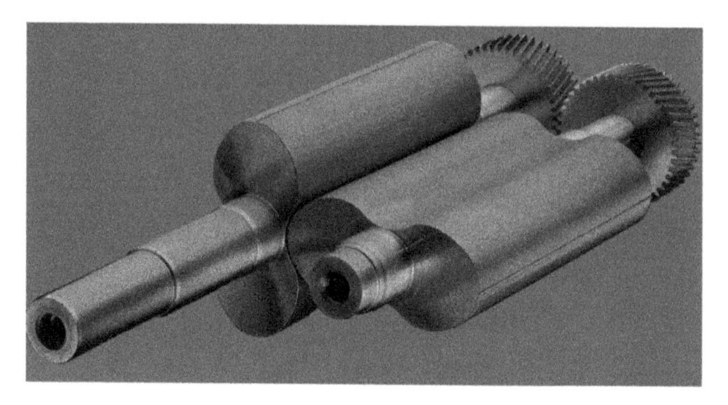

Figure 8.4 Synchronous gear transmission. Source: Aerzener.

Because there is a small play between the rotating parts and the housing, it must absolutely be avoided that any contaminations enter the compressor. A filter at the inlet is necessary.

The frictionless operation guarantees a high mechanical efficiency; only in the bearings and the synchronous gear transmission does friction occur.

Thanks to the contactless operation, it is possible to make the compressor oil-free. There is no contamination of the transported gas: it is free of lubricating oil or wearing parts.

Because of the frictionless operation the rotors can rotate at high speed (1500 [rpm] and more) via a direct coupling with an electromotor.

Because of the high speeds, a high flow can be reached. Dependent on the sizes of the roots blowers to 80.000 [m^3/h]. The gap (some tenths of a millimeter) between the rotor and housing, and between the rotors themselves, allows gas to flow back from the discharge the suction side; this leak flow increases as the pressure difference between the suction and discharge side increases (Figure 8.5). The effective flow is 50–95% of the theoretical flow. This, of course, has a negative influence on the necessary power.

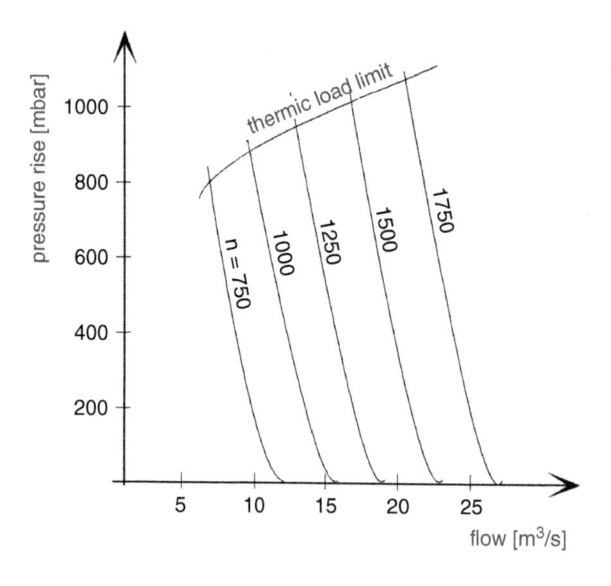

Figure 8.5 Flow. Source: after Aerzener.

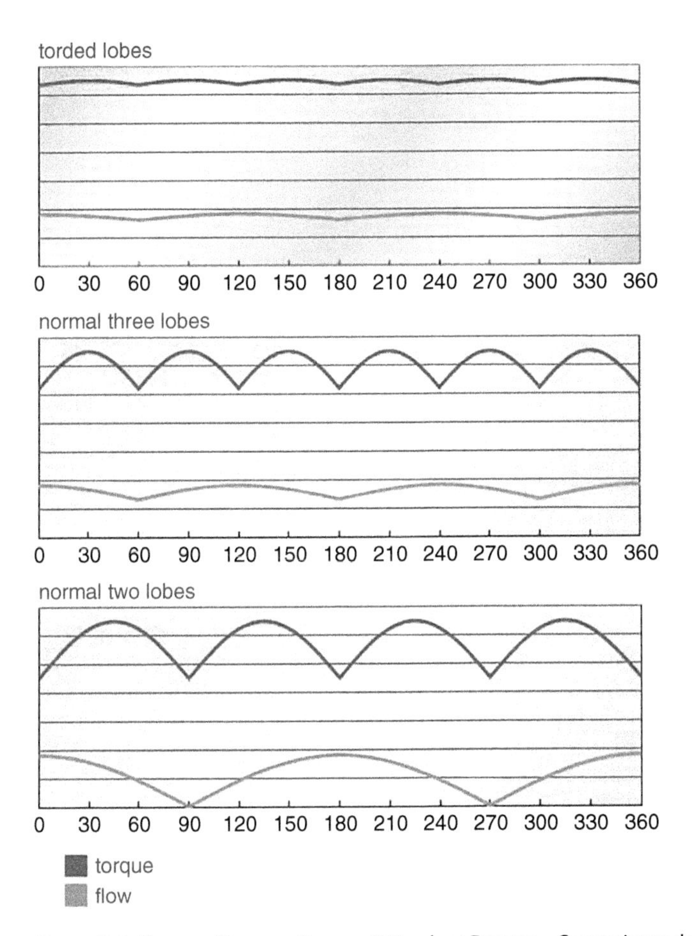

Figure 8.6 Flow and torque. Source: © Gardner Denver – Compair used with permission.

The flow is somewhat irregular. This can be improved by using a three-lobe blower (Figure 8.6).

Oil-free compressors need the gas to be separated from the oil in the gearbox. This can be done with a labyrinth seal (Figure 8.7). The gas flows outside via narrow annular slits. Thereby the gas keeps the oil that wants to enter inside, outside. There is no friction between the sealing and the shaft.

A small problem with roots compressors is that the flow is not pulsation free. This causes noise. One solution is to use more than two lobes, for instance three lobes (Figure 8.8). Another, better, solution is to twist the rotors, so that the flow is nearly continuous (Figure 8.9).

8.2 Vane Compressor

8.2.1 Operation

The vane compressor (Figures 8.10 and 8.11) consists of a rotor of radial slots where vanes can move in and out. The centrifugal force pushes the vanes against the housing.

Figure 8.7 Labyrinth seal. Source: Kaeser.

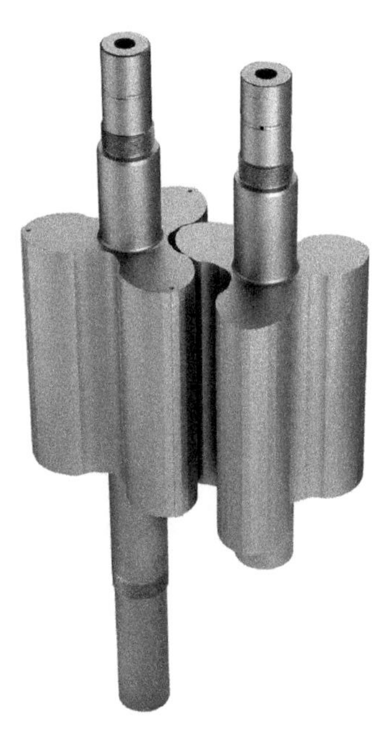

Figure 8.8 Three-lobe oil-free roots blower. Source: Kaeser.

And the housing can push the vanes back in their slots. Because of the eccentric mounting of the rotor in the pump housing, chambers are created during rotation that can become bigger and smaller. Like a roots compressor this compressor has no valves but ports. But the vanes compressor has an *internal compression.*

When on the inlet side chamber I (Figure 8.10) with volume V_I presents itself, it is filled with gas on suction pressure p_1. The gas is then locked up in the rotor in the space limited by rotor, pump housing, vanes, and cover. This space decreases so that the gas is compressed.

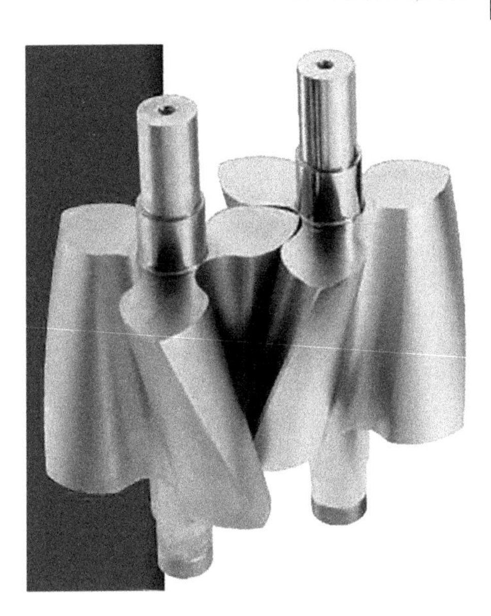

Figure 8.9 Twisted rotors.Source: © Gardner Denver, used with permission.

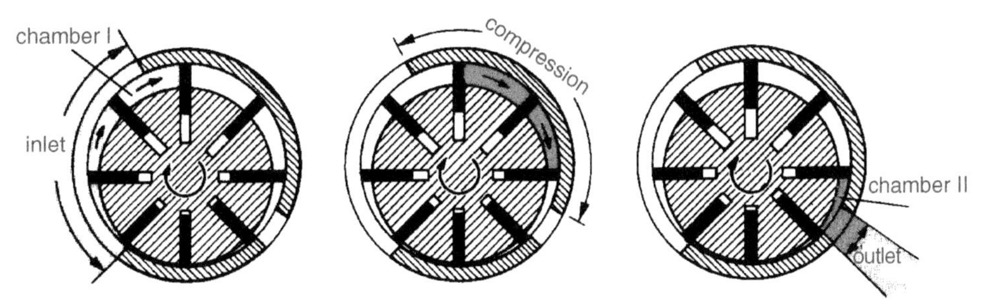

Figure 8.10 Operation vane compressor.

Figure 8.11 Three-dimensional sight vane compressor.

Finally, the gas is confined in the smallest chamber II with volume V_{11}. The pressure that the gas has in that moment is called the *design pressure* p_2.

If the rotor rotates now a little bit further then the gas will present itself on pressure p_2 before the discharge port. The gas is then pressed through the discharge line to the press vessel (if that exists). The pressure in the press vessel is the *compression pressure* p_3.

If the gas is compressed polytropic (with exponent n) from p_1 to p_2 from chamber I to chamber II:

$$\frac{p_2}{p_1} = \left(\frac{V_I}{V_{II}}\right)^n$$

The volumes V_I and V_{II} are constant because of constructive reasons, so the compression ratio is well determined. The compressor has an *internal compression ratio*.

The *design pressure* p_2 is determined by the construction but doesn't have to be equal to the vessel pressure p_3.

One and another differs from the piston compressor. The last one has spring-loaded valves that open automatically when there is a small pressure difference between the cylinder and the pressure vessel. The piston compressor delivers the gas when the pressure in the cylinder p_2 is a little bit greater that the pressure p_3 in the discharge line, so that in the first approximation the discharge valve opens when $p_2 = p_3$ (Figure 8.12).

Three cases can be distinguished on the *indicator diagram* (Figure 8.13; suppose no dead volume). Or the *design pressure* is equal, greater, or smaller that the *compression pressure*.

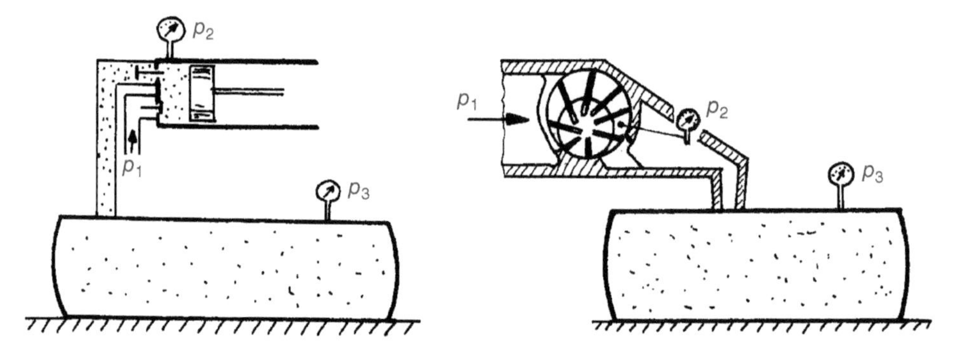

Figure 8.12 Difference between piston and vane compressor.

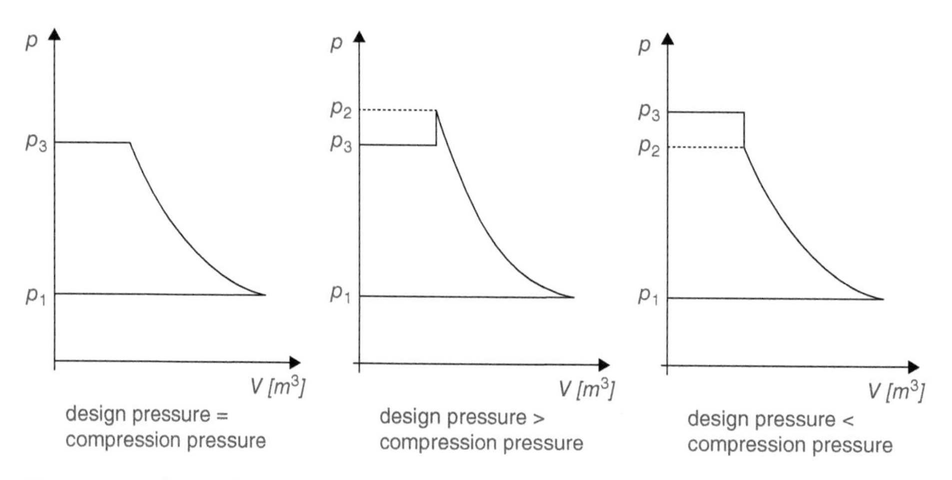

Figure 8.13 Indicator diagram of a vane compressor.

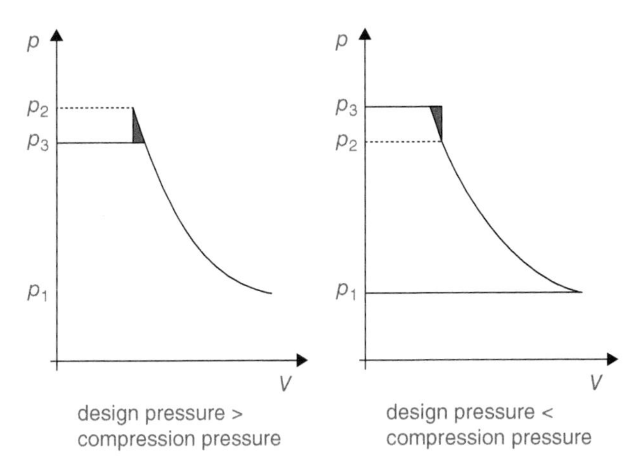

Figure 8.14 Extra energy with a vane compressor.

When the design pressure is not equal to the vessel pressure, the vane compressor will need an extra energy compared to the piston compressor, that is shaded in Figure 8.14.

8.2.2 Properties

When the discharge pressure is more than 2 [bar] abundant lubrication is necessary. The injected oil mist provides the sealing between vanes and housing. On the other hand, the oil will eliminate an important part of the compression heart.

The compressor delivers a flow from 100 to 10 000 [m^3/h]

The discharged pressure attains 8 [bar] in a single-stage design and 15 [bar] in a double-stage compressor with intercooling.

Two-stage vane compressors are possible (Figures 8.15 and 8.16).

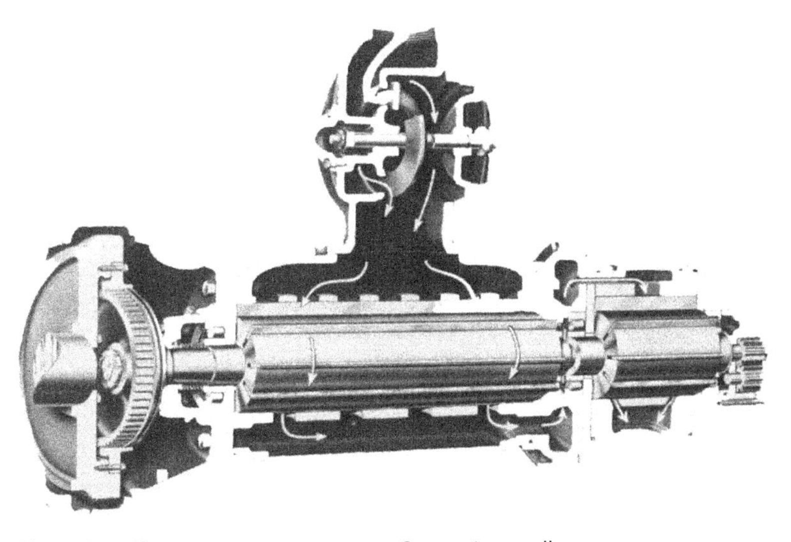

Figure 8.15 Two-stage vane compressor. Source: Ingersoll.

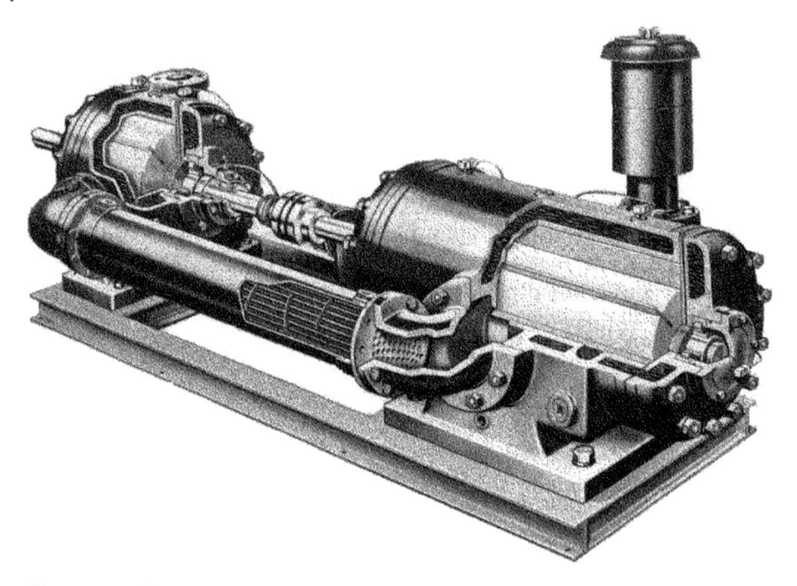

Figure 8.16 Two-stage vane compressor with intercooling. Source: Ingersoll.

8.3 Screw Compressor

8.3.1 Operation

The screw compressor consists of two screw-shaped rotors with different profiles that can grip one another (Figures 8.17–8.19). They rotate in opposite directions.

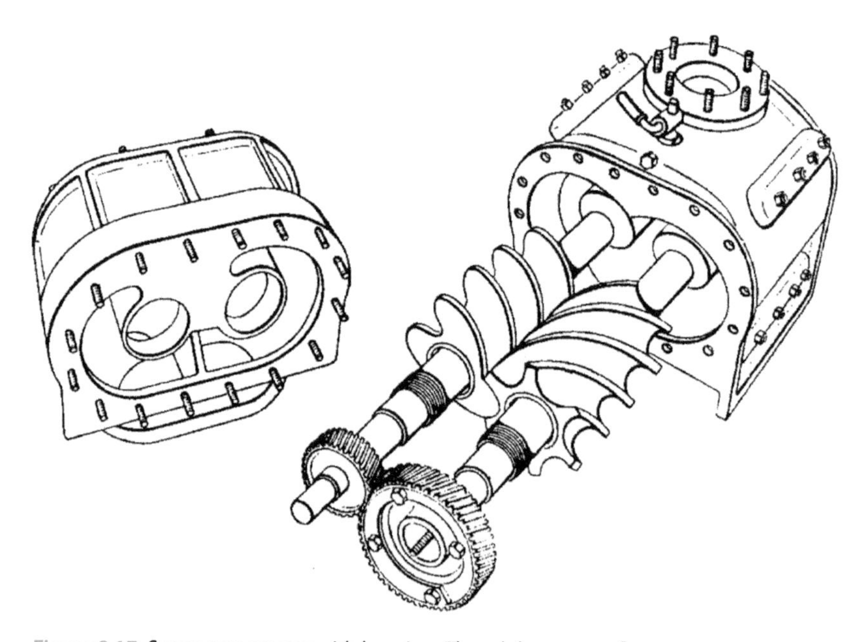

Figure 8.17 Screw compressor with housing. Three lobes versus five recesses.

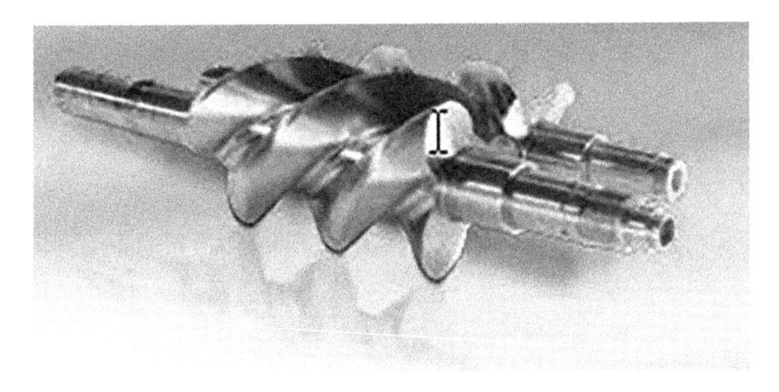

Figure 8.18 Three- to four-screw compressor, 1 [bar]. Source: Aerzener.

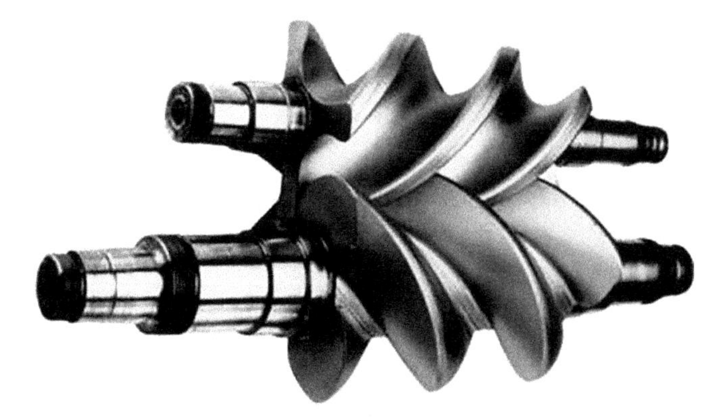

Figure 8.19 3/2 compressor. Source: Ingersoll.

They can have different numbers of lobes. Mostly the main rotor (male rotor) has four lobes, while the female rotor or side rotor has six. When the rotors do not have the same number of lobes, they must be driven at a different speed.

The flow through the compressor is *axial*.

The operation mode is as shown in Figures 8.20 and 8.21. First the grooves or recesses in both rotors are filled with gas from the suction line because these chambers present themselves before the suction port. By rotating the chambers become bigger. The gas cannot leak to the outlet port that is situated at the other side of the cylinder block, because at the time of compression the outlet port is blocked by the screw.

When the rotors rotate the female rotor is first locked from the inlet port and the male rotor gets free and is filled with gas. Then this couple of chambers will be completely locked off from the inlet port. In the chambers the gas is now at inlet pressure. The male lobe grips then into the female recess and this leads to the chamber becoming smaller and an *internal compression* of the gas takes place.

When the male rotor has traveled a whole rotation then the holes of the profiled ends of the rotors present themselves before the discharge port and the compressed gas is displaced. After that the same thing happens with the other chambers.

Two-stage versions are possible but because most often use is made of standard implementations the most favorable intermediate pressure is not reached. This can, however, be the case if the implementation is in one casing, like in Figure 8.22.

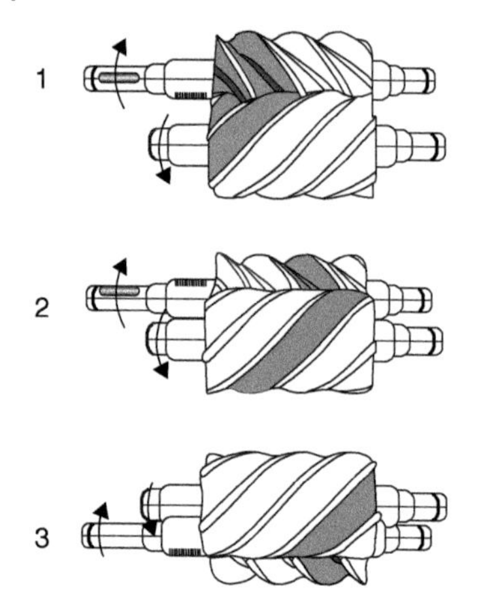

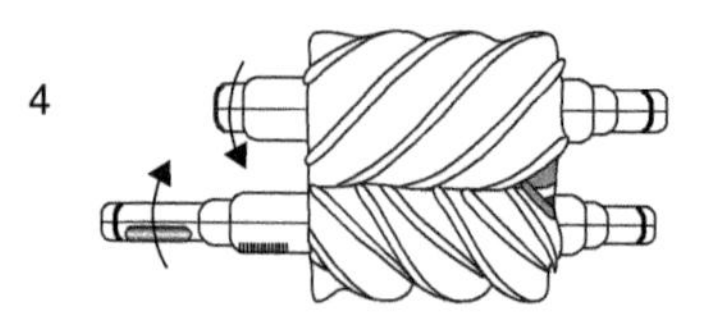

1 Suction: Air inlet through gas inlet opening in the open screw threads on the suction side of the rotors

2/3 Compression process: Continuous turning of the rotors closes the gas inlet opening, the volume decreases, the pressure rises

4 Discharge: The compression is complete, the discharge pressure has been reached. Expulsion, exhaust gas begins.

Figure 8.20 Operation screw compressor.

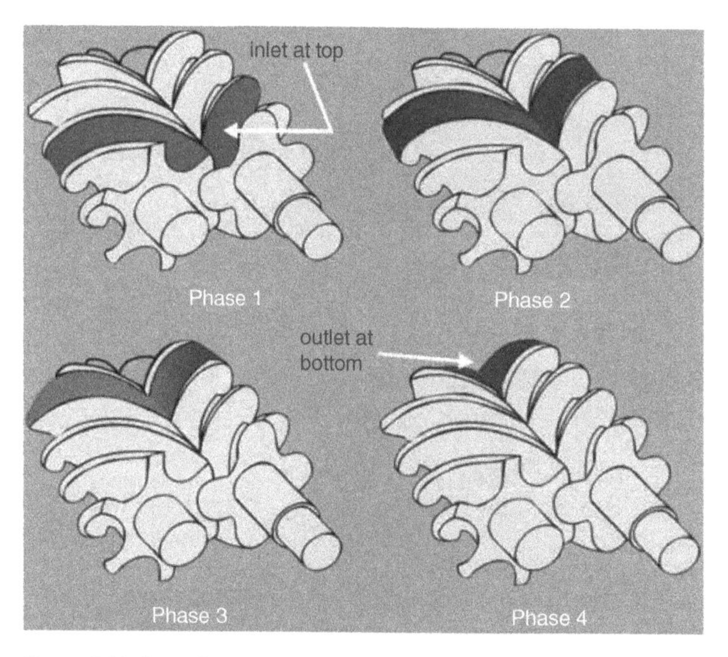

Figure 8.21 Operation screw compressor. Source: ALUP.

Oil-free screw compressors have a temperature limit of 220 °C (higher temperatures can be reached with a special screw, to 260 °C); the maximum temperature of oil-injected compressors is 120 °C.

There are also water-injected screw compressors. The water cools the compressed gas (Figure 8.22), and in this way the adiabatic compression becomes partly an isothermal compression, with power gain as a result (Figure 8.23).

The two-stage version of Figure 8.24 has a water curtain after the first stage.

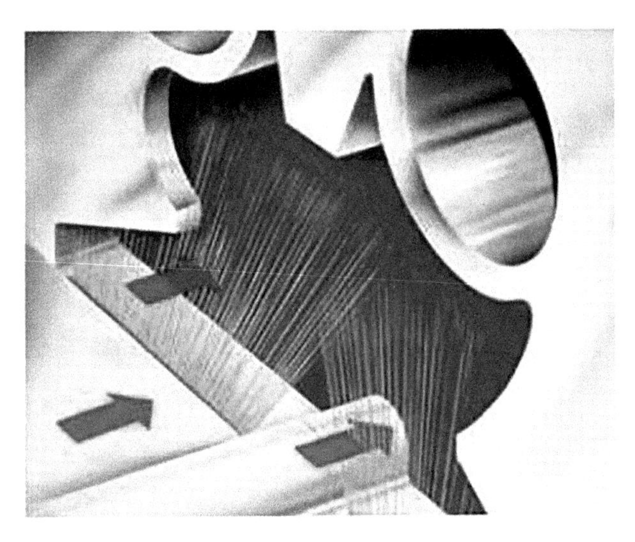

Figure 8.22 Water cooling curtain after the first stage. Source: Ingersoll.

Figure 8.23 Energy gain in a two-stage implementation screw compressor. Source: Ingersoll.

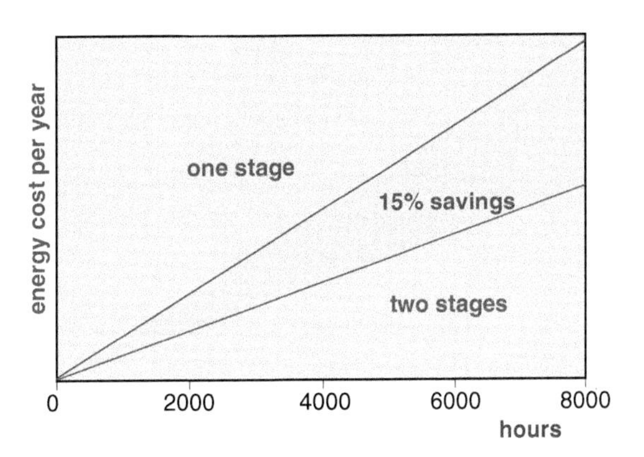

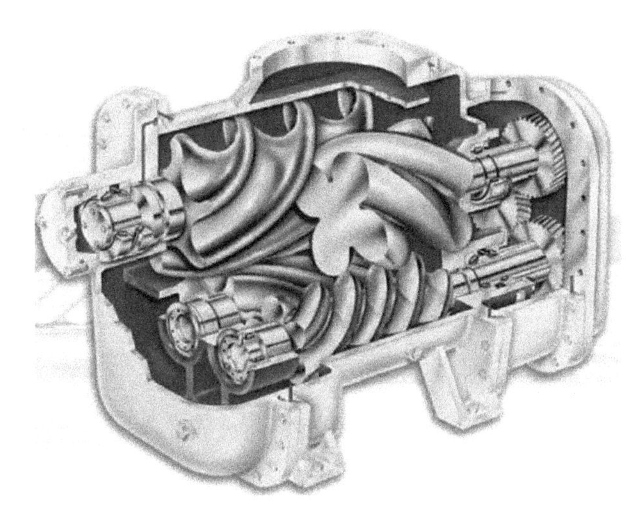

Figure 8.24 Two-stage implementation of a screw compressor. Source: Ingersoll Rand.

8.3.2 Properties

Oil-injected screw compressors are also available. With the oil-free compressors, the rotors don't touch each other, nor do they make contact with the housing. Wear is excluded. But because of the necessary clearance there is a small leak flow. To decrease this leak flow proportionately the flow and thus the rotation speed is higher than with oil-injected compressors. With a circumferential velocity of 40 [m/s] for oil-injected compressors and 80 [m/s] for oil-free compressors the rotation speed attains, dependent on the rotor diameter, from 3000 to 25 000 [rpm]. There are designs to 10 000 [m^3/h].

The rotors of oil-free compressors are synchronized with a gear-box drive (Figure 8.25).

What is the influence of the number of lobes? The ratios 3/5 and 3/2 are used for low pressures but high flows use a 4/6 ratio, and the most efficient is between 4 and 14 [bar] (best compromise between cost and efficiency). With oil-free compressors the female rotor is driven by the male rotor.

Radial forces arise. With a 6/4 compressor the radial pressure is divided over six recesses and that gives a lower contact pressure, less wear, and less bending of the female rotor.

It was described that the screw compression works with an *internal* compression ratio, determined by the constructive sizes. The compressor delivers all the time the same *design* pressure, independent of the rotation speed or press vessel pressure.

The remarks about vane compressors are valid here too. The delivered pressures attain 2–13 [bar] for single-stage designs.

In Figures 8.26–8.28 a screw compressor is represented in various ways.

Oil-free screw compressors tolerate a temperature of up to 220 °C (high temperature elements 260 °C); the maximum temperature with oil-injected screw compressors is only 120 °C. There are also oil-free water-injected screw compressors. The water cools down the compressed gas. This way what in principle would be an adiabatic compression becomes an isothermal compression, with a consequent power gain.

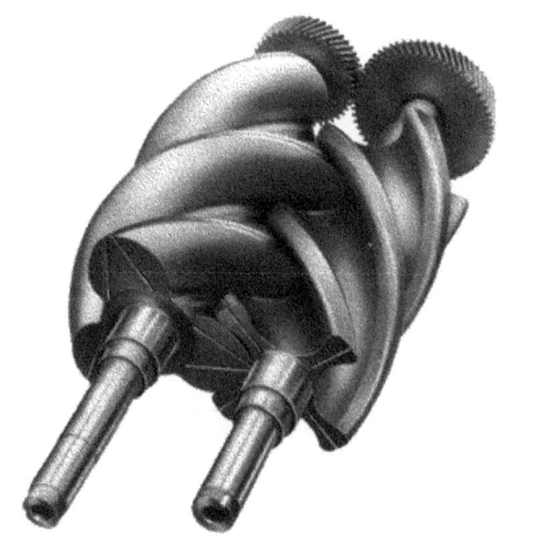

Figure 8.25 Synchronous gearbox.

Figure 8.26 Screw compressor: internal sight. Source: Aerzener.

Figure 8.27 Screw compressor of Figure 8.26.

Figure 8.28 Screw compressor of Figure 8.26 with soundproofing housing. Source: Aerzener.

When two stages are used the gas is led after the first stage through a water curtain to cool down. Also the shaft sealing makes no contact with the shaft. There wear is excluded and non-lubricating oil can penetrate the compression space. Oil-free compressors are available from 75 to 2000 [kW].

Oil-free compressors are more expensive because of the tolerances on the rotors.

With oil-injected compressors a one-stage implementation can attain 13 [bar]. A few tens of liter/min oil are injected before the compression.

Oil-injected compressors are available from 4 to 750 [kW].

8.3.3 Regulation

The capacity regulation with steering slide (Figures 8.29 and 8.30) is based on the steering of the effective rotor length. The shorter the rotor, the lower the flow. This is realized by means of a hydraulic steering slide that is a part of the housing. When the steering slide is regulated by a bypass, it will present itself and make a connection to the suction side of the compressor. The effective groove volume decreases.

By means of this regulation the flow can be regulated from 10 to 100% continuously.

Figure 8.31 shows the operation of the capacity steering slide.

The internal volume ratio V_I/V_{II} is the ratio between the chamber volume at the inlet port, presented at the beginning of the compression, to the one of the chamber at the outlet, presented at the end of the compression. According to the law of Poisson for a polytropic state change the volume ratio determines the internal compression ratio and thus the design pressure. By working on the start volume V_I of the chamber the internal compression ratio V_I/V_{II} is changed. If there is no pressure vessel this end pressure will also be the pressure in the outlet port. But the design pressure is determined by the volume I. And this design pressure does not have to be equal to the pressure in the discharge vessel. This leads to an energy loss. The design pressure is thus regulated

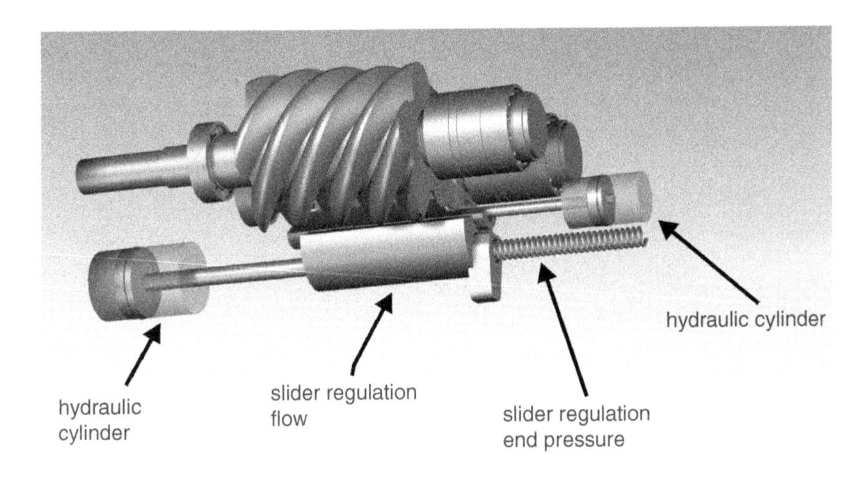

Figure 8.29 Regulation screw compressor. Source: Grasso.

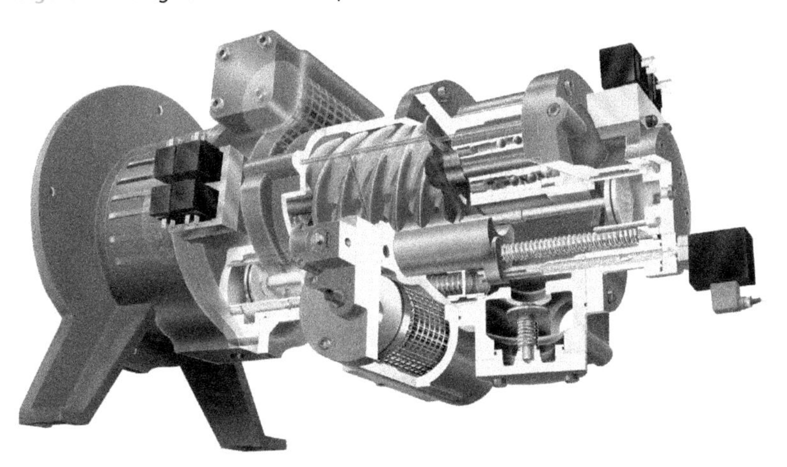

Figure 8.30 Regulation screw compressor. Source: Grasso.

Figure 8.31 Regulation pressure ratio. Source: After Grasso.

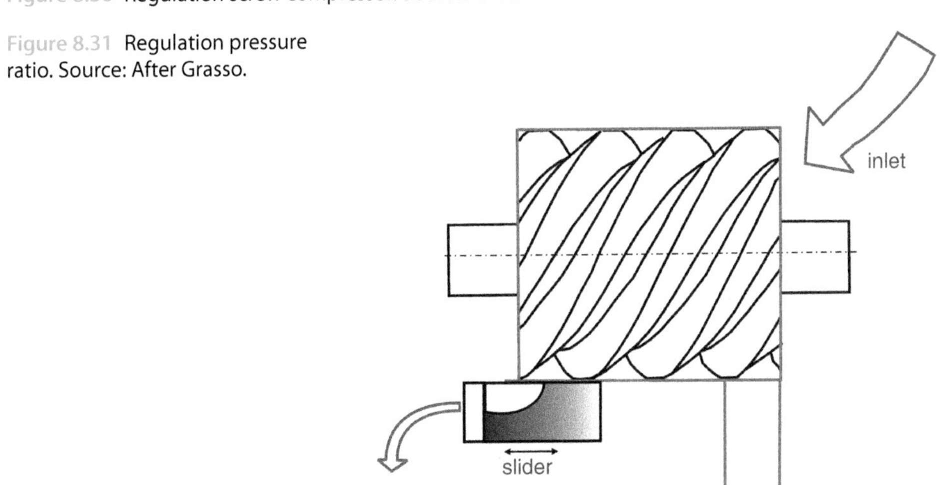

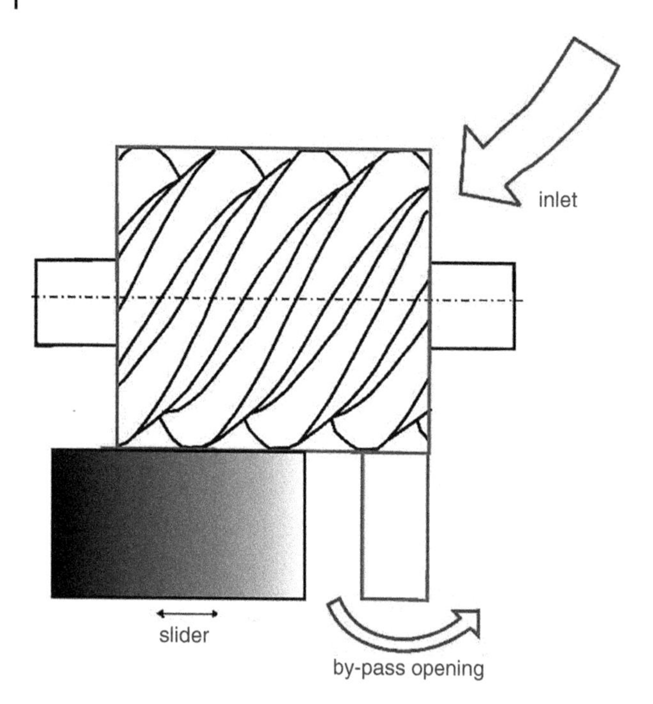

Figure 8.32 Regulation flow. Source: After Grasso.

by placing a valve at the inlet, that regulates the volume I, and so the pressure ratio (Figure 8.32). The valve is driven by the desired end pressure.

At the radial outlet is a second steering slide; its position determines the position of the outlet port (Figure 8.32). If the slider is open some gas is by-passed to suction side. The compressor delivered then less flow.

The steering slides work together, the first one on the inlet, the second one at the outlet. This way pressure ratio and flow can be regulated together.

The most energy-friendly flow regulation is a speed regulation VSD (variable speed drive), mostly based on a frequency converter.

8.3.4 Refrigerant Compressors

Screw compressors for refrigerant purposes have powers from 15 to 1200 [kW]. Like mono-screw compressors (see later) they are provided with a regulation that allows unloaded startup and a regulation of the desired pressure and flow. The latest trend is to continuously regulate Energy = speed to make it economically efficient (Figure 8.33).

8.4 Mono-screw Compressor

8.4.1 Operation

A mono-screw compressor (Figure 8.34) uses one screw gripped by two gears. The screw is driven by a motor. The gears are of metal or composite material. The gas is locked up at the inlet side in the groove of the screw and then axially transported to the outlet side.

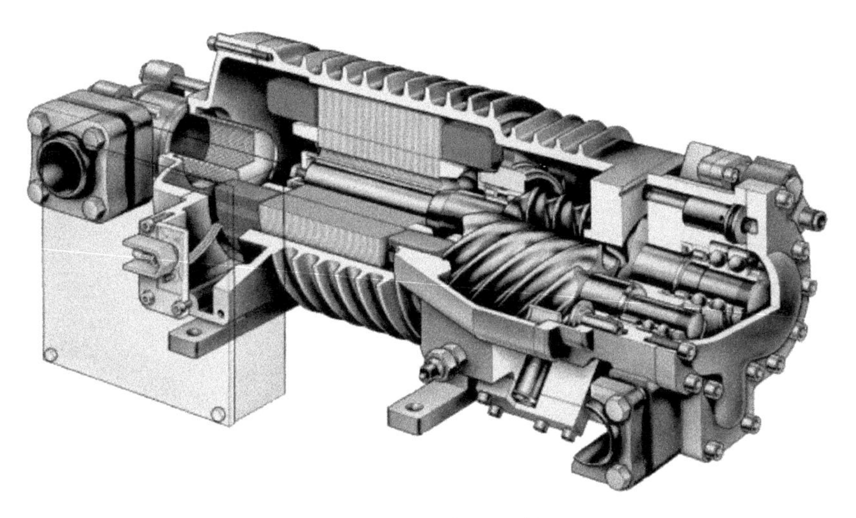

Figure 8.33 Semi-hermetic screw compressor. Source: Bitzer.

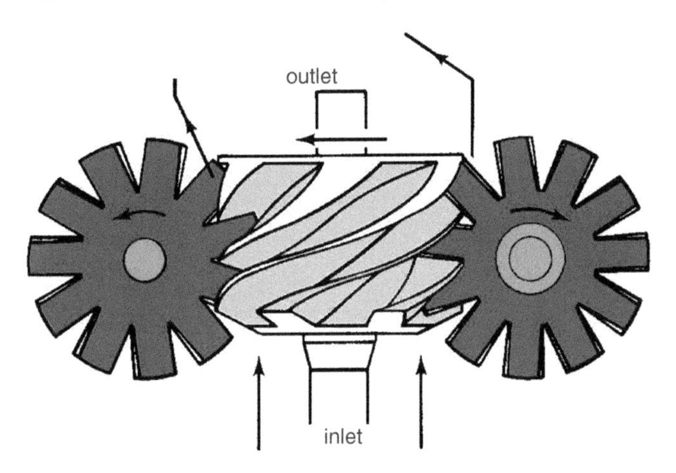

Figure 8.34 Mono-screw compressor.

During the compression, there is a volume reduction like with the classic two-rotor screw compressor (Figures 8.35–8.40).

8.4.2 Properties

- A cycle from suction to outlet is finished after 160° of the screw rotor. The speed of the rotor can thus be kept low while the pressure builds up quickly. Because of this fast build-up of pressure, the leak flow is very low, resulting in a high pump efficiency.
- The build-up of the pressure takes place on both sides of the screw rotor. The radial load on the screw rotor is then compensated and balanced.
- The force originating from the compression pressure is perpendicular on the gate rotors. In the rotation sense they don't experience any load from this pressure. This means that there is no friction between the gate rotors and the screw rotor; the only resistance that the gate rotors experience comes from the rolling bearing.

inlet
Atmospheric gas flows in the compression element and fills the grooves of the main rotor. The rotors grip in the grooves and form two compression chambers at the top and bottom of the main rotor

Compression
The rotors follow automatically the main rotor. The volume in the rotor decreases and theg gas is compressed. Purified water is injected in the compression element and takes care of the sealing and the lubrication.

Outlet
The mix of water and gas is evacuated by the compression element and flows to a water separator. Because of the low end temperature of the gas an aftercoller is not needed.

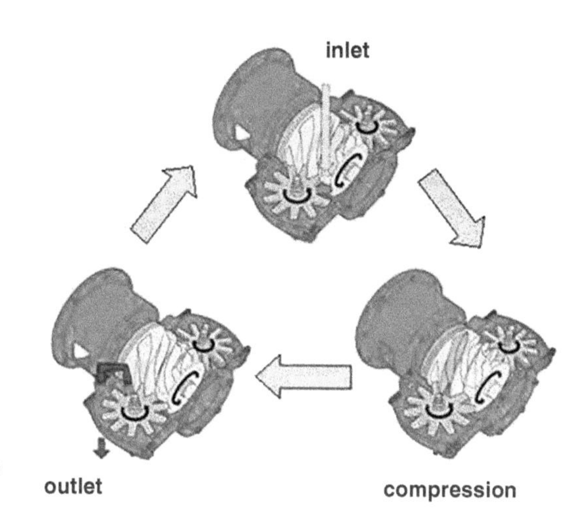

Figure 8.35 Operation mono-screw compressor with water injection.

Figure 8.36 Elements mono-screw: a screw rotor and two gate rotors. Source: © Gardner Denver – Compair used with permission.

Figure 8.37 Mono-screw compressor. Source: Daikin.

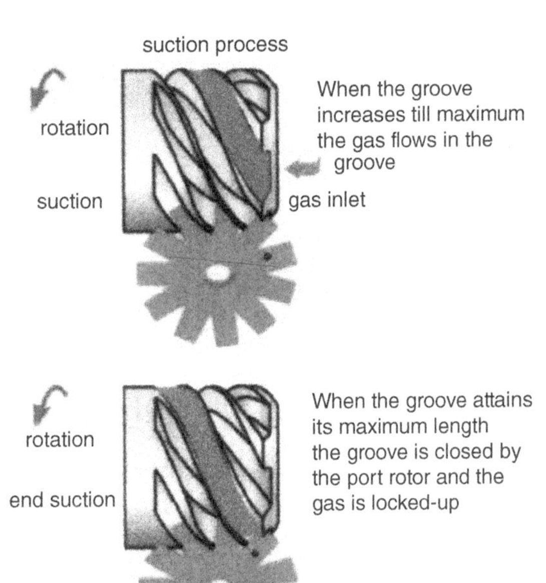
suction process
rotation
suction
When the groove increases till maximum the gas flows in the groove
gas inlet
rotation
end suction
When the groove attains its maximum length the groove is closed by the port rotor and the gas is locked-up

Figure 8.38 Operation suction process. Source: Daikin.

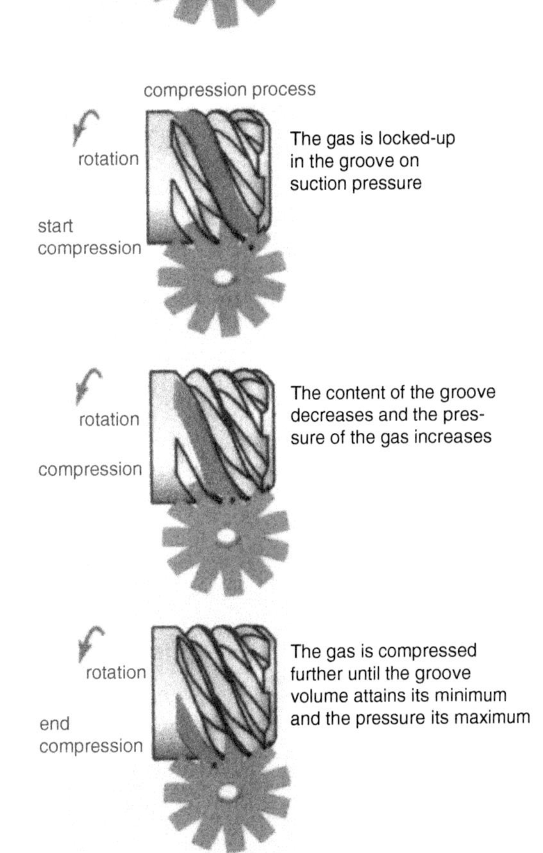
compression process
rotation
start compression
The gas is locked-up in the groove on suction pressure
rotation
compression
The content of the groove decreases and the pres-sure of the gas increases
rotation
end compression
The gas is compressed further until the groove volume attains its minimum and the pressure its maximum

Figure 8.39 Operation compression. Source: Daikin.

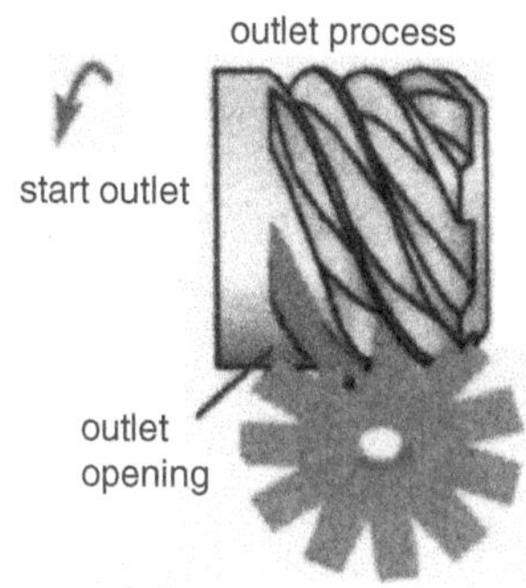
outlet process
start outlet
outlet
opening
The gas that is locked-up
is now at the outlet port.
The gas is displaced.

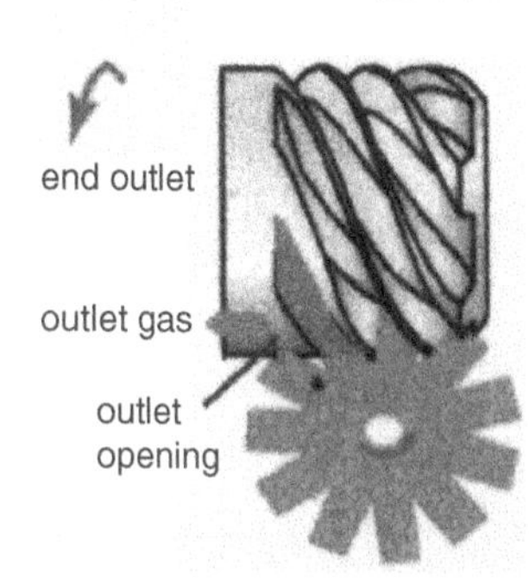
end outlet
outlet gas
outlet
opening
All the locked up gas
is removed.

Figure 8.40 Outlet. Source: Daikin.

- The pressure build-up happens 12 times in one revolution of the screw rotor (2 × 6). This minimizes the gas pulsation, vibration, and noise. Because of the quick pressure build-up the rotation speed can be kept low and this had advantages for its efficiency and lifecycle.
- The material of the gate rotors is a very special plastic; there is no metal-on-metal contact. And because of the lack of resistance and the balance between the compression forces, there is nearly no noise and vibration.
- Because of the high efficiency of the screw compressor, it can be run at low speeds, so there is no need for a gearbox or special transmission. The screw rotor can be coupled directly to the motor.

For refrigerant compressors there is a need for regulation of flow and pressure in the function of the cooling load.

8.4.3 Regulation

The operation is as follows (Figures 8.41–8.44). As with the vane compressor, the inlet volume of the groove is V_I and the volume at outlet is V_{II}. A polytropic compression with factor n gives:

$$\frac{p_2}{p_1} = \left(\frac{V_I}{V_{II}}\right)^n$$

where p_1 is the suction pressure (mostly the atmospheric pressure) and p_2 the design pressure. This is the pressure by which the gas enters the discharge space, which will eventually become pressure p_3. Flow regulation is possible by placing a disc at the inlet so that only a part of the groove volume is allowed and compressed. This implies that less flow is delivered, but the pressure p_2 will decrease because the volume V_I is decreased.

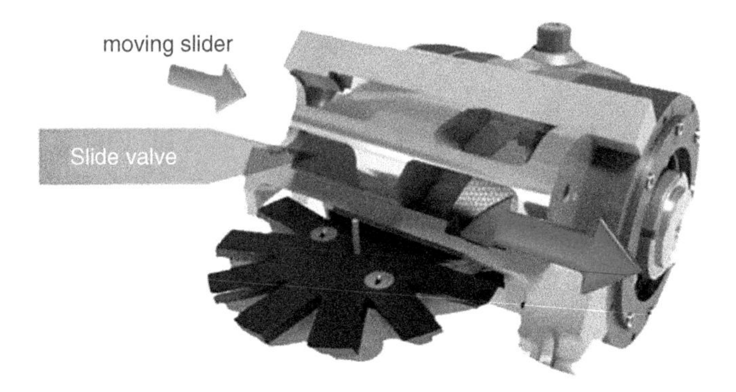

Figure 8.41 Regulation flow with slider. Source: Daikin.

regulation volume ratio (pressure control)
regulation capacity (flow control)
The outlet port opens early in the cyclus so that only a moderate compression occurs
The volume gas is in the groove.
The outlet port opens later in the cyclus so that a great volume reduction takes place and thus a greater pressure is achieved
Only 50% of the groove volume is allowed and compressed.
Here the slider valve is in the maximum position(left) so that the maximum pressure at the outlet is achieved.
Only 20% of the groove volume is used.

Figure 8.42 Regulation of flow and pressure – separately. Source: Daikin.

If there is a pressure vessel at pressure p_3 the gas pressure will rise until it reaches that pressure p_3 (by backflow). But in the case of refrigeration there is no vessel.

By placing a slider at the outlet side, or a by-pass to the suction side, the volume V_{II} will also decrease. This has an implication for the pressure p_2. It will increase when V_{II} decreases. Working on both sliders at the same time can regulate the flow and the outlet pressure.

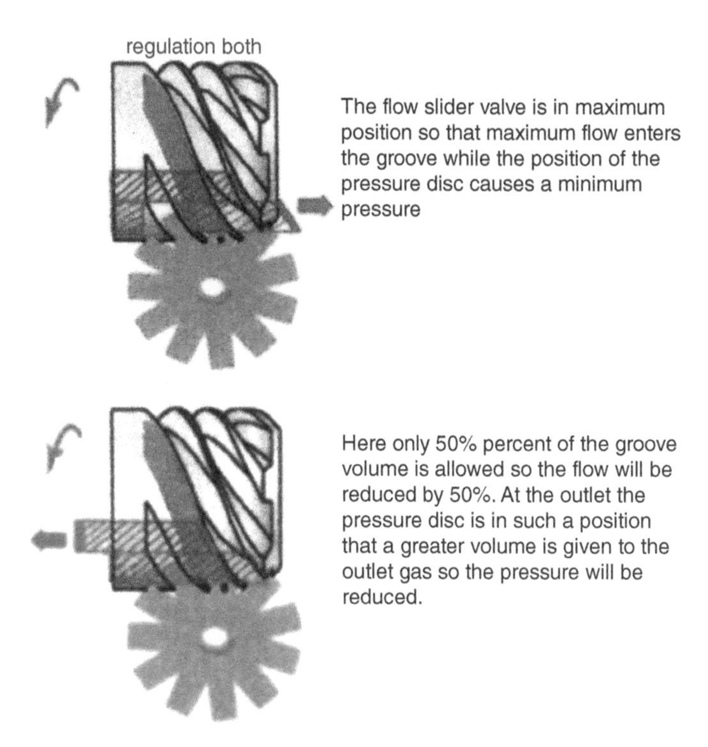

regulation both
The flow slider valve is in maximum position so that maximum flow enters the groove while the position of the pressure disc causes a minimum pressure
Here only 50% percent of the groove volume is allowed so the flow will be reduced by 50%. At the outlet the pressure disc is in such a position that a greater volume is given to the outlet gas so the pressure will be reduced.

Figure 8.43 Regulation of flow and pressure – together. Source: Daikin.

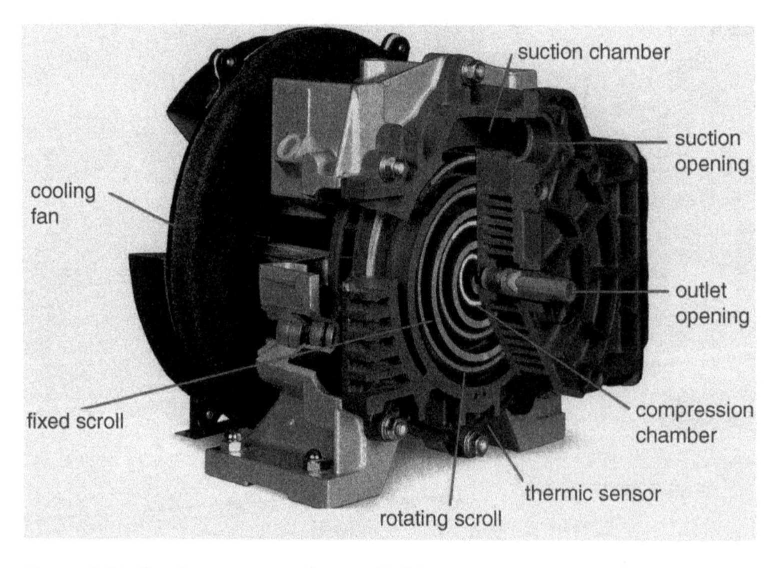

Figure 8.44 Scroll compressor. Source: Daikin.

These methods to regulate pressure and flow together are prone to efficiency loss; that's why nowadays increasingly speed control (VFD) is used to regulate the flow.

8.5 Scroll Compressor

Another displacement compressor that is used mainly for small capacities is the *scroll compressor*. It consists of a fixed spiral. Another spiral, the rotor, makes an eccentric movement in the stationary spiral. Figures 8.45 and 8.46 show the operation for compressed air applications. Scroll compressors are often used for refrigeration applications and heat pumps. The scroll compressor is oil-free because there is no metal-on-metal contact between the compression spirals. Furthermore, it is driven directly instead of a gearbox drive, and so the compressed gas is absolutely oil-free.

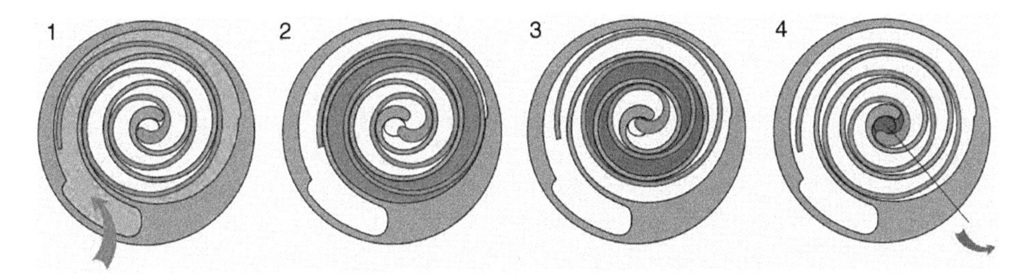

Figure 8.45 Operation scroll compressor – internal compression. Source: Bitzer.

Figure 8.46 Rotor.

Figure 8.47 Spiral.

The delivered pressure attains 10 [bar(g)]. The flow in FAD (free air delivery) is as high as 70 [m^3/h].

Because of the rotation speeds and a practically continuous flow the scroll compressor is noted for its noiseless action (50–75 [dBA]).

Figures 8.46–8.49 show the parts of a hermetic scroll compressor.

8.6 Tooth Rotor Compressor

The tooth rotor compressors consist of two claws that rotate in opposite directions. There are designs with one or two claws (Figures 8.50 and 8.51); their form can be the same or different. Nowadays there is a trend toward two claws because of higher and more uniform flow, and also a symmetric radial load on the rotors.

The main purpose is that the rotors can cover axial ports or open them and that the claws in their movement compress the sucked gas during their rotating movement. So, there is internal compression.

Just like the mechanism found in a roots pump, the rotors of a claw pump are also synchronized by a gear. In order to attain an optimum seal, the clearance between the rotor inside the casing wall is very small, in the order of magnitude of some 10 [μm].

So, in principle the compressor is oil-free. Nominal pressure: to 10 bar(g) and flows to 2000 [m^3/h]. Influx and discharge of the gas is performed during two half periods. Each rotor turns twice during a full work cycle.

8.7 Rolling Piston

8.7.1 Operation "Rotary"

Small cooling installations for Airco often use the simple rotary compressor. The suction disc, eccentrically secured on the shaft end of the motor shaft, rolls within the inner wall of a cylindrical housing where a suction and discharge opening are bored; between the

Figure 8.48 Stator.

Figure 8.49 Rotor and Scroll.

two ports stands a plate that by means of a spring pushes against the suction disc. The suction disc and plate divide the space into two chambers. When the suction disc goes from state a to b, the gas is sucked off (Figure 8.52). The gas in it is compressed until some, in the discharge line build valve, opens and the gas can flow to the discharge line. The process then begins again.

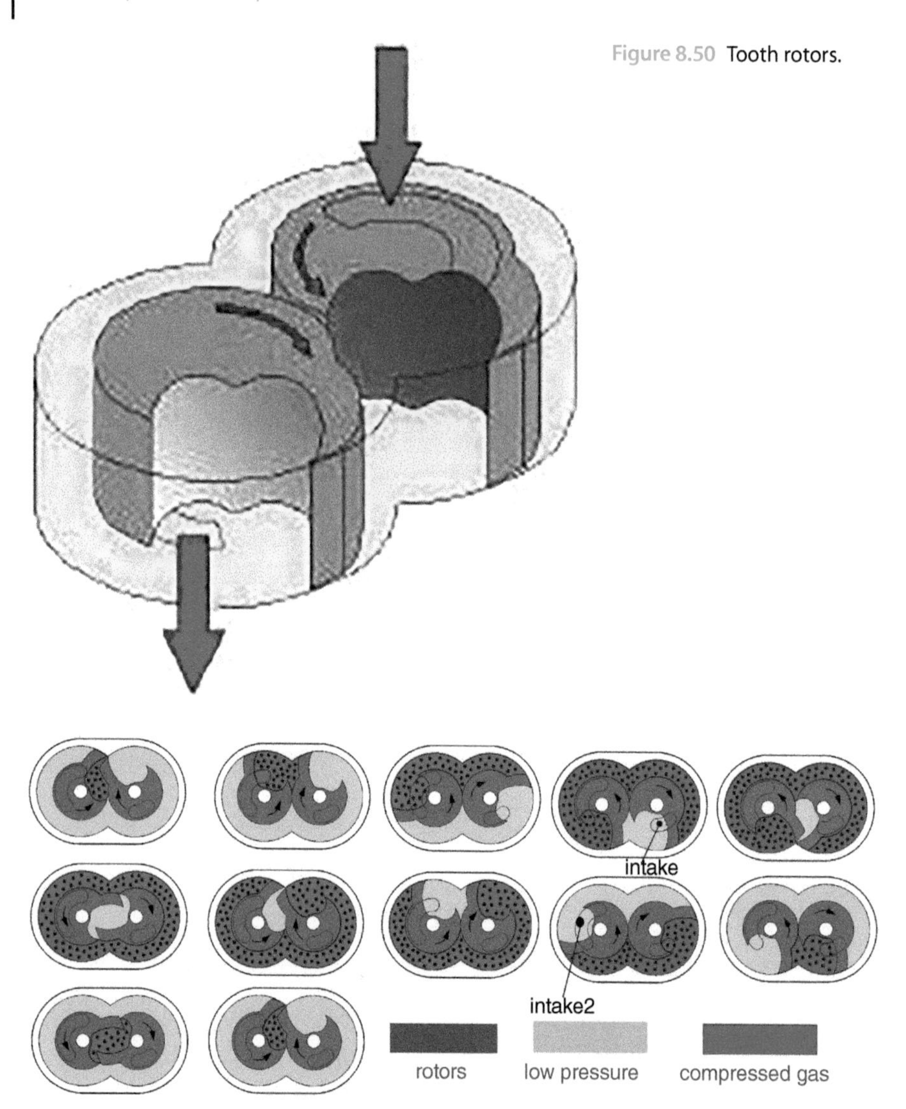

Figure 8.50 Tooth rotors.

Figure 8.51 Operation principle of tooth compressor: two cycles.

Rotary swing compressors of up to 7 [kW] are used for Airco applications.

The different parts of a rotary compressors are represented in Figures 8.53 and 8.54.

8.7.2 Swing Compressor

The rotary compressor has a drawback: the discharge and suction side are connected via a clearance of the plate. Back in the old days, when refrigerant R20 was used, it was not very harmful because the pressure stayed below the 20 [bar].

The alternative, ozone-friendly refrigerants like R407C and R410C use pressure of up to 40 [bar], and that is problem. The solution was very simple: the plate is attached to the rotor and the plate slides and turns in a joint in the housing (Figures 8.55–8.58).

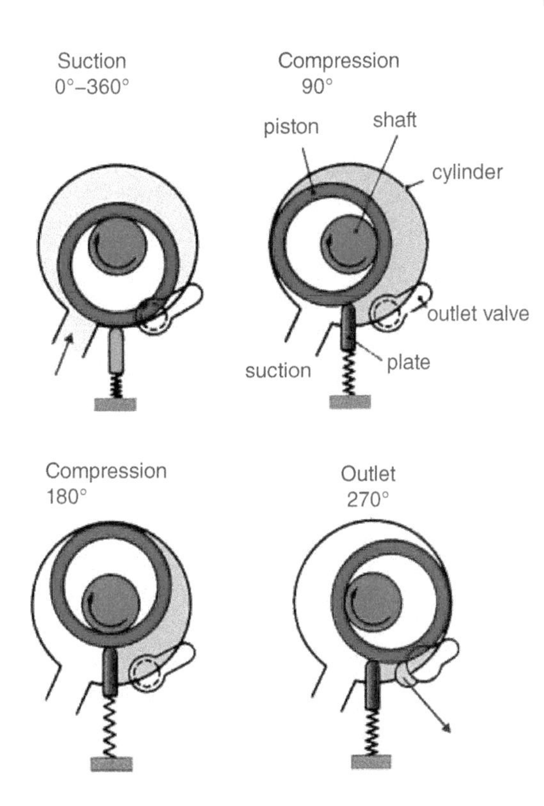

Figure 8.52 Operation rolling piston.

Figure 8.53 Parts rotary compressor.

Figure 8.54 Stator.

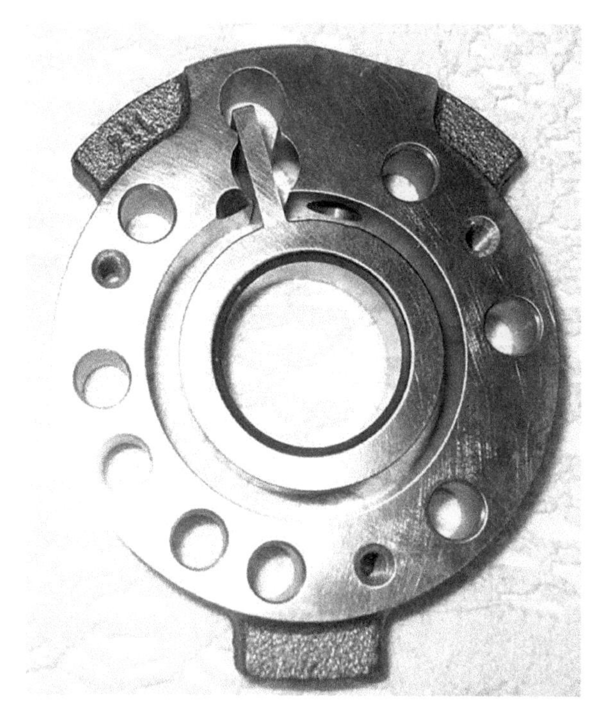

Figure 8.55 Swing compressor. Source: Daikin.

Figure 8.56 Half-moons as hinges. Source: Daikin.

Figure 8.57 Parts.

Figure 8.58 Swing compressor.

8.8 Liquid Ring Compressor

8.8.1 Operation

A rotor with vanes (Figures 8.59–8.62) is eccentrically positioned in a cylindrical housing. When the housing is partly filled with liquid and the wheel rotates quickly, a

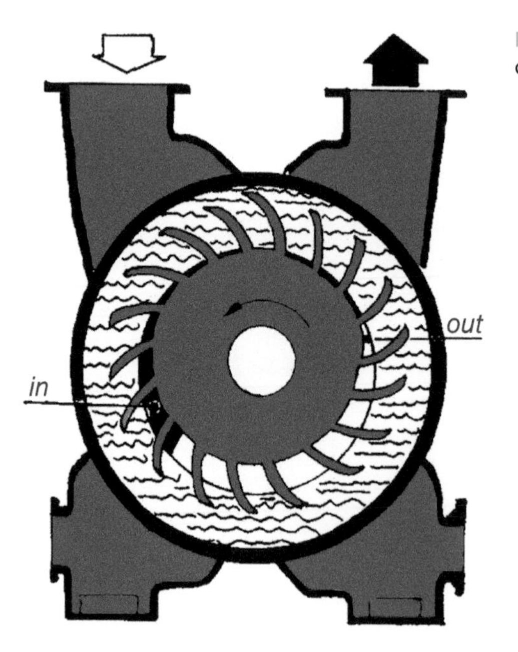

Figure 8.59 Operation of a water ring compressor.

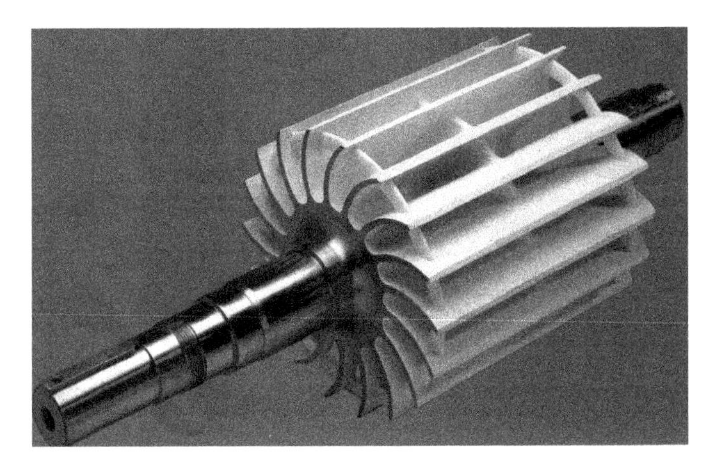

Figure 8.60 Water ring compressor rotor. Source: Hibon.

Figure 8.61 Water ring compressor with housing. Source: Hibon.

liquid ring caused by centrifugal force will be formed. Between the inner surface of the liquid ring and the outer surface of the hub of the wheel, chambers appear that change size cyclically: on the suction side they become greater and on the discharge side they become smaller, so that an internal compression takes place.

The inlet and outlet lines are axial; there are no valves.

A two-stage compressor is shown in Figure 8.63.

8.8.2 Properties

Because the compressed gas is in very tight contact with the liquid ring, that has a great heat storage capacity, the temperature rise in the gas is very small. The gas is compressed

Figure 8.62 Sight on the housing and inlet port. Source: Hibon – Ingersoll.

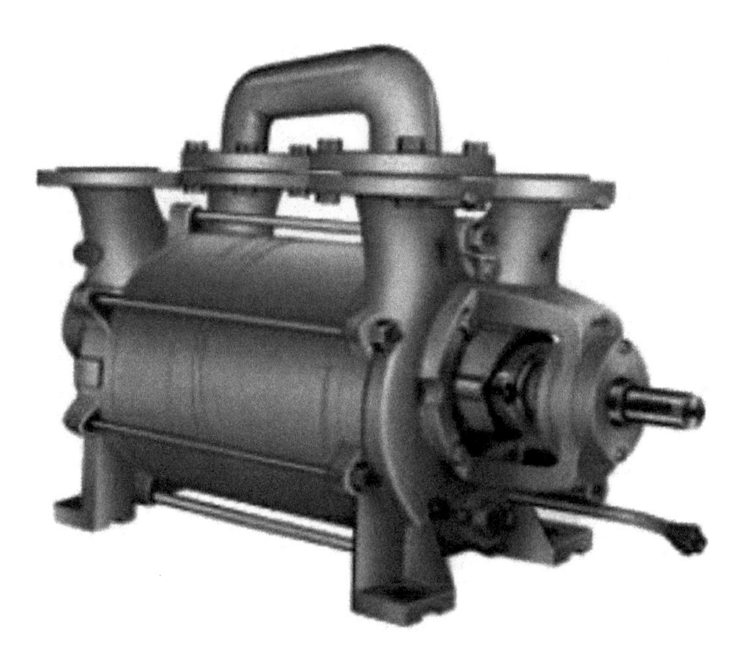

Figure 8.63 Two-stage compressor.

nearly isothermally. These compressors are used in processes where the temperature rise must be limited.

The liquid will need to be regularly supplied and drained in order to exchange the heat taken during the compression received form the gas. There are three possibilities:

- The liquid is drained to the sewage.
- A part of the liquid is recuperated.
- The hot liquid flows through a cooler and then flows back to the compressor.

A part of the liquid in the liquid ring is in any case received by the compressed gas and must eventually be separated from it.

Mostly, the liquid is water, but in function of the application other fluids can be used, for instance when a substance of the gas must be taken up by absorption by the liquid or when the compressor must be protected against the corrosive or chemical aggressive gasses or vapor.

Although the compression is in principle isothermal, this type of compressor needs more energy than a piston (or other) compressor: there are extra losses caused by friction of the fluid against the pump housing and the movement of the vanes in the liquid. There are oil-free water ring compressors on the market.

8.9 Regulation Displacement Compressors

First of all, it must be noted that many compressors start unloaded so that the approach torque is not too high. But also, when stopping, an unloaded compressor is wanted. Take, for instance, a screw compressor: the rotor has a considerable mass inertia; the speed is high. Laying the motor down during action (under load) would produce a shock load for the bearing.

8.9.1 Blow Off (Figure 8.64)

The compressor works continuously, the surplus flow is blown off to the atmosphere. This is, of course, a simple, but not an energy economical, regulation. It has only sense if the gas may be discharged to the air. This is possible with an oil-free air compressor. For compressed air applications a valve is driven by the compressed air itself. Other applications ask for an automatic regulation.

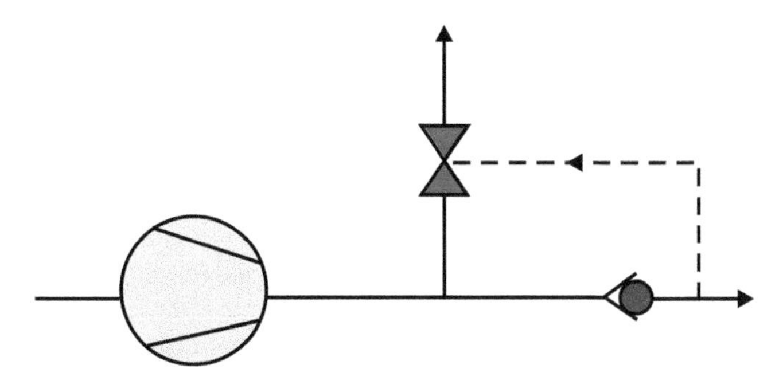

Figure 8.64 Blow-off to the atmosphere.

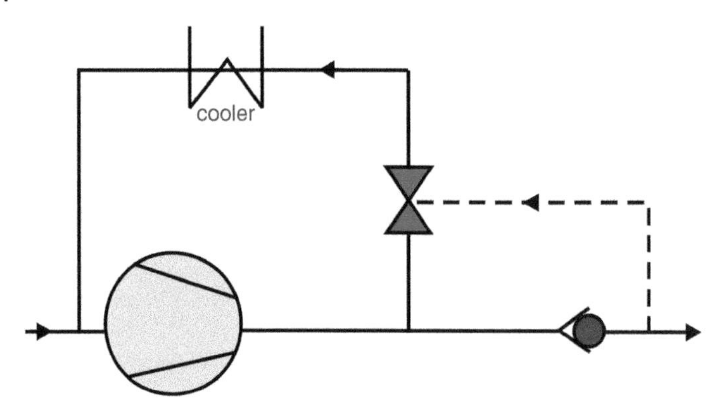

Figure 8.65 Bypass regulation.

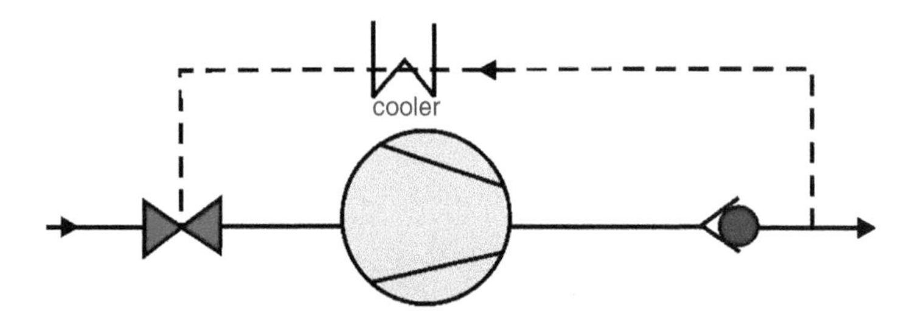

Figure 8.66 Throttling the suction line.

8.9.2 Bypass Regulation (Figure 8.65)

Here the redundant gas will flow back via a bypass to the suction line, after pressure reduction in a valve, and cooling. This regulation, of course, consumes energy.

8.9.3 Throttling the Suction Line (Figure 8.66)

This method is discussed in Section 7.9.2, in Chapter 7. The operation is identical for other compressors. It is a continuous regulation that is energy consuming because in the valve energy is spoiled.

8.9.4 Start–Stop Regulation (Figure 8.67)

This method is applied with small machines (<10 [kW]). When no compression gas is necessary the motor is shut off. But the number of times per minute that a electromotor may start is limited and becomes smaller as the motor power increases. The repeated run-up currents (about 10 times the nominal current) would burn the windings.

How can that be prevented? Take as an example a compressed air application with pressure vessel. A delay relay is used. Suppose a motor that may be started 10 times per hour. The delay relay is set to six minutes. If the desired pressure in the pressure vessel is attained after two minutes the compressor will still work for four minutes. The pressure in the vessel will be higher than normal, and that costs energy.

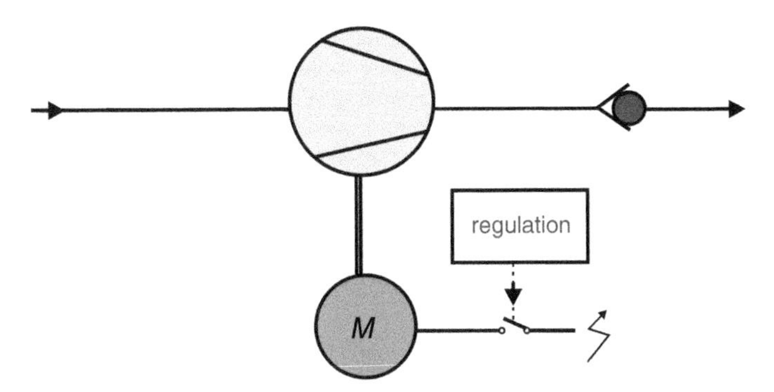

Figure 8.67 Start–stop regulation.

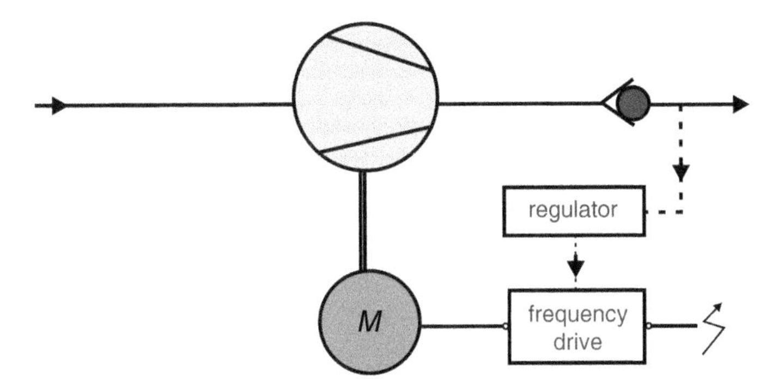

Figure 8.68 Frequency regulation.

A similar method can be used for compressors in refrigeration. There is no pressure vessel, but the compressed refrigerant goes to the evaporator and the cooling cycle just continues. In other words, there will be too much cooling.

8.9.5 Full Load–No Load Regulation

When switching from nominal load to no load the suction is closed and the compressor blows off to the atmosphere or the method of the bypass is used. There is no delivery of gas. In zero load the energy consummation is 25% of full load.

8.9.6 Speed Control with a Frequency Regulator (Figure 8.68)

This is the most energy-friendly regulation. The rotation speed of the compressor is regulated continuously dependent on the demand.

8.10 Refrigerant Compressors

Most compressors discussed in this chapter are also used in air-conditioning and refrigeration applications (screw Figure 8.69, mono-screw, scroll, rolling piston).

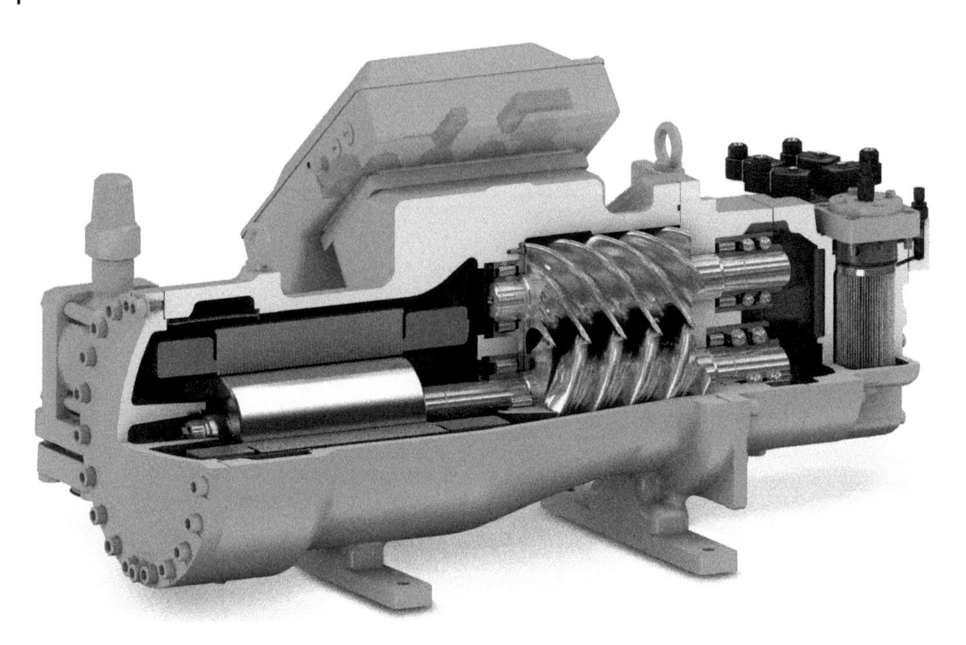

Figure 8.69 Semi-hermetic refrigerant compressor. Source: Bitzer.

9

Turbocompressors

CHAPTER MENU

9.1 Centrifugal Fans

9.1.1 General

The operational principle of a centrifugal fan (aka a radial turbo fan) accords with that of a centrifugal pump, but the fluid is different, now it's a gas. It is built up out of an impeller with vanes that rotates fast in an abutting volute (Figures 9.1–9.4). The rotating movement of the impeller produces a pressure rise on the locked-in gas; the gas is propelled out of the impeller and caught in the volute. An underpressure arises in the inlet and this cause the gas to be sucked into the suction line.

As far as that is concerned the resemblance with a centrifugal pump is complete. But there are differences, for instance the delivered pressure will be much smaller.

Indeed, in a first approximation the manometric feed pressure of a radial turbo fan can be written as:

$$p_{man} \cong \rho \cdot u_2^2$$

where ρ is the specific mass (sometimes called the *absolute density* and u_2 the peripheral velocity of the impeller (at the outlet).

For a liquid like water ρ is of a magnitude of 1000 [kg/m^3]. A gas like air at atmospheric condition has a specific mass of 1.3 [kg/m^3].

Pumps and Compressors, First Edition. Marc Borremans.

Companion website: www.wiley.com/go/borremans/pumps

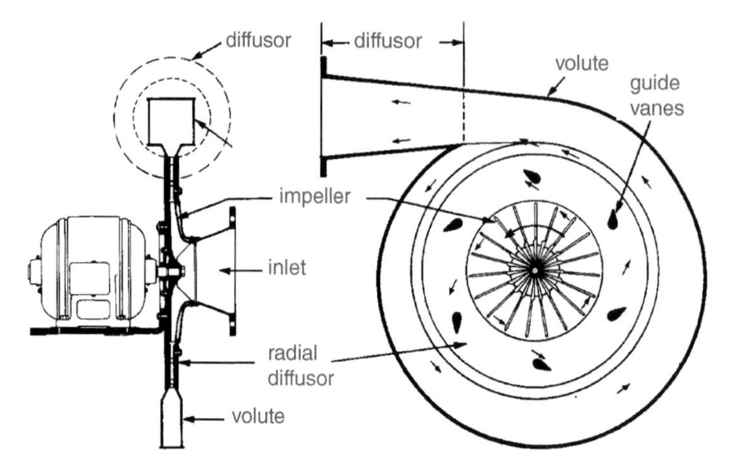

Figure 9.1 Centrifugal fan.

Figure 9.2 Centrifugal fan for air treatment, forward-curved vanes.

Figure 9.3 Plastic fan (corrosion resistant).

Figure 9.4 Fan with two-sided entry. Source: Howden.

Where a centrifugal pump can deliver about 5 [bar] a centrifugal fan at the same speed will deliver 5 [mbar] [hPa].

For liquid one tolerates peripheral speeds of about ±35 [m/s]. Higher speeds would have no sense because this would imply very high liquid velocities and then the friction losses in the pump would become inadmissible (intolerable) high.

For gasses the friction losses are much smaller and so higher peripheral speeds of the impeller can be allowed.

Above that centrifugal fans are aimed for a limited pressure rise, to transport the gas (for instance ventilation). The fan pressure is only meant to overcome the downstream friction losses.

The volumetric flows can attain: 100 000 [m^3/h]

9.1.2 Static and Dynamic Pressure

Let's look at the order of magnitude of the pressures for a fan. It is supposed that the law of Bernoulli is applicable, where it is supposed that the specific mass ρ stays constant when flowing through the fan because the pressure rises are very small. The pressure rises are just a few tens of pascals.

In the fan the geodetic pressure rise will be of no importance; suppose an impeller diameter of 30 [cm], the geodetic is $\Delta h = 0.3$ [m]. The corresponding geodetic pressure rise would be:

$$p_{geo} = \rho \cdot g \cdot \Delta h$$

At atmospheric pressure, it can be determined that the specific mass $\rho \cong 1.2$ [kg/m^3]. if the ideal gas law is applied, so that:

$$p_{geo} = 1.2 \cdot 9.81 \cdot 0.3 \cong 3\,[\text{Pa}].$$

This term is negligible.

The dynamic pressure rise is another question entirely. With a velocity at the outlet of the fan that can attain easily 20 [m/s] the dynamic pressure rise would be:

$$p_{dyn} = \rho \cdot \frac{c^2}{2} = 1.2 \cdot \frac{20^2}{2} = 240\ [\mathrm{Pa}]$$

In comparison with the total feed pressure the dynamic pressure is not negligible. The total pressure of the fluid is then:

$$p_{tot} = p_s + p_{dyn}$$

where p_s is the static pressure of the gas.

With an opening against the stream (the left-hand side of Figures 9.5 and 9.6) the total pressure is measured. Crosswise on the stream (right) the static pressure is measured. The pressure difference is the dynamic pressure.

This pressure difference can then be measured on the manometer. Because only small height differences are measured (vertically), the measure tube is placed obliquely, then the measure scale is more sensitive (see a course on fluid mechanics). If the ρ is given, one can calculate the mean velocity of the flow. With the knowledge of the section A of the inner side of the channel the volumetric flow Q_V can be determined.

This principle can be applied in channels where other measuring devices can't be applied because of the high temperature or the aggressivity (fuel gas channels).

The velocity can also be determined with a wing wheel anemometer, comparable to a small windmill. The rotational speed of the wheel is proportional to the gas velocity. Anemometers for measurements in gas channels are very compact and can be glided in an opening of 12 [mm]. The velocity can be measured at different places and the device calculates the mean velocity.

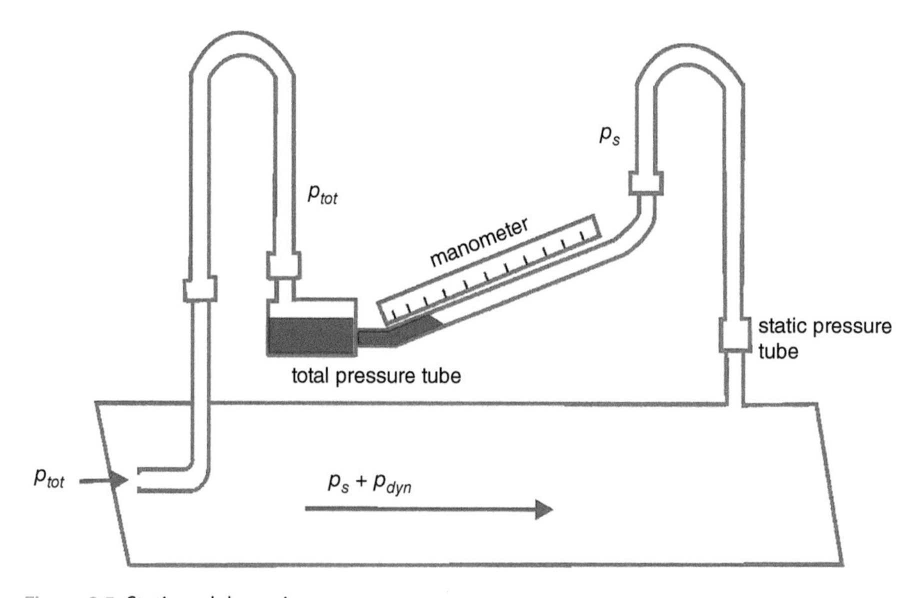

Figure 9.5 Static and dynamic pressure.

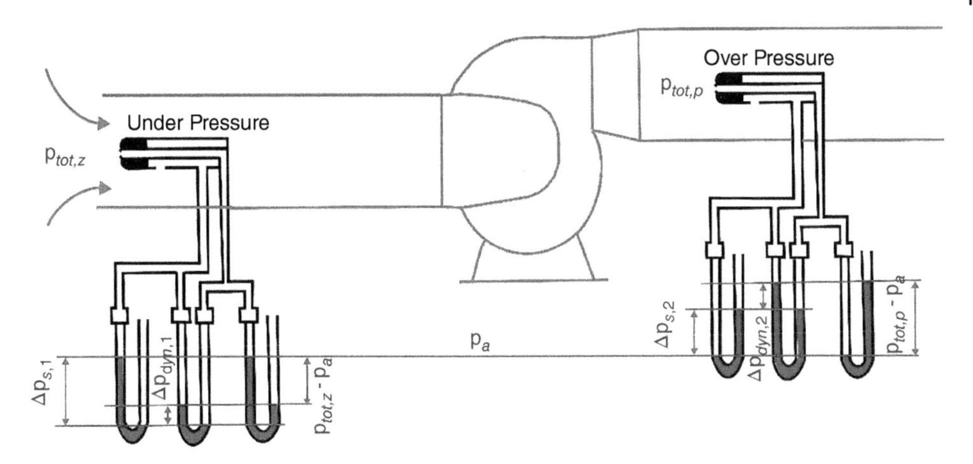

Figure 9.6 Determination of the total pressure of a fan.

Figure 9.6 should elucidate how the total feed pressure p_{tot} delivered by the fan is determined. The feed pressure is given by:

$$p_{tot} = p_{tot,p} - p_{tot,s} = p_{dyn} + p_s$$

9.1.3 Types of Vanes

9.1.3.1 Forward-curved Vanes

These are characterized by a big number of vanes that are slightly curved forwards (Figures 9.7 and 9.8). The gas leaves the impeller at high velocity and requires therefore a diffusing volute to transfer the high velocity into usable static energy. The efficiency is lower than that of an aerodynamic or simple backwards curved fan. The advantage lies in the fact that this type may rotate at the lowest speed of all kinds of fans. And at low speeds the fan produces accordingly less noise. The use of it is discouraged for dusty gasses or fuel gasses that have the tendency to stick to the short curved vanes, producing imbalance. Cleaning is difficult.

Applications: low pressure heating, ventilation, air conditioning.

9.1.3.2 Aerodynamical Vanes

These are backward-curved vanes but with an aerodynamic profile. They are what called *airfoil fans*. Because of their profile these fans have the best efficiency, make less noise, function softly, but are also the costliest.

Application: in big installations where the efficiency and not the initial price is of primary importance.

9.1.3.3 Backward-curved Vanes

These have fewer vanes than fans with forward-curved vanes (Figures 9.7 and 9.9).

This type can eventually be made without a volute. It is better, of course, taking into account the efficiency, to have one. They are available with a single-sided or double-sided entry. Their efficiency is lower than that of an aerodynamic type. The blades are often straight with constant thickness. More expensive types have a more adequate profile.

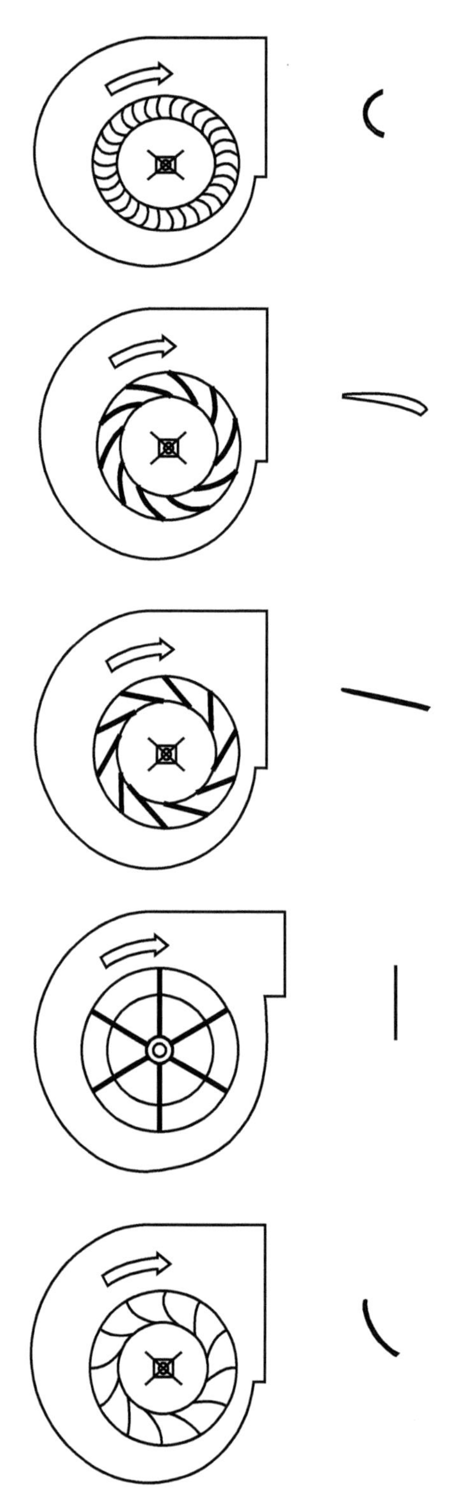

Figure 9.7 Types of fans.

Figure 9.8 Backward-curved vanes.

Figure 9.9 Forward-curved vanes.

Application: use on big installations because of their good efficiency and in corrosive or erosive environments.

9.1.3.4 Radial Vanes

The vanes are straight. This type has the worst efficiency of them all (Figures 9.7 and 9.10). A problem with centrifugal fans with forward- or backward-curved vanes is that underpressure zones with dust agglomeration can be formed. This dust can cause imbalance, vibrations, and bearing damage, and increases maintenance demands (cleaning). That's why they are less adequate for dusty air, or require a filter. Fans with straight vanes don't have underpressure zones and therefore are suitable for dusty air.

Application: for moderate to high pressures.

9.1.3.5 Radial Tip Vanes

The vanes end radially but begin backward curved (Figure 9.7). The efficiency is higher than that of the pure radial fan.

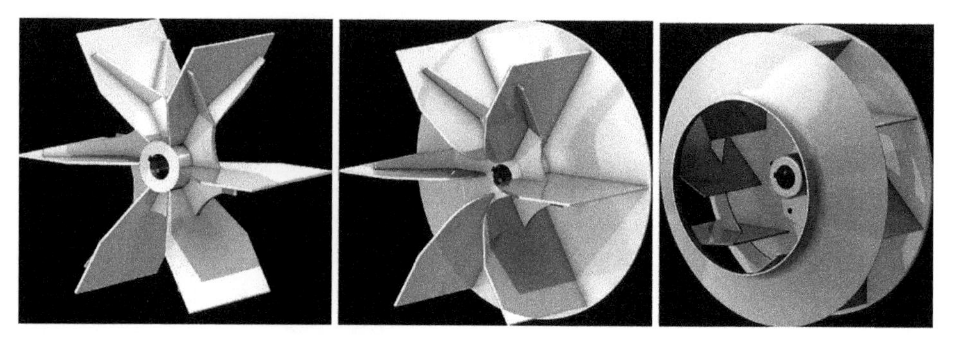

Figure 9.10 Impeller forms - radial vanes. Source: © 2018 Greenheck Fan Corporation.

Application: applied for its wear resistance in erosive air flows.

Open impeller: are suitable for most industrial applications. Applications: abrasive dust, suction (grinding and polishing), transport if granular materials (sawdust, wooden splinters), fuel gas outlet, and high temperature air conditioning.

Half-open impeller: is applied for long fibers and string material. Applications: transport of wooden abrade and paper clips. The efficiency is higher than that of an open impeller for granular material.

Closed impeller: for the treatment of pure air and light material transport. Applications: fuel gas and warm gas outlet, corrosive and heavy fuel gasses, and light dust.

9.1.4 Behavior of the Different Impeller Types

9.1.4.1 Backward-curved Vanes

At free outflow (maximum flow) and at closed channel (minimum flow) the power is smaller than at the operating point (Figure 9.11). The motor can thus be selected on the basis of the operating point. At occasional opening of the air conditioning cabinet the motor will not be surcharged. The power of the motor should be 15–20% higher than the impeller power, in order to take into account the loss in the drive, reading errors, influences of temperature...

9.1.4.2 Forward-curved Vanes

As the shaft power is proportional to the multiplication of flow and pressure, it will increase strongly with increasing flow (Figure 9.12).

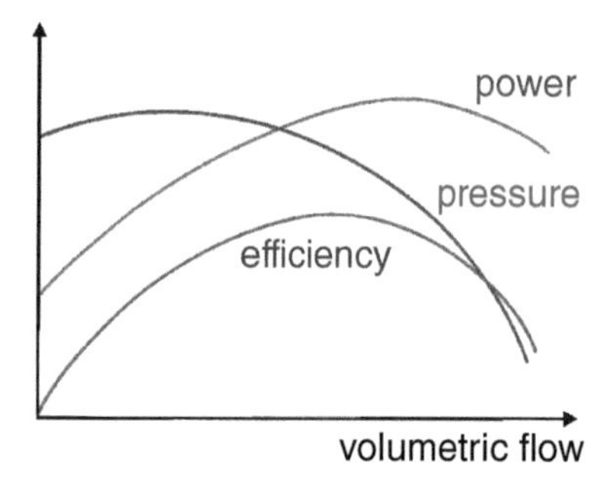

Figure 9.11 Characteristics for backward-curved vanes.

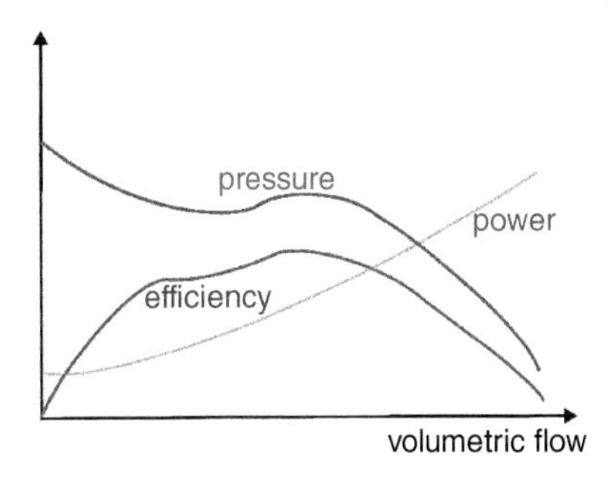

Figure 9.12 Characteristics for forward-curved vanes.

Above the operating point the shaft power increases (and thus the electric power) to two to three times the nominal power. This makes it necessary to do the following:

- Correct the setting of the electrical motor's current limit.
- To over dimension the electric motor (200% or 3× too big).
- Indeed, at a constant rotational speed of the shaft, an increasing power is possible by an increasing torque, so don't let the door of the air treatment cabinet open.

Radial-curved vanes (Figure 9.13) have a characteristic between those of forward- and backward-curved vanes.

9.1.5 Study of the Characteristics

As discussed before, fan characteristics are comparable with the characteristic of a centrifugal fan. It gives the feed pressure (or feed head) in function of the volumetric flow, and this at different speeds and at $\rho = 1.2$ [kg/m^3] and a temperature of 20 [°C] Besides that one can also read (Figure 9.14):

- the shaft power P
- the peripheral velocity u_2
- the efficiency, thus also the highest efficiency
- the outlet velocity c_2' of the gas
- dynamic, static, and/or total pressure.

Outlet velocity and dynamic pressure are based on the surface of the outlet channel.

Both axes often have a logarithmic scale that transforms the system curves to straight lines. Parabola become then straight lines because the log of y^2 transfers it to $2 \cdot y$.

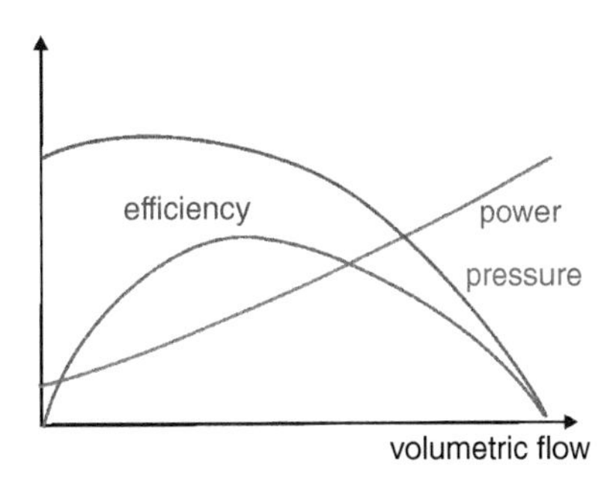

Figure 9.13 Characteristics for radial-curved vanes.

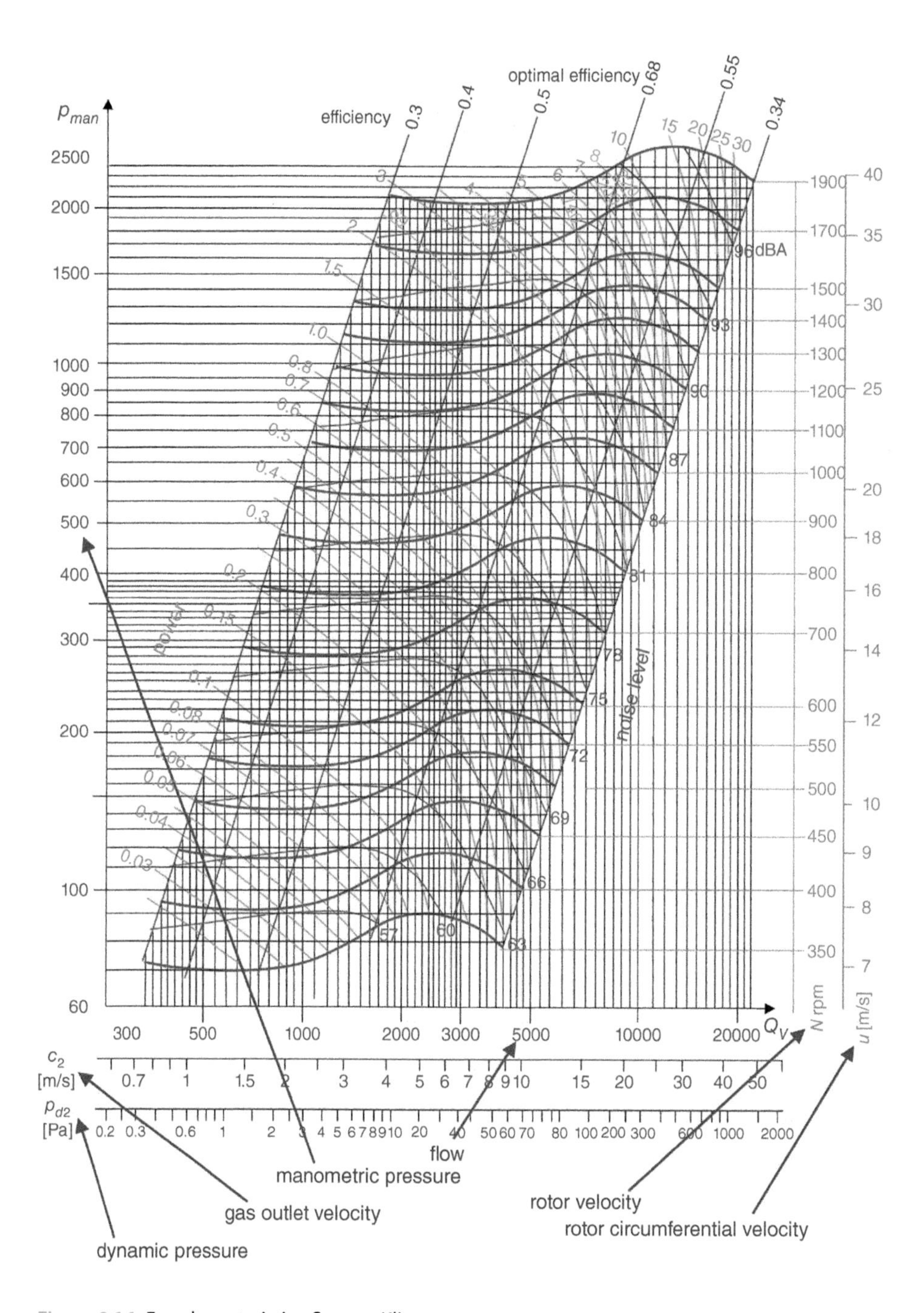

Figure 9.14 Fan characteristics. Source: Klima.

9.1.6 Selection of a Fan

A number of criteria for the selection of a fan are:

- Choice of impeller diameter, single or double sided entry.
- The noise level decreases as the peripheral velocity u_2 is lower.
- The efficiency is dependent on the type of fan and on the operating point.
- In installations with a large volumetric flow and small feed head (pressure) forward-curved vanes are more adequate; in the other case one prefers backward-curved vanes.
- Fan with backward-curved vanes are more expensive.

Now we treat an example of such a selection. The characteristics of the company Klima are used. There are two options:

- A fan series FGS (forward-curved vanes, single-sided entry) and FGD (forward-curved vanes double-sided entry).
- The fan series BGS (backward-curved vanes, single-sided vanes, single-sided suction) and BGD (backward-curved vanes, double-side entry).

The first section takes place on the base of *preset diagrams* (Figures 9.15 and 9.16) that sets the static pressure difference Δp_{st} against the volumetric flow Q_V and this for the whole offer of fans.

Example 9.1 A single entry fan must be selected for a flow of 5000 [m^3/h] and a static pressure difference of 300 [Pa]. From the preset diagram the following fans are considered:

- The type FGS-315 to 450 (these numbers point on the impeller diameters).
- The type BGS-355 to 560.

The selection is on the base of the criterion that the appropriate fan must be situated at the right of the area where lies the η_{max} curve.

As an example take a fan with forward-curved vanes, the fan type FGS-400 and compare with a fan with backward-curved vanes having the same impeller diameter, the BGS-400.

From Figure 9.15–9.18 it follows that:

	Unity	FGS-400	BGS-400
Air flow	m^3/h	5000	5000
Discharge velocity	m/s	11	11
Static pressure difference	Pa	330	330
Dynamic pressure	Pa	71	71
Rotational speed	rpm	760	1630
Peripheral velocity	m/s	16	34.8
Shaft power	kW	0.854	0.854
Noise level	dB(A)	76	80
Efficiency	%	65	65

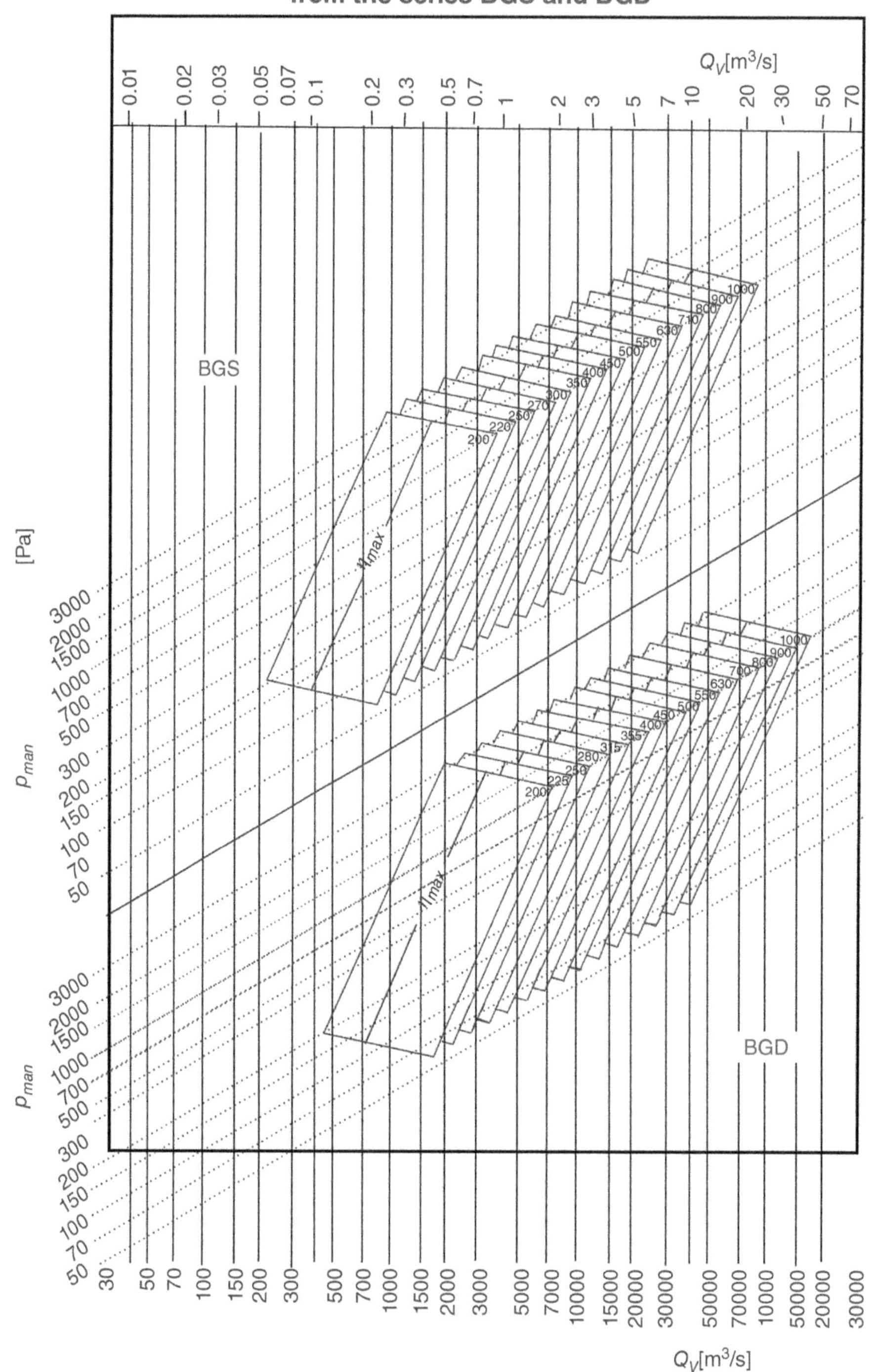

Figure 9.15 Preset diagram for BGS and BGD. Source: Klima.

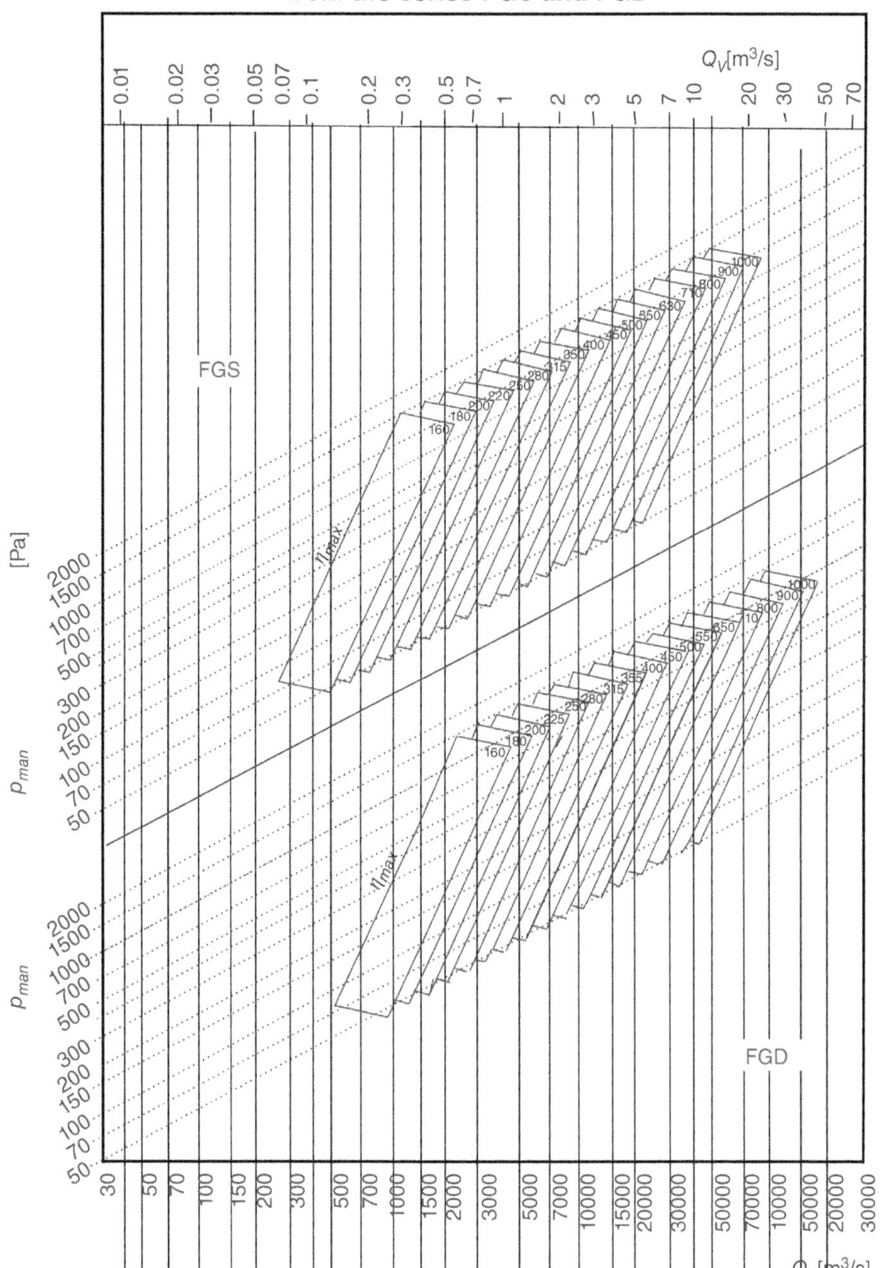

Figure 9.16 Preset diagram FGD and FGS. Source: Klima.

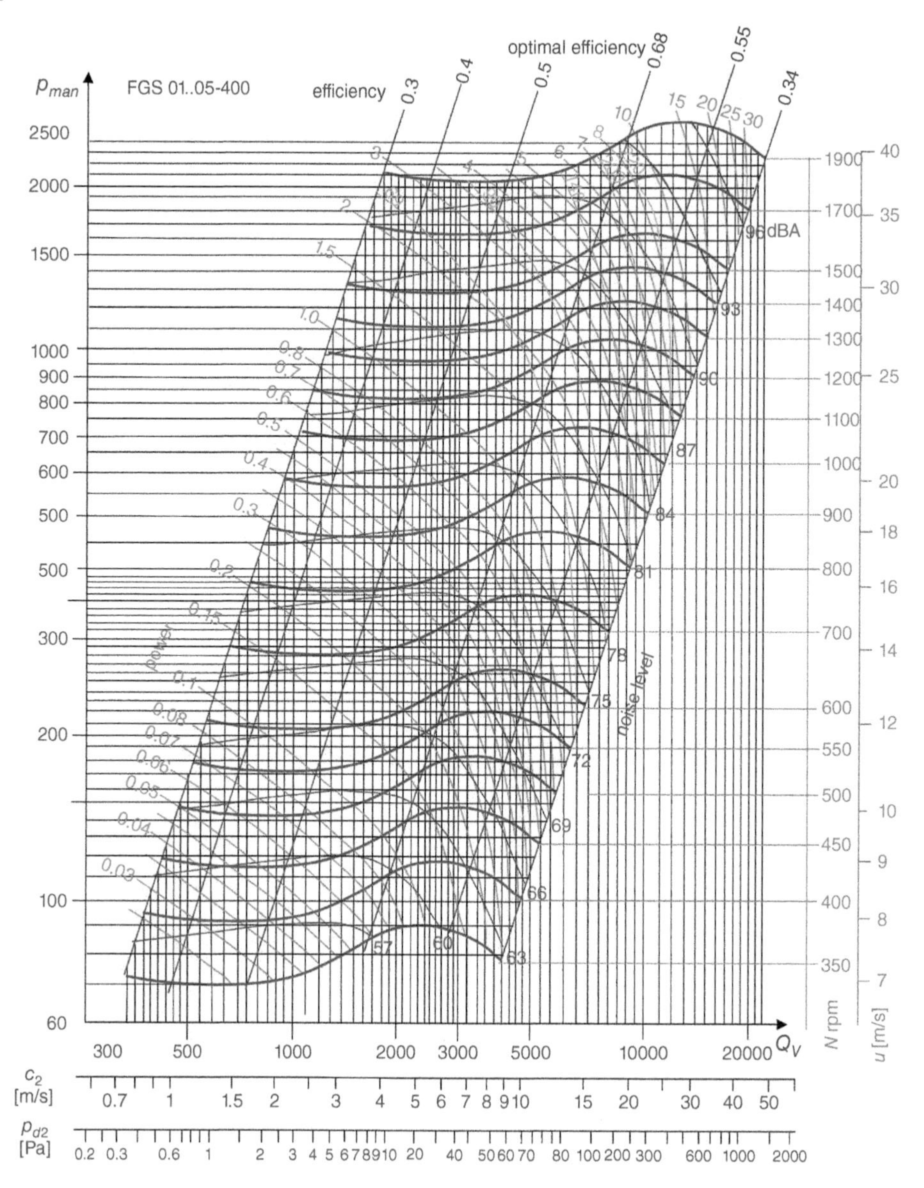

Figure 9.17 FGS-400. Source: Klima.

The fan with backward-curved vanes will have to turn faster. The noise level is then higher. Because the efficiency and the shaft power are in both case the same, the choice is easy: the fan with forward-curved vanes. This one is cheaper and less noisy.

Example 9.2 A flow of 5000 [m^3/h] and a static pressure rise of 929 [Pa] is wanted this time. Take the same fans as above.

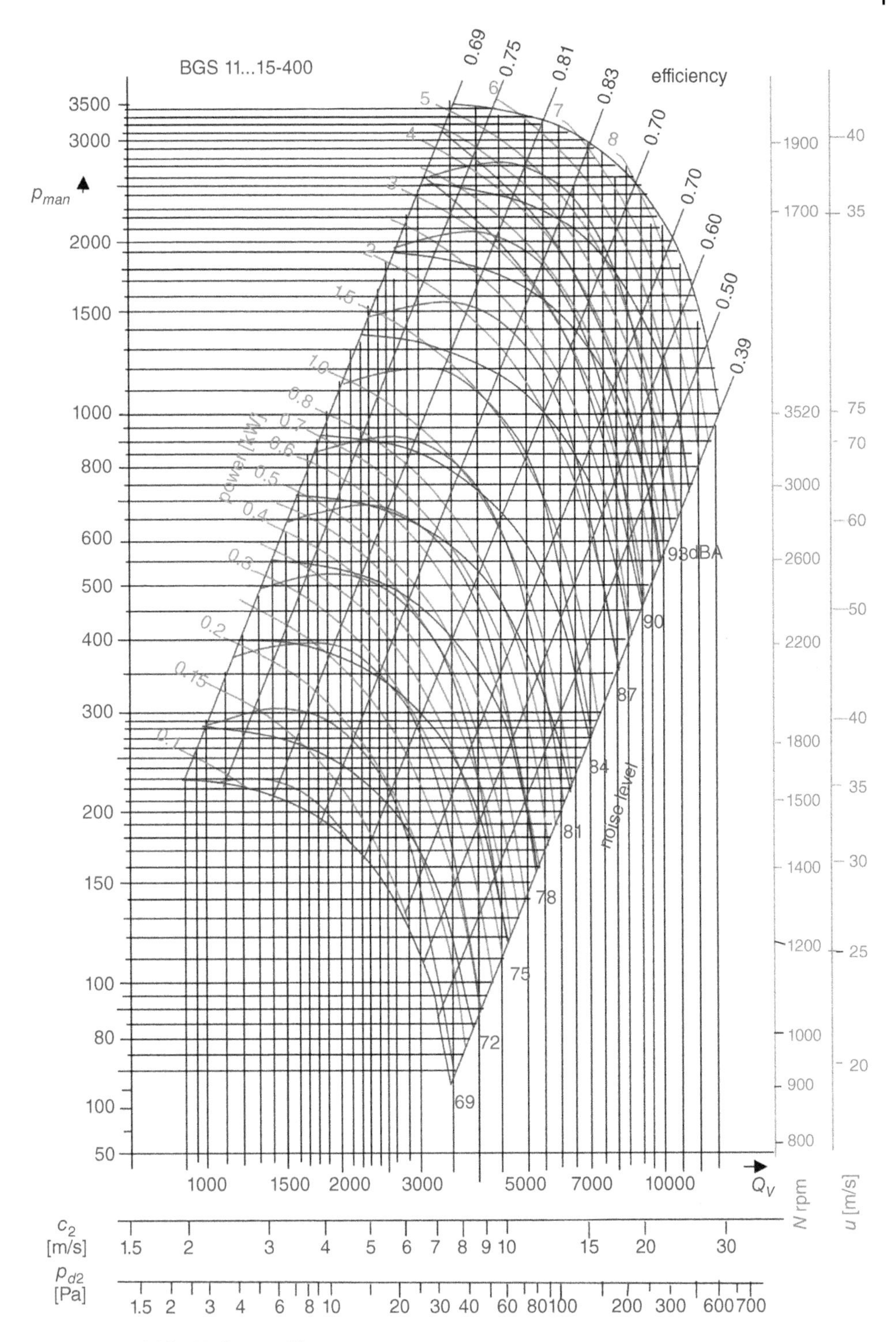

Figure 9.18 BGS-400. Source: Klima.

	Unity	FGS-400	BGS-400
Air flow	m^3/h	5000	5000
Discharge velocity	m/s	11	11
Static pressure difference	Pa	1000	1000
Dynamic pressure	Pa	25.8	45
Rotational speed	rpm	1230	2130
Peripheral velocity	m/s	25.8	45
Shaft power	kW	2	1.7
Noise level	dB(A)	83	85
Efficiency	%	69.4	81.6

The noise level for the fan with backward-curved vanes is here also higher. The efficiency is now significantly higher. Above that a big over-dimensioning of the electrical motor is not necessary. The preference will be the fan with backward-curved vanes.

9.2 Cross-stream Fans

The cross-stream fan (Figures 9.19–9.21) is composed out of a cylindrical rotor with forward-curved vanes.

During the rotation gas streams through a part of the rotor blades toward the inner side of the rotor. This creates turbulence (a vortex) that bends to the other section of the rotor, through the rotor blades and on to the outlet of the fan housing. These fans have the advantage that they work with low gas velocities in comparison with centrifugal fans. High velocities mean high noise.

9.3 Side Channel Fans

Side channel pumps are discussed above. Side channel fans (Figures 9.20–9.24) suck gas and increase the pressure by a series of twirls that are created in a side channel by

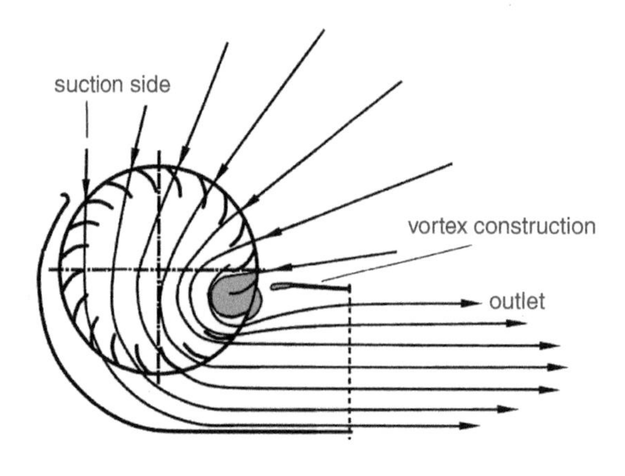

Figure 9.19 Cross-stream fan. Source: Ebm-Papst.

Figure 9.20 Cross-stream fan vortex. Source: Ebm-Papst.

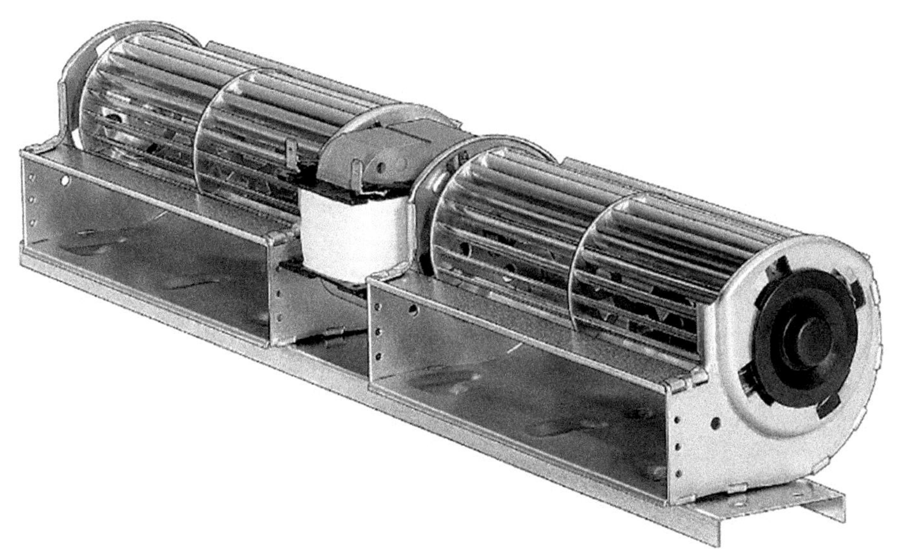

Figure 9.21 Implementation. Source: Ebm-Papst.

centrifugal force. The rotation of the impeller makes the gas rotate in the separate cells during which the created centrifugal force pushes the gas out of the side channel. This creates spiral formations that help to compress the gas again. This leads to an increase in pressure along the pad of the side channel.

The efficiency of such a fan might not be good, but is sufficient for some applications.

Almost 300 [mbar] can be attained with a single-stage implementation, and 600 [mbar] with a two-stage implementation. Flows go up to 500 [m^3/h].

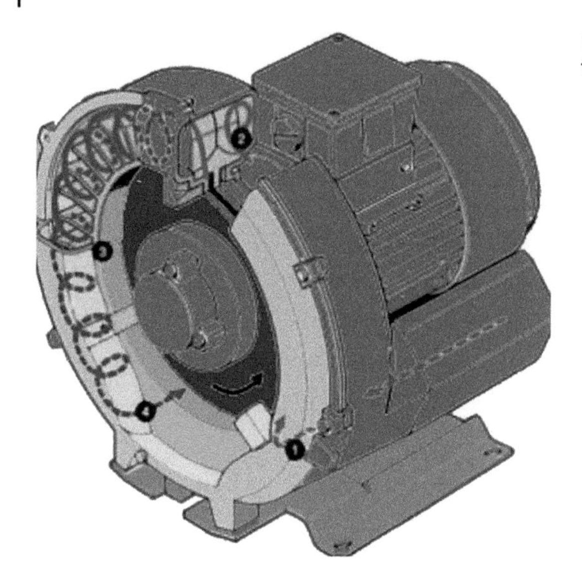

Figure 9.22 Operation of a side channel fan.

Figure 9.23 Sight on rotor. Source: GUT mbH.

9.4 Turbo Fan

In extreme circumstance the peripheral velocity attains 400 [m/s]. For air this gives a pressure rise of ±:

$$p_{man} \cong \rho \cdot u_2^2 \cong (1.3) \cdot 400^2 \cong 2 \cdot 10^5 \text{ [Pa]} \cong 2 \text{ [bar]}$$

Figure 9.24 Implementation side channel fan. Source: GUT mbH.

Here the specific mass ρ was given the value of 1.3 [kg/m^3], which is the magnitude at atmospheric conditions.

The above expression is only an approximation because the specific mass in fans and compressors is not constant, as with pumps.

The *pressure rise* is then about 2 [bar], what with a suction pressure of 1[bar] this gives a *pressure ratio* of 3 [bar]. This all implies of course enormous speeds of 10 000–500 000 [rpm]. The impellers are milled out of one block of material with CNC (computer numerical control) techniques. Implementations are given in Figures 9.25–9.28.

Figure 9.25 Centrifugal fan.

Figure 9.26 Turbocompressor in a diesel engine.

Figure 9.27 Turbocompressor (Source: Atlas Copco)

9.5 Centrifugal Compressor

When higher pressures are wanted multiple impellers are placed on one shaft. Such a multistage compressor (Figures 9.29–9.34) has 6–10 stages and a pressure rise of 6–10 [bar], but there are implications that go as high as 30 [bar].

There are designs with intercooling (Figure 9.35 and 9.36).

The operating speeds by turbomachines are very high compared with other types of compressors. For airplane applications, where weight is of paramount importance, turbocompressors have speeds of 50–100 000 [rpm]. Commercial machines work at 20 000 [rpm] or lower.

Small airplanes or helicopter motors limit the number of stages mostly to two.

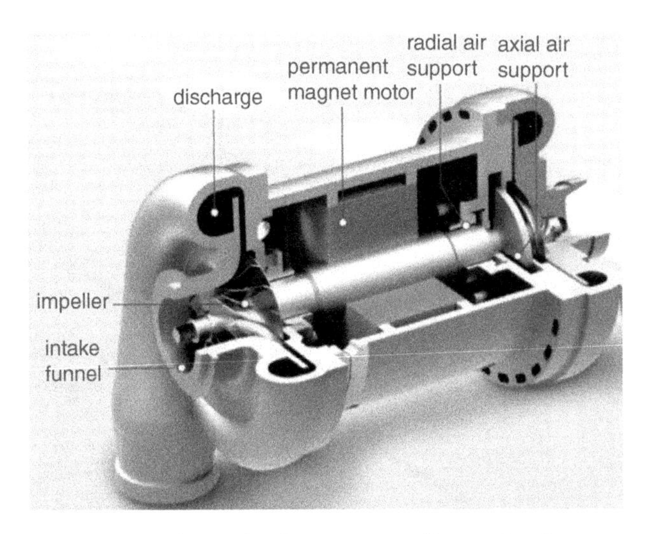

Figure 9.28 High speed turbocompressor (Source: Boge)

Figure 9.29 Rotor of a centrifugal compressor (Source: Hitachi)

9.6 Refrigerant Turbocompressor

Centrifugal compressors are used for very large refrigeration installations (cooling power from 350 [W] to 3.5 [kW] per compressor); see Figure 9.37.

9.7 Axial Fans

9.7.1 General

The operating principle of axial fans accords to that of an axial pump, apart from the use of fluid (Figure 9.38). The fan is applied for high volumetric flows and low

Figure 9.30 Multistage centrifugal with half stator and full rotor. Source: ACEC.

Figure 9.31 Upper sight on a multistage open centrifugal compressor during revision I.

pressures; the achievable pressure ratio is 1.05 to 1.8. Examples are given in Figure 9.39 and 9.40.

The regulation of the pressure occurs mostly by variable voltage, often in multiple steps, in modern systems with an electronic drive.

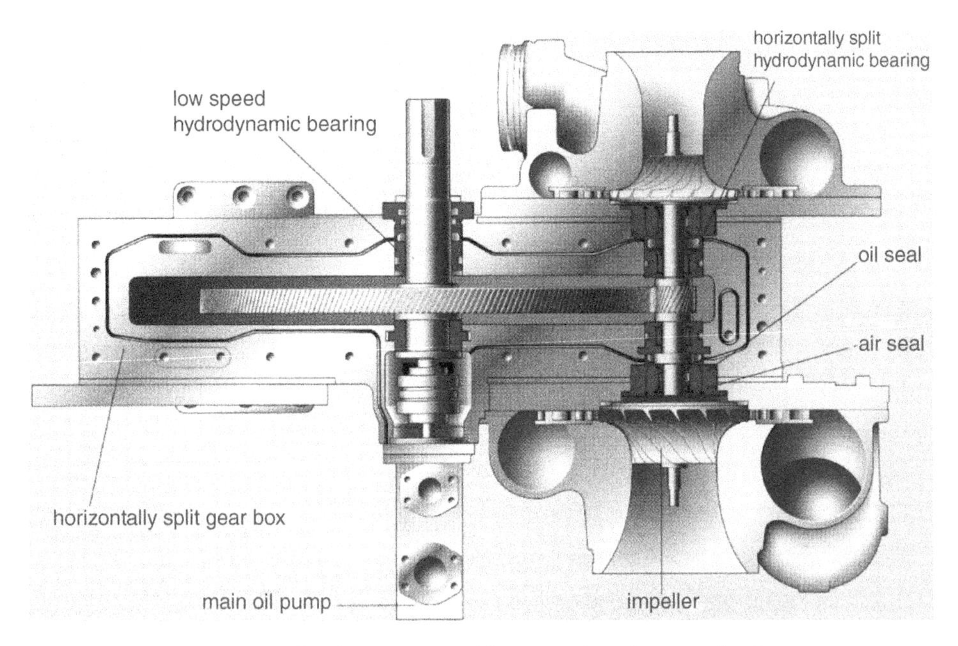

Figure 9.32 Two stage centrifugal compressor with intercooling. Source: Atlas Copco.

Figure 9.33 Implementation three stage compressor. Source: Atlas Copco.

Regulation of pressure and flow often occurs with variable voltage, often in more steps, in modern systems with an electronic device.

Figure 9.41 shows the characteristic of such a fan at various values of the applied voltage.

For high pressure fans sometimes half-axial fans are chosen (Figure 9.42). These are used with a guide wheel that is placed after the rotor (Figures 9.42 and 9.43, yellow).

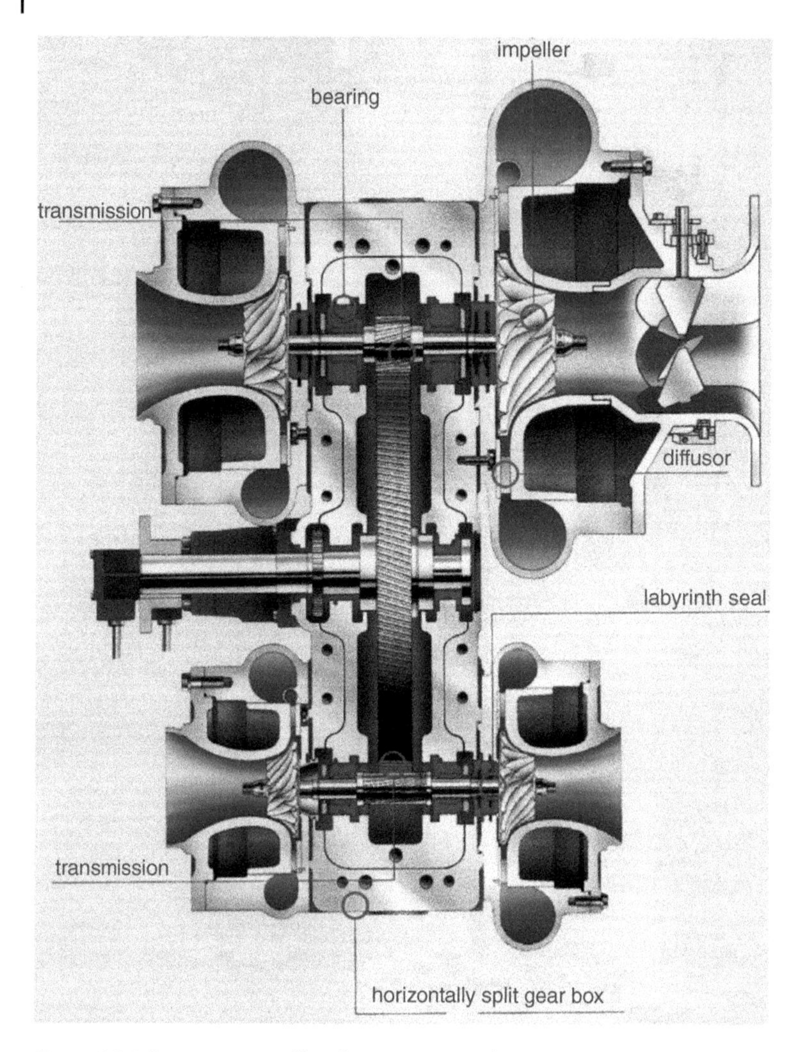

Figure 9.34 Four stage centrifugal compressor with intercooling. Source: Atlas Copco.

These are two-stage implementations. The possibilities are:

- To work with the first fan and shut off the second one.
- To work with the second one, which is smaller, and turn off the first one.
- Work with the two fans at the same time.

Figure 9.44 represents the characteristic of such a two-stage fan.

9.7.2 Reaction Degree Axial Fan

9.7.2.1 Definition

According to thermodynamics, a fan is an *open system*. This is a device where mass flows in, undergoes a process, and flows out.

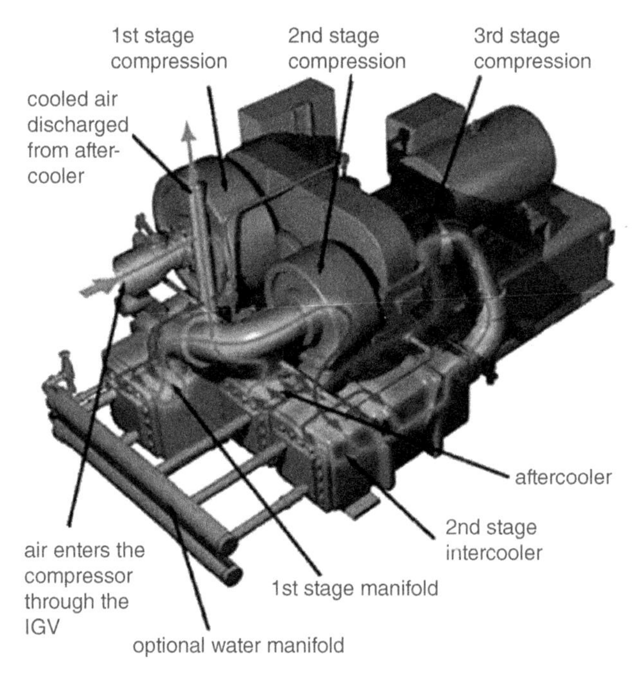

Figure 9.35 Isothermal compression with a centrifugal compressor. Source: Cooper.

Figure 9.36 Flow of compressed air through coolers at a three stage implementation. Source: Atlas Copco.

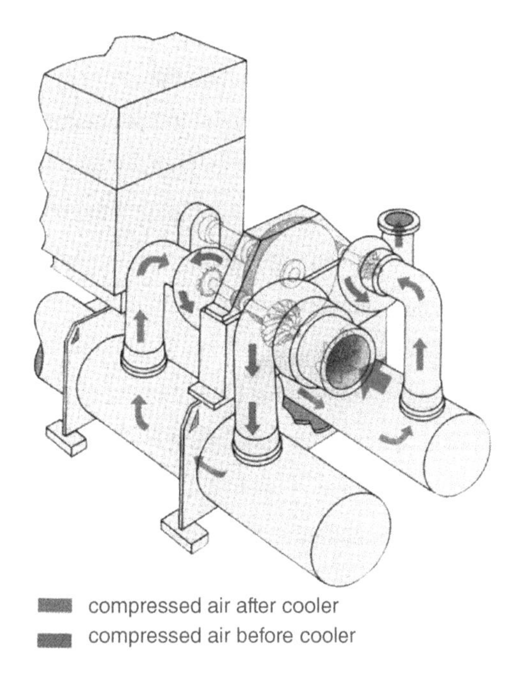

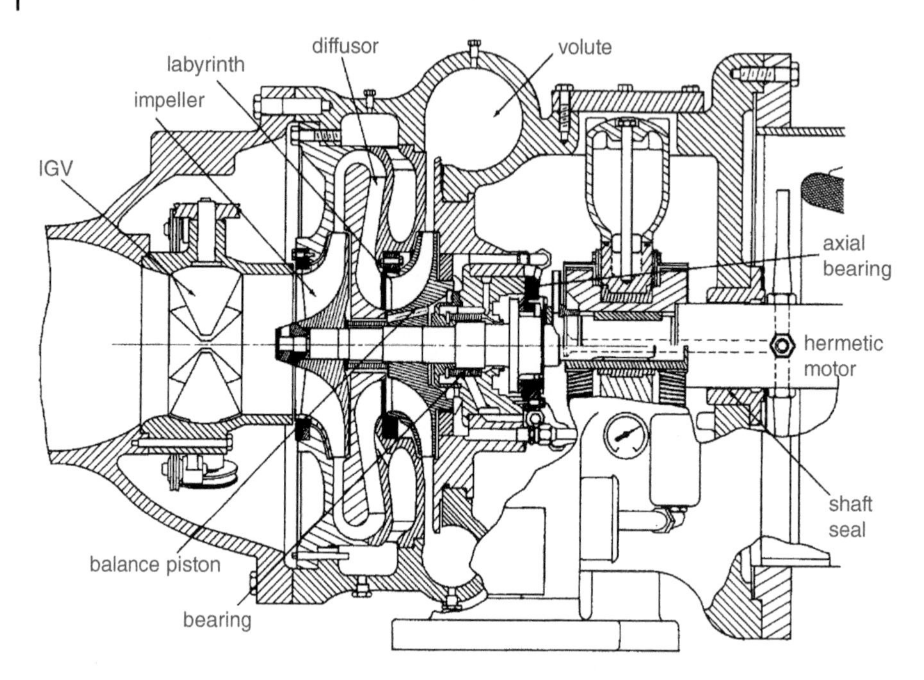

Figure 9.37 Centrifugal refrigeration hermetic compressor.

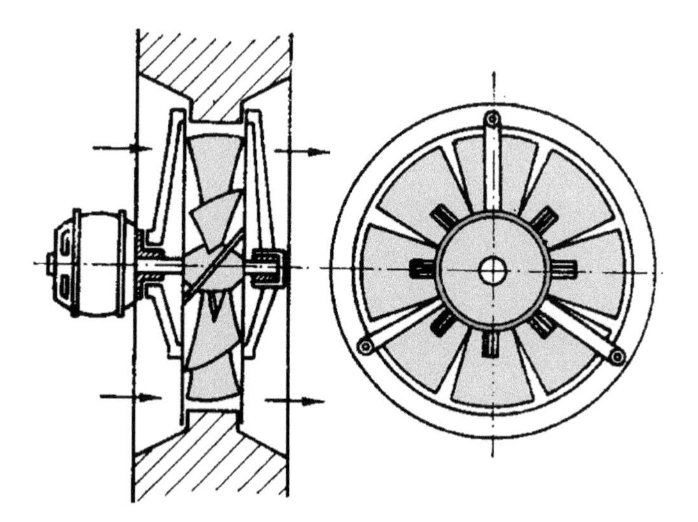

Figure 9.38 Principle of an axial fan.

State 1 points on the inlet of the device, state 2 on the outlet. Neglect the importance of the potential energy of the gravitation field. The law of conservation of energy for such an open system, for the change of state 1–2 is then:

$$(h_2 - h_1) + \frac{1}{2} \cdot (c_2^2 - c_1^2) = w_{t12} + q_{12}$$

Figure 9.39 Axial fan. Source: Systemair.

Figure 9.40 Axial fan. Source: Systemair.

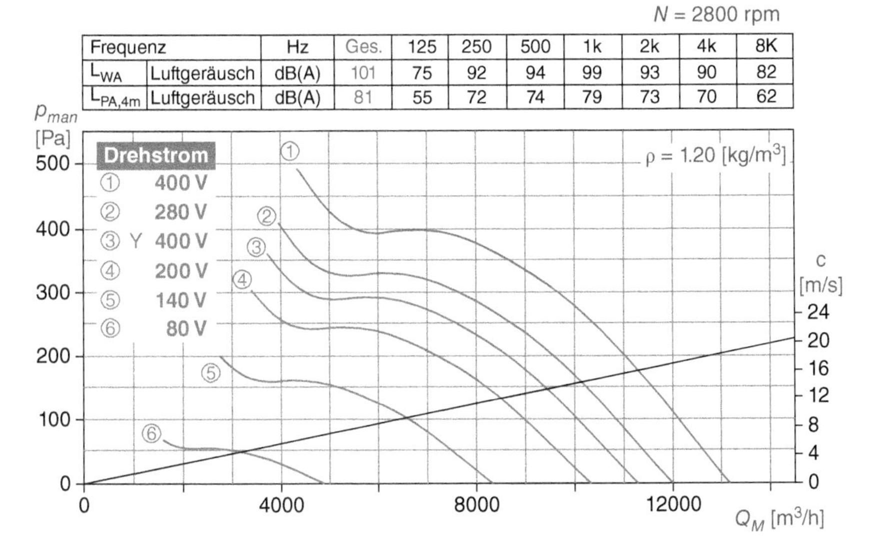

Frequenz		Hz	Ges.	125	250	500	1k	2k	4k	8K
L_{WA}	Luftgeräusch	dB(A)	101	75	92	94	99	93	90	82
$L_{PA,4m}$	Luftgeräusch	dB(A)	81	55	72	74	79	73	70	62

Figure 9.41 Characteristic of a fan. Source: Helios.

Figure 9.42 Two-stage axial fan. Source: Helios.

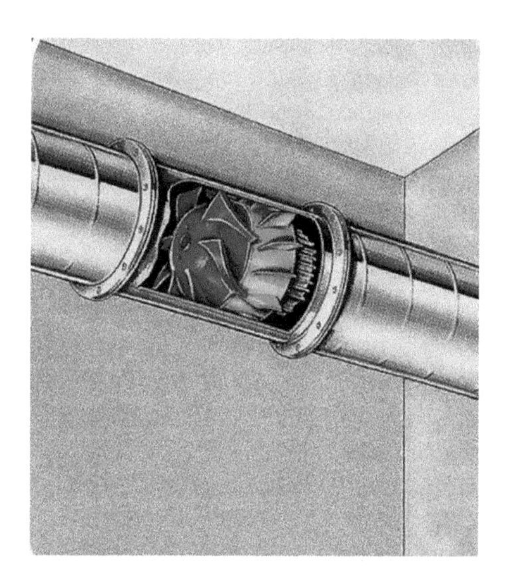

Figure 9.43 Mounting. Source: Helios.

where:

h_1 : specific enthalpy of state 1 [J/kg]

h_2 : specific enthalpy of state 2 [J/kg]

c_1 : velocity at entrance, at state 1 [m/s]

c_2 : velocity at discharge, at state 2 [m/s]

w_{t12} : specific technical work delivered by the surrounding of the device (work of the shaft of the device) [J/kg]

q_{12} : specific heat provided by the surrounding [J/kg].

Note that this formulae is valid with and without friction!

Z-VAR.. 280/4

N = 1450 rpm

Frequenz		Hz	Ges.	125	250	500	1k	2k	4k	8K
L_{WA}	Luftgeräusch	dB(A)	77	56	68	70	73	71	63	53

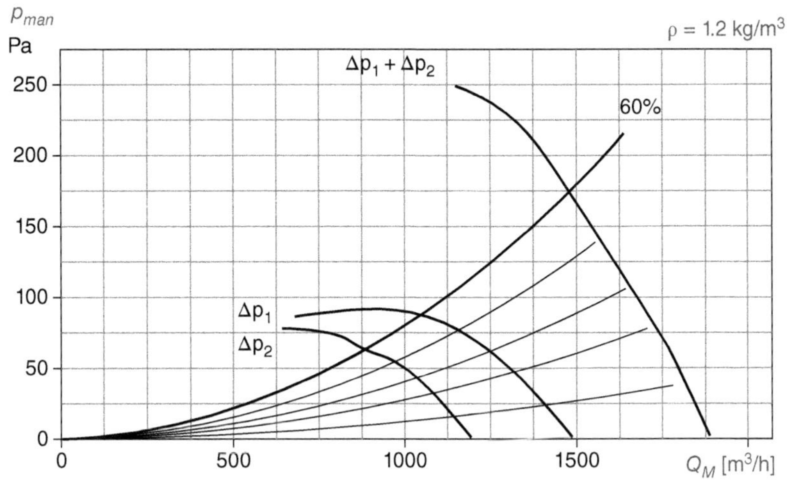

Figure 9.44 Characteristic of a two-stages axial fan. Source: Helios.

What is *enthalpy*? In ancient times, when the animals could speak, the word *enthalpy* was not used. h was written i and was called the *heat content* of a substance per kilogram. Better to say that the specific enthalpy of a substance is the heat applied to it from the beginning state, where nothing moves, at constant pressure, to the present state.

Interesting formulae:

$$(h_2 - h_1) = q_{12} + \int_1^2 v \cdot dp + w_{f,12}$$

Further, for an isobaric change from state 1 to 2 for a gas (always):

$$(h_2 - h_1) = c_p \cdot (T_2 - T_2)$$

where c_p is the mean value in the interval $(T_2 - T_1)$, because it is, in fact, dependent on the temperature. Mostly, one considers it constant in a good approximation of 10–15%.

If we pose that the change of state is adiabatic, which is true because the velocities involved are so high that the gas has no time to exchange heat with the surrounding, thus $q_{12} = 0$:

$$(h_2 - h_1) + \frac{1}{2} \cdot (c_2^2 - c_1^2) = w_{t12} \tag{9.1}$$

Consider now the *rotor channels*.

From the standpoint of the traveling observer in the channel, with velocity w, because no centrifugal acts on the flow (the stream is axial, the gas cannot move along the radius):

$$(h_2 - h_1) + \frac{1}{2} \cdot (w_2^2 - w_1^2) = w_{t12}$$

When the observer sits in the vane channel all he sees is a simple flow process. So, seen relatively, no technical work is exercised to or from the gas.

$$(h_2 - h_1) + \frac{1}{2} \cdot (w_2^2 - w_1^2) = 0$$

Interpret this now. In the rotating channel the passage is diffusing. This means that the relative velocity w decreases. Thereby a static pressure strike (better: an enthalpy strike) takes place. The specific energy that is involved is equal to:

$$(h_2 - h_1) = (w_1^2 - w_2^2) \tag{9.2}$$

The *reaction degree* Γ is defined as the ratio of the delivered *static* energy in the fan to the *total energy*:

$$\Gamma = \frac{(h_2 - h_1)}{w_{t12}}$$

Making use of Equations (9.1) and (9.2):

$$\Gamma = \frac{w_1^2 - w_2^2}{(w_1^2 - w_2^2) + (c_2^2 - c_1^2)}$$

This way the reaction degree is expressed with only velocities as variables, and that is handy because problems can easily be resolved with the velocity triangles.

CASE 1: $\Gamma < 1$ (Figure 9.45)

For the reaction degree:

$$\Gamma = \frac{w_1^2 - w_2^2}{(w_1^2 - w_2^2) + (c_2^2 - c_1^2)}$$

To be smaller than 1 is necessary that: $c_2 > c_1$

Presume that the gas enters the fan axially, then the velocity triangles are as those shown in Figure 9.45.

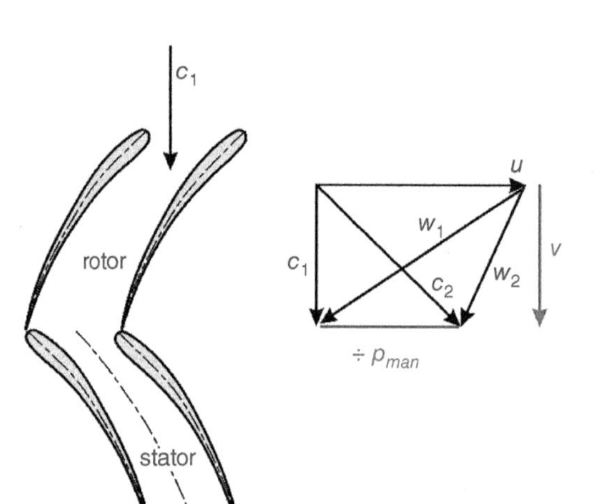

Figure 9.45 $\Gamma < 1$.

The *total* pressure rise (see Section 3.2.3, in Chapter 3) is:

$$p_{man} = \rho \cdot u \cdot |\overline{w_1} - \overline{w_2}| = \rho \cdot u \cdot c_{2u}$$

where c_{2u} is the projection on the vector $\overline{c_2}$ on the vector $\overline{u}$ (see Section 3.2.3). Because $\overline{c_2}$ does not exit the fan axially a guide wheel (that does not rotate – be careful with the name!) is placed that reorients the velocity c_2 in the axial direction.

CASE 2: $\Gamma = 1$ (Figure 9.46)

In that case: $c_2 = c_1$, from where it follows that only static pressure is created in the impeller. The $\overline{c}$ vectors cannot be axial, because then they would coincide and the $\overline{w}$ vectors. But if both the $\overline{w}$ vectors are equal there is no static pressure rise in the impeller (see theory axial pumps). Thus, the $\overline{c}$ vectors are oblique, and that means there should be a guide wheel preceding the rotor.

CASE 3: $\Gamma > 1$ (Figure 9.47)

In this case c_1 must be greater than c_2. Take as an example that c_2 stands perpendicular on u, and thus exits axially. This case gives directly the smallest value for c_2. This implies

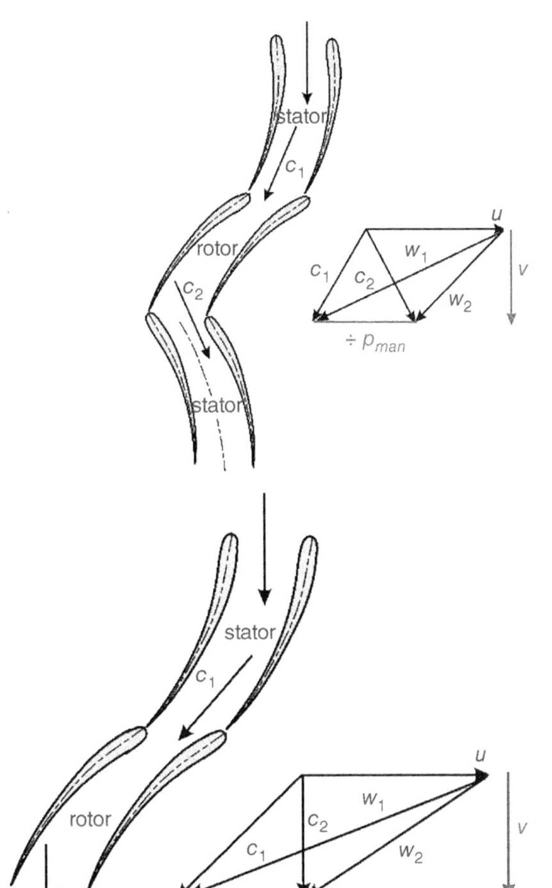

Figure 9.46 $\Gamma = 1$.

Figure 9.47 $\Gamma > 1$.

that then also c_1 cannot be axially directed (it must be greater than the perpendicular one). From there it follows that, if out of symmetry considerations it is desired that the gas enters and exits axially, it is an obligation to place a guide wheel before the rotor, or else the velocity c_1 would be principally axial.

In Figure 9.47 this situation is drawn, where, as in the preceding paragraphs, the same values for u, v, and p_{man} are retained.

CASE 4: $\Gamma = 0.5$

In this case (Figure 9.48): $w_1 = c_2$ and $c_1 = w_2$. Further investigation shows that $\alpha_2 = \beta_1$ and $\alpha_1 = \beta_2$. This holds that the impeller vanes have the same profile as the guide wheel vanes, but they are each other's mirrors.

Comparing all these different cases one can conclude that in the case $\Gamma > 1$ high values of the relative velocities w occur. This leads to high friction losses. Above that an extra guide wheel is needed.

More interesting then are the cases 1 and 4. But in case 4 an extra guide wheel is necessary; the first case is chosen for simple fans.

However, for multistage compressors (discussed later) preference is given to case 4. Here the friction losses are minimal. And the same vane profile scan be used for all vanes, as well as impeller vanes as guide wheel vanes. This last argument is, however, relative, because vanes are designed with CFD (computational fluid dynamics) programs and the complex vanes that result from them are automatically milled with five-axis CNC machines via CAD-CAM (computer-aided design and computer-aided manufacturing) programs.

9.7.3 Contrarotating Axial Fans

An implementation of two axial fans that are placed right after each other, but rotating in different directions, are perhaps one of the most attractive designs. This way (Figure 9.49) a guide wheel is redundant, and the total pressure is the result of a two-stage compression. The only drawback is the drive, where two rotors have to rotate in opposite directions. In Figure 9.49 it can be seen that the incoming air enters the fan axially with

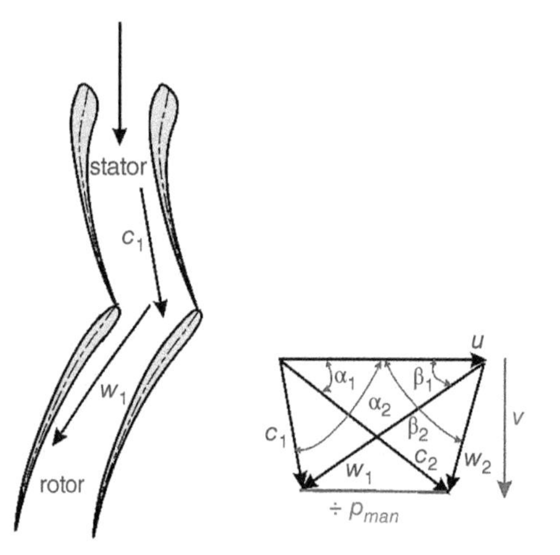

Figure 9.48 $\Gamma = 0.5$.

Figure 9.49 Velocity diagram contrarotating fans.

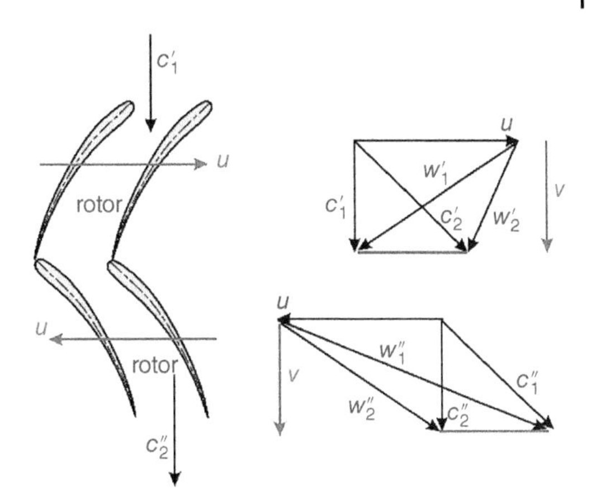

velocity c_1' and the outgoing velocity c_2' is equal to the incoming velocity c_1'' of the second rotor.

Fans of this type are rare. In the aviation sector, especially propeller airplanes (Figures 9.50–9.52), prototypes of contrarotating rotors were build. Also some attack helicopters (e.g. Black Hawk) make use of contrarotating rotors for the tail rotor, in order to produce significantly less noise in order to surprise the enemy. The sound waves of both rotors are in contra phase (180° displaced) and destroy each other.

Figure 9.50 Lockheed XFV-1.

Figure 9.51 Lockheed XFV-1.

Figure 9.52 Antonov AN-70.

9.7.4 Variable Pitch Axial Fan

In some cases, use is made of fans with variable vane angles to regulate the flow. With some versions the vane angle can only be varied at standstill; others can do it continuously.

The flow increases or decreases with the vane angle, but the efficiency is only optimal for one angle. Alternatively, for the use of vanes with variable pitch there are frequency invertors, but also here there is only one determined optimum for a combination vane angle/peripheral velocity, thus when the velocity of the rotor changes, the efficiency decreases. Frequency invertors are, however, more expensive.

Figure 9.53 shows a fan with adjustable vane angles, at a standstill. A typical example of the characteristics of such a fan is shown in Figure 9.54. Figures 9.55–9.60 show some implementations.

Figure 9.53 Fan with variable pitch. Source: Howden.

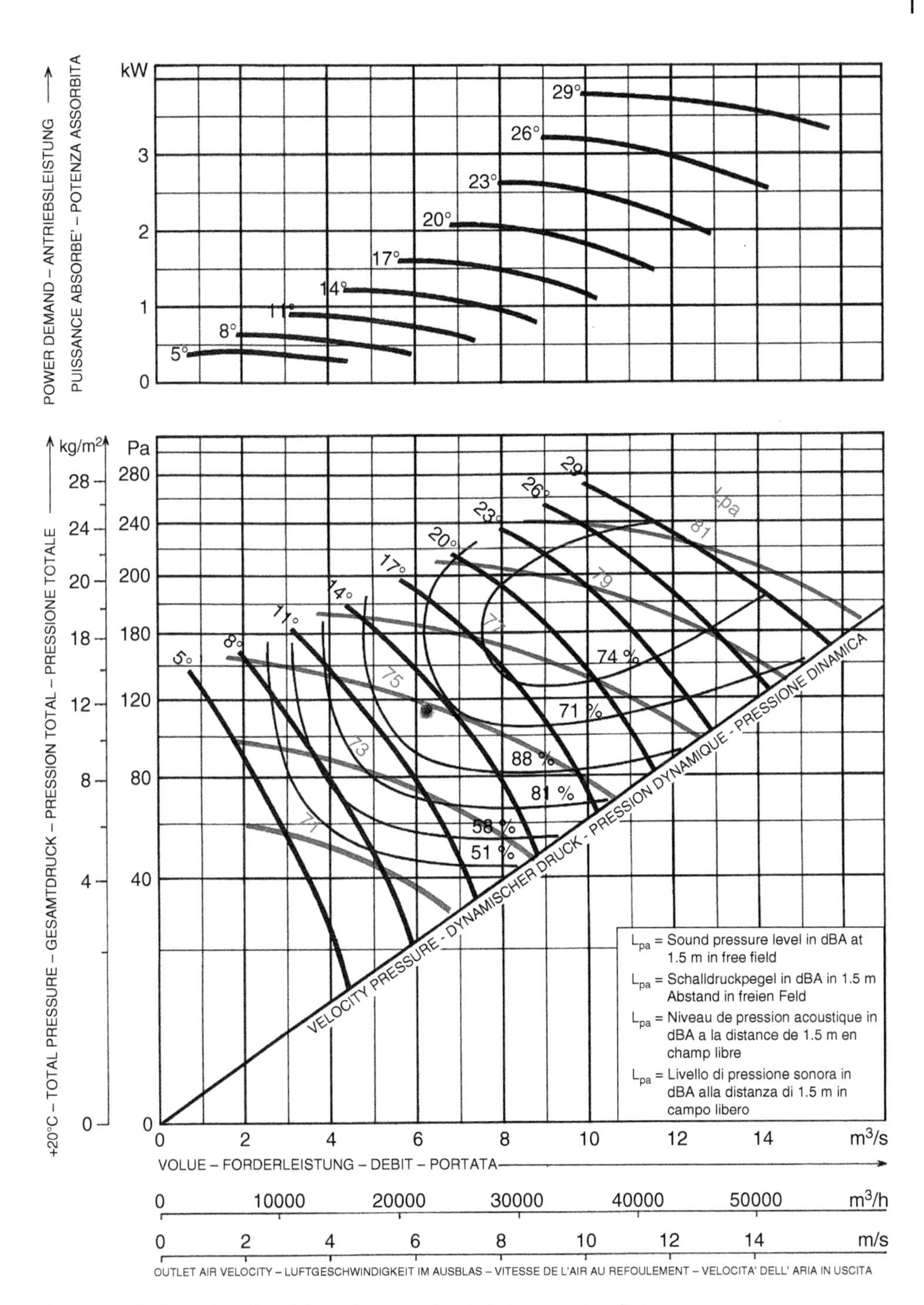

Figure 9.54 Characteristics of fans with variable pitch. Source: Howden

Figure 9.55 Variable vane. Source: Howden.

Figure 9.56 Trim mechanism (hydraulic). Source: Howden.

9.8 Axial Compressor

The pressure rise of a single stage axial turbocompressor is limited and amounts to 0.05–0.8 [bar], depending on the version. If higher pressures are desired, multiple stages

Figure 9.57 Hyper fan for refrigeration installation. Source: Howden.

Figure 9.58 Cooling tower. Source: Howden.

are to be placed on one shaft. After each rotor comes a stator (guide wheel). The job of the rotor is to decrease the relative velocity of the gas in order to increase the pressure. The stator has the following functions (Figures 9.61 and 9.62):

- Change the flow direction in order to lead the gas to the entrance of the following rotor.

Figure 9.59 Transport of vanes. Source: Howden.

Figure 9.60 Curvature of vanes. Source: Howden.

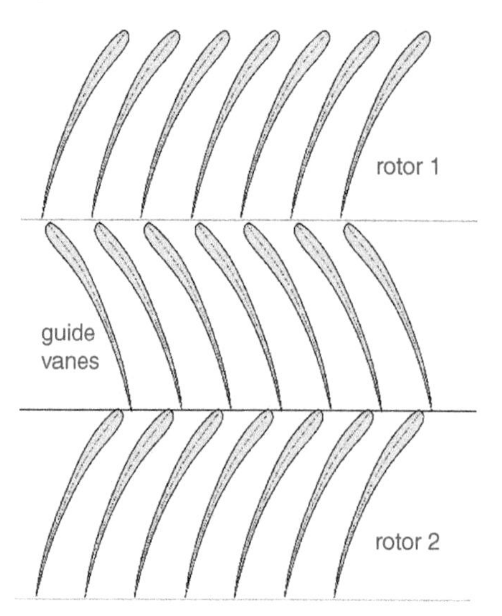

Figure 9.61 Vane channel axial compressor.

Figure 9.62 Axial compressor. Source: Les Turboréacteurs, Eyrolles, Paris.

- Decrease the absolute velocity of the gas at the discharge side of the rotor so that the static pressure increases. Therefore the channel passage has to be diverging.

9.9 Calculation Example

Consider a multistage compressor from which in Figure 9.63 two stages are drawn, together with the used symbols for angles and velocities. The variables that have a relation with the second stage have one accent, those in the third stages two accents, and so on… As we saw in the previous section, it is interesting to work with a reaction degree $\Gamma = 0.5$. Do not forget that the changes of state are supposed to be adiabatic, and this is practically true.

Take a so-called periodic compressor with reaction. The peripheral velocity $u = 250$ [m/s] and the vane angles are typically $\beta_1 = 30°$ and $\beta_2 = 50°$ (Figure 9.64).

Consider a periodic compressor, where:

$$\overline{c_1'} = \overline{c_1''} = \overline{c_1'''} = \ldots \text{ and } \alpha_1' = \alpha_1'' = \alpha_1''' = \ldots$$

With $\boldsymbol{\Gamma} = 0.5$ the following expressions are valid:

$$\alpha_1' = \beta_2' \; en \; \alpha_2' = \beta_1'$$

The same goes for the second rotor:

$$\alpha_1'' = \beta_2'' \; en \; \alpha_2'' = \beta_1''$$

And for a periodic compressor:

$$\beta_1' = \beta_1'' \; en \; \beta_2'' = \beta_2''$$

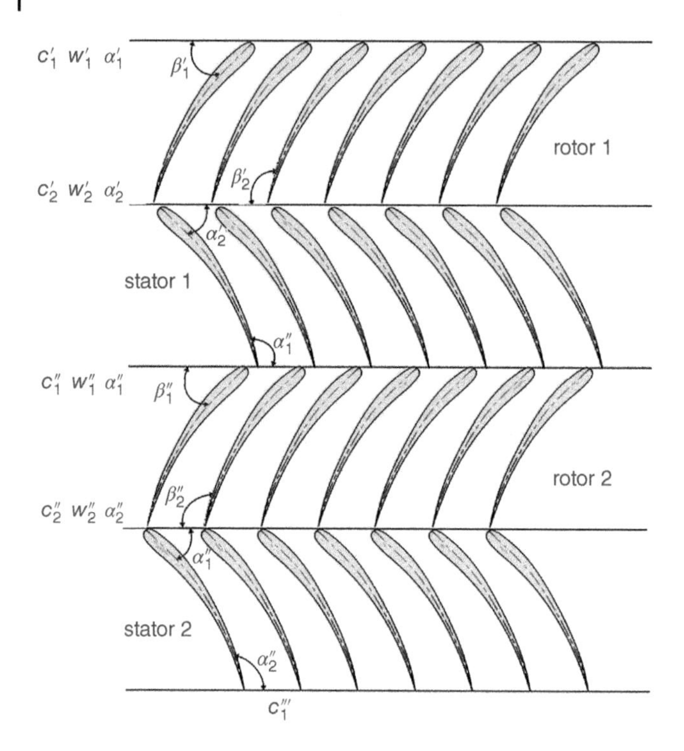

Figure 9.63 Multistage axial compressor.

In such case all profiles of the vanes are equal, because the velocity triangles repeat themselves in each stage. So:

$$c_1'' = c_1'''$$

$$w_1' = w_1'' \text{ and } w_2' = w_2''$$

What are the enthalpy drops in the various rotors and stators?

- In rotor 1: $(h_2' - h_1') = \frac{1}{2} \cdot (w_1'^2 - w_2'^2) = \frac{1}{2} \cdot (c_2'^2 - c_1'^2)$ because $\boldsymbol{\Gamma} = 0.5$ see definition of $\boldsymbol{\Gamma}$)
- In stator 1: $(h_1'' - h_2') = \frac{1}{2} \cdot (c_2'^2 - c_1''^2) = \frac{1}{2} \cdot (c_2'^2 - c_1'^2)$
- In rotor 2: $(h_2'' - h_1'') = \frac{1}{2} \cdot (w_1''^2 - w_2''^2) = \frac{1}{2} \cdot (c_2''^2 - c_1''^2)$ because $\Gamma = 0.5$
- And: $(h_1'' - h_1') = \frac{1}{2} \cdot (c_2''^2 - c_1''^2) = \frac{1}{2} \cdot (c_2''^2 - c_1'^2) = \frac{1}{2} \cdot (w_1''^2 - w_2''^2) = \frac{1}{2} \cdot (w_1'^2 - w_2'^2)$

$$= \frac{1}{2} \cdot (c_2'^2 - c_1'^2)$$

- So: $\frac{1}{2} \cdot (c_2'^2 - c_1'^2) = \frac{1}{2} \cdot (c_2''^2 - c_1''^2)$ from where: $c_2'^2 = c_2''^2$
- In stator 2: $(h_1' - h_2'') = \frac{1}{2} \cdot (c_2''^2 - c_1'''^2) = \frac{1}{2} \cdot (c_2''^2 - c_1'^2) = \frac{1}{2} \cdot (c_2'^2 - c_1'^2)$

The velocity diagram for a pressure stage of such a compressor is shown in Figure 9.64: From the sine rule:

$$\frac{w_1}{\sin 50°} = \frac{c_1}{\sin 30°} = \frac{u}{\sin(180° - 80°)} = \frac{u}{\sin 100°}$$

Figure 9.64 Velocity triangles example.

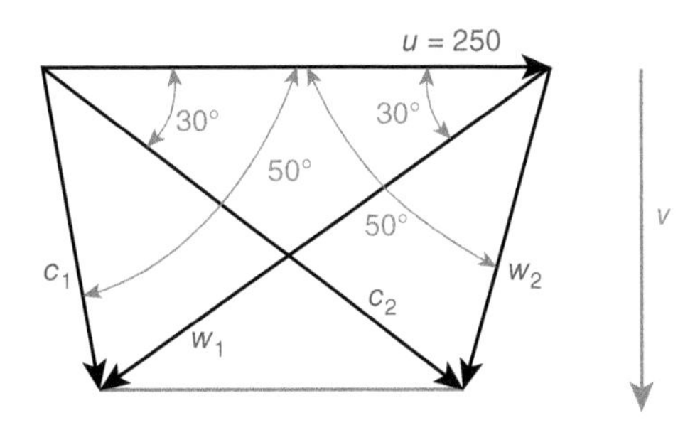

With $c_1 = w_2$.
From where:

$$w_1 = 194\,[\text{m/s}] \quad w_2 = 127\,[\text{m/s}] \quad (h_2 - h_1) = \frac{1}{2}(w_1^2 - w_2^2) = 10853\,[\text{J/kg}]$$

The *flow component* (Greek letter): $\nu = w_1 \cdot \sin 30° = 97$ [m/s].

Note: in accordance with logic the axial component of the absolute velocity would best be noted c_{ax}, as is the case with centrifugal machines for the component of c in radial direction, c_{2r}. However, it is so that in most handbooks the Greek notation ν is employed, and therefore this symbol is used in this book. Don't, however, confuse it with the v for *specific volume*, because the two letters are very similar. Anyway, when we talk about the axial component we don't mention anything about the properties of a gas, so there is no real problem.

Because the change of state is adiabatic and friction is negligible, which is not at all true in practice with these velocities, the thermodynamically expression reduces to:

$$(h_2 - h_1) = \int_1^2 v \cdot dp + q_{12} + w_{f,12}$$

$$(h_2 - h_1) = \int_1^2 v \cdot dp$$

For a small pressure rise in one stage where the substance is considered incompressible (an approximation, what else do you propose?), thus the specific volume $(h_2 - h_1) \cong v \cdot (p_2 - p_1) \cong v \cdot p_1 \cdot \left(\frac{p_2 - p_1}{p_1}\right)$.

Using the ideal gas law:

$$p_1 \cdot v = \frac{R}{M} \cdot T_1$$

then:

$$(h_2 - h_1) \cong \frac{R}{M} \cdot T_1 \cdot \left(\frac{p_2 - p_1}{p_1}\right)$$

From where:

$$\frac{p_2 - p_1}{p_1} = \frac{M}{R \cdot T_1} \cdot (h_2 - h_1)$$

The total static pressure rise Δp in a stage – this is a rotor plus a stator – is $2 \cdot (p_2 - p_1)$.

And so, finally, one arrives at:

$$\left(\frac{\Delta p}{p_1}\right) = \frac{2 \cdot (h_2 - h_1) \cdot M}{R \cdot T_1}$$

Consider air on inlet, with $M = 29$ [kg/kmol], the universal gas constant $R = 8315$ [kg/kmole.K], and an inlet temperature of 300 [K], then the relative pressure rise is:

$$\frac{\Delta p}{p_1} = \frac{2 \cdot 10853 \cdot 29}{8315 \cdot 300} = 0.252$$

With a start pressure of $p_1 = 1$ [bar], $p_2 = p_1 + \Delta p = 1.252$ [bar]
The pressure ratio for the first stage is then in a first approximation: 1.252.
For a ten-stage axial compressor the pressure ratio is:

$$(1.252)^{10} = 9.46$$

9.10 Surge Limit

Consider a compressor that presses gas to a pressure vessel via a discharge line (Figure 9.65). Consumers use compressed gas delivered by the vessel. The operating point on the characteristic is represented by point P. When consumption decreases, in the worst-case scenario becomes zero, then the compressor will deliver a higher flow than the press vessel delivers to its consumers. This means that the pressure in the vessel rises. The operating point shift to point P′. If the consumption stays too low the point P′ will shift toward point P″ and so on., until point P⁗ is reached. In the supposition that the consummation still stays lower than that of the compressor the following situation occurs: the pressure in the vessel keeps on increasing but the compressor is not able to produce this. The flow of the compressor then becomes zero and the operating point of the compressor shifts on its curve to point R. The operating point on the system curve is now S. The pressure in S is, however, higher than the pressure in R. In other words, a flow back from vessel to compressor takes place!

And so the pressure in the vessel decreases. Above that this pressure might also decrease because of consumption. At a certain moment the compressor is again able to deliver so that the operating point is again P‴.

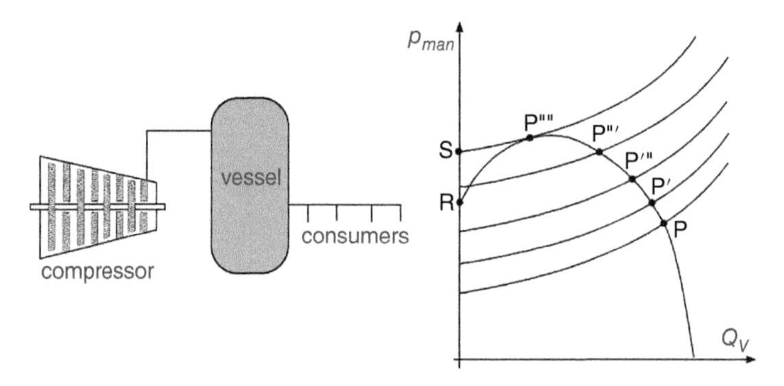

Figure 9.65 Surge limit.

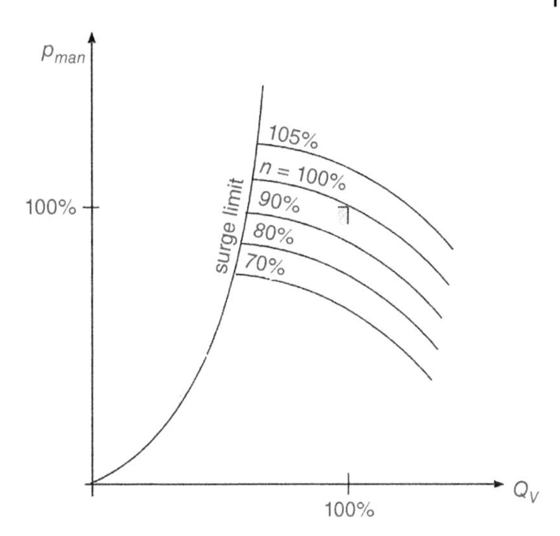

Figure 9.66 Characteristic.

This cycle can be repeated many times. The phenomenon is called *pumping the compressor*, or *surge*. Of course, this little game is harmful for the compressor and therefore care must be taken that the operating point is always situated on the right side of the compressor curve. This, in fact, means that the flow of the compressor must be big enough. Eventually the extra part of it must be blown off, as itis in airplane compressors.

The surge limit is situated between 45 and 90% of the nominal flow depending on the design of vanes, the number of stages, the gas, the speed, etc. The greater the number of stages, the smaller the stable area.

In Figure 9.66 the surge limit in function of the speed is represented.

9.11 Choke Limit (Stonewall Point)

9.11.1 Introducing Nozzles

9.11.1.1 Calculation of the Discharge Speed

Consider a tube which is first converging, then diverging (Figure 9.67). Enter the tube, which at its entry point is well rounded, so that the velocity there can be supposed to be zero. What is the discharge velocity?

The change of state is adiabatic and friction is negligible.

For such a state of change the law of conservation of energy is:

$$h + \frac{c^2}{2} = h_1 + \frac{c_1^2}{2}$$

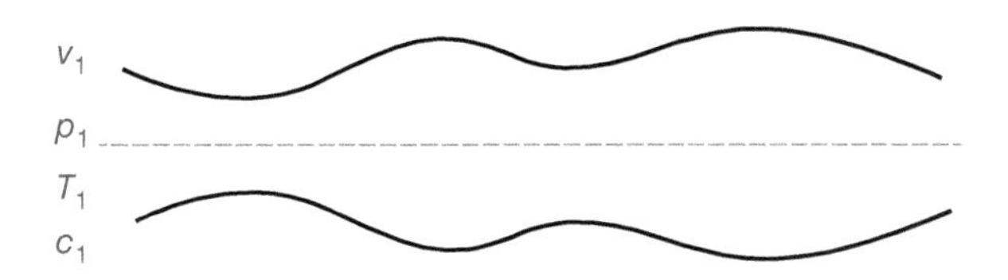

Figure 9.67 General case of a tube

where:

h : specific enthalpy on a random place in the tube

c : its velocity there

h_1 and c_1: the variables at the inlet.

The velocity on a random place can be written as:

$$c^2 = c_1^2 + 2 \cdot (h_1 - h)$$

For an ideal gas:

$$c^2 = c_1^2 + 2 \cdot c_p \cdot (T - T_1)$$

$$c^2 = c_1^2 + 2 \cdot c_p \cdot T_1 \cdot \left(1 - \frac{T}{T_1}\right)$$

Making use of the laws of Poisson:

$$\frac{T}{T_1} = \left(\frac{p}{p_1}\right)^{\frac{\gamma-1}{\gamma}}$$

So:

$$c^2 = c_1^2 + 2 \cdot c_p \cdot T_1 \cdot \left(1 - \left(\frac{p}{p_1}\right)^{\frac{\gamma-1}{\gamma}}\right)$$

With the ideal gas law: $p_1 \cdot v_1 = \frac{R}{M} \cdot T_1$ the variable T can be eliminated:

$$c^2 = c_1^2 + 2 \cdot c_p \cdot \frac{p_1 \cdot v_1}{R} \cdot M \cdot \left(1 - \left(\frac{p}{p_1}\right)^{\frac{\gamma-1}{\gamma}}\right)$$

It is common that instead of c_p the isentropic factor γ is used:

$$c_p = \frac{\gamma \cdot R}{(\gamma - 1) \cdot M}$$

$$c^2 = c_1^2 + \frac{2 \cdot \gamma}{\gamma - 1} \cdot p_1 \cdot v_1 \cdot \left(1 - \left(\frac{p}{p_1}\right)^{\frac{\gamma-1}{\gamma}}\right)$$

If the counterpressure p decreases, the velocity c increases.
It may be supposed that the incoming velocity c_1 may be negligible.
So finally:

$$c^2 = \frac{2 \cdot \gamma}{\gamma - 1} \cdot p_1 \cdot v_1 \cdot \left(1 - \left(\frac{p}{p_1}\right)^{\frac{\gamma-1}{\gamma}}\right)$$

9.11.1.2 Calculation of the Flow

In this section the mass flow Q_M will be expressed as a function of the ratio $\frac{p}{p_1}$.

$$Q_M = \rho \cdot c \cdot A = \frac{c \cdot A}{v}$$

Or, replacing c with its value:

$$Q_M = \frac{A}{v} \cdot \sqrt{\frac{2 \cdot \gamma}{\gamma - 1} \cdot p_1 \cdot v_1 \cdot \left(1 - \left(\frac{p}{p_1}\right)^{\frac{\gamma-1}{\gamma}}\right)}$$

According to Poisson:

$$v = \left(\frac{p_1}{p}\right)^{\frac{1}{\gamma}} \cdot v_1$$

$$Q_M = A \cdot p^{\frac{1}{\gamma}} \cdot p_1^{-\frac{1}{\gamma}} \cdot v_1^{-1} \cdot \sqrt{\frac{2 \cdot \gamma}{\gamma - 1} \cdot p_1 \cdot v_1 \cdot \left(1 - \left(\frac{p}{p_1}\right)^{\frac{\gamma-1}{\gamma}}\right)}$$

Bring everything under the root, except A:

$$Q_M = A \cdot \sqrt{p^{\frac{2}{\gamma}} \cdot p_1^{-\frac{2}{\gamma}} \cdot v_1^{-2} \cdot \frac{2 \cdot \gamma}{\gamma - 1} \cdot p_1 \cdot v_1 \cdot \left(1 - \left(\frac{p}{p_1}\right)^{\frac{\gamma-1}{\gamma}}\right)}$$

Or:

$$Q_M = A \cdot \sqrt{\frac{2 \cdot \gamma}{\gamma - 1} \cdot p_1^{-\frac{2}{\gamma}} \cdot v_1^{-1} \cdot p^{\frac{2}{\gamma}} \cdot \left(1 - \left(\frac{p}{p_1}\right)^{\frac{\gamma-1}{\gamma}}\right)}$$

Bring the variable p into the round brackets:

$$Q_M = A \cdot \sqrt{\frac{2 \cdot \gamma}{\gamma - 1} \cdot p \cdot v_1^{-1} \cdot \left(\left(\frac{p^{\frac{2}{\gamma}}}{p_1}\right) - \left(\frac{p}{p_1}\right)^{\frac{\gamma+1}{\gamma}}\right)}$$

Conclusion: the mass flow can be expressed as a function of the counter pressure p. Write this expression as:

$$\psi = \sqrt{\frac{2 \cdot \gamma}{\gamma - 1} \cdot \left(\left(\frac{p^{\frac{2}{\gamma}}}{p_1}\right) - \left(\frac{p}{p_1}\right)^{\frac{\gamma+1}{\gamma}}\right)}$$

Ψ is the flow function.

$$Q_M = A \cdot \psi \cdot \sqrt{\frac{p_1}{v_1}}$$

9.11.1.3 The Flow Function ψ

The flow function is represented graphically in Figure 9.68. It becomes zero for the ratio p/p_1 equal to zero and to 1. The function has a maximum that can be calculated by partially deriving the function by p:

$$\frac{\partial \sqrt{\frac{2 \cdot \gamma}{\gamma-1} \cdot \left(\left(\frac{p^{\frac{2}{\gamma}}}{p_1}\right) - \left(\frac{p}{p_1}\right)^{\frac{\gamma+1}{\gamma}}\right)}}{\partial p} = 0$$

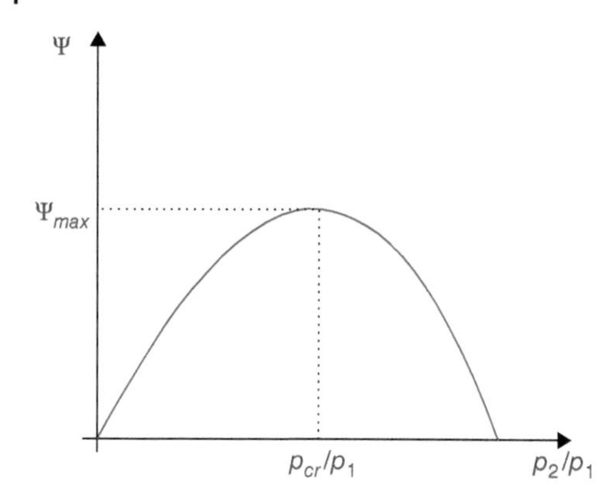

Figure 9.68 Flow function ψ.

It is easy to find out that this function has maximum for:

$$\psi = \psi_{max} \text{ with p} = \left(\frac{2}{1+\gamma}\right)^{\frac{\gamma}{\gamma-1}} \cdot p_1$$

For air: $\psi_{max} = 0.685$ at a pressure ratio of $\frac{p_2}{p_1} = 0.528$ (Figure 9.68).

9.11.1.4 The Critical Pressure

Now let's express the section A in function of the static pressure p.

$$Q_M = A \cdot \psi \cdot \sqrt{\frac{p_1}{p_2}}$$

Or:

$$A = \frac{Q_M}{\psi \cdot \sqrt{\frac{p_1}{v_1}}}$$

The section A is a function of p via the function Ψ. When is A minimal? Therefore should:

$$\frac{\partial A}{\partial p} = 0 \text{ with } Q_M \text{ constant}$$

This will happen when the denominator is minimum or if Ψ maximum. The value of the pressure where the section A is minimal is called the *critical pressure* p_{cr}.

But that has already been calculated, so:

$$p_{cr} = \left(\frac{2}{1+\gamma}\right)^{\frac{\gamma}{\gamma-1}} \cdot p_1$$

If one takes into account that for γ the following values are valid:

1.41 dry air

1.3 superheated steam

1.135 dry saturated steam

$1.035 + 0.1 \cdot x$ wet steam (approximation), x = vapor content

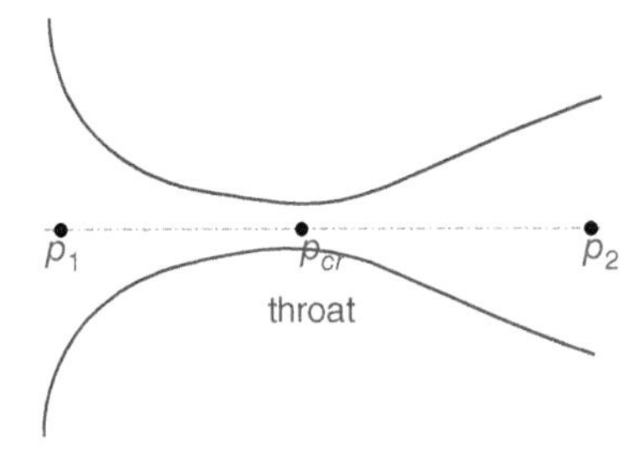

Figure 9.69 Converging–diverging nozzle.

then, in general (Figure 9.69):

$$p_{cr} = 0.528 \cdots 0.6\, p_1$$

If for air, the counterpressure p_2 at the discharge side of the tube is lower than 0.528 p_1, a converging diverging nozzle should be used; otherwise, a converging nozzle will suffice.

It can be proved that the velocity in the throat c_{cr} is the speed of sound in the specific gas, at that point.

Indeed, for an ideal gas, it can be proven that the velocity of sound is equal to:

$$c_{sound} = \sqrt{\frac{\gamma \cdot R \cdot T}{M}}$$

The speed of sound is dependent on the temperature. But in the nozzle the temperature changes all the time. It is only in the nozzle that the velocity of the gas is exactly equal to the speed of sound at that point.

Example 9.3

Design a nozzle where air with a flow of 5 [kg/s] flows adiabatically and expands from a start pressure of 10 [bar] and a start temperature of 15 °C until a counterpressure of 1 [bar].

Given: $\gamma = 1.4$, $c_p = 1$ [kJ/kg.K] and M = 29 [kg/kmol].

Solution

$$p_{cr} = \left(\frac{2}{1+\gamma}\right)^{\frac{\gamma}{\gamma-1}} \cdot p_1 = 10 \cdot \left(\frac{2}{2.4}\right)^{\frac{1}{0.4}} = 5.28\ [\text{bar}]$$

$$T_{cr} = T_1 \cdot \left(\frac{p_{cr}}{p_1}\right)^{\frac{\gamma-1}{\gamma}} = 600 \cdot \left(\frac{5.28}{1}\right)^{\frac{0.4}{1.4}} = 500\ [\text{K}]$$

$$h_1 - h_{cr} = c_p \cdot (T_1 - T_{cr}) = \frac{1}{2} \cdot (c_{cr}^2)$$

$$c_{cr} = \sqrt{2 \cdot c_p \cdot (T_1 - T_{cr})} = \sqrt{2 \cdot c_p \cdot (600 - 500)} = 447\ \left[\frac{\text{m}}{\text{s}}\right]$$

$$v_{cr} = \frac{R \cdot T_{cr}}{M \cdot p_{cr}} = \frac{8315 \cdot 500}{29 \cdot 5.28\ 10^5} = 0.2714\ \left[\frac{\text{m}^3}{\text{kg}}\right]$$

$$T_2 = T_1 \cdot \left(\frac{p_2}{p_1}\right)^{\frac{\gamma-1}{\gamma}} = 600 \cdot \left(\frac{1}{10}\right)^{\frac{0.4}{1.4}} = 310.6\ [\text{K}]$$

$$c_2 = \sqrt{2 \cdot c_p \cdot (T_2 - T_2)} = \sqrt{2 \cdot 1000 \cdot (310.6 - 600)} = 760.8$$

$$v_2 = \frac{R \cdot T_2}{M \cdot p_2} = \frac{8315 \cdot 310.6}{29 \cdot 10^5} = 0.89 \left[\frac{\text{m}^3}{\text{kg}}\right]$$

$$Q_M = \rho_2 \cdot c_2 \cdot A_2 = \rho_{cr} \cdot c_{cr} \cdot A_{cr}$$

$$A_{cr} = \frac{Q_M}{\rho_{cr} \cdot c_{kr}} = \frac{Q_M \cdot v_{cr}}{c_{cr}} = \frac{0.2714 \cdot 5}{29 \cdot 10^5} = 0.00303\ [\text{m}^2]$$

$$D_{cr} = \sqrt{\frac{4 \cdot A_{cr}}{\pi}} = \sqrt{\frac{4 \cdot 0.00303}{\pi}} = 6.21\ [\text{cm}]$$

$$A_2 = \frac{Q_M}{\rho_2 \cdot c_2} = \frac{v_2 \cdot Q_M}{c_2} = \frac{0.89 \cdot 5}{760.8} = 0.005585\ [\text{m}^2]$$

$$D_2 = \sqrt{\frac{4 \cdot A_2}{\pi}} = \sqrt{(4 \cdot 0.00585)/\pi} = 8.63\ [\text{m}^2]$$

The dimension L of the diverging part (Figure 9.70).

$$L = \frac{r_2 - r_1}{tn\, 5^\circ} = \frac{1}{2} \cdot \frac{(8.63 - 6.21)}{tn\, 5^\circ} = 13.8\ [\text{cm}].$$

9.11.2 Behavior at Changing Counter Pressure

Consider only a converging nozzle (Figure 9.71). At the inlet the pressure is p_1.The *counter pressure* p_3 in the chamber where the nozzle discharges can be regulated with a valve S. The gas leaves the nozzle with a pressure p_2. On first sight p_2 should always be equal to p_3,but situations will arise where that will not be the case.

On a graphic representation (Figures 9.72 and 9.73) the behavior of the static pressure will be followed. It is assumed that the gas enters the nozzle at subsonic speed. Different cases are possible:

- *Situation a*: The valve is completely closed; $p_3 = p_1$.
- *Situation b*: Let now by varying the opening of the valve by the switch S decrease the pressure p_3 a little until situation b is reached (see Figures 9.71 and 9.72). The converging nozzle behaves as a converging nozzle where the gas expands subsonically. The flow through the nozzle is given by:

$$Q_M = A \cdot \psi \cdot \sqrt{\frac{p_1}{p_2}}$$

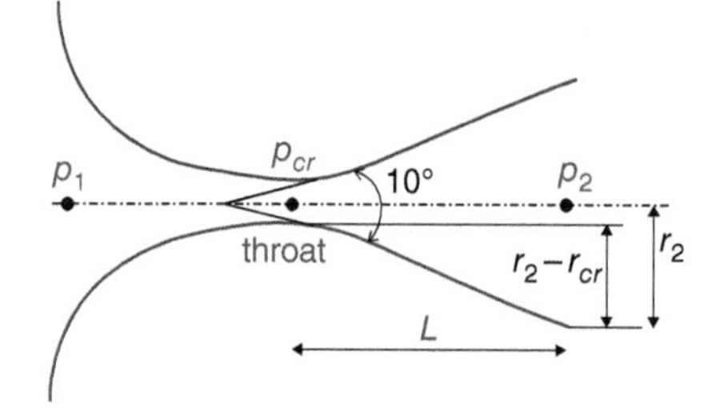

Figure 9.70 Calculation example of a nozzle.

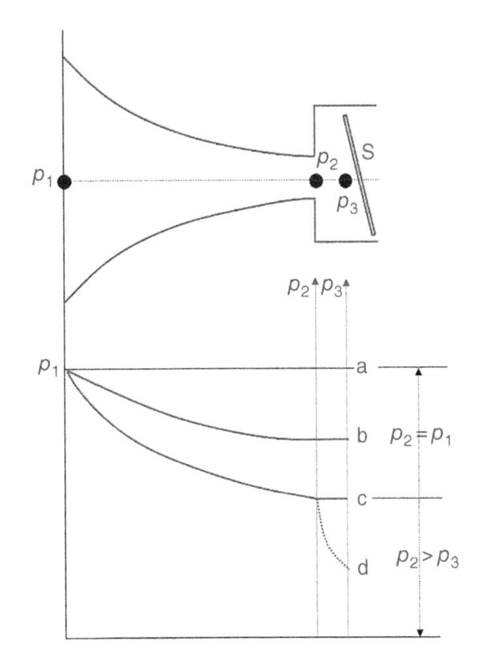

Figure 9.71 Various situations of counterpressure.

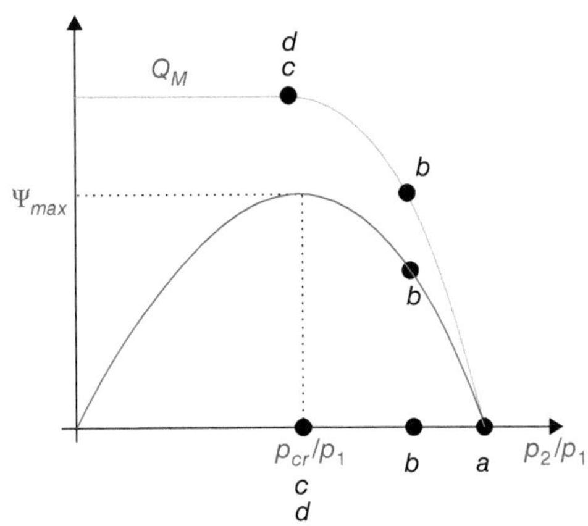

Figure 9.72 Progress of mass flow with situations marked in function of p_2.

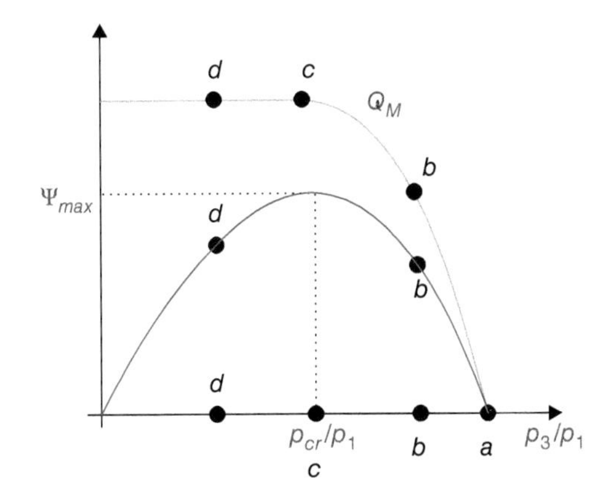

Figure 9.73 Progress of mass flow with situations marked in function of p_3.

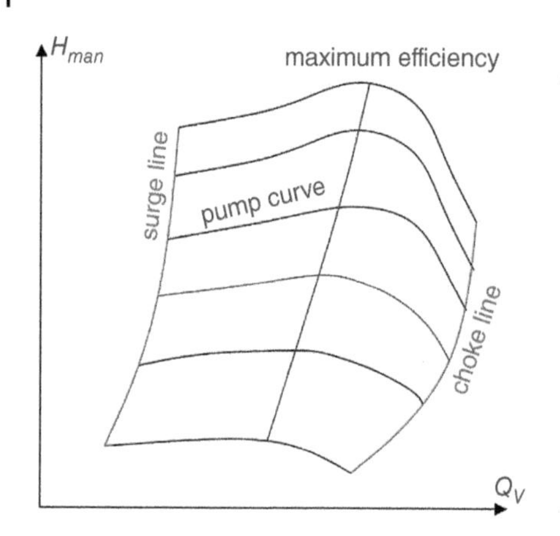

Figure 9.74 Choke limit line on compressor characteristic.

With:

$$\psi = \varphi\left(\frac{p}{p_1}\right)$$

- *Situation c*: Let p_3 decrease some more until the value of the critical pressure is reached. The flow Q_M has no maximum value. In this situation the nozzle is said to be *saturated*, or *choked*. At the outlet of the nozzle the speed of sound is reached. The outlet is, in fact, the *throat* of a imaginable converging–diverging nozzle.
For all situations between *a* and *c*:

$$p_3 = p_2$$

- *Situation d*: When the pressure p_3 decreases still more, one would expect that the mass flow and the velocity would further increase. However, a higher velocity than the speed of sound is not possible in a converging nozzle; therefore, you need a diverging part, as discussed in the Chapter 6. The result is that situation *d* corresponds with situation *c*: the maximum mass flow holds on for all:

$$p_3 < p_2$$

Once arrived in the chamber the gas beam will break up by means of expansion from p_2 *to* p_3.

In Figures 9.72 and 9.73 the progress of the flow is represented as a function of the outlet pressures.

Such a situation of a choked passage can happen in a turbocompressor. The passage becomes smaller and smaller as the gas flows though the vanes. At low counter pressures and high flows stone walling, or choking, can happen; this can damage the compressor. In Figure 9.74 the limit for choking is shown in the characteristic of a turbocompressor.

9.12 Comparison Axial/Radial Compressor

The characteristic of an axial compressor progresses much steeper than that of a centrifugal pump (Figure 9.75). This means that the stable zone, the working area

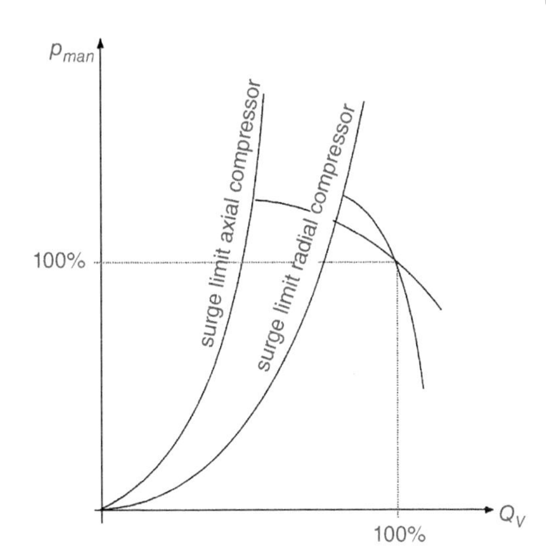

Figure 9.75 Comparison of characteristics.

(application field) with an axial compressor, is much smaller than its counterpart. The flow with an axial compressor can then deviate little from the nominal flow, so this compressor is generally good when working at a constant flow.

The lack of a *centrifugal dimension* leads to the fact that axial compressors have a lesser high value of pressure rise per stage than a centrifuge compressor.

On the other hand, because the flow resistances are limited by axial compressors, efficiencies of 85% are reachable, against 80% for centrifugal compressors. For the definitions of *efficiency* see the end of this chapter. Piston compressors score even better, but the drawback is a heavy construction and a bigger and less flexible installation. Because of the smaller radial dimensions on one side, and the higher section that axial compressors disposed at the inlet on the other side, they can attain higher flows (Figure 9.76).

Axial compressors mostly don't have cooling, centrifugal more often. (Figure 9.77).

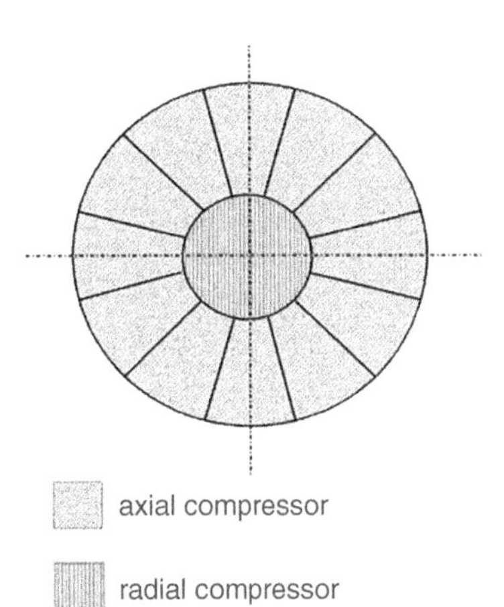

Figure 9.76 Flow passage.

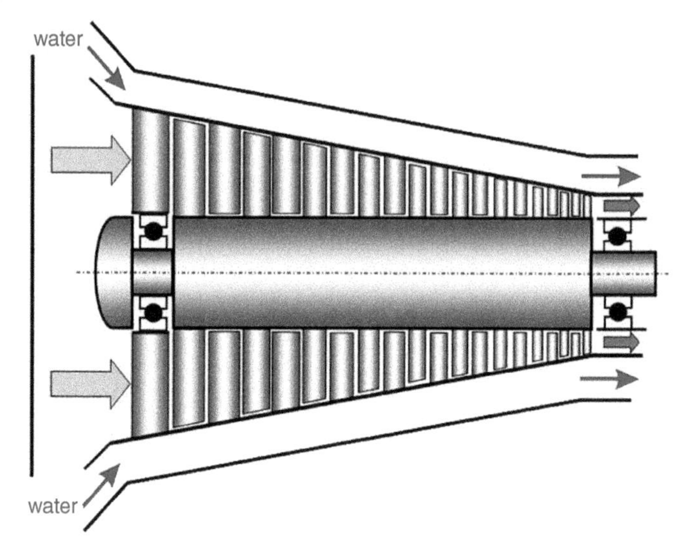

Figure 9.77 Cooling axial compressor.

9.13 Regulation of Turbocompressors

9.13.1 Rotation Speed

A first way to regulate the flow of a turbocompressor is by means of a regulation of the speed (Figure 9.78). Every time the compressor speed is altered, the operating point shift on the compressor characteristic.

The danger is that the operating point lands in the surge zone. If, however, a smaller flow is desired then the only way is to blow off gas or shut off the compressor temporarily.

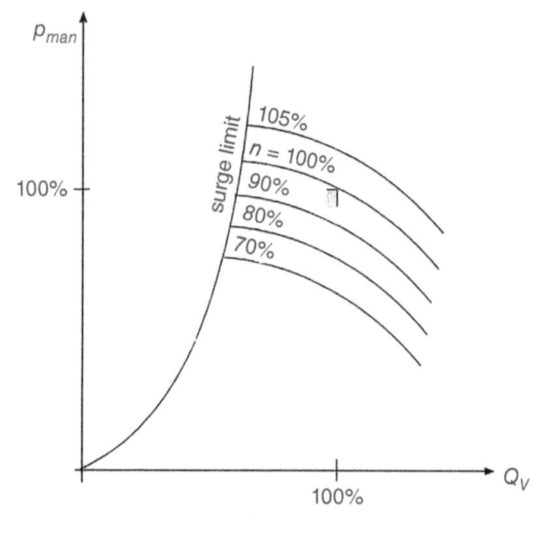

Figure 9.78 Regulation volumetric flow via rotation speed.

9.13.2 Throttling

A common method is throttling the suction line. This method is customary with machines that rotate at a constant speed. The manometric pressure and the flow decrease.

9.13.3 Variable Guide Vanes

9.13.3.1 Axial Compressor

A third method is making use of *variable guide vanes*.

Consider first the axial compressor.

The first question is to look at the problem that arises on the left part of the compressor characteristic. What happens exactly?

In Figure 9.79 the compressor curve is represented. It is explained qualitatively as follows.

Start from the operating point P. In the velocity diagram of Figure 9.80 the velocities c_1, w_1, c_2, w_2, u, and v are drawn.

Let the flow rise, at *constant speed*, thus at constant magnitude of u.

Figure 9.79 Characteristic.

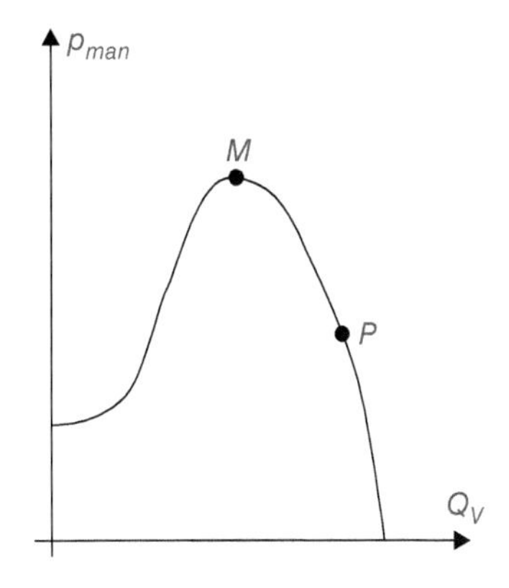

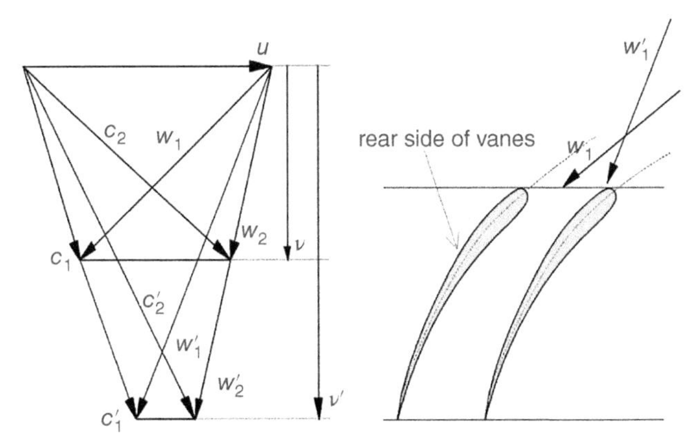

Figure 9.80 Flow increase from v to v'.

In the velocity diagram the increase in flow expresses itself as an increase of the flow component v to the new value v'. The new velocities are noted c_1', w_1', c_2', w_2'.

Suppose that the direction of the velocity c_1' at the inlet is the same as that of c_1. There is no reason it should change.

Build the new velocity diagram (Figure 9.80, at given value of v'. The new relative velocity w_1' will not have the same direction as w_1. That means that the gas will not enter the compressor bump free: there is a collision with the vane.

Practically this means that the incoming gas has a bad angle of attack and will bump on the rear of the vane, so that the compressor works partly as a turbine. After some time in the channel it might be presumed that the gas is adapted at the channel, so that at outlet of the vane channel the w_2' has the same direction as w_2.

Knowing this the velocity diagram can be redrawn. The velocities now have accents (').

Once that is done, it can be guessed that the delivered pressure will decrease. Indeed, the difference vector $\overline{w_1} - \overline{w_2}$ shrinks (see Section 3.2.3, Chapter 3).

The delivered manometric head of the compressor, according to the theory, is determined by the absolute value of this vector:

$$p_{man} \div \|\overline{w_1} - \overline{w_2}\|$$

On the base of: $\|\overline{w_1'} - \overline{w_2'}\| < \|\overline{w_1} - \overline{w_2}\|$ it can be concluded that when the flow increases, the delivered pressure will decrease.

But from point M the representation with velocity triangles is just a pure geometrical approximation. It doesn't explain everything. There are, indeed, aerodynamical aspects that cannot be explained by a purely geometrical model. It is so that from a certain angle of incidence the gas layers cannot follow the vanes profile (stall). This leads to a strong pressure decrease. This is represented in the left part of the characteristic.

Consider now the case as the flow decreases. In Figure 9.81 the corresponding velocity diagram is shown. It follows that the delivered pressure by the compressor will increase, indeed:

$$\|\overline{w_1'} - \overline{w_2'}\| < \|\overline{w_1} - \overline{w_2}\|$$

The use of variable guide vanes at the entrance of the first rotor stage of the fan or compressor comes down to the gas is presented in a way that it will always enter the first rotor stage bump less, independent of the value of the flow. This is realized as follows.

In Figure 9.82 a vane channel of the rotor is represented. The rotor is now preceded by a guide wheel. This is a stator, provided with so-called guide vanes. In the first case (the left-hand side of Figure 9.82) the gas flowing through the guide vanes channels is

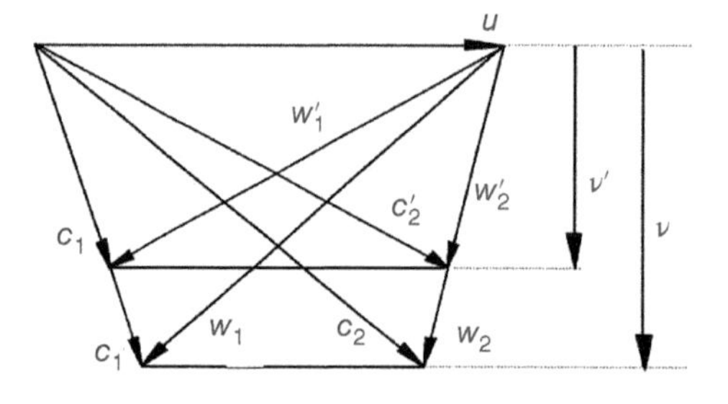

Figure 9.81 Velocity triangles in the case that the flow decreases.

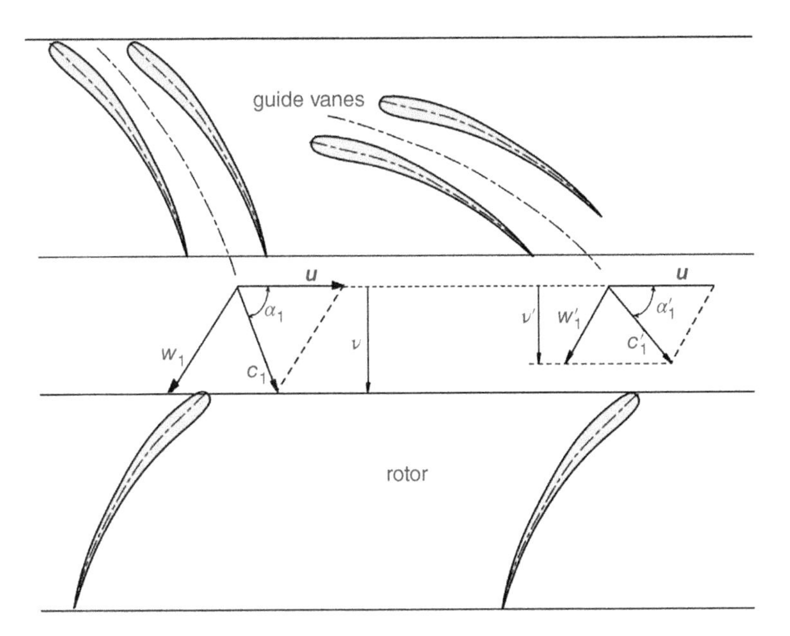

Figure 9.82 Effect of orientable guide vanes.

presented on the inlet of the rotor with an absolute velocity $c_1$and therefrom a relative velocity w_1 that flows tangentially to the rotor, without any bump losses.

When the flow now, for one reason or another, decreases, for instance because of throttling the suction line, (as illustrated in Figure 9.82), then the velocity w_1 swill not flow bump less tangentially to the rotor vanes anymore.

By now rotating the guide vanes the velocity c_1, that on its turn the direction of w_1 will alter. The aim is to have a rotor inlet angle that fits perfectly the incoming angle of the fluid. This is illustrated in Figure 9.82, on the right-hand side of the figure.

This way the surge limit is moved to the left on the fan curve.

The rotation of the guide vanes has, however, a side effect: the more the vanes are rotated, the more they throttle the gas. The final fan characteristic is given in Figure 9.83. On one hand, the normal surge limit is shown. On the other side, there is the normal surge limit. Look at the nominal rotational speed: when the guide vanes aren't rotated at all the fan sits in its nominal operating point. If these vanes are turned the pump curve moves to the left. At the same time, however, the pressure decreases because of the throttling action of these vanes.

The variable guide vanes are mounted over the whole circumference of the casing (Figure 9.84). Via a rod system they are attached with one another and driven by a hydraulic or electric servomotor.

For multistage axial compressors a few stages are provided with those orientable guide vanes; these are called *variable stator vanes*. See Figure 9.85 for an implementation.

9.13.3.2 Centrifugal Compressor

In the same way as with an axial compressor IGVs (inlet guide vanes) can be placed at the entrance of a centrifugal compressor (Figures 9.86 and 9.87).

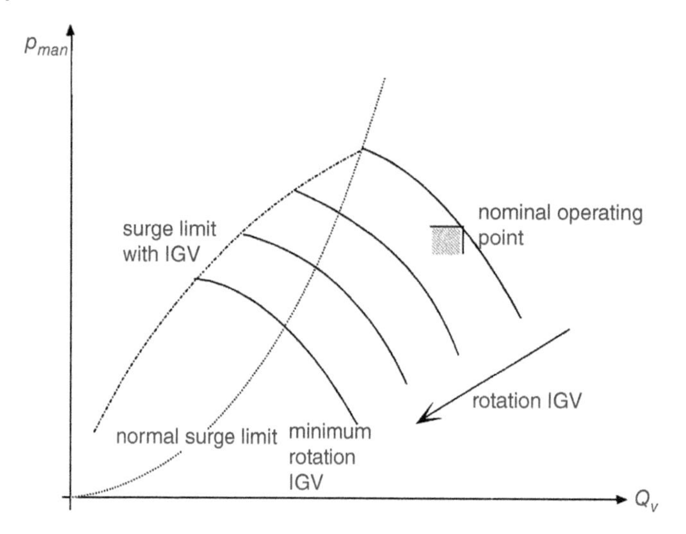

Figure 9.83 Characteristic with IGV.

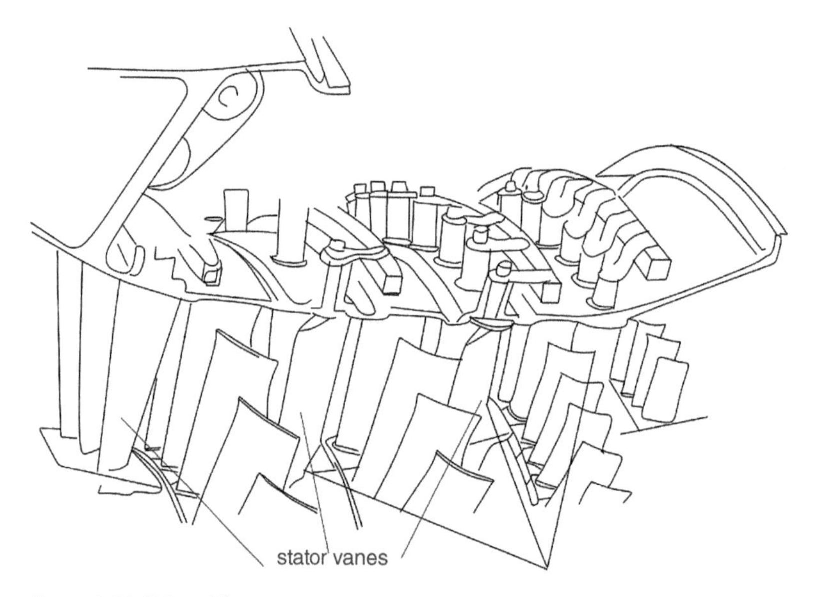

Figure 9.84 Orientable stator vanes.

Is should be clear that variable stator vanes in centrifugal compressors like in axial compressors cannot be realized constructively.

9.14 Efficiency of Turbocompressors

In this section only turbocompressors will be handled. The compression is always adiabatic, because the gas flows with such a speed (100–200 [m/s]) that it has no time to

Figure 9.85 Implementation of variable stator vanes

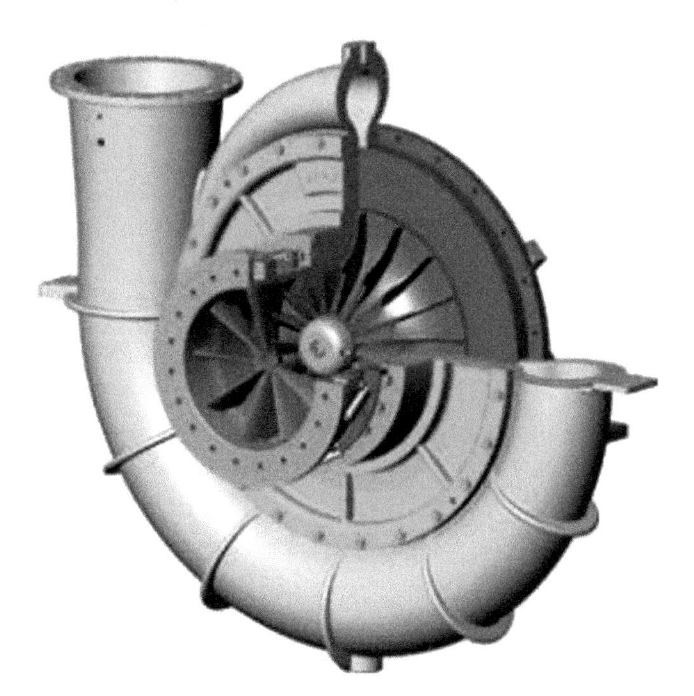

Figure 9.86 IGV. Source: Dresser.

exchange heat with the surroundings. On the other hand, with such speeds, there is certainly friction.

In the case of an adiabatic system, and neglecting potential and kinematic energies, the law of conversion of energy becomes, no matter if there is friction or not:

$$h_2 - h_1 = w_{t12}$$

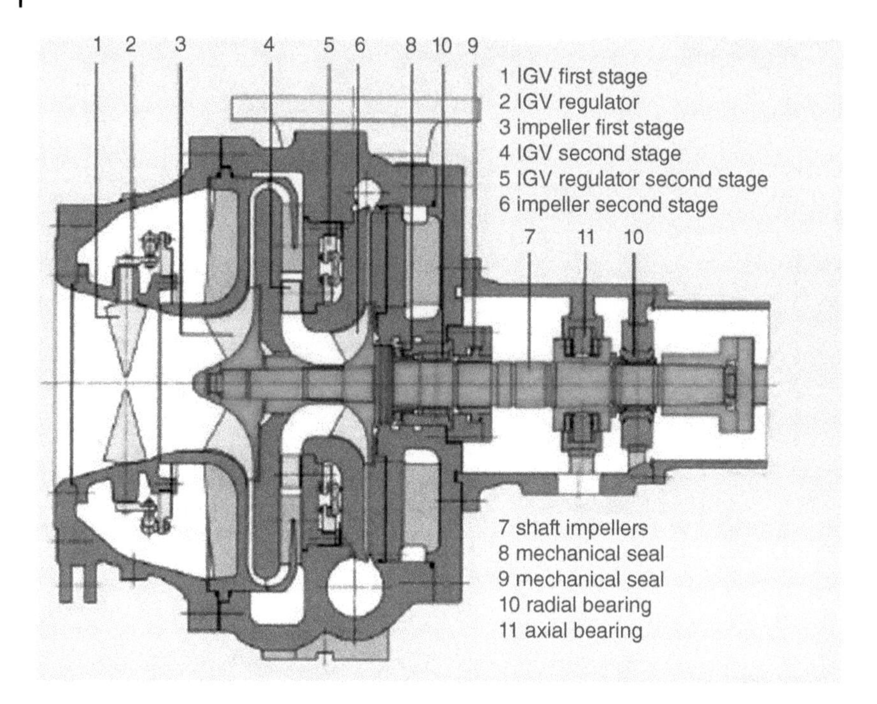

Figure 9.87 IGV and a centrifugal compressor. Source: Friotherm.

For an adiabatic compression from pressure p_1 to p_2 in a turbocompressor that includes friction, we note the specific technical work as $w_{t,12}$, the change of state is $1 \rightarrow 2$. We compare with an adiabatic compression to the same end pressure, without friction $1 \rightarrow 2'$ (Figures 9.87–9.89).

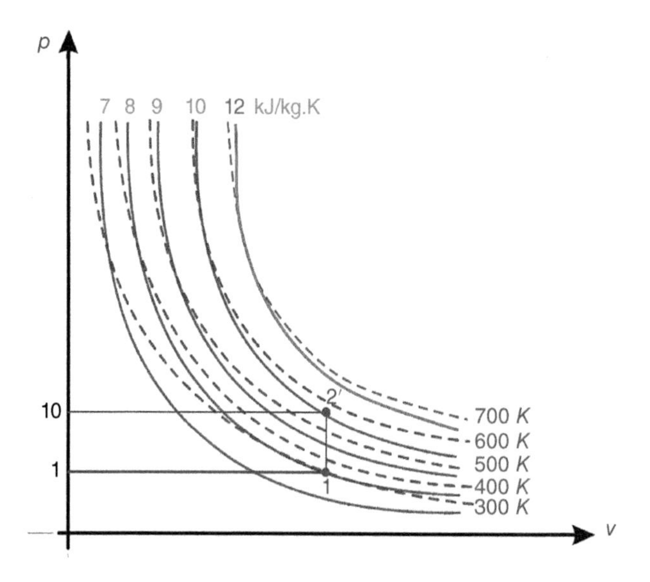

Figure 9.88 *pv* diagram.

For both compressions:

$$w_{t12'} < w_{t12}$$

The definition of the isentropic efficiency is then:

$$\eta_s = \frac{w_{t,12'}}{w_{t12}}$$

which for an ideal gas is:

$$\eta_s = \frac{T_{2'} - T_1}{T_2 - T_1}$$

In Figures 9.88 and 9.89, a *pv* diagram and a *Ts* diagram, entropy is used. The general formulae for the definition of entropy is a differential definition:

$$ds = \frac{dq}{T} + \frac{dw_f}{T}$$

where:

s : specific entropy [J/K]
T : absolute temperature [K]
dw_f : a small part of friction work in the system, not a differential from work! [J]

dw_f is always positive. It is created. It cannot be given back.

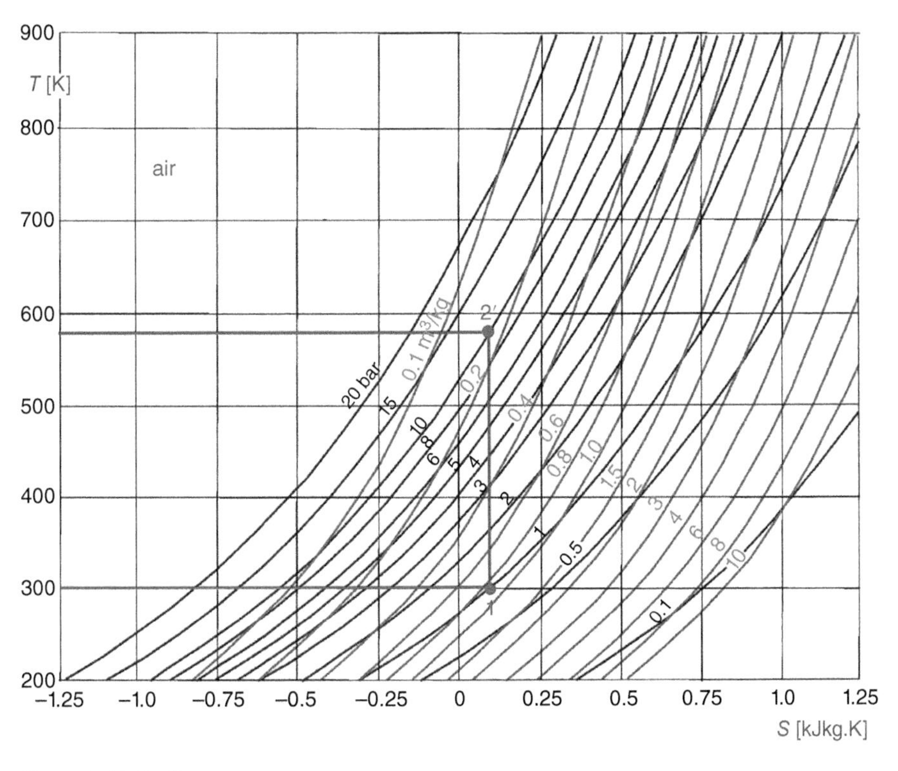

Figure 9.89 *Ts* diagram.

It must be interpreted as follows. The entropy doesn't care if the heat developed in the system is the result of heat that comes from the surroundings or heat created by friction.

When a system is adiabatic ($dq = 0$) and there is no friction ($d_{wf} = 0$), the change of entropy is zero. The entropy is then constant, and in such a case we follow for the change of state a curve with constant entropy, and in a *Ts* diagram a vertical line.

These changes of state are drawn in the diagrams for a compression from 1 to 10 [bar], in the case of a constant entropy change. On the diagram we can read that the final temperature will be 580 [K].

But this can be calculated. Let's do it for air.

Poisson said (he died on April 25, 1840):

$$\frac{T_2'}{T_1} = \left(\frac{p_2}{p_1}\right)^{\frac{\gamma-1}{\gamma}}$$

From where:

$$T_2' = T_1 \cdot \left(\frac{p_2}{p_1}\right)^{\frac{\gamma-1}{\gamma}}$$

For air, with $\gamma = 1.4$, a compression from 1 to 10 [bar], and a start temperature of 300 [K], this gives:

$$T_2' = 579\ [\mathrm{K}]$$

Example 9.4

Suppose the compressor works with an isentropic efficiency of 85%, a value that was defined by earlier measurements. What will be the resulting temperature of the compression?

Solution

$$\eta_s = \frac{T_2' - T_1}{T_2 - T_1}$$

From where:

$$T_2 = \frac{(T_2' - T_1)}{\eta_s} + T_1 = \frac{1}{0.85} \cdot (579 - 300) + 300 = 628\ [\mathrm{K}]$$

Draw now on a paper the change of state 1→ 2. Why did the temperature rise when there was friction? Why does the entropy rise?

10

Jet Ejectors

CHAPTER MENU

According to DIN 24 290 Jet Ejectors are classified as follows:

Gas jet pumps	Steam jet pumps	Liquid jet pumps
Gas jet fan	*Steam jet fan*	*Liquid jet fan*
Gas jet compressor	*Steam jet compressor*	*Liquid jet compressor*
Gas jet vacuum pump	Steam jet vacuum pump	Liquid jet liquid vacuum pump
Gas jet liquid pump	Steam jet liquid pump	Liquid jet liquid pump
Gas jet solid pump	Steam jet solid pump	Liquid jet solid pump

The fans and compressors discussed in this chapter are in italics.

10.1 Steam Ejector Compressor

10.1.1 General

The aim is to suck a gas, mostly air, in a suction chamber (Figure 10.1).

A driving or motive fluid with a high pressure of a few bars is used to suck the gas. Depending on the type of motive fluid, there are three types:

- liquid ejectors
- steam ejectors
- gas ejectors.

The final pressure of the gas determines whether the jet is a fan or a compressor.

In Figure 10.1 the case is represented for a steam jet compressor. The steam acts as a motive fluid and gas is sucked as a result. The gas will be compressed. The steam on high pressure is sent through a converging–diverging nozzle. The nozzle emerges in a

Pumps and Compressors, First Edition. Marc Borremans.

Companion website: www.wiley.com/go/borremans/pumps

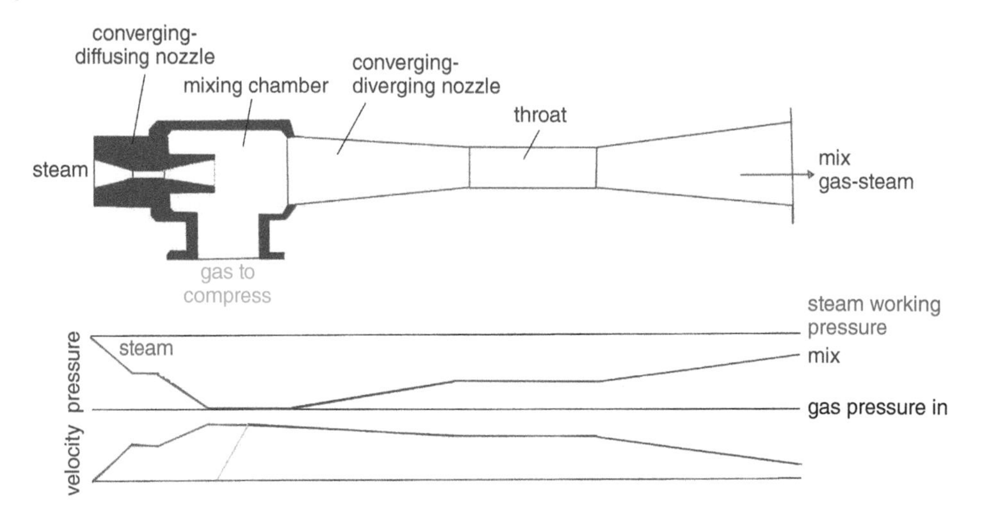

Figure 10.1 Principle of a jet ejector.

suction chamber, where the gas to be compressed can flow in. The steam expands until the pressure in the suction chamber is lower than the final pressure of the gas at the outlet of the jet pump.

From thermodynamic courses it is known that the expansion of the steam in the nozzle is accompanied with an increase of its velocity. In the throat of the first nozzle the sonic velocity is reached (speed of sound, or Mach 1) and in the diverging part of it the flow is supersonic (see also Chapter 6).

In the suction chamber the steam is mixed with the gas to be compressed and the mix flows further into a converging–diverging nozzle. But this time the entrance velocity is supersonic! This means that the velocity of the mix will decrease and the static pressure of the mix will increase.

The ratio $\frac{p}{p_0}$, where p is the *counterpressure* (pressure at the discharge side) and p the static pressure of the motive fluid, is the *compression ratio* of the motive fluid. A single stage has a maximum ratio of 20 : 1. The maximum attainable ratio is dependent on the suction pressure and the pressure of the motive fluid. The higher it is (the *expansion ratio*), the higher the compression ratio.

The converging part acts like a subsonic nozzle but the diverging part is a *supersonic nozzle*. The steam enters the nozzle and its pressure decreases while its velocity increases, in the throat the steam reaches Mach 1, then in the diverging part the velocity continues to increase and is thereby supersonic, while the static pressure continues to decrease. Because the pressure of the steam decrease so much, it will become vapor, and finally it will condense. Meanwhile, the gas, entrained by the motive steam, will be compressed. An implementation is given in Figure 10.2.

An example. Suppose a suction pressure of 0.1 [mbar(a)] and a static pressure of the motive steam to be 8 [bar(a)], the expansion ratio is 80 000 : 1. In that case one can reach a compression ratio of 20. But with a suction pressure of 1 [bar(a)] and a steam pressure of 8 [bar(a)], this is with an expansion ratio of van 8 : 1, only a compression ratio of 3 can be reached.

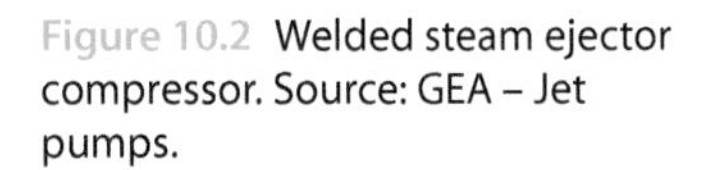

Figure 10.2 Welded steam ejector compressor. Source: GEA – Jet pumps.

10.1.2 Jet Pumps with Mixing Heat Exchangers

A mixing heat exchanger is one in which the cold and hot media have contact (Figure 10.3).

The condensed steam can be separated from the compressed gas by a heat exchanger. In Figure 10.4 two jet pumps (1 and 2) are used, to attain a higher end pressure of the gas. The gas enters the suction chamber at A. The steam enters the jet pump via B. If it condenses a little bit the condensed water is immediately drained via D. The gas–steam mix enters the heat exchanger I at the bottom, moves upwards, and hits the obstructions, where droplets are formed that fall down by gravity and are evacuated via I. Then follows a second jet pump. The drain of the droplets is via II. Then follows a third jet pumps 4, a heat exchanger and a fifth jet pump. Via C cooling water is supplied that comes in close contact with the steam and enhances the forming of condensate. It is also drained via I, II, and III.

10.1.3 Jet Pump with Three Surface Heat Exchangers

In a surface heat exchanger the cooling steam and the mix of motive steam and gas do not have any contact. The heat is evacuated through the surface of the heat exchanger (Figures 10.5–10.11).

What does such an exchanger look like? Well, a shell and tube heat exchanger is represented in Figures 10.6–10.8.

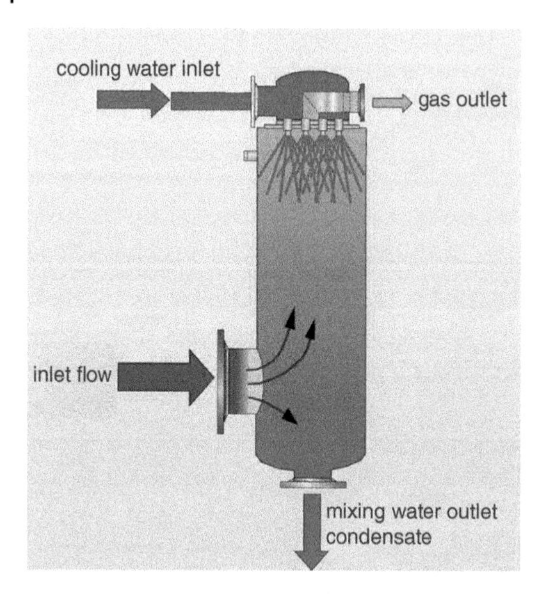

Figure 10.3 Mixing heat exchanger. Source: Körting Hannover AG.

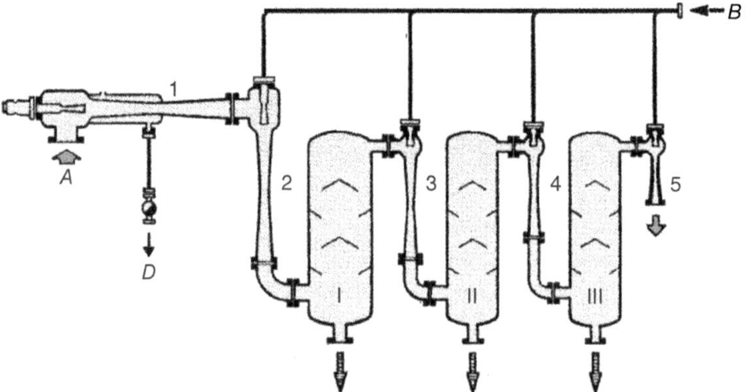

Figure 10.4 Single-stage steam jet fan ejector, using two 5 ejectors and three contact heat exchangers. Source: GEA.

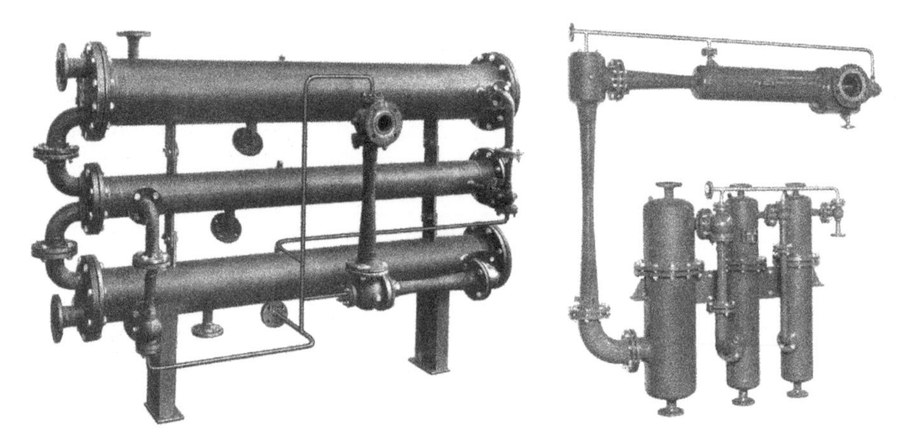

Figure 10.5 Single-stage steam jet fan ejector using three surface heat exchangers. Source: GEA.

Figure 10.6 Surface heat exchanger.

Figure 10.7 Tubes of a shell and tube heat exchanger.

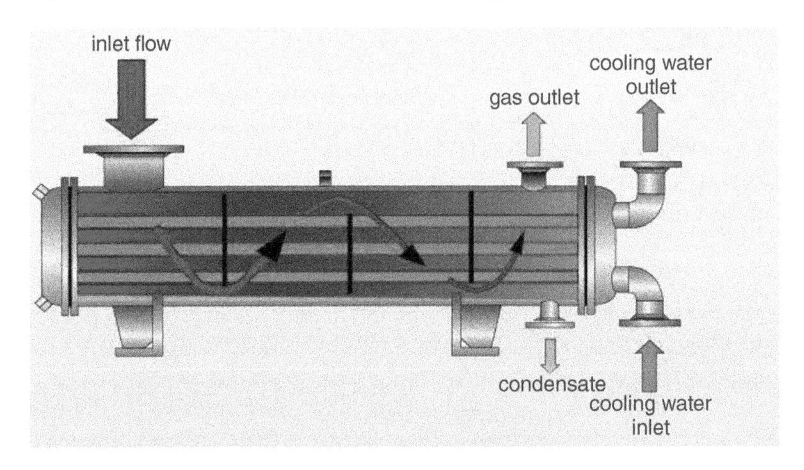

Figure 10.8 Drawing of shell and tube heat exchanger. Source: Körting Hannover AG.

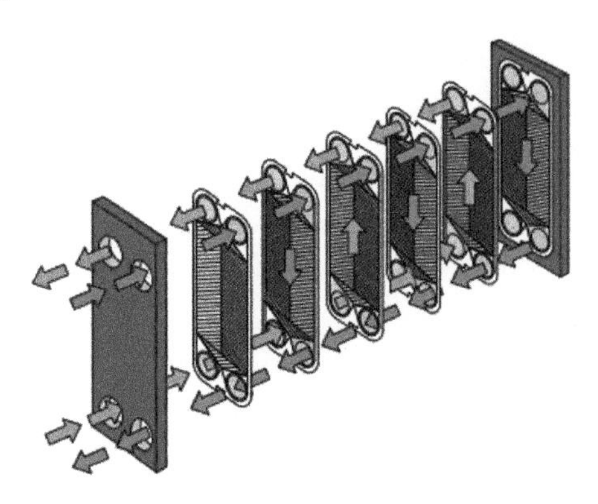

Figure 10.9 Plate heat exchanger. Source: Alfa Laval.

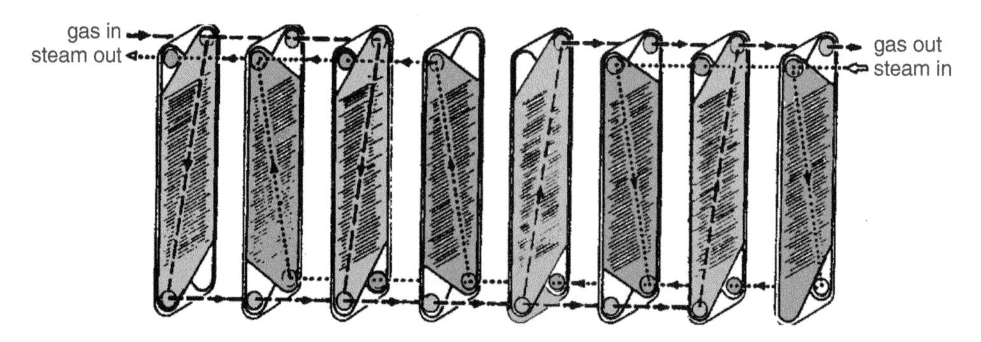

Figure 10.10 Drawing of plate heat exchanger. Source: Alfa Laval.

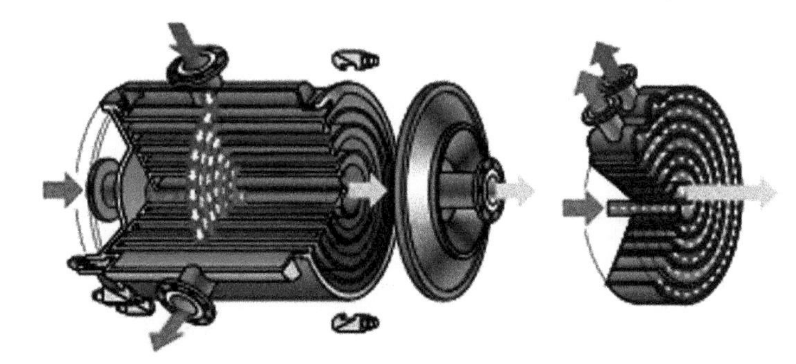

Figure 10.11 Section view of spiral heat exchanger. Source: Alfa Laval.

The steam enters the heat exchanger via chest 2, and exchanges heat with the gas that flows in the pipes. There is no contact, just heat exchange. There are five plates in the exchanger with holes in it where the pipes fit exactly in. But the platen fills only one half of a circle. So the steam can pass from one chamber of the exchanger to the next one, and travels across through the shell. This way the steam has a passage through the shell. It flows from the left to the right, and leaves the exchanger via 5. The gas enters the pipes via the chest 6. Then it flows through the pipes of the heat exchanger. The best result is attained in a counter-wise exchanger, that is the directions of cooling water and gas are opposite.

Figure 10.12 Spiral heat exchanger. Source: Alfa Laval.

Other types of heat exchanger exist. Shell and tube heat exchangers are very expensive. Alternatives are plate exchangers and spiral heat exchangers; see Figures 10.9 and 10.10 for plate heat exchangers and Figures 10.11 and 10.12 for spiral heat exchangers.

10.2 Gas Jet Ejector

The flow chart of a gas jet fan is given in Figure 10.13. An application of a gas jet compressor is given in Figure 10.14.

Sometimes a liquid is sucked (Figure 10.15). The motive gas creates an underpressure in the liquid jet ejector that sucks the air out of the suction pipe line, so that water and mud from the swamp will be sucked and discharged via the outlet.

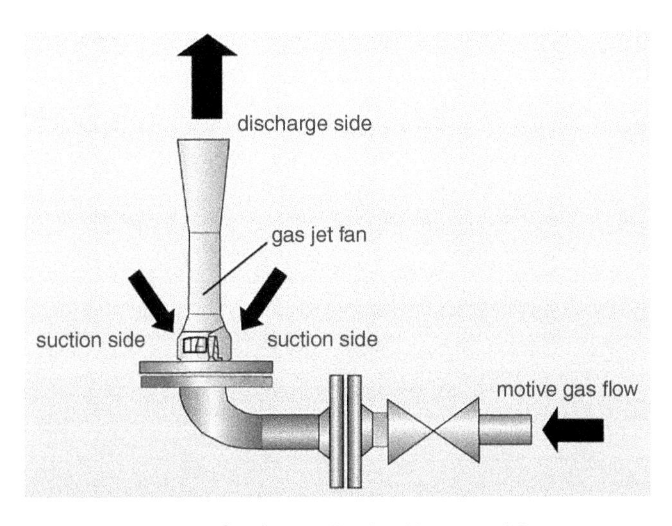

Figure 10.13 Gas jet fan. Source: Körting Hannover AG.

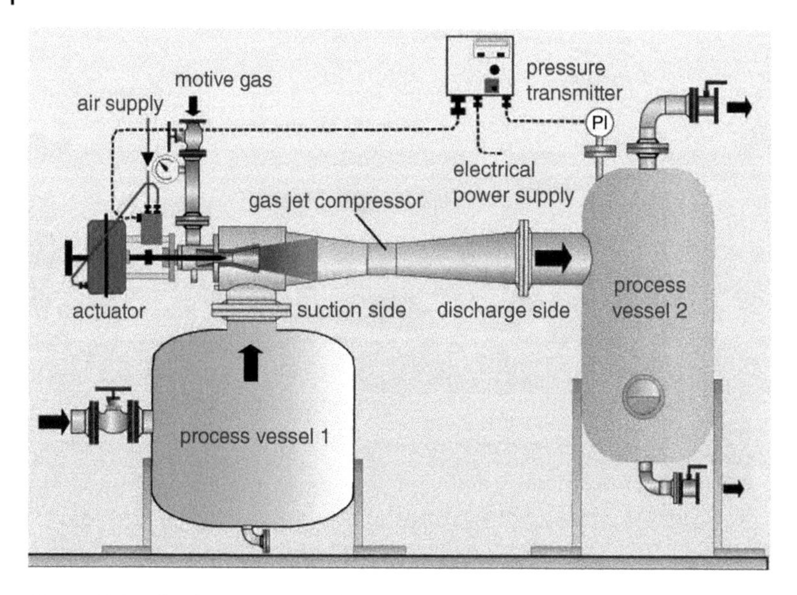

Figure 10.14 Gas jet compressor. Source: Körting Hannover AG.

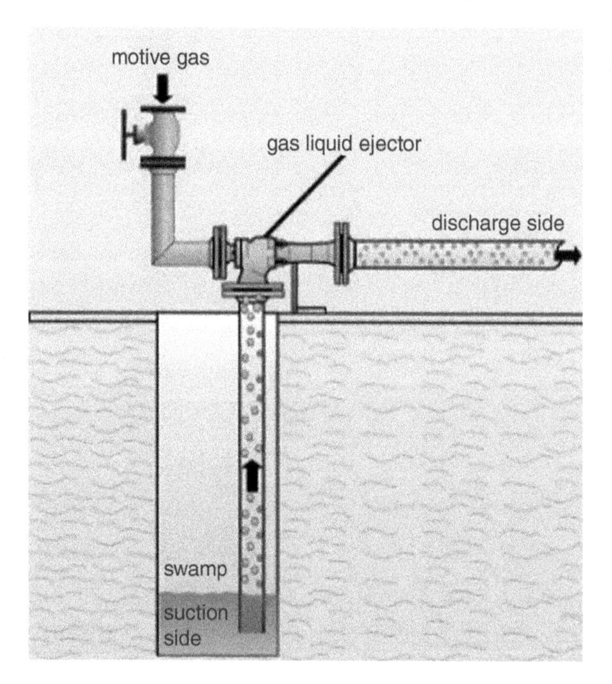

Figure 10.15 Gas jet liquid ejector. Source: Körting Hannover AG.

10.3 Applications

10.3.1 Application 1

Figure 10.16 shows an application of a steam fan ejector. Exhaust gasses that still contain fuel and air are sucked by the steam ejectors together with natural gas. Steam not only works as a motive gas but also as a catalyst for a burning process.

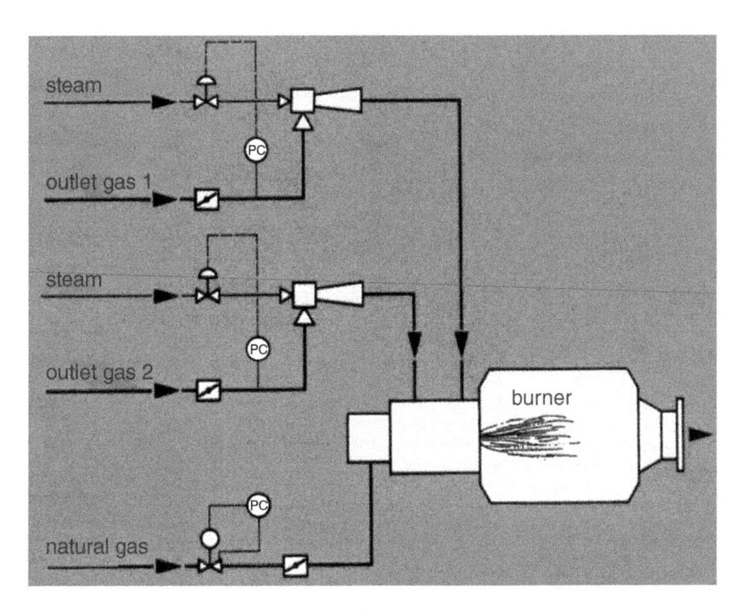

Figure 10.16 Application steam jet fan. Source: GEA.

In function of the pressure of the exhaust gasses the quantity of motive steam is regulated with a pressure control (PC) regulator (PC).

10.3.2 Application 2

In Figure 10.17, the same sort of application is used. The driving gas A is an exhaust gas that still contains enough air for a burning process. B is just natural gas that stands on a delivered pressure of 5 [bar]. It is expanded in reduction. C is the mixed gas. 1 is the fan of the burner.

Figure 10.17 Application gas jet fan. Source: GEA.

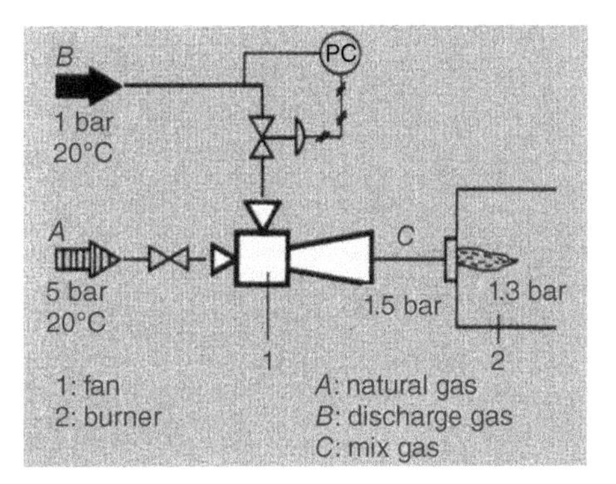

11

Vacuum Pumps

CHAPTER MENU

11.1 Vacuum Areas

11.1.1 Kinetic Gas Theory

In kinetic gas theory one considers a gas as a set of molecules that are locked up in a vessel (Figure 11.1). They are point masses that are all identical. The gas molecules move in all directions, across each other, at all possible velocities.

The absolute temperature T [K] of the gas is a measure of the mean kinetic energy of the molecules.

They collide with each other and against the wall of the vessel. It is supposed that these collisions are *elastic*, which means that no kinetic energy is lost in friction. The collisions against the wall lead to the notion of *pressure p* of the gas, that is the force that the molecules exercise on the wall per unity surface.

The notion of *mean free distance* λ should then introduced (Figure 11.2). This is the average distance that a gas molecule covers before there is a collision against another molecule or the wall.

It can be proved that, at constant temperature T, for every gas:

$$\lambda \cdot p = \text{constante } C^*$$

For every gas the constant is different (see Table 11.1) at 20 °C.

The pressures in SI units are given in [mbar] (hPa).

Sometimes an old unit of pressure is used: the *Torr*. 1 [Torr] = 133.32 [Pa] ≅ 100 [Pa] = 1 [hPa] = 1 [mbar].

Pumps and Compressors, First Edition. Marc Borremans.

Companion website: www.wiley.com/go/borremans/pumps

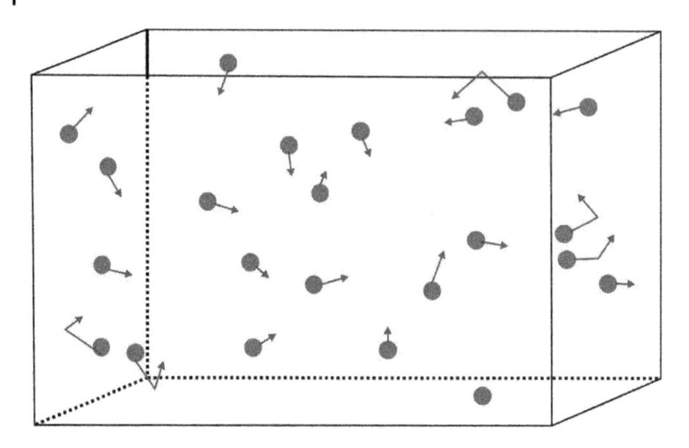

Figure 11.1 Kinetic gas theory.

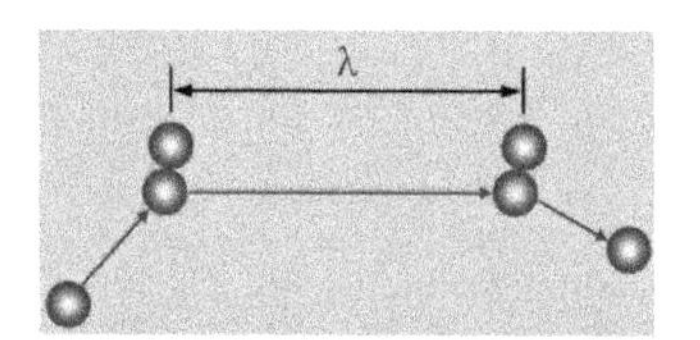

Figure 11.2 Mean free distance. Source: Pfeiffer.

Table 11.1 Mean free distance at 20 °C.

gas	$C^* = \lambda \cdot p$ [cm · mbar]
H_2	$12.00 \cdot 10^{-3}$
He	$18.00 \cdot 10^{-3}$
Ne	$12.30 \cdot 10^{-3}$
Ar	$6.40 \cdot 10^{-3}$
Kr	$4.80 \cdot 10^{-3}$
Xe	$3.60 \cdot 10^{-3}$
Hg	$3.05 \cdot 10^{-3}$
O_2	$6.50 \cdot 10^{-3}$
N_2	$6.10 \cdot 10^{-3}$
CO_2	$3.950 \cdot 10^{-3}$
air	$6.67 \cdot 10^{-3}$

In vacuum technique one has to deal with a very extensive pressure area, namely 16 power of 10. Therefore this vast area will be divided into partitions:

- *Rough vacuum*: 1000–1 [mbar]
- *Medium vacuum*: $1–10^{-3}$ [mbar]
- *High vacuum*: $10^{-3}–10^{-7}$ [mbar]
- *Ultrahigh vacuum*: $10^{-7}–10^{-14}$ [mbar]
- *Kosmos*: $<10^{-14}$ [mbar]

In reality the collisions of the molecules with the walls are not elastic. When the particles collide with the vessel wall some of them stick with the wall. This is named *adsorption. Desorption* is the process where they come loose again and regain their freedom. The result is that the diffraction is quite independent of the direction and velocity: the particles forget the prehistory. They leave the wall with a velocity distribution dependent on the temperature of the wall. Some preference direction is absent.

When pressures are low the theory or elastic collisions is not valid anymore, then comes the theory of wall adsorptions. But outside that pressure area the insight of molecule collisions must be adapted. This is discussed below.

In Figure 11.3 applications for vacuum areas are represented.

11.1.2 Formation Time

At high vacuum it is so that some of the gas molecules fly around in the free space and some molecules are adsorbed by the walls. The time that is necessary to reach a mono atomic layer on the surface of the walls is called the *formation time* τ.

In the high and ultra-high vacuum the properties of the walls are of critical importance. Below 10^{-3} [mbar] there will be more gas molecules on the surfaces of a sphere vessel with a volume of 1 [l]; the ratio of the number of adsorbed gas molecules to the number of free molecules in the space will be:

- at 1 [mbar]: 10^{-2}
- at 10^{-6} [mbar]: 10^{+4}
- at 10^{-11} [mbar]: 10^{+9}

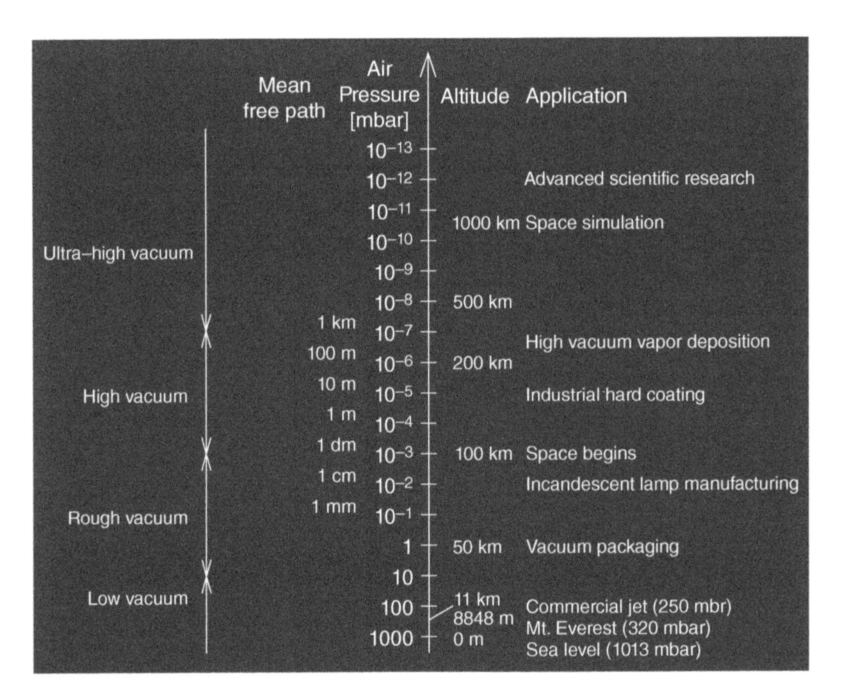

Figure 11.3 Applications on vacuum areas.

This leads to the fact that the mono layer formation time τ will be used in making the difference between ultra-high vacuum and high vacuum. That's why surfaces free of atoms can only be attained in ultra-high vacuum and high vacuum. In high vacuum the formation time τ will just be a fraction of a second; in the ultra-high vacuum this is in the order of minutes or hours. That's why surfaces that are free of atoms can only be reached under ultra-high vacuum circumstances.

It speaks for itself that the vacuum pumps and pressure-measuring devices will be of a different character depending of the vacuum area that is under study.

11.2 Measuring Devices

11.2.1 Introduction

The pressures measured in vacuum technology today cover a range from 1013 [mbar] to 10^{-14} [mbar], i.e. over 17 orders of magnitude.

Measuring instruments designated as vacuum gauges are used for measurement in this broad pressure range. Since it is impossible for physical reasons to build a vacuum gauge which can carry out quantitative measurements in the entire vacuum range, a series of vacuum gauges is available, each of which has a characteristic measuring range that usually extends over several orders of magnitude.

Depending on the pressure indication of the type of gas, a distinction must be made between the following vacuum gauges:

1. Instruments that by definition measure the pressure as *the force* which acts on an area, the *direct* or *absolute vacuum gauges*. According to the kinetic theory of gasses, this force, which the particles exert through their impact on the wall, depends only on the number of gas molecules per unit volume (number density of molecules) and their temperature, but not on their molar mass. The reading of the measuring instrument is independent of the type of gas. Such units include liquid-filled vacuum gauges and mechanical vacuum gauges.
2. Instruments with *indirect pressure measurement*. In this case, the pressure is determined as a function of a pressure-dependent (or more accurately, density-dependent) property (thermal conductivity, ionization probability, electrical conductivity) of the gas. These properties are dependent on the molar mass as well as on the pressure. The pressure reading of the measuring instrument depends on the type of gas.

The scales of these pressure-measuring instruments are always based on air or nitrogen as the test gas. For other gasses or vapors correction factors, usually based on air or nitrogen, must be given. For precise pressure measurement with indirectly measuring vacuum gauges that determine the number density through the application of electrical energy, it is important to know the gas composition.

11.2.2 Bourdon Measuring Devices

Bourdon tube pressure gauges are used for the measurement of relative pressures from 10 [mbar] to 7000 [bar]. They are classified as mechanical pressure-measuring instruments and thus operate without any electrical power.

The interior of a tube bent into a circular arc (the Bourdon tube) is connected to the vessel to be evacuated (Figure 11.4). Through the effect of the external air pressure the end of the tube is deflected to a greater or lesser extent during evacuation and the attached pointer mechanism is actuated. Since the pressure reading depends on the external atmospheric pressure, it is accurate only to approximately 10 [mbar], provided that the change in the ambient atmospheric pressure is not corrected.

It may be of any interest to situate the pressure areas (Figure 11.5). A *manometer* measures the effective pressure or overpressure. This device is necessary for pumps and compressors.

In vacuum techniques there is a need for a *vacuum meter* that gives the relative pressure with reference to the atmospheric pressure.

Finally, there are measuring devices available that can measure in a vacuum as well as in the overpressure area. Figure 11.2 shows three such measuring devices (in the classic analog implementation). It goes without saying that these Bourdon measuring devices cannot operate satisfactorily in a high vacuum area.

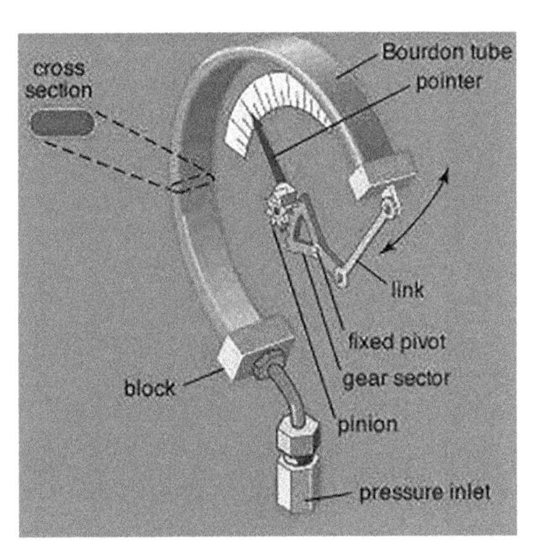

Figure 11.4 Bourdon measuring device.

Figure 11.5 Bourdon measuring devices.

11.2.3 Pirani Devices

All electronic gauges are subject to shifts in reading due to the "drift" inherent in electronic controllers. Both the electronic gain and the zero set point are adjustable. Direct gauges measure the pressure independently of the composition of the gas being measured. Indirect gauges are dependent on the composition of the gas being measured, such as the thermal conductivity, electrical conductivity, or ionization capability.

Thermocouple and Pirani gauges are classified as indirect gauges that typically operate from atmosphere to 0.1 [mbar]. Thermocouples and Pirani gauges differ a small amount depending on the manufacturer, but fundamentally they function in the same manner. Pirani gauges have a slightly wider range and provide better resolution than thermocouple gauges. The basic theory of operation is the measurement of heat lost from a wire (Figure 11.6). At pressures below 0.1 [mbar] there is so little gas present that the gauge it cannot provide an accurate reading.

One of the thin wire filaments of a standard *Wheatstone bridge* circuit is placed inside the vacuum system and the other side in a reference gas. The gauge electronics measures the temperature loss of the wire inside the vacuum system and compares it to that of the reference wire, and then it displays the vacuum reading. When large amounts of gas are present near the wire, a large amount of heat is removed from the wire, and a large amount of current is consequently required to maintain the temperature of the wire. The current required to maintain the temperature is directly proportional to the heat being conducted away by the gas present near the wire inside the vacuum system. As the amount of gas near the wire decreases, less heat is removed from the wire. The composition of the gas significantly affects the reading (accuracy) of these gauges. Typically, the gauges are calibrated for air (78% nitrogen) and a factor must be applied to the gauge's reading for argon, helium, or hydrogen. Argon absorbs heat much faster than nitrogen, while nitrogen is much faster than helium or hydrogen. As the pressure decreases the offset due to the gas composition decreases as well.

Application of Pirani and thermocouple gauges are inexpensive and used primarily for measuring the vacuum in noncorrosive and nonreactive chambers. Their use in semiconductor systems is limited because hot wires and reactive gasses are generally not compatible.

11.2.4 Thermocouple gauges

A thermocouple gauge works very similarly to a Pirani gauge. The difference is that the temperature of the wire is measured precisely by the gauge, which is attached

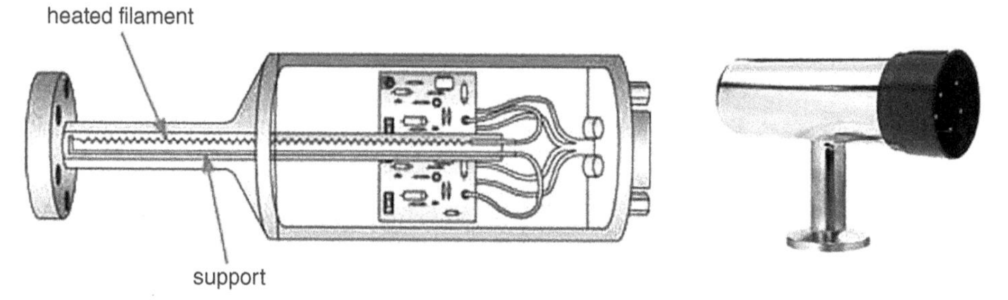

Figure 11.6 Pirani gauge.

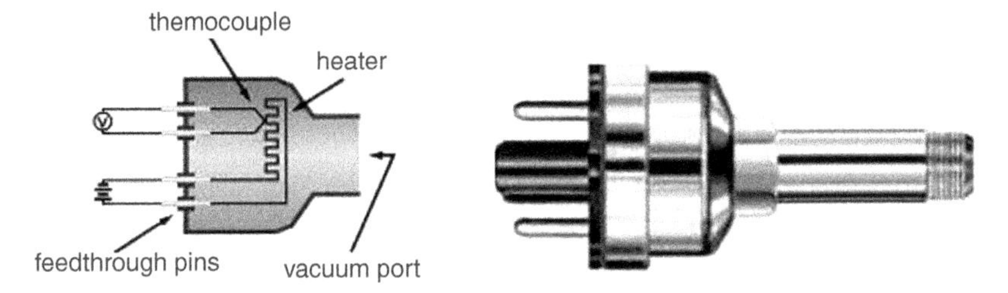

Figure 11.7 Thermocouple gauge.

to the wire (Figure 11.7). The current is determined based on the resistance. This gauge is normally used for comparison purposes and the sensitivity varies based on the pressure and the strength of the current. The reading is on a mini-voltmeter calibrated to show pressure, but it must be calibrated for each gas other than air and nitrogen. Another disadvantage is that it is not marked in a linear order. At low pressures, the scale markings are spread apart, and in higher ranges, the marks are closer together. For the most part, thermocouple gauges have the same advantages and disadvantages as Pirani gauges, although the thermocouple gauge is considered less expensive and more user friendly. Figure 11.7 is an example of a typical thermocouple gauge.

11.2.5 Capacity Membrane Gauge

This direct pressure gauge typically operates from atmosphere to 10^{-4} [mbar]. The gauge contains a small metal diaphragm with one side of the diaphragm exposed to the vacuum chamber and a known pressure on the rear (Figure 11.8). The pressure in the vacuum chamber either compresses or allows the expansion of the metal diaphragm. The gauge's electronics measures the movement of the metal diaphragm as a function of capacitance of the metal diaphragm with a fixed *parallel* electrode. When the diaphragm and the housing are constructed from stainless steel, this type of gauge is very resistant to corrosive and reactive gasses.

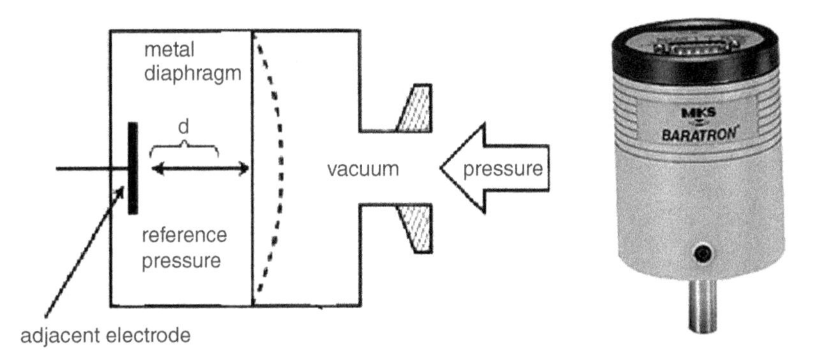

Figure 11.8 Capacity membrane gauge.

11.2.6 Ionization Gauges

Every modern vacuum furnace capable of operating in high vacuum relies on some form of ionization gauge for pressure measurements under 10^{-3} [mbar]. There are two competing ionization gauge technologies to choose from which are viable means for pressure measurements between 10^{-2} and 10^{-10} [mbar]. They sense pressure indirectly by measuring the electrical ions produced when gas is bombarded with electrons. Fewer ions will be produced by lower-density gasses.

A filament is used to emit electrons which are attracted to a positively charged grid. Inside the grid is a negatively charged collector (Figure 11.9). The electrons collide with gas molecules around the grid and ionize them. The positively charged ions are attracted to the collector and create an ionic current.

As in all vacuum gauges dependent on the type of gas, the scales of the indicating instruments and digital displays in the case of thermal conductivity vacuum gauges also apply to nitrogen and air. Within the limits of error, the pressure of gasses with similar molecular masses, i.e. O_2, CO, and others, can be read off directly.

11.2.7 Cathode Gauges

A *hot cathode ionization gauge*, or Penning gauge, typically operates from 10^{-2} to 10^{-12} [mbar]. It is composed mainly of three electrodes (Figure 11.10) consisting of three electrodes (cathode, anode, and ion collector) where the cathode is a hot cathode. The three electrodes are a collector or plate, a filament, and a grid. Electrons emitted from the filament move several times in back and forth movements around the grid before finally entering the grid. The hot cathode is a very high-yield source of electrons. The electrons are accelerated in the electric field and receive sufficient energy from the field to ionize the gas in which the electrode system is located. The positive gas ions formed are transported to the ion collector, which is negative with respect to the cathode, and give up their charge there. The ion current thereby generated is a measure of the gas density and thus of the gas pressure. Since the gaseous molecule density is proportional to the pressure, the pressure is estimated by measuring the ion current.

A cold cathode is an electrode that emits electrons, which is not electrically heated by a filament. There are two types of cold cathode ionization gauges: the Penning gauge and the inverted magnetron, also known as the *redhead gauge*.

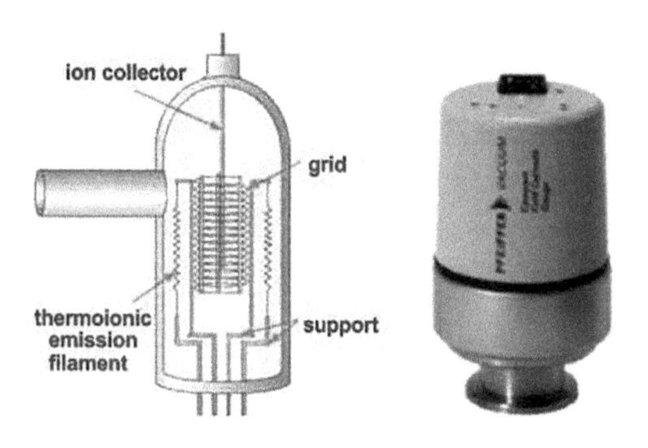

Figure 11.9 Ionization gauges.

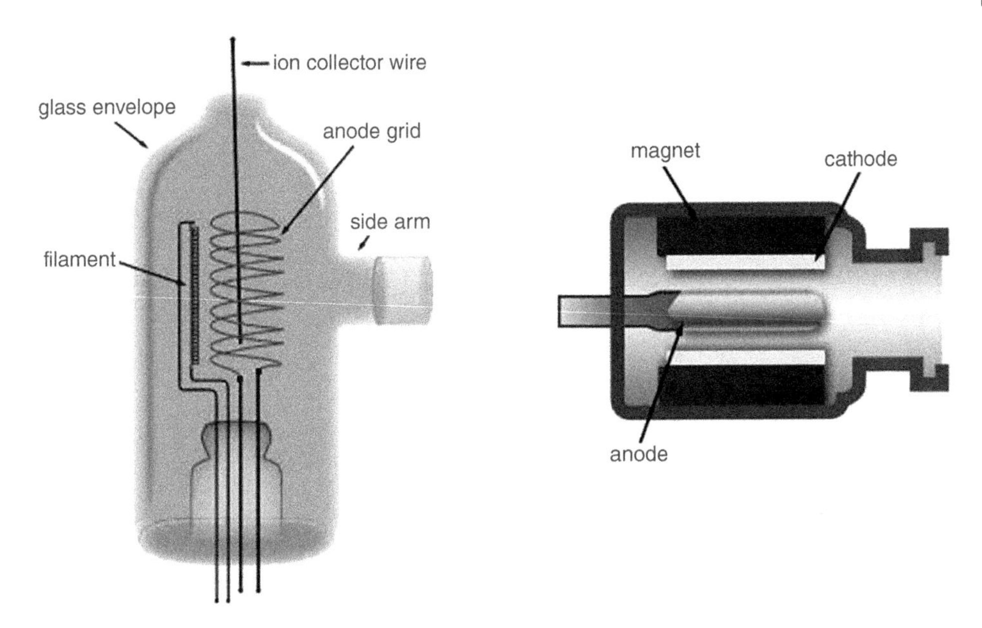

Figure 11.10 Hot cathode gauge and cold cathode gauge.

This gauge, like the one in Figure 11.10, makes use of the fact that the rate of ion production by a stream of electrons in a vacuum system is dependent on pressure and the ionization probability of the residual gas.

Two parallel connecting cathodes and the anode are placed midway between them. The cathodes are metal plates or shaped metal bosses. The anode is a loop of flattened metal wire, the plane of which is parallel to that of the cathode. A high-voltage potential is maintained between the anodes and the cathodes. In addition, a magnetic field intensity is applied between the elements by a permanent magnet, which is usually external to the gauge tube body.

Emitted electrons travel in helical paths (owing to the magnetic field), eventually reaching the anode, thus increasing the amount of ionization occurring within the gauge. Normally, the anode is operated at about 2 [kV], giving rise to a direct current caused by the positive ions arriving at the cathode. The pressure is indicated directly by the magnitude of the direct current produced. The pressure range covered by this gauge is from as low as 10^{-7} [mbar]. It is widely used in industrial systems because it is rugged and simple to use.

11.3 Types of Flow

There are three types of flow in vacuum techniques: the *viscous flow*, the *molecular flow*, and the transition between both, the *Knudsen flow* (Figure 11.11).

In the rough vacuum one has to deal with viscous flow. This has been studied before. The viscosity of a gas is caused by the collision between molecules whereby they exchange kinetic energy. When an upper layer of gas is moved, the molecules get a steering direction and energy. They exchange kinetic energy with the underlaying layer.

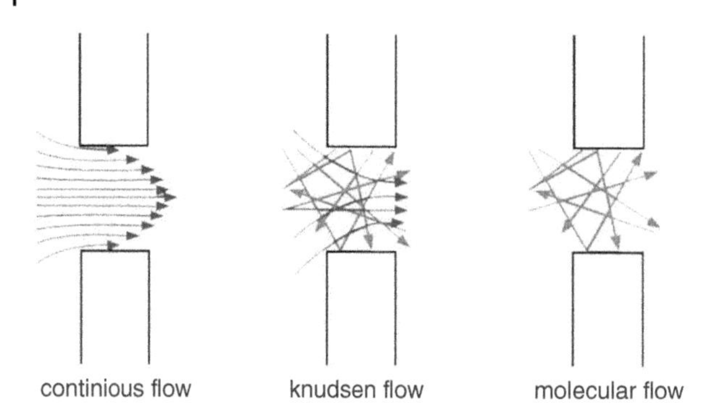

Figure 11.11 Types of flow. Source: Pfeiffer.

The underlaying layer will break the upper layer. This is called *friction*, better still, *viscosity*. As you know, in the viscous flow the distinction can be made between laminar and turbulent flow.

In this vacuum area it is so that if D is a characteristic dimension of the vessel (diameter, height), the following is valid:

$$\lambda \ll D$$

In other words, the molecules collide all the time with each other and sporadically with the wall of the vessel where they are locked up. *Molecular* flow is active in the high and ultra-high vacuum area. In these regimes the molecules can freely move, without interacting with each other. A molecular flow is characterized by:

$$\lambda \ll D$$

In other words, the molecules will practically not collide any more with each other, only with the wall surface. The transition area between viscous and molecular flow is the Knudsen flow. This one is characterized by:

$$\lambda \approx D$$

In the viscous flow area the preference direction of the molecules will be identical with the macroscopic movement of the gas. This is necessary because the gas molecules are so closely packed together that, caused by the collisions, they forward their preference direction to each other. The macroscopic velocity of the gas is thus a "group" velocity and has nothing to do with the thermal velocity of the molecules, which one superposes itself on the group velocity In the molecular flow area, however, the collisions of the molecules with the surfaces (or pipe line) are predominant. When a gas molecule collides against a wall it is kept there for some time and afterwards fired back into the space. But this isn't an elastic collision anymore, the gas molecule can go in any direction. It is then not any longer justified to speak about a *flow*, in the macroscopic sense.

In fact, it is so that the distribution of the vacuum areas that was made on the base of pressures does not need to correspond with the type of flow, because also the dimension of the pipe line or vessel is of importance. The distribution is based on vessels under laboratory conditions.

The volumetric flow of a vacuum pump is called the *pumping speed*. It is the flow at the *entrance* of the pump.

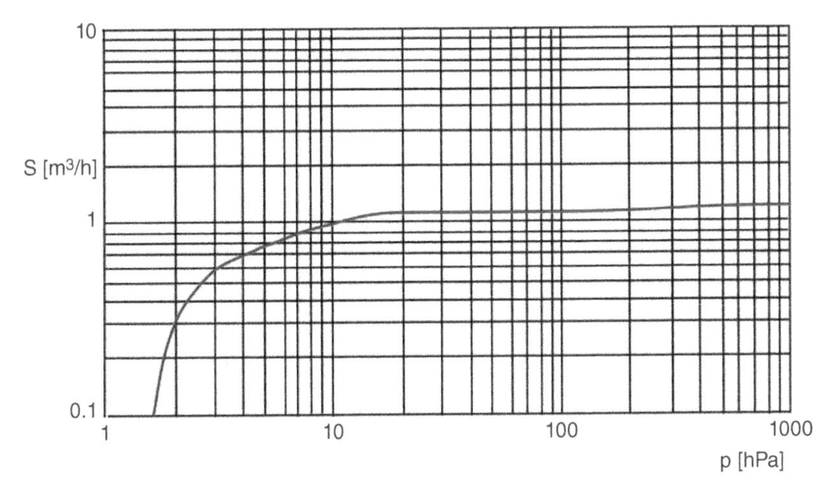

Figure 11.12 Characteristic of a diaphragm vacuum pump. Source: Pfeiffer.

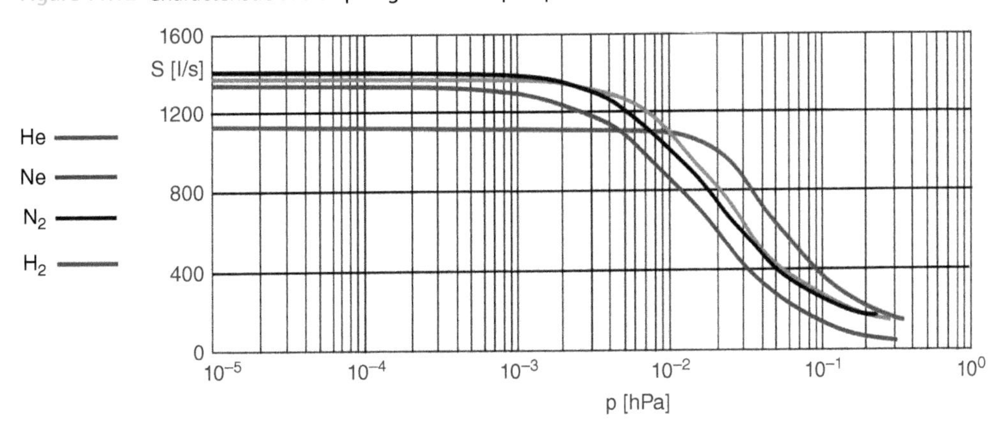

Figure 11.13 Characteristic of a turbomolecular vacuum pump.

For vacuum pumps in the low and middle vacuum area the characteristic is not dependent on the type of molecule (Figure 11.12). The characteristic in Figure 11.12 shows that this vacuum pump works well between 10 [mbar] and 1 [bar].

For vacuum pumps that work in the high and ultra-high vacuum area the characteristic is dependent on the type of molecule pumped (Figure 11.13). In Figure 11.13 it can be seen that this vacuum pump works well between 10^{-5} and 10^{-2} [mbar] ([hPa]). In other words, the pump must be followed by another pump that can overcome the vacuum area between 10^{-2} [mbar] and 1 [bar].

11.4 Rough Vacuum (1000–1 [mbar])

11.4.1 Membrane Pumps

Piston compressors can be used as vacuum pumps. The attainable pressure is 100 [mbar] for a single-stage and 10 [mbar] for a two-stage implementation. Often a membrane

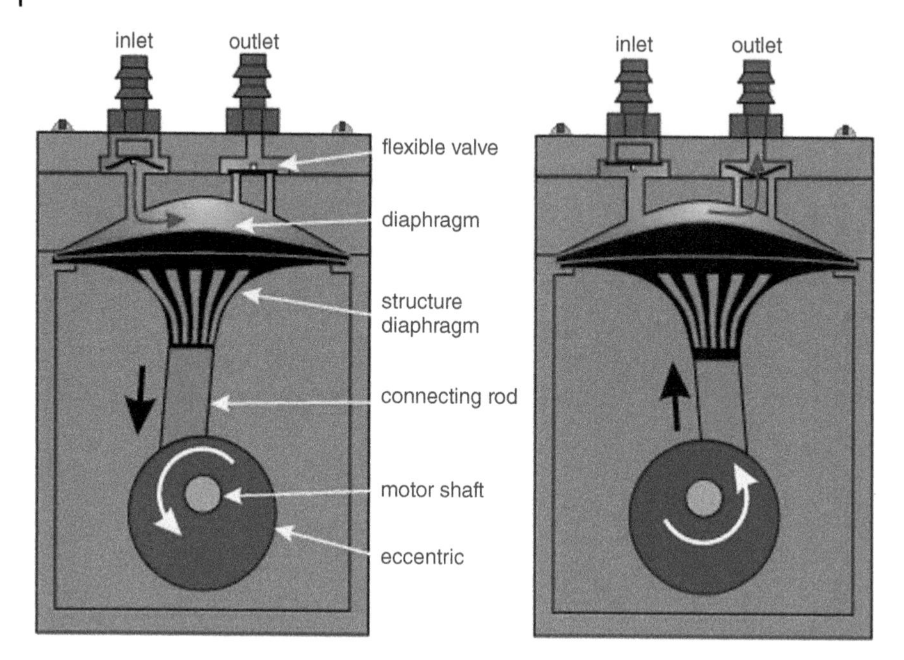

Figure 11.14 Diaphragm piston vacuum pump. Source: KNF.

compressor is used. These ones can be designed with a small dead volume, so that vacuum pressures up to 80 [mbar] can be reached for a single stage and 10 [mbar] for a two-stage implementation (Figure 11.14). Three-stage implementations go as high as 2 [mbar] and four-stage implementations to $5 \cdot 10^{-1}$ [mbar].

The so-called pump speed (normalized after DIN 28400) goes to as much as 30 [m^3/h].

The big advantage of those membrane pumps is, in comparison with, for instance, vane pumps, the oil-free character of the discharged gas. One speaks of *dry vacuum pumps*.

11.4.2 Steam Jet Vacuum Pumps

In case the reader hasn't read the sections on ejectors, the working principle is repeated here (Figure 11.15). At the inlet, steam, the motive fluid, is fed to the head (1) of the ejector where the high steam pressure is converted to a vacuum pressure in exchange for velocity. The steam now has a supersonic velocity, but because of the very low pressure in

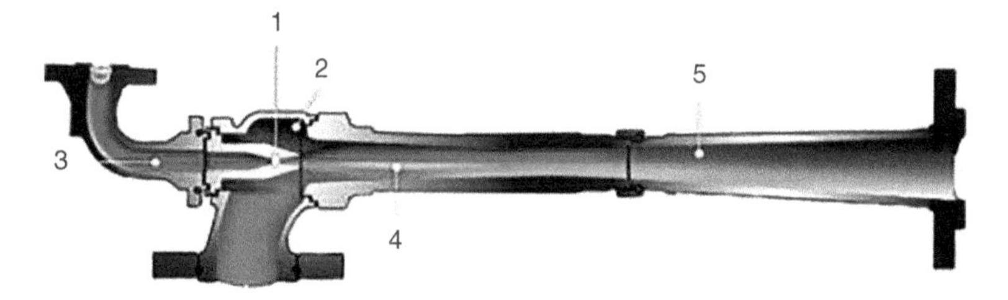

Figure 11.15 Operating principle. Source: Körting Hannover AG.

the chamber (2) gas can be sucked from a vessel (on vacuum pressure). After that the mix steam/gas will enter the chamber (4) supersonically. A converging passage in the chamber leads to a velocity increase and a pressure decrease. Then the velocity of the mix will, in chamber (5) in the diverging part, be converted to a subsonic velocity and the pressure increased to the atmospheric counterpressure in order to dump the sucked gas from the vacuum vessel. When the ejector is used as a vacuum pump the whole device is designed so that the pressure at the discharge of the ejector is the atmospheric pressure.

A single-stage ejector attains a compression ratio of 1/10, so the vacuum pressure will be 100 [mbar(a)]. If a deeper vacuum is wanted refuge must be taken to multistage ejectors. So a five-stage ejector will attain a vacuum pressure of 0.01 [mbar(a)].

Between two ejectors a condenser is placed to condense the motive fluid (the steam) and drain it away. This is done with a heat exchanger with cooling water but is only possible when the temperature of the cooling water is below the condensation temperature of the steam.

In Figures 11.16 and 11.17 a single-stage ejector is shown. Such a single-stage ejector reaches 100 [mbar(a)], in the supposition that the counter pressure is the atmospheric pressure. A two-stage ejector without intercooling attains 40 [mbar(a)] and is applied for low suction flows. For higher flows at 40 [mbar(a)] a two-stage or three-stage ejector must be used with intercooling. A four-stage ejector can reach 0.4 [mbar(a)] and with a five-stage device 0.01 [bar(a)].

Now the cooling job. Suppose a heat exchanger is placed after the ejector, with a cooling temperature of 25 °C. If the cooling water increases by 10 °C in the heat exchanger the final temperature will be 35 °C.

Consider an ideal heat exchanger on one hand and on the other hand that the final temperature of the water–air mix is also 35 °C. If the steam flow in the ejector is significantly higher than the flow of the sucked air, then the partial pressure of the steam will be equal to the total pressure of the mix. The steam would condensate at a temperature of 35 °C, and a look in the saturated steam tables leads to a saturation pressure of 56 [mbar(a)].

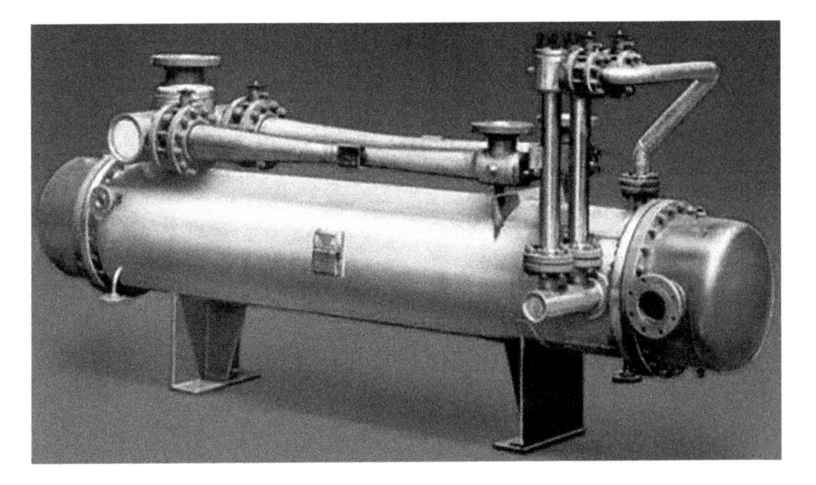

Figure 11.16 Single-stage jet pump. Source: Scam Torino.

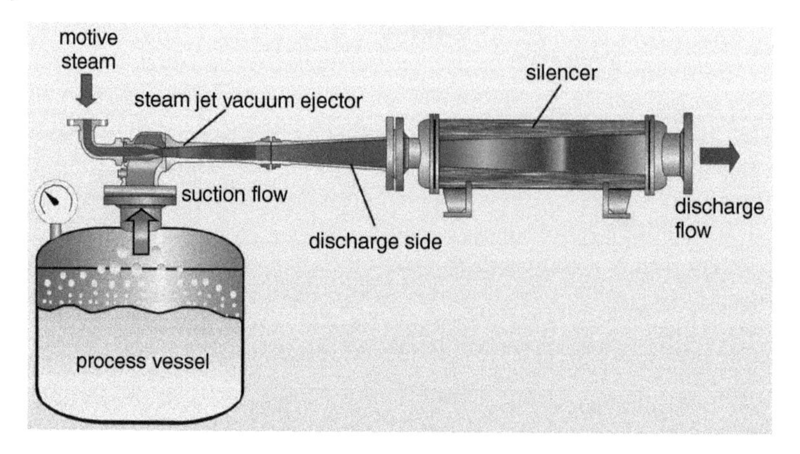

Figure 11.17 Single-stage steam jet vacuum ejector. Source: Körting Hannover AG.

The deeper the vacuum pressure wanted, the lower the condensation temperature will have to be. This means that it is necessary to have a lower cooling temperature, and eventually a higher cooling water flow.

From the steam tables (saturated steam):

°C	Pressure [mbar(a)]
0	6112
35	56 280

But if the total pressure of the mix that flows out of the ejector is lower than 6 [mbar(a)] then its temperature is 0 °C, and that means that the steam would freeze. In the ejector an ice layer would build up, something that is absolutely not desired. How to prevent this? By heating the ejector, with steam (Figure 11.18).

A five-stage steam jet pump (Figure 11.19) works for suction pressures of 1 [mbar] but there are designs for 0.01 [mbar].

In the first stage the pump sucks vapor and gas from the process vessel to an end pressure of, for instance, 2 [mbar]. Stage 2 compresses the motive steam and the sucked vapor and gas to 55 [mbar]. Then follows a surface heat exchange where the vapor and gas is cooled. This way the steam condenses to water, and is drained way. A lesser mass flow goes to stage 3. This has the advantage that for the compression in stage 3 less

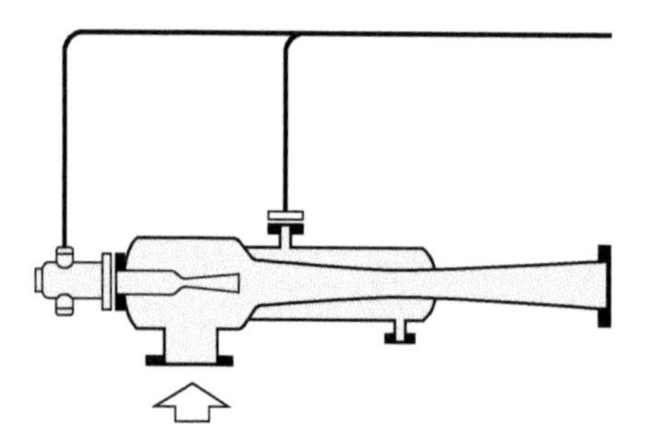

Figure 11.18 Heating ejector with steam. Source: GEA.

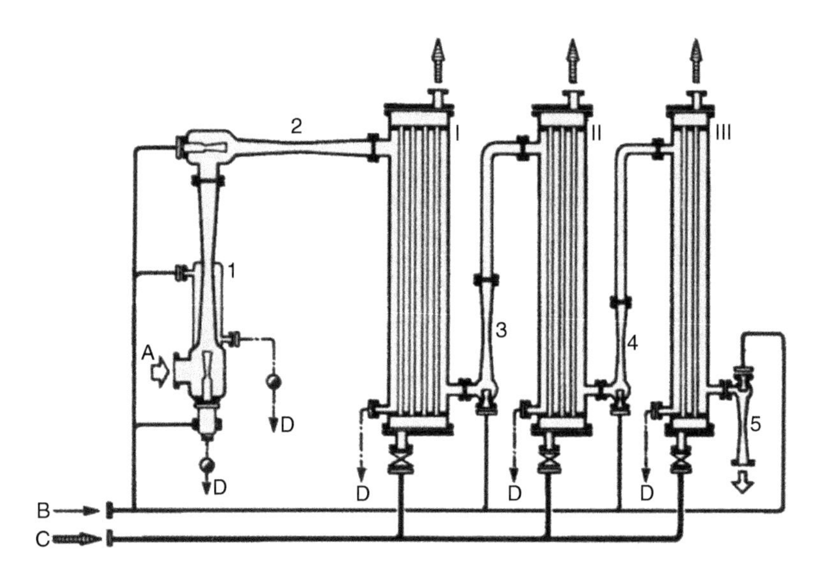

Figure 11.19 Five-stage steam jet pumps. Source: GEA.

motive steam is necessary. A vacuum pressure of 160 [mbar] is attained. Then follows heat exchanger II. In stage 4 the pressure will become 400 [mbar]. Then follows surface heat exchanger III and finally stage 5 that compresses everything against the atmosphere.

For suction pressures below 6 [mbar] the head of the mixing tube of the first stage is heated. In this way ice formation is avoided. When not all the air but also other substances with high melting points are sucked, higher stages have to be heated too.

This type of heat exchange, a surface heat exchange, is applied where the sucked medium may not come into contact with the cooling water or where the condensed steam can't be recuperated. Steam is a very expensive fluid because it needs water treatment before getting heated.

In principle steam jets and condensers are executed in steal or bronze. For aggressive substances there are designs in porcelain, PTFE, graphite, and glass.

Figures 11.20–11.23 show a two-stage construction. Figure 11.24 shows the performance of multistage steam jet pumps.

11.4.3 Liquid Vacuum Ejector Pump

These very simple ejectors use water as a motive fluid.

The attainable absolute vacuum pressure is dependent on the temperature of the water (Figures 11.25 and 11.26).

11.4.4 Gas Jet Vacuum Pump

Gas, and mostly compressed air, is used as a motive fluid here.

In Figures 11.27 and 11.28 two applications are represented. Figure 11.27 shows an ejector that is used for making a suction pipe airless for a centrifugal pump. In Figure 11.27 an ejector is used to suck a liquid.

Figure 11.20 Two-stage ejector with twin element surface condenser. Source: Scam Torino.

Figure 11.21 Three stage ejector, 3 surface condensers. Source: Scam Torino.

Figure 11.22 Two-stage jet pump with heat exchangers. Source: Scam Torino.

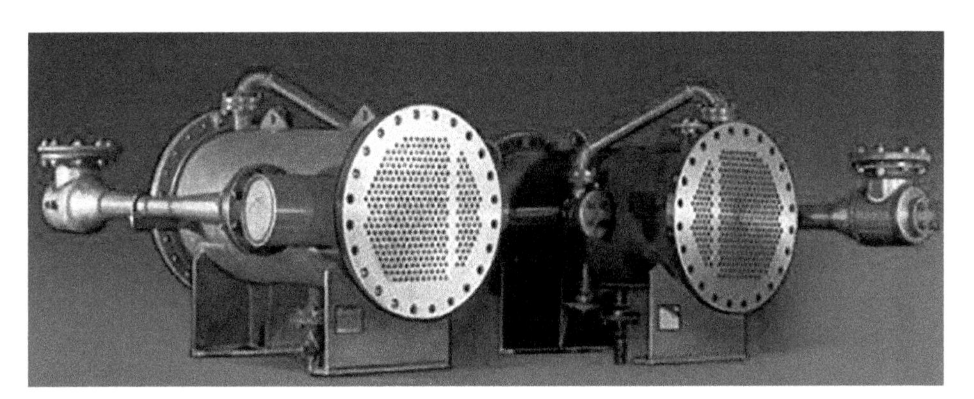

Figure 11.23 Two-stage jet pump with shell and tube heart exchangers : open shell-plate. Source: Scam Torino.

Gas ejectors are often used in combination with a liquid ring pump (Figure 11.28). The ejector delivers pressures from 50 to 100 [mbar] to the water ring pump and can attain a vacuum pressure of 5 [mbar].

11.4.5 Centrifugal Vacuum Pumps

These ones deliver high flows to 50 000 [m^3/h] at moderate vacuum pressure of 500 [mbar(a)].

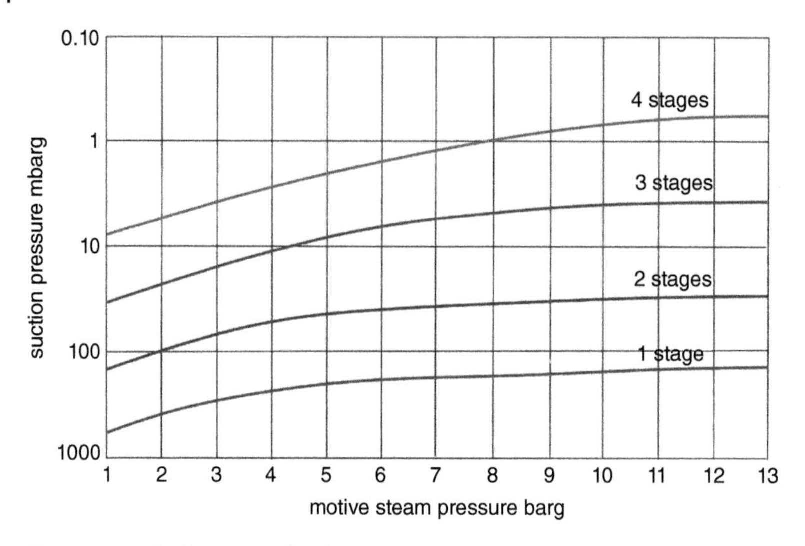

Figure 11.24 Performance of multistage steam jet pumps.

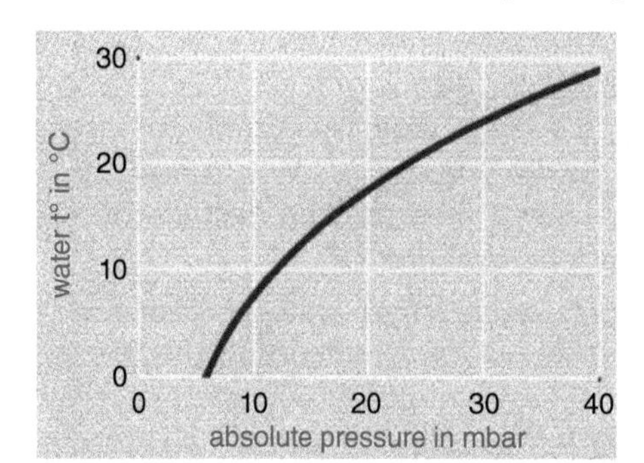

Figure 11.25 Dependence of vacuum pressure on temperature of the water. Source: GEA Wiegand.

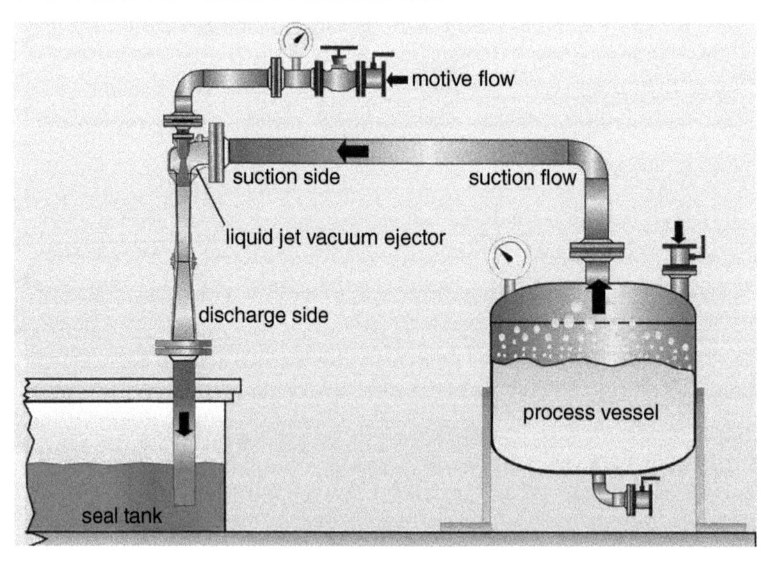

Figure 11.26 Application of a liquid vacuum ejector pump. Source: Körting Hannover AG.

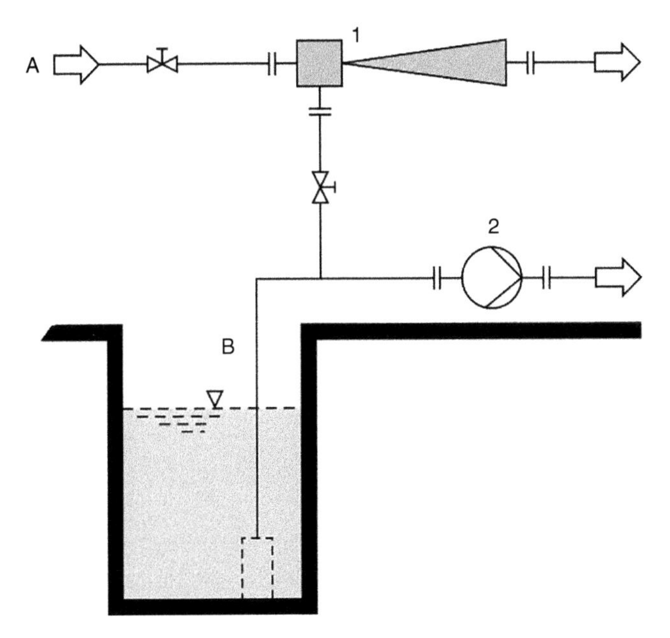

Figure 11.27 Application of a gas jet vacuum pump. Source: GEA.

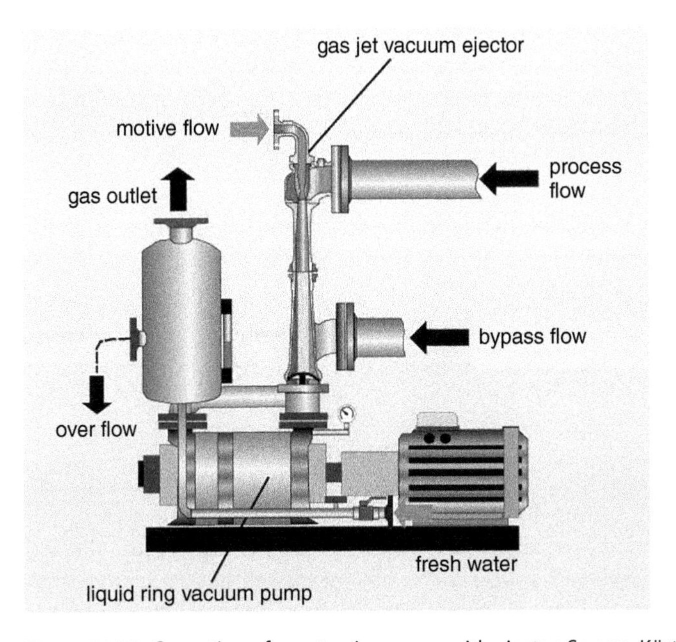

Figure 11.28 Operation of a water ring pump with ejector. Source: Körting Hannover AG.

11.4.6 Liquid Ring Pumps

Liquid ring pumps, which were discussed before, are applied a lot in vacuum techniques (Figure 11.29). They are applied for pressures of 30 [mbar(a)] and flows to 40 000 [m^3/h].

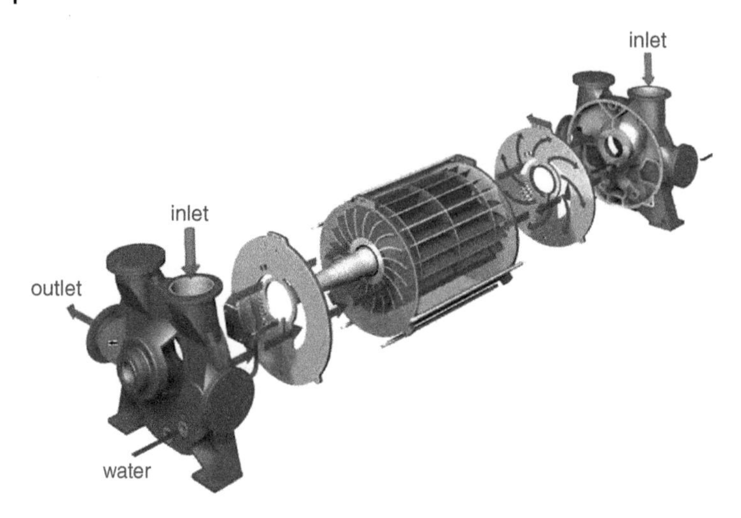

Figure 11.29 Liquid ring pump with two-sided entry. © Gardner Denver – Nash-Elmo used with permission.

11.5 Medium Vacuum (1–10^{-3} [mbar])

11.5.1 Vane Pump

There are also vane pumps with only two or three vanes. These ones are exclusively used in vacuum, and for low pressures.

A distinction must be made between oil-lubricated and "dry" vacuum pumps. By using a sealing oil film it is possible to reach ratios, in a single-stage high compression, of up to 10^5. Without oil the internal leak between the press and suction side will be so great that the achievable pressure ratio will only reach a value of 10!

Figure 11.30 shows a section view of a vane pump with three vanes. There is plentiful oil lubrication. The lubrication is being taken care of by the difference in pressure between the press and the suction side.

They can work against the atmosphere and attain vacuum pressures of $<2.5 \cdot 10^{-2}$ [mbar].

Lubricated vane pumps with three vanes reach 275 [m^3/h]. Nonlubricated pumps reach 1200 [m^3/h], but reach pressures of only $<5 \cdot 10^{-1}$ [mbar].

The more modern two-vane pump is represented in Figures 11.31 and 11.32. The oil is supplied by a gear pump.

There are single-stage and two-stage implementations. With two-stage vane pumps (Figure 11.33), lower pressures are reachable ($<1 \cdot 10^{-4}$ [mbar]) than can be reached with single-stage versions. The reason is that in the case of single-stage pumps oil inevitably comes into contact with the atmosphere in the vessel where the gas is sucked from. In other words, oil escapes to the vacuum side, and if this must be avoided as much as possible, the minimum attainable pressure must be limited. The higher the difference in pressure between the outside atmosphere and the atmosphere at the discharge side of the pump, the more oil escapes to the vessel. Why doesn't this happen with a two-stage pump? Because the oil is first supplied to the second stage and then diffuses to the first stage.

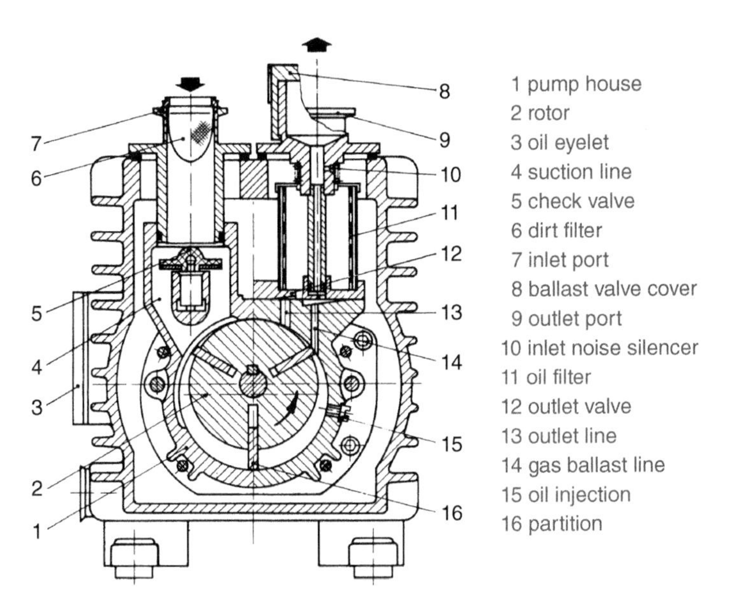

Figure 11.30 Vane pump with three vanes. Source: Leybold GmbH.

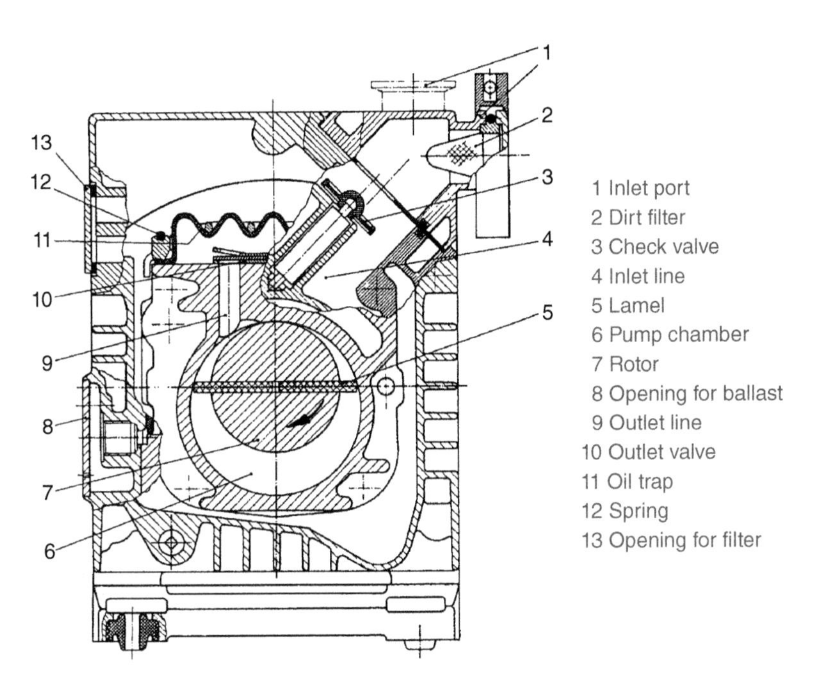

Figure 11.31 Two-vane pump. Source: Leybold GmbH.

11.5.2 The Gas Ballast

The *gas ballast* as used in the vane pump and the rotating piston pump (see later) allows not only permanent gasses but also high quantities of condensable vapors to be pumped.

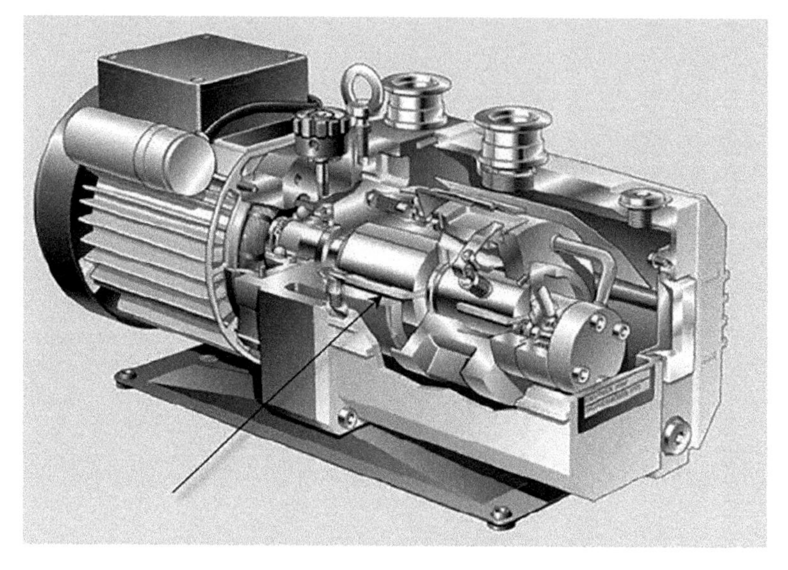

Figure 11.32 Two-vane vacuum pump. Source: Leybold GmbH.

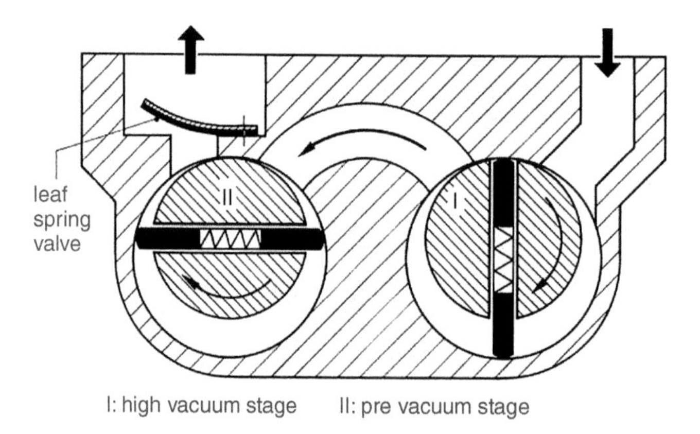

Figure 11.33 Two-stage vane pump. Source: Leybold GmbH.

The gas ballast (Figure 11.34) avoids condensation of vapors in the pumping room of the vacuum pump. When these condensable vapors are compressed until their saturation temperature, at the governing temperature of the pump, liquid is formed. This is the case with the water in the air. This vapor is in principle in a gas state, in the form of superheated steam. Consider a pump temperature of 70 °C, then this superheated steam can only be compressed to a pressure of 312 [mbar] (as one can look up in the steam tables; see Table 11.2) without becoming liquid. In the case that the steam becomes liquid one consequence is that the pressure in the chamber will stay at this saturation value and the pressure will not be able to increase anymore. The outlet valve does not open! However, the water will stagnate in the pump and form an emulsion with the oil. The lubricating properties of the oil are gone and the pump can strand (i.e. lock up because of lack of oil (metal faces heat up and weld together).

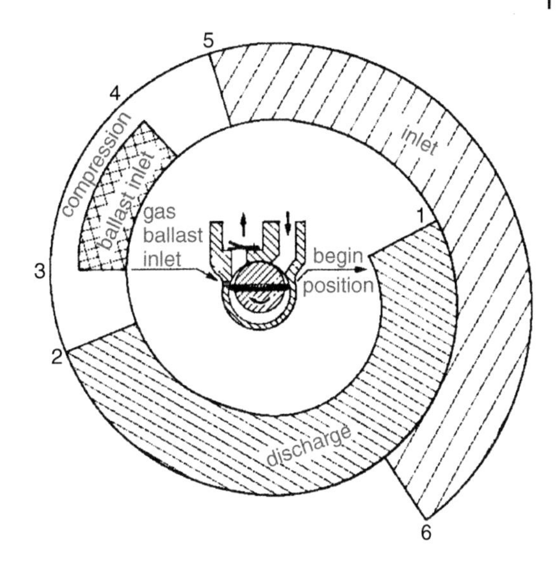

Figure 11.34 Gas ballast. Source: Leybold GmbH.

Table 11.2 Steam table saturated steam.

T (°C)	p [bar]	v' [m³/kg]	v'' [m³/kg]
0	0.0061	0.0010002	206.3
10	0.0123	0.0010002	106.4
20	0.0234	0.0010017	57.84
30	0.0424	0.0010043	32.94
40	0.0738	0.0010078	19.56
50	0.1233	0.0010121	12.05
60	0.1992	0.0010172	7.682
70	0.3116	0.0010229	5.048
80	0.4736	0.0010293	3.410
90	0.7011	0.0010363	2.361
100	1.0132	0.0010438	1.673

That's why an operative part must be provided to counteract that. It works as follows (Figure 11.34). Before compressing the gas, a determined quantity of air, the *gas ballast*, is allowed in the pump chamber of the pump. The quantity is determined in a way that the compression ratio of the pump is reduced to 10 : 1 maximum. In this way condensable gasses that were sucked by the pump are compressed together with the gas ballast, and will never reach their condensation point, and in this way will be discharged.

In Figure 11.35 the processes with and without gas ballast are shown schematically. On the left is the operation without gas ballast. In a1 the pump is connected with the vessel that contains a little bit of air: the pressure is 70 [mbar]. In a2 the pump is blocked from the vessel: the compression can start. In a3 the compression cycle is already so far

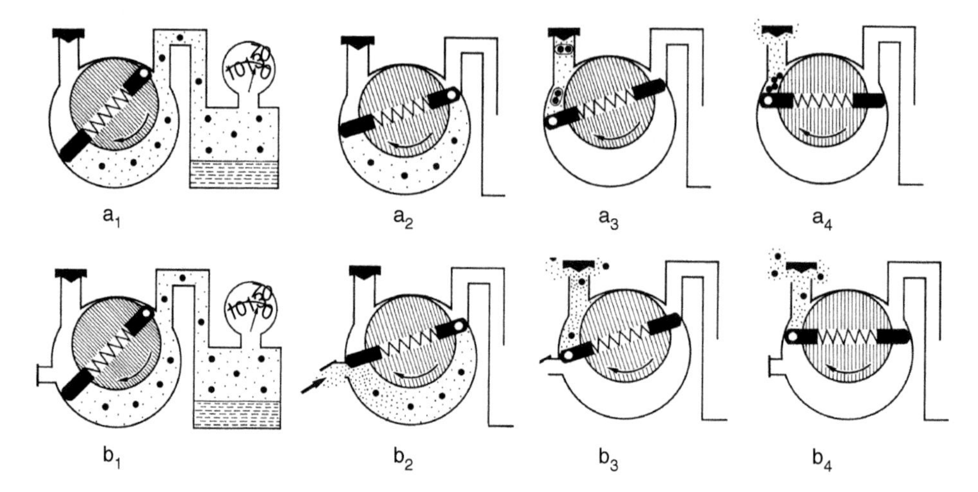

Figure 11.35 Process in a vane pump with and without gas ballast. Source: Leybold GmbH.

that the vapor condenses: droplets are not yet formed. That happens in a4, where the residual air comes in overpressure and the outlet valve opens: the droplets stay behind in the pump.

In b1 the same state is shown as in a1. But in b2 it can be seen that, as the pump chamber is locked from the chamber, the gas ballast valve (an overpressure valve) opens and air from the outside comes in. In b3 the outlet valve opens and the gas with its condensable vapor leaves the pump. The overpressure required to open the valve is attained very quickly and this is due to the suppletory quantity of gas, the gas ballast. In b4 the process is continued.

11.5.3 Screw Vacuum Pumps

Oil-free screw pumps can also be applied as vacuum pumps (Figures 11.36–11.38). They reach vacuum pressures of 10^{-2} [mbar] against the atmosphere, and flow to 750 [m^3/h].

Figure 11.36 Screw vacuum pump. Source: Ilmvac.

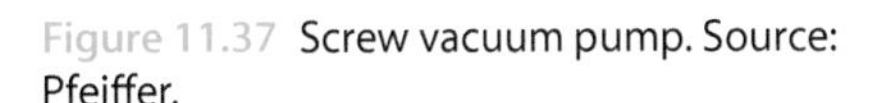
Figure 11.37 Screw vacuum pump. Source: Pfeiffer.

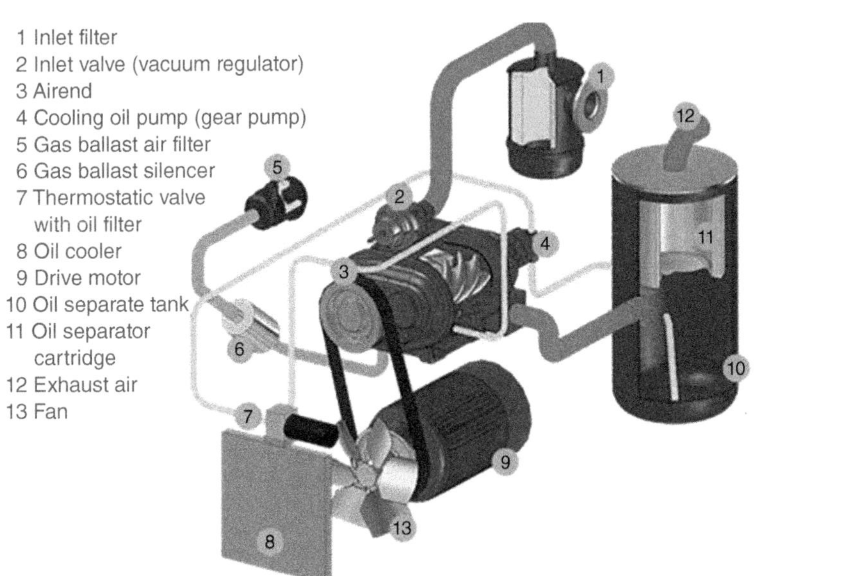

Figure 11.38 Screw vacuum pump. Source: Kaeser.

11.5.4 Scroll Vacuum Pump

The lowest attainable vacuum pressure is 10^{-2} [mbar] against the atmosphere. The flow reaches 50 [m^3/h].

11.5.5 Rolling Piston

In Figures 11.39 and 11.40 a vacuum pump with rolling piston is shown as a section drawing. A piston (2) is moved by an eccentric (3) in the direction of the arrow and moves therefore along the wall of the chamber. The gas flows to pump via the inlet port (11) in the pump. It passes along a suction valve (12) and arrives in the pump chamber (14). The suction valve forms one part with the piston. The bearing makes a rotating and gliding movement and periodically releases thereby the inlet opening. Finally, the gas arrives in the compression chamber (14). During the rotation of the rotor the gas is compressed until finally it is pressed through the oil secured valve (5). As with a vane pump the oil reservoir serves for lubrication, sealing, and cooling.

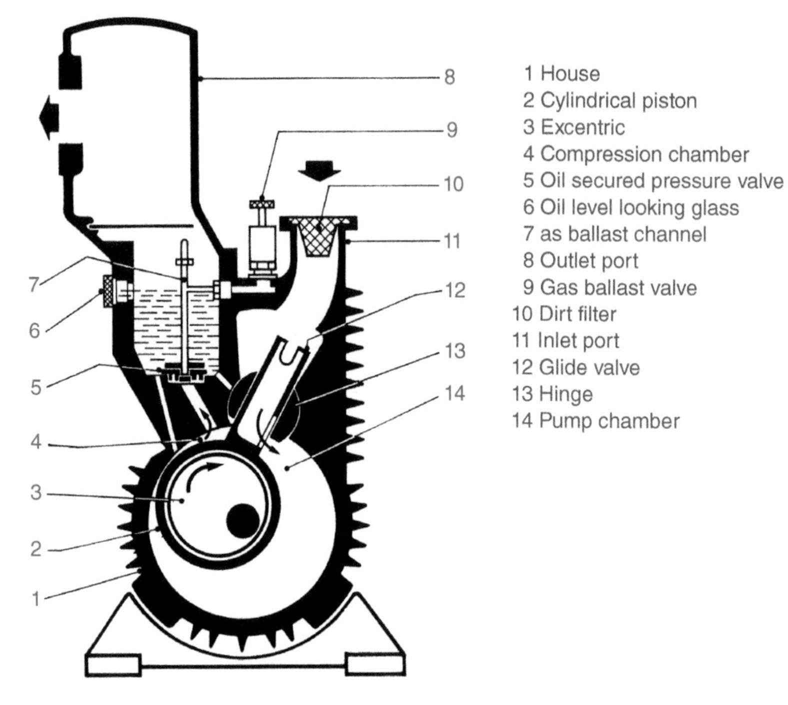

Figure 11.39 Rolling piston. Source: Leybold GmbH.

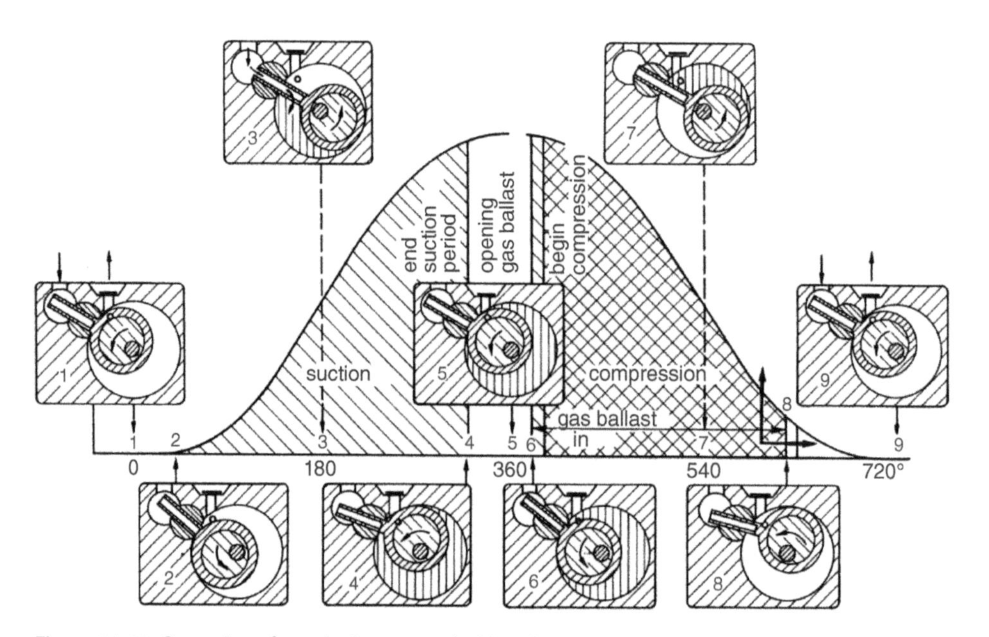

Figure 11.40 Operation of a cycle. Source: Leybold GmbH.

Operating pressures without gas ballast reach $4 \cdot 10^{-2}$ [mbar] and deliver flow of up to 250 [m^3/h].

11.5.6 Claw Pump

Isochoric compression, which also limits the temperature ultimately attained during compression, especially in the stage on the side of the atmosphere, and which ensures protection against internal explosions, is performed by venting the pumping chamber with cold gas from a closed refrigerating gas cycle. In Figure 11.41, 1 indicates the start of the intake process by opening the intake slot through the control edge of the right rotor. The process gas then flows into the intake chamber, which increases in size. The intake process is caused by the pressure gradient produced by increasing the volume of the pumping chamber. The maximum volume is attained after 3/4 of a revolution of the rotors (Figure 11.41, 2). After the end of the intake process, the control edge of the left rotor opens the cold gas inlet and at the same time the control edge of the right rotor opens the intake slot (Figure 11.41, 3) once more. In 4, the control edge of the left rotor terminates the discharge of the gas which has been compressed to 1000 [mbar] with the cold gas; at the same time the control edge of the right rotor completes an intake process again.

Normally, the pump has a few stages. The attainable vacuum pressure can then be 1 [mbar] (Figure 11.42). In the following stages the vacuum pressure goes from 1 to 1000 [mbar].

- First stage: from 1 to 3 [mbar]
- Second stage: from 3 to 15 [mbar]
- Third stage: from 15 to 150 [mbar]
- Fouth stage: from 150 to 1000 [mbar]

Pump speeds achieve 800 [m^3/h].

11.5.7 Roots Vacuum Pumps

Vacuum application roots pumps (Figures 11.43 and 11.44) differ a bit from the classic roots blowers. They have a smaller suction flange than the press flange. On the press side there are three openings. There are two long pipe lines on both sides of the press side. They serve as an inlet for fresh air.

Step 1

The upper rotor encloses a volume of gas.

Step 2

The upper roots rotor has completely enclosed the volume suction gas.
This gas is in a vacuum state.

Step 3

The vacuum is rapidly filled with atmospheric air.
The upper rotor presses the gas to the press side.

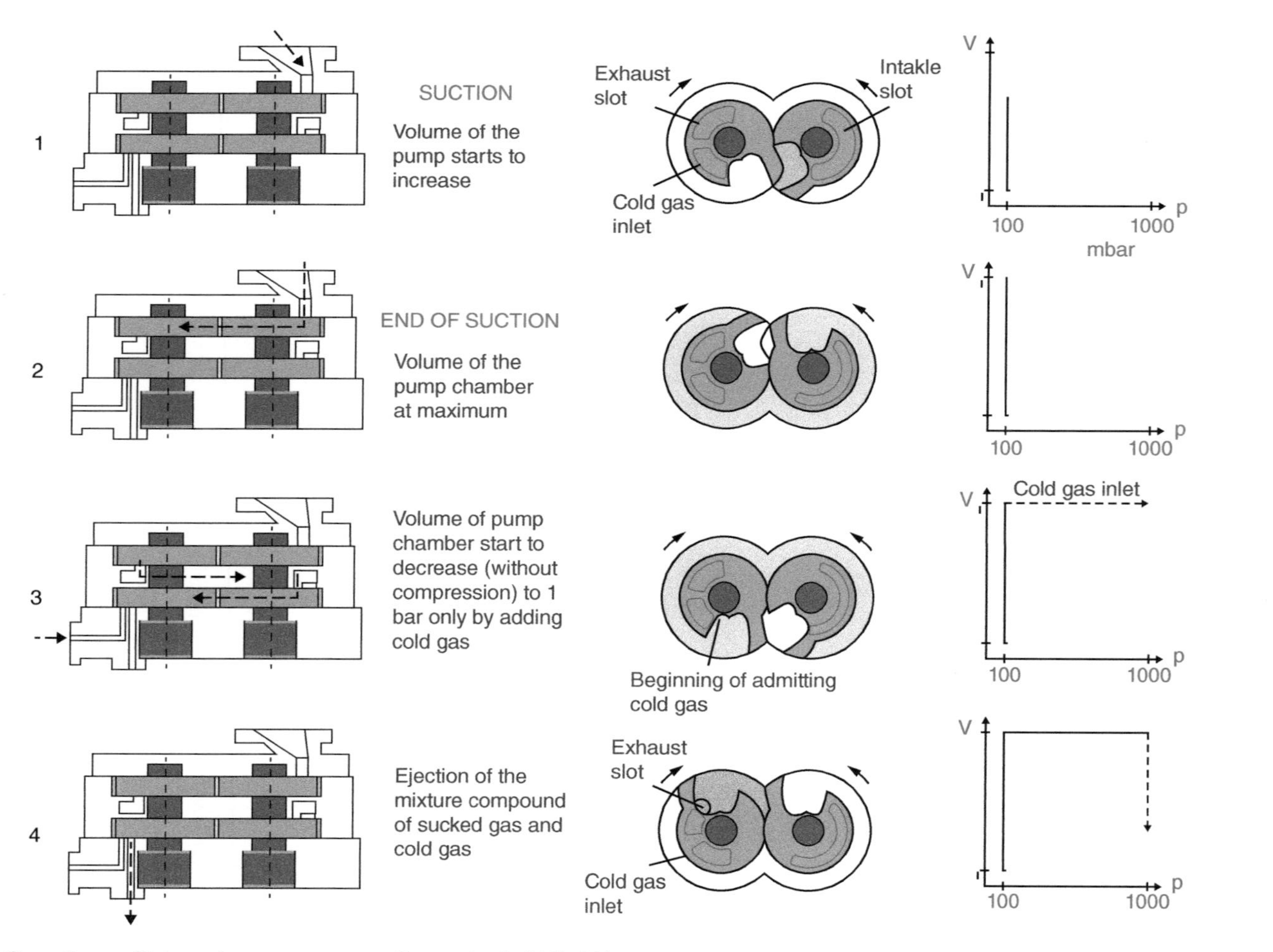

Figure 11.41 Operation multistage claw vacuum pump. Source: Leybold GmbH.

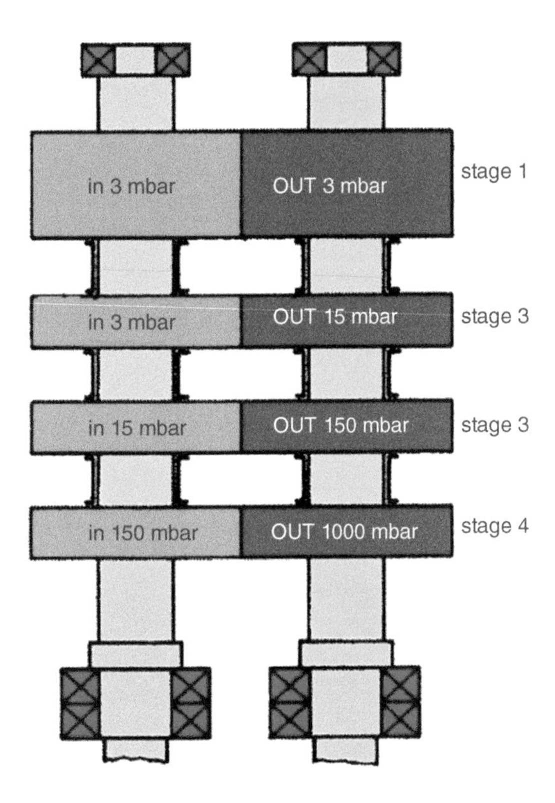

Figure 11.42 Performance of multistage claw pumps . Source: Leybold.

Figure 11.43 Roots vacuum pump. Source: Hibon.

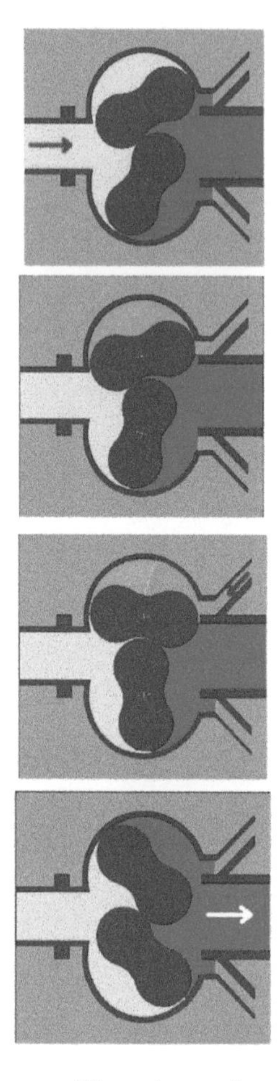

Figure 11.44 Roots vacuum pump. Source: Hibon.

The mix was heated by the compression but fresh air takes care of its cooling.

Step 4

Normally, the end compression temperature is about 100 °C with a vacuum pressure of 400 [mbar(a)].

Figure 11.45 shows such a roots vacuum pump, with a sound silencer on the suction and press side.

The counterpressure may not be too high; otherwise, the rotors of the compressor will become too hot; they have an initial clearance of 0.5 [mm] and would expand too much. That's why a safety valve is provided (Figure 11.45).

Since roots pumps do not have internal compression or an outlet valve, when the suction chamber is opened its gas volume surges back into the suction chamber and must then be re-discharged against the outlet pressure. As a result of this effect, particularly in the presence of a high pressure differential between inlet and outlet, a high level of energy dissipation is generated, which results in significant heat-up of the pump at low gas flows that only transport low quantities of heat. The rotating roots pistons are relatively difficult to cool compared to the housing, as they are practically vacuum-insulated.

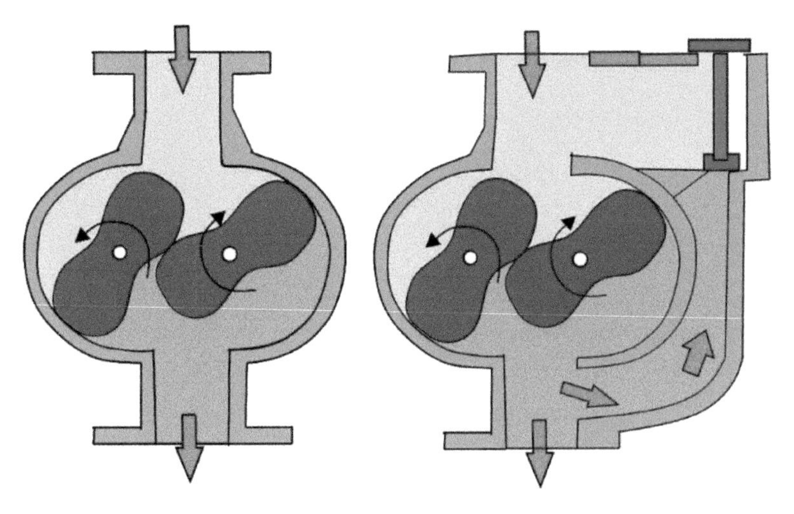

Figure 11.45 Roots compressor as vacuum pump. Source: Pfeiffer.

Consequently, they expand more than the housing. To prevent contact or seizure, the maximum possible pressure differential, and so also the dissipated energy, is limited by an overflow valve. It is connected to the inlet side and the pressure side of the pump through channels. A weight-loaded valve plate opens when the maximum pressure differential is exceeded and allows a greater or lesser portion of the intake gas to flow back from the pressure side to the inlet side, depending on the throughput. Owing to the limited pressure differential, standard roots pumps cannot discharge against atmospheric pressure and require a backing pump. However, roots vacuum pumps with overflow valves can be switched on together with the backing pump even at atmospheric pressure, thus increasing their pumping speed right from the start. This shortens evacuation times.

Usually, roots pumps cannot counteract the atmospheric pressure unless they are cooled; they are used as a first stage and are followed by a vacuum pump of another type like a piston pump, a vane pump, a screw compressor, or a water liquid ring pump. Or they are designed as a multistage, like in Figure 11.46.

One can reckon that every roots compressor has a compression ratio of 10 : 1; if two roots pumps are placed after each other followed by another vacuum pump vacuum pressures of up to 10^{-5} [mbar] can be reached!

And flows of up to 100 000 [m^3/h].

11.6 High Vacuum (10^{-3}–10^{-7} [mbar])

First of all some words that will be used are defined:

Sublimation: the change of state whereby a solid evaporates directly without going through the liquid phase.

Sorption: is a physical and chemical process by which one substance becomes attached to another. Specific cases of sorption are adsorption and absorption.

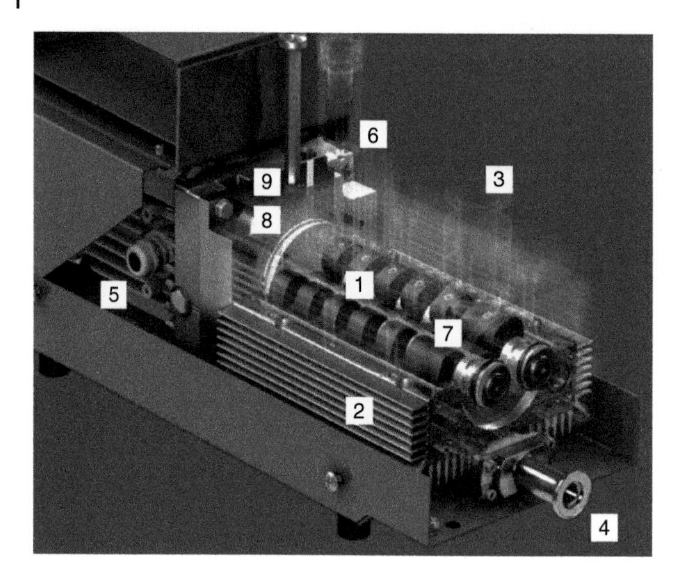

Figure 11.46 Multistage roots vacuum pump. Source: Pfeiffer.

Adsorption: the incorporation of a substance in one state into another of a different state (e.g. liquids being absorbed by a solid or gasses being absorbed by a liquid).
Absorption: the physical adherence or bonding of ions and molecules onto the surface of another phase (e.g. reagents adsorbed to a solid catalyst surface).
Diffusion: the movement of particles without a driving force; the movement is just a result of the thermal energy.
Precipitation: deposit.
Fractioning: a separation process in which a certain quantity of a mixture (gas, solid, liquid) is divided into a number of smaller quantities.
Hydrophile character: the character of a molecule to attract water.

11.6.1 Diffusion Pumps

Diffusion pumps (Figure 11.47) belong to the group of flow-driven pumps. Distinction is made between ejector pumps as *waterjet pumps* (17 [mbar] $< p <$ 1013 [mbar]), *vapor jet pumps* ($10^{-3} < p < 10^{-1}$), and finally the *diffusion pumps* ($p < 10^{-3}$ [mbar]). They all work with a rapid moving motive substance (water, water vapor, oil, or mercury vapor). The pumping mechanism of all these devices is the same. The gas molecules are removed from the vessel and finish up in the motive fluid, that then expands after they have passed a narrowing passage. The molecules of the motive fluid bring over their impulse to the sucked gas molecules in the direction of the flow. In this way the gas is pumped to the space on higher pressure. With a diffusion pump a paradoxical phenomenon appears that gasses at a very low pressure can be pumped with a fluid flow from essentially low pressure to a place of high pressure! This paradox is due to the fact that the vapor flow is initially free of the gas to be pumped and that a gas on a higher partial pressure can diffuse to an area of lower partial pressure (the vapor flow). In Figure 11.41 the principle of a diffusion pump is represented. An electric heating element (1) takes care of the evaporation of oil in the boiler (2). The oil vapors are then expanded in four consecutive

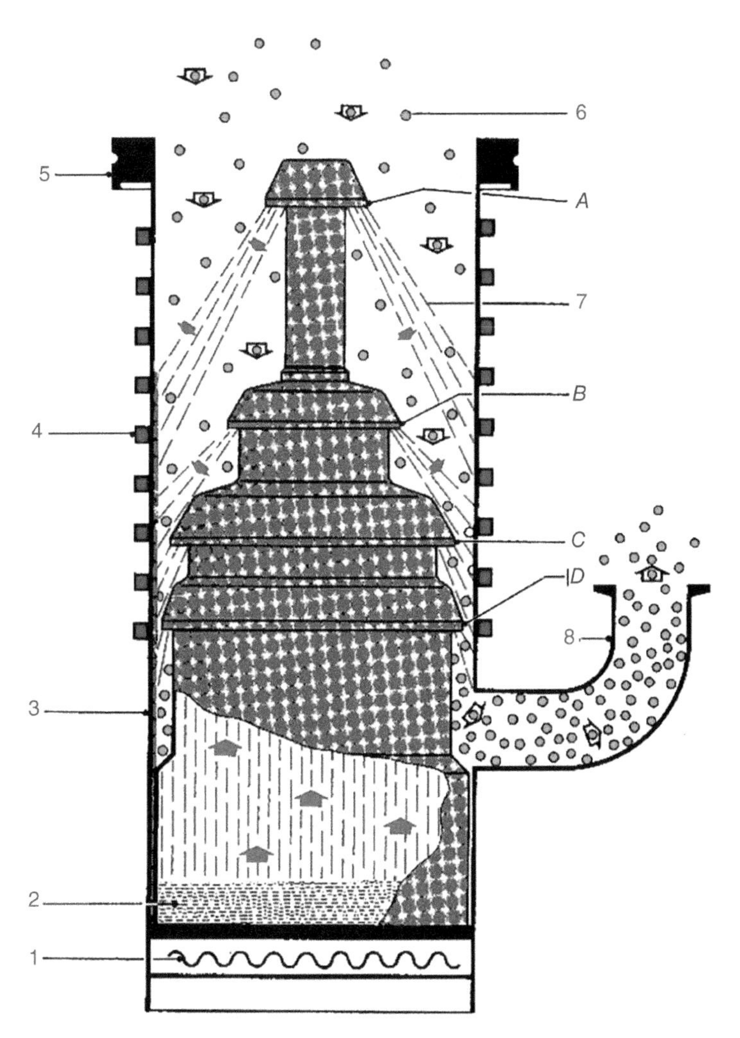

Figure 11.47 Diffusion pump.

stages A, B, C, and D in nozzles and flow with supersonic velocity outwards (7). The result is a mist of oil in the form of a cone.

The vapor pressure is, say, now 1 [mbar]. The air is sucked out of the vessel (6) by diffusion. At the outlet of the pump (8) care musty taken for a lower pressure than 1 [mbar], for instance 10^{-2} [mbar]. This can be done with a mechanical vacuum pump. Note that the compression ratio that is reachable with those diffusion pumps is very high; suppose a vacuum pressure of 10^{-9} [mbar] at the inlet port and a counterpressure of 10^{-2} [mbar], then the gas will be compressed with a factor 10^{7}.

To avoid oil being carried away to the outlet, the walls of the pump housing (3) are cooled (4). Mostly this is water cooling that keeps the housing at 30 °C (for oil); sometimes forced air cooling suffices for smaller versions. The liquid droplets are caught in the boiler.

The oil that is used is not monomolecular. It is composed out of various molecules, each with its own boiling point and vapor pressure. The oil vapors are split in more or less homogeneous vapor rays by a split method that is called *fractioning*. This works as follows (Figure 11.48).

The nozzle (1) with the highest vacuum is fed with the oil fraction that has the lowest vapor pressure. This assures a final fall of the condensed oil in the boiler at the outside of it. The oil is fed to the nozzle (3) at the outlet of the pump. The most volatile fractions of the oil evaporate here. The oil flows in the boiler to the middle part, which will feed the nozzle (2). Finally, the oil flows to the center of the boiler, where the heaviest fractions will evaporate. These have the lowest vapor pressure.

Figure 11.49 shows a diffusion pump with a *cold trap*, a conic cover that stands above the first stage. It is cooled by conduction or by water and avoids that oil vapor diffuses to the low pressure side.

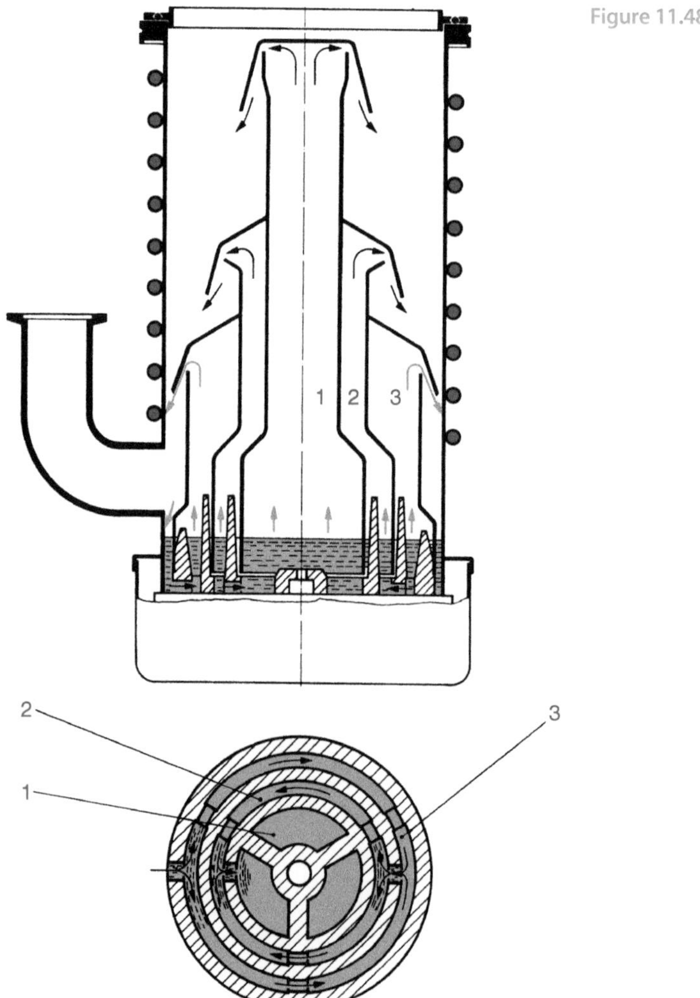

Figure 11.48 Fractioning.

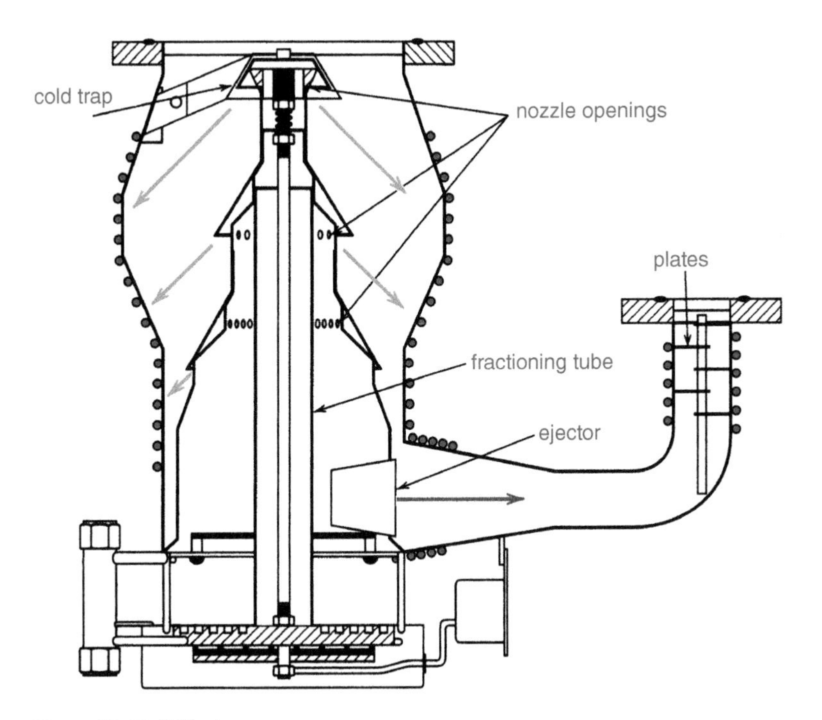

Figure 11.49 Diffusion pump.

Furthermore, it must be prevented that oil vapors diffuses to the counter pressure side. This is realized by an ejector diffusor along which the vapors and gasses flow and that flow in an extra piece that is cooled. It consist of a plate heat exchanger. This construction avoids oil vapors flowing to the high pressure side.

Oil diffusion pumps attain pump speeds of 50 000 [l/s] and pressures of up to $7.5 \cdot 10^{-11}$ [mbar].

11.6.2 Diffusion Ejector Pumps (Booster Pumps)

The diffusion pump (Figure 11.50) needs a counterpressure in the order of 1–10 [Pa], and this is mostly realized with a roots pump. Another possibility for a so-called booster pump is a diffusion ejector pump. It does not belong to the pumps for high vacuum but now that the diffusion pumps has been discussed this is the right moment to explain the operating principle of this pump. In fact, this pump is based on the design of a steam ejector but it uses stable carbon hydrogen oil on a high boiler pressure instead of steam.

11.6.3 Turbomolecular Pump

Molecular pumps are, like *vapor jet pumps,* based on the principle of exchange of impulse. In this case it is, however, not a fast vapor jet but a fast-moving wall that will realize this directed impulse exchange.

A fast-rotating propeller in a free molecular gas gives the molecules an extra velocity component above the normal thermal velocity distribution (Figures 11.51–11.54).

Meanwhile, the molecules collide with a steady wall: the stator.

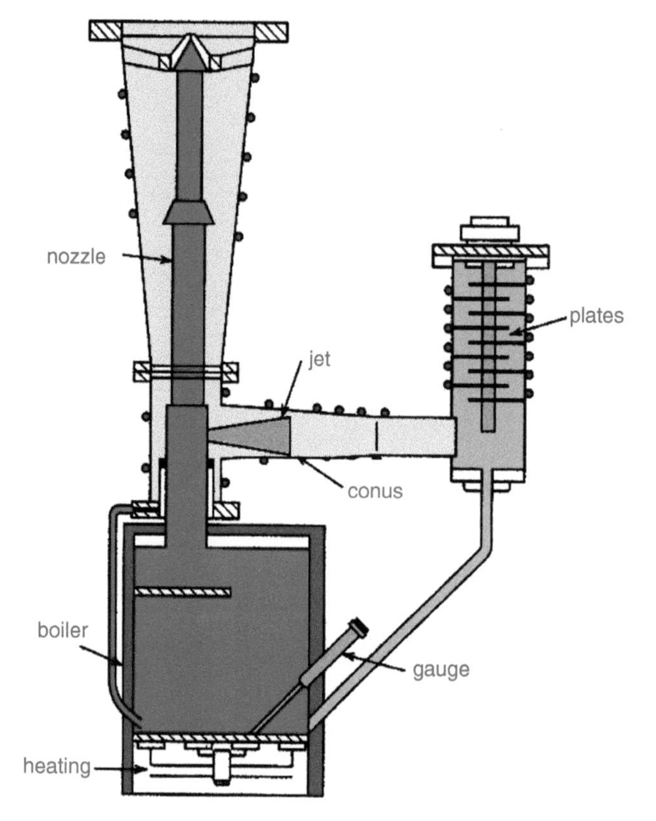

Figure 11.50 Booster pump. Source: Leybold GmbH.

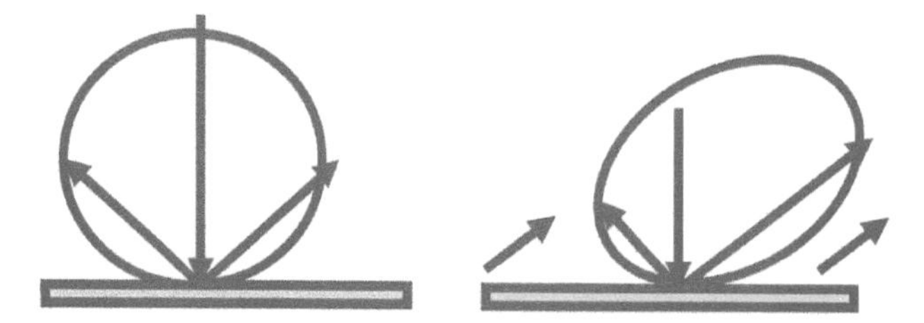

Figure 11.51 Distraction of molecules on a steady plate and a moving plate.

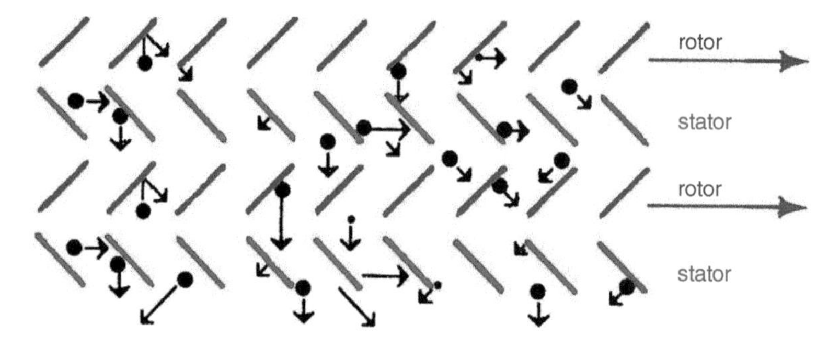

Figure 11.52 Movement of molecules through the pump.

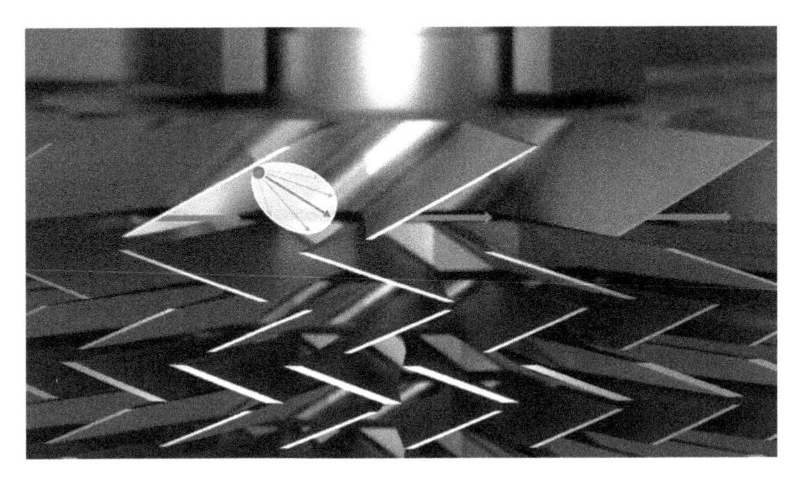

Figure 11.53 Principle of a molecular pump. Source: Pfeiffer.

Figure 11.54 Turbomolecular pump. Source: Leybold GmbH.

Table 11.3 Mean thermal velocity molecules.

Gas	Molar mass M [kmol/kg]	Mean thermal velocity [m/s]
H_2	2	1761
He	4	1245
H_2O	18	587
Ne	20	557
N_2	28	471
O_2	32	440
CO_2	44	375

A pumping effect can, however, only be significant if the peripheral velocity of the rotor blades is greater than the mean thermal velocity of the molecules that have to be pumped. According to kinetic gas theory, this is dependent on the molar mass M of the molecule (see Table 11.3).

To prevent this, turbomolecular pumps have to rotate very quickly, as much as 72 000 [rpm]. This demands a lot of the bearings.

The pump speed for air ($M = 29$) is 10 000 [l/s]. The vacuum is 10^{-5} [mbar], but there are designs that reach $7.5 \cdot 10^{-11}$ [mbar]. But that is already an ultra-high vacuum.

11.7 Ultrahigh Vacuum (10^{-7}–10^{-14} [mbar])

11.7.1 Sorption Pumps

The term *sorption pumps* comprises all devices destined for the removal of gasses or vapors from a vessel by means of sorption methods. What does that mean? The pumped gas particles are bound to the surfaces or its inside, by means of physical temperature dependent adsorption forces (Van der Waals forces), chemisorption, absorption, or by being imbedded during the course of the continuous formation of new sorption surfaces.

When operating principles are compared, these pumps can be distinguished:

- *adsorption pumps*, where the sorption of gasses simply takes place by temperature-controlled adsorption processes, this is the catching of molecules by a steady surface.
- *getter* pumps, where the sorption and retention of gasses essentially is caused by the formation of chemical compositions of *getter materials.*

11.7.2 Adsorption Pumps

Zeolite (Figure 11.55) is an aluminum silicate that is available in both its natural and its synthetic form. The zeolite has a three-dimensional, porous structure (Figure 11.56). It consists of silicon, aluminum, and oxygen atoms. The silicon ions are neutrally charged in the crystal structure. The aluminum ions are charged negatively. To keep the charge in balance here are counterions, cations, like (Na^+, K^+, ...), or a proton (H^+) in the pores.

Figure 11.55 Zeolite crystals.

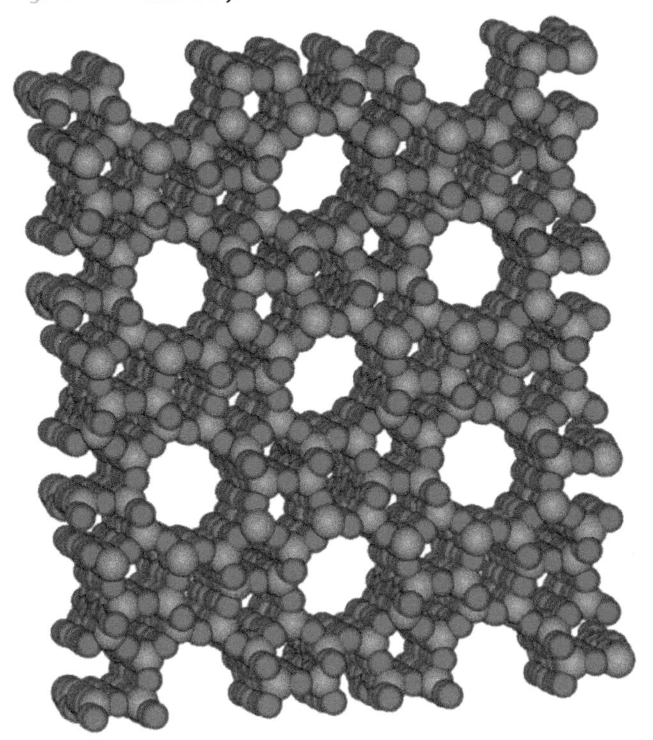

Figure 11.56 Zeolite with a porous structure.

One type of zeolite has pores of the same diameter in the whole crystal structure. The crystal structure is partly determined by the dimension of the rings of the crystal structure.

If the ratio of aluminum and silicon is changed, this has an effect on the dimension of the pores as well as the type of the counterion. All natural zeolites contain aluminum and have a hydrophile character. These materials form a good absorbent for polar substances (e.g. water and water-soluble substances).

Natural zeolites	Formula
Analcime	$NaAl(Si_2O_6) \cdot (H_2O)$
Limonite	$CaAl_2Si_4O_{12} \cdot 4(H_2O)$
Stilbite	$NaCa_4Al_8Si_{28}O_{72} \cdot 30(H_2O)$
Natrolite	$Na_2Al_2Si_3O_{10} \cdot 2(H_2O)$

Zeolite is available with pores of magnitude 0.3–3 [nm]. This pore dimension is uniform for one type of zeolite. A zeolite will not absorb molecules that are greater than the dimension of the pores. Also, molecules that have no affinity for zeolite will not be absorbed.

The pores diameter of zeolite 13X is nearly 13 [Å] (10^{-10} [m] = 10 [nm]). This is in the order of magnitude of the dimension of water vapor, oil vapor, and bigger molecules. These molecules are caught on the surface of the pores and form there a mono-atomic layer. Hydrogen and other noble gasses like neon and helium, however, have a relatively small diameter compared with the pore diameter and are therefore only moderately adsorbed.

The adsorption of gasses on the surface depends on the temperature and the pressure. This is represented in Figure 11.57.

How does one practically proceed? The adsorption pump is connected via a valve with the vessel to be sucked (Figure 11.58). Only when the pump casing is dipped in liquid

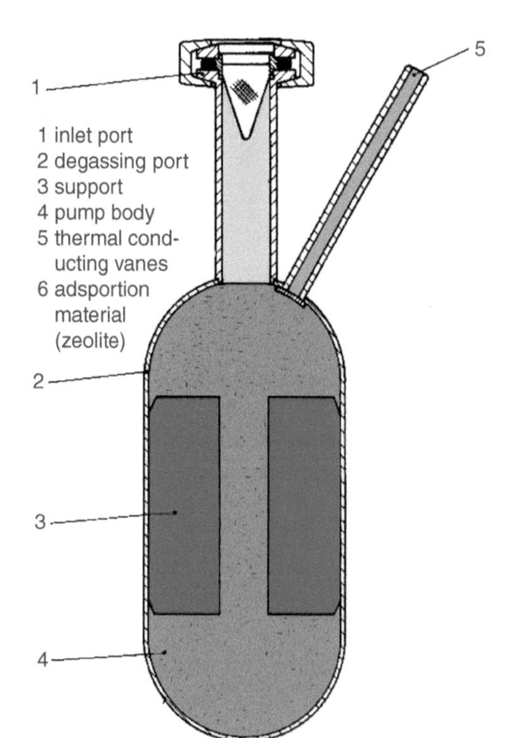

Figure 11.57 Section adsorption pump. Source: After Leybold GmbH.

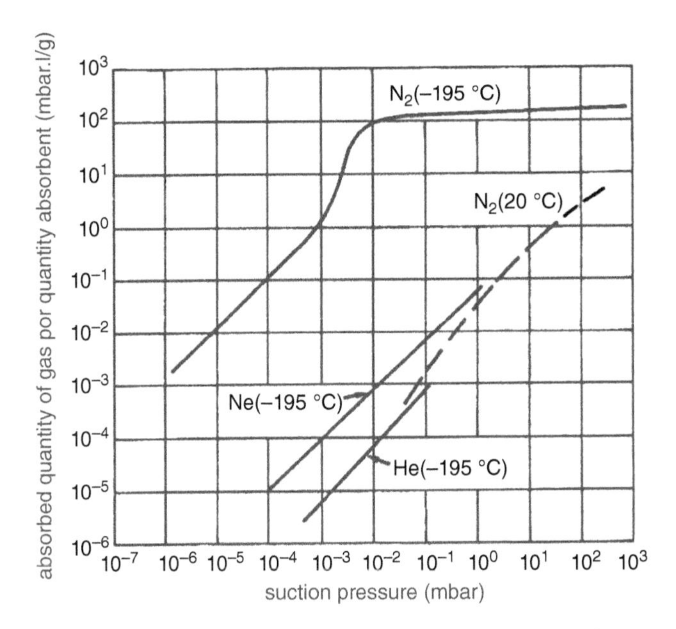

Figure 11.58 Adsorption isotherms for zeolite 13X. Source: After Leybold GmbH.

nitrogen is the sorption effect technically feasible. Because of the different pump speeds, the eventual pressure will be different for the different partial gasses in the vessel. The best values are attained for carbon dioxide, water vapor, and carbon hydrogen gasses. Light noble gasses are hardly pumped.

The more pumping, the more the zeolite will be saturated with molecules and the pump speed will decrease.

The attainable end pressure is 10^{-2} [mbar] and 10^{-3} [mbar] in as far as there is no neon or helium present in the gas mix. This pump does not work in the *ultra-high vacuum*, as the title of this paragraph suggests. During the pump process, the pump must be heated to room temperature to give the adsorbed gasses the chance to escape from the zeolite, and in this way regenerate the zeolite for a new application. If air with a lot of water vapor was sucked the pump must be heated to 200 °C.

11.7.3 Sublimation Pump

Sublimation pumps (Figures 11.59 and 11.60) are sorption pumps whereby the *getter* material evaporates and precipitates on a cold wall. In this way a stable getter film of gas molecules is formed. Then the active getter film is recreated by evaporation of the getter material. This process is repeated.

Mostly titanium is used as sublimation material. The titanium is evaporated from a wire of titanium that is heated with an electric current of 50 [A].

Those sublimation pumps are often used as booster pumps for sputter pumps (see later) and turbomolecular pumps.

The operation area runs from 10^{-1} to 10^{-11} [mbar].

Figure 11.59 Sublimation pump. Source: LewVac.

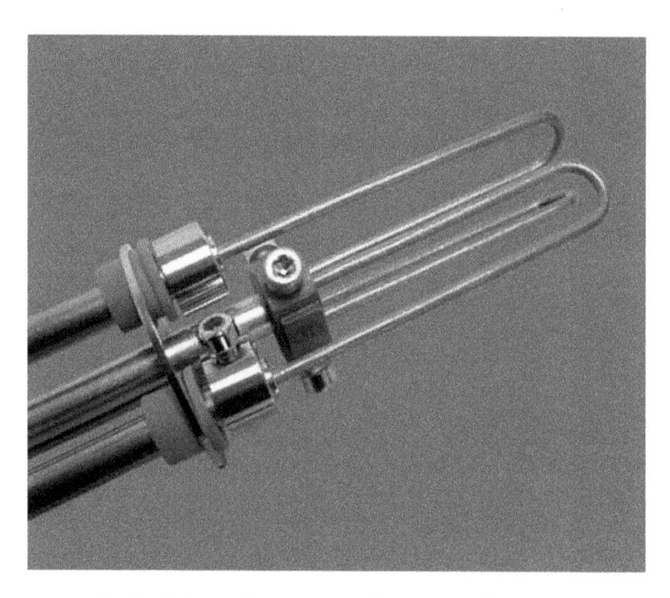

Figure 11.60 Sight on titanium wires. Source: LewVac.

11.7.4 Ion Getter Pump

This ion pump uses a cold gas discharge, called a *Penning discharge*. What is that? Well, I'm going to tell you.

The dispositive consists of a cylindrical anode (Figure 11.61) and two cathode plates (Figures 11.62 and 11.63) at the end of the anode tube. The whole is placed in a uniform magnetic field B. When the pressure is lower than 1 [Pa] and a voltage of 5 [kV] is applied on the anode, the electrons are no longer attached to the titanium cathodes and leave them. The electrons are electrical negative charges that care repelled by the cathode. On

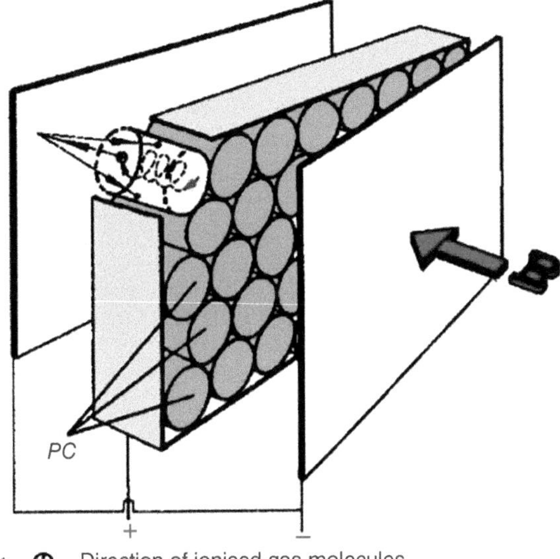

Figure 11.61 Operational principle of an ion pump. Source: After Leybold GmbH.

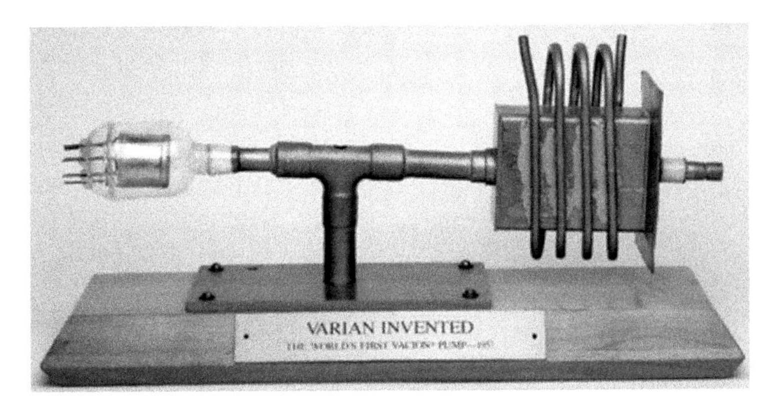

Figure 11.62 First ever ion pump (1957). Source: Agilent.

the other side, they are attracted by the positive anode, but that happens relatively slowly. The field $\overline{B}$ is the cause of a force $\overline{F}$ that acts on the moving charges q with velocity $\overline{c}$:

$$\overline{F} = q \cdot \overline{v} \times \overline{B}$$

With × the cross-product or vectoral product.

The result is a rotating movement of the electrons round the magnetic field.

Electrons can also be bent by the cathode in the direction from where they came from. In this way a spiral movement is created. Finally, the electrons will arrive at the anode.

Figure 11.63 Ion pump: sight on cathodes. Source: Agilent.

Other electrons collide with the gas molecules. This collision sets another electron free in the gas and this one shoots away. The created ion (gas) moves to the nearest cathode and the two electrons in the opposite direction.

In this way a cloud of free electrons is caught by the anode tube. The density of these electrons is high enough to maintain sufficient ionizing collisions. Even at pressures as low as 10^{-3} [mbar] this goes on. Without a magnetic field this wouldn't be the case. This discharge is a Penning discharge.

As far as concerns the air ion, this precipitates on the cathode and is absorbed by it. At the impact titanium parts can be shot away (*sputtering*). This cloud of dust precipitates on the anode and the cathode on the opposite side. This forms again a new cathode layer. In the case of titanium the cathode is renewed all the time. This layer is, so to speak, a getter pump. The electrons that are shot away by the collisions cause new collisions with gas molecules or move to the anode. Achievable pressure: $7.5 \cdot 10^{-12}$ [mbar], pumping speed: 50 000 [l/s].

Appendix A

The Velocity Profile and Mean Velocity for a Laminar Flow

Consider a cylindrical tube. In the tube with length L there will be a pressure decrease p_f caused by viscous friction.

The shear stress is given by:

$$\tau = \eta \cdot \frac{dc}{dr}$$

Let's try to find another expression for the shear stress. Suppose that the fluid flows through the tube with a constant mean velocity.

To find that, isolate a quantity of liquid, forming a little cylinder; with as axis the axis of the tube, length L and a radius r (Figure A.1).

On the liquid cylinder the following forces apply:

- At the left the pressure force of the fluid: $p \cdot A$ with $A = \pi \cdot r^2$.
- At the right the pressure force of the liquid: $(p - p_f) \cdot A$.
- On the sideways surface of the cylinder, there are two equal shear forces (symmetry):

$$\tau \cdot L \cdot 2 \cdot \pi \cdot r$$

When the liquid flows on a constant (mean)velocity, then the sum of all forces that act on the small cylinder is zero, according to Newton' law:

$$-\tau \cdot L \cdot 2 \cdot \pi \cdot r + p \cdot \pi \cdot r^2 - p \cdot \pi \cdot r^2 + p_f \cdot \pi \cdot r^2 = 0$$

From where:

$$\tau = \frac{r \cdot p_f}{2 \cdot L}$$

From this expression it is clear that the shear stress is line air proportional to the distance of the tube.

The velocity distribution can now be found by identifying the two expressions we have for the shear stress, provided that a minus sign is introduced because, in fact:

$$\frac{dc}{dr} < 0$$

$$\eta \cdot \frac{dc}{dr} = -\frac{r \cdot p_f}{2 \cdot L}$$

$$c = \int dc = \int -\frac{p_f}{2 \cdot L \cdot \eta} r.dr$$

Pumps and Compressors, First Edition. Marc Borremans.

Companion website: www.wiley.com/go/borremans/pumps

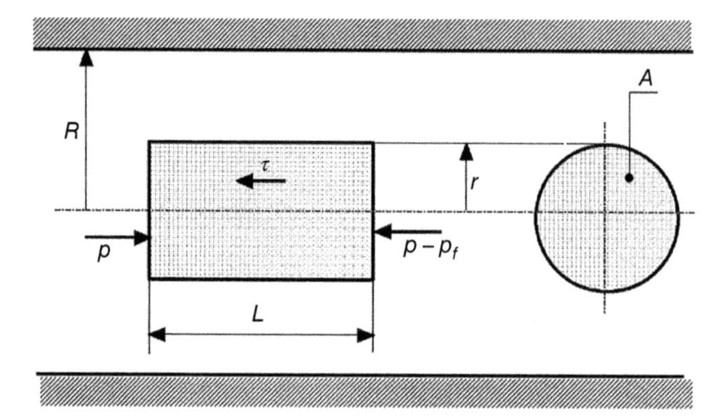

Figure A.1 Elementary volume of liquid.

$$c = -\frac{p_f \, . \, r^2}{4 \cdot \eta \cdot L} + C_1$$

The integration constant C_1 can be found by:

$$r = R \Rightarrow c = 0$$

$$r = R \Rightarrow 0 = -\frac{p_f . \, R^2}{4 \cdot \eta \cdot L} + C_1$$

So that:

$$c = \frac{p_f . (R^2 - r^2)}{4 \cdot \eta \cdot L}$$

The velocity changes thus parabolically with the distance r to the perimeter (Figure A.2).

Now that the velocity profile is known, it is possible to calculate the flow that circulates through the tube. For this, we isolate a very small ring surface on distance r from the origin and with a thickness dr. The local velocity there is c. The infinitesimal flow that flows through that infinitesimal area is (Figure A.3):

$$dQ_V = c \cdot dA$$

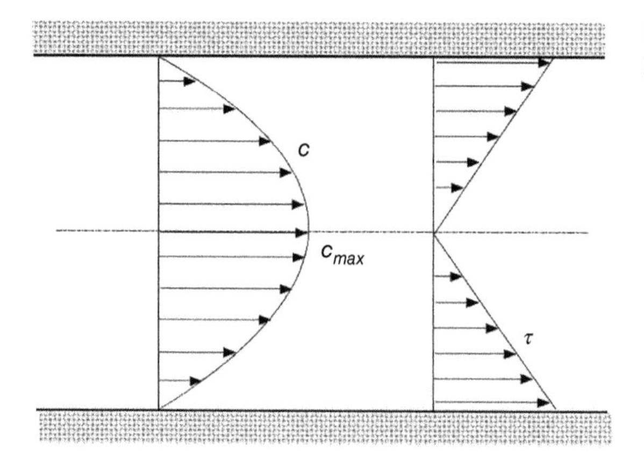

Figure A.2 Velocity and shear stress profile.

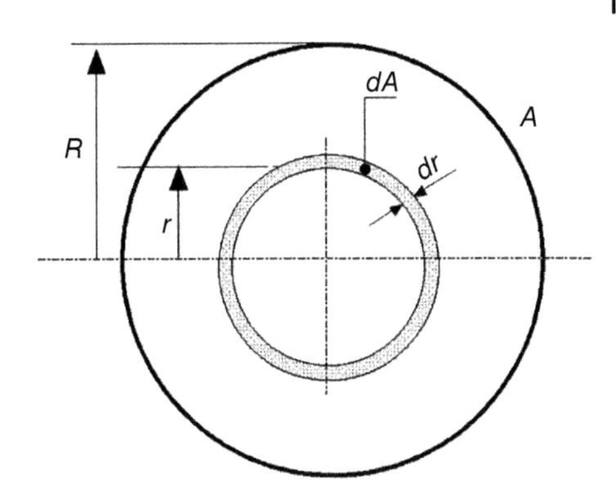

Figure A.3 Infinitesimal ring surface.

With:

$$dA = \pi \cdot (r + dr)^2 - \pi \cdot r^2 \cong 2 \cdot \pi \cdot r \cdot dr$$

So:

$$Q_V = \int_0^R dQ_V = \frac{\pi \cdot p_f}{2 \cdot \eta \cdot L} \cdot \left[\frac{R^2 \cdot r^2}{2} - \frac{r^4}{4}\right]_0^R$$

Or:

$$Q_V = \frac{\pi \cdot p_f \cdot R^4}{2 \cdot \eta \cdot L}$$

The mean velocity is:

$$c_m = \frac{Q_V}{A}$$

Or:

$$c_m = \frac{p_f \,.\, R^2}{8 \cdot \eta \cdot L}$$

Appendix B

Calculation of λ for a Laminar Flow

The (experimental) formula of Darcy–Weisbach for the pressure loss in a tube is given by:

$$p_f = \lambda \cdot \frac{L}{D} \cdot \rho \cdot \frac{c_m{}^2}{2}$$

And, for a laminar flow:

$$p_f = \frac{\eta \cdot 8 \cdot L}{R^2} \cdot c_m = \frac{\eta \cdot 32L}{D^2} \cdot c_m$$

Dividing both equations by p_f:

$$\frac{\eta \cdot 32 \cdot L}{D^2} \cdot c_m = \lambda \cdot \frac{L}{D} \cdot \rho \cdot \frac{c_m{}^2}{2}$$

Or:

$$\frac{64 \cdot \eta}{D} = \lambda \cdot \rho \cdot c_m$$

Or:

$$\lambda = \frac{64}{\frac{\rho \cdot c_m \cdot D}{\eta}} = \frac{64}{\frac{c_m \cdot D}{\nu}} = \frac{64}{Re}$$

Pumps and Compressors, First Edition. Marc Borremans.

Companion website: www.wiley.com/go/borremans/pumps

Index

Pumps and Compressors, First Edition. Marc Borremans.

Companion website: www.wiley.com/go/borremans/pumps